Texts and Monographs in Computer Science

Editors

David Gries
Fred B. Schneider

Texts and Monographs in Computer Science

(continued after index)

A Logical Approach to Discrete Math

David Gries
Fred B. Schneider

With 25 Illustrations

Springer

David Gries
Department of Computer Science
Cornell University
Upson Hall
Ithaca, NY 14853-7501
USA

Fred B. Schneider
Department of Computer Science
Cornell University
Upson Hall
Ithaca, NY 14853-7501
USA

Library of Congress Cataloging-in-Publication Data
Gries, David.
    [1st ed]
    A logical approach to discrete math / David Gries and Fred B.
Schneider.
        p.   cm. — (Texts and monographs in computer science)
    Includes bibliographical references and index.

    1. Mathematics.   I. Schneider, Fred B.   II. Title.   III. Series.
QA39.2.G7473   1994
510—dc20   93-27848

Printed on acid-free paper.

ISBN 978-1-4419-2835-1

Printed in the United States of America.   (SBA)

Springer-Verlag is a part of Springer Science+Business Media

springeronline.com

To the women in our lives,

Elaine and Mimi

Preface

This text attempts to change the way we teach logic to beginning students. Instead of teaching logic as a subject in isolation, we regard it as a basic tool and show how to use it. We strive to give students a skill in the propositional and predicate calculi and then to exercise that skill thoroughly in applications that arise in computer science and discrete mathematics.

We are not logicians, but programming methodologists, and this text reflects that perspective. We are among the first generation of scientists who are more interested in using logic than in studying it. With this text, we hope to empower further generations of computer scientists and mathematicians to become serious users of logic.

Logic is the glue

Logic is the glue that binds together methods of reasoning, in all domains. The traditional proof methods —for example, proof by assumption, contradiction, mutual implication, and induction— have their basis in formal logic. Thus, whether proofs are to be presented formally or informally, a study of logic can provide understanding.

But we want to impart far more than the anatomy of the glue —proof theory and model theory. We want to impart a skill in its use. For this reason, we emphasize syntactic manipulation of formulas as a powerful tool for discovering and certifying truths. Of course, syntactic manipulation cannot completely replace thinking about meaning. However, the discomfort with and reluctance to do syntactic manipulation that accompanies unfamiliarity with the process unnecessarily forces all reasoning to be in terms of meaning. Our goal is to balance the tendency to reason semantically with the ability to perform syntactic reasoning. Students thereby acquire understanding of when syntactic reasoning is more suitable, as well as confidence in applying it.

When we teach the propositional and predicate calculi, students are put in a syntactic straightjacket. Proofs must be written rigorously. This is an advantage, not a disadvantage, because students learn what it means for a proof to be rigorous and that rigor is easily achieved. We also describe principles and strategies for developing proofs and go over several proofs for the same theorem, discussing their advantages and disadvantages. Gradually, students learn how the shape of formulas can help in discovering proofs. The students themselves develop many proofs and, with time and practice, begin to feel at ease with the calculi. We also relate formal logic to informal proofs and to various well-known proof methods. This allows students to put what they have learned in context with their past experiences with proofs. In the end, students have a firmer understanding of the notion of proof and an appreciation for rigor, precision, brevity, and elegance in arguments.

Teaching logic as a tool takes time. It is a terrible mistake to skim over logic in one or two weeks. Five to eight weeks are needed to digest and master the material. Time spent on logic early in the game can be gained back many times over in later courses. By mastering this material early, students will have an easier time with subsequent material in mathematics and computer science.

An equational logic

We need a style of logic that can be used as a tool in every-day work. In our experience, an equational logic, which is based on equality and Leibniz's rule for substitution of equals for equals, is best suited for this purpose.

- Equational logic can be used in class, almost from the first day, to solve in a simple fashion problems that otherwise seem hopelessly complex. Students see right from the beginning that logic is *useful*.

- Proofs in an equational logic are a nice *alternative* to reasoning in English, because they rarely parrot informal English arguments in a formal way. Formal logic is more than just a language in which English arguments are to be couched. Moreover, equational proofs are frequently shorter, simpler, and easier to remember than their counterparts in English or in other formal styles (e.g. Hilbert or natural deduction).

- The equational style is already familiar to students, because of their earlier experiences with high-school algebra.

- The equational style has wide applicability. This text is evidence for this claim. We have used the equational style to reason about sets, sequences, relations, functions, the integers, combinatorics, recurrence relations, programs, and graphs.

Teacher's manual and answer book

The departure of this text from the traditional method of teaching logic and discrete math may initially present difficulties for instructors. To help them make the switch, we have written a teacher's manual, which contains general guidelines on presenting the material as well as detailed suggestions for each chapter. The teacher's manual, which includes answers to all the exercises in the text, can be obtained by writing the authors at Cornell.

Notation

Where possible, we use conventional mathematical notation. In a few places, however, our standards for consistency, uniformity, unambiguity, and ease of syntactic manipulation compel us to depart from tradition. The first departure is our notation for function application, where we use an infix period, as in $f.b$. Those who have been practicing formal manipulation have found that eliminating unnecessary parentheses helps in making the structure of formulas clear. When the structure of the function application dictates the use of parentheses, as in $f.(b+2)$ and $g.(a,b)$, we abbreviate by eliminating the period, as in $f(b+2)$ and $g(a,b)$.

We use a single notation for quantifications (e.g. $(+i \mid 0 \le i < n : i^3)$ and $(\wedge i \mid 0 \le i \le 20 : b[i] = 0)$) over any symmetric and associative operator. Having a single notation cleans up what until now has been a rather muddled affair in mathematics, and it enables students to see semantic similarities that are obscured by the standard notations.

We also depart from standard English usage in a few ways. For example, we use the logical (rather than the traditional) approach to placing stops relative to quote marks. In this, we follow Fowler and Gower in their "Dictionary of Modern English Usage". Traditionalists would have us place the period of the previous sentence before the quote mark (but if the sentence ended in an exclamation point it would be after!). The logical approach puts the stop where it belongs. When the period is not part of the quote, it appears outside the quote marks.

We place a space on one side of an em dash —here are examples— in order to help the reader determine whether the em dash begins or ends a parenthetical remark. In effect, we are creating two symbols out of one. In longer sentences—and we do write long sentences from time to time—the lack of a space can make it difficult to see the sentence structure—especially if the em dash is used too often in one sentence. Parenthetical remarks delimited by parentheses (like this one) have a space on one side of each parenthesis, so why not parenthetical remarks delimited by em dashes?

Historical notes

Students generally find math texts to be rather dry affairs. This reputation is well deserved. But the creators of mathematics and computer science are every bit as colorful as the characters one finds lurking behind other disciplines. Mathematics and its notation were shaped by personalities, cultures, and forces of history. To give students a taste for our history and culture, this text contains 30-odd historical notes (and some are very odd indeed). These historical notes introduce mathematicians and computer scientists as well as discuss topics such as the history of symbols for equality, the golden ratio, and software patents. We hope that these notes help convince students that our field is not sterile, but vibrant and living.

The facts for the historical notes were gleaned from a number of sources, all of which appear in the list of references beginning on page 473. Three sources were particularly helpful: Eric T. Bell's *Men of Mathematics* [3], Florian Cajori's *A History of Mathematical Notations* [7], and the *Encyclopædia Britannica* [12].

Selecting topics for different audiences

This book contains far too much material for a typical one-semester, freshman or sophomore course, and the instructor is advised to select a subset. The core of the book —Chaps. 1–5.1, 8, and 9— belongs in every course. At Cornell, this material takes five weeks, but there is nothing wrong with spending as much as seven or eight weeks on it. We usually mix parts of Sec. 5.1 on applications with the presentation of Chap. 3, thereby providing motivation. Chaps. 11, 12, and 14 on sets, induction, and relations are also central to almost any course. Finally, Chap. 13 provides a wealth of exercises on proving theorems by induction, outside the domain of the natural numbers.

Thereafter, there is ample opportunity for divergence.

- Computer-science oriented courses will want to cover Chaps. 10 (concerning programming), 16 (combinatorial analysis), 17 (recurrence relations), or 19 (graph theory).

- Math-oriented courses may tend towards Chaps. 15 (integers), 18 (modern algebra), and 20 (infinite sets).

- A logic-oriented course would cover Chaps. 6 and 7 thoroughly.

This text is also suitable for students seeking some exposure to mathematics. The material in Chaps. 1–5.1, 8, and 9 constitutes an effective alternative to calculus as the introduction to rigorous thinking that is required by most colleges. After all, the notion of proof is important in almost all areas, and not only in scientific and engineering ones. We believe that the material on propositional and predicate logic can be learned by *every* student, as long as the teacher paces the material to the mathematical aptitude and maturity of the students.

A reference text

We have organized most of the chapters for reference, as well as study. Theorems are grouped in boxes, to assist the student or scholar in finding a pertinent theorem when necessary. This is not only for the logic chapters but also for the chapters on sets, sequences, relations, and integers. A list of theorems for the propositional and predicate calculi has been placed at the end of the text for easy reference.

Acknowledgements

Our views on the use of logic have been strongly shaped by research over the past 20 years in the formal development of programs. This work led to the development of the proof style and the form of propositional calculus of Chap. 3. The work of Edsger W. Dijkstra has been most influential, and our propositional calculus is modeled on one developed by Dijkstra and Carel Scholten in the late 1980s [10]. The influence of Dijkstra and his group at Eindhoven on this text will be clear to all who know them.

The less formal styles of proof presentation discussed in Sec. 4.2, as well as the proof presentations in Chap. 6, were based on work by Robert Constable and his group on Nuprl and later work by Leslie Lamport.

Several people critiqued drafts of this book. Robert Constable, Edsger W. Dijkstra, Tom Henzinger, Dexter Kozen, Doug McIlroy, Jay Misra, and Richard Zippel made major contributions. Navindra Gambhir proofread answers to exercises and developed some answers. We also received help from Eric Aaron, Vicki Almstrum, Betty Cheng, Nell Dale, Alan Demers, Sofoklis Efremidis, Alan David Fekete, Konrad Gries, Paul Gries, Juris Hartmanis, Rick Hehner (and his text [21]), Tony Hoare, Jamey Leifer, Andrew Malton, Paul MacMillan, Paul Myers, Bill McKeeman, Gil Neiger, Walt Potter, Arnd Potz-Hoeffner, Juris Reinfelds, Jan van de Snepscheut, Phil Wadler, Stan Warford, and Tony Wojcik. William Rucklidge took pictures for the morph images on the cover, and he and Harry Barshatsky were helpful in working with the various systems to produce the images.

We also thank the students in courses that were subjected to drafts of this text at Cornell and Texas at Austin for their help in making the text better. Tom Henzinger and Jay Misra were brave enough to teach courses using a draft.

We are grateful for our funding for research and education over the years, which has enabled us to become educated and informed users of logic. Gries has been supported by the NSF and DARPA (through the ONR). Schneider acknowledges support from the NSF, ONR, DARPA, and Digital Equipment Corporation.

The people at Springer-Verlag, New York, have been most helpful: Rüdiger Gebauer, Hal Henglein, Jacqueline Jeng, and Karen Kosztolnyik.

The text was created and typeset using LaTeX[28], which is built on TeX [27]. Typing and editing were done using Emacs on various machines with the Unix operating system and at times on a Macintosh powerbook. The morphs on the back cover were produced using the software package *Morph* on the Macintosh. Drafts were printed using LaserWriters, which use the language PostScript. Take the capability of any of these systems away, and this book would have been less complete and polished.

But more has to be acknowledged than the authors of these individual systems. The first author remembers well using computers and punch cards in typesetting his first book, *Compiler Construction for Digital Computers*, almost twenty-five years ago. Although better than using a typewriter, it was still a painful process —as reading the final text of that book was! The remarkable development of all the systems we used in developing this book was made possible by basic, foundational research (and hacking) in many areas of computer science and engineering, including formal languages, programming languages, compiler construction, operating systems, machine architecture, VLSI design, AI, and even theory of computation. It is computer scientists and engineers *en masse* that deserve a round of applause for their work and achievements in the past twenty-five years.

Computer Science Department
Upson Hall
Cornell University
Ithaca, New York 14853
May 1993

David Gries
gries@cs.cornell.edu

Fred B. Schneider
fbs@cs.cornell.edu

---

[0] LaserWriter, Macintosh, and Powerbook are trademarks of Apple Computer. Morph is a trademark of Gryphon Software Corporation. PostScript is a trademark of Adobe Systems, Inc. TeX is a trademark of the American Mathematical Society. Unix is a trademark of AT&T.

Contents

Chapter 0

Using Mathematics

$\mathbf{M}$ athematics can be used to represent, or model, the world. This is because mathematics provides a way to represent relationships that is concise, precise, and well-suited to manipulations for revealing insights about the objects being modeled. For example, the equation

$$e = m \cdot c^2$$

was Albert Einstein's way of expressing a belief about the relationship between energy e and mass m (c denotes the speed of light). The laws of planetary motion, or at least models of these laws, are used in launching satellites and keeping them in orbit. Social scientists employ mathematics, especially statistics, in understanding, analyzing, and making predictions about the behavior of society. Mathematical models help in anticipating the stock market and the weather. Since all areas of science and engineering employ mathematical models of one kind or another, it is not surprising that much time is spent building, manipulating, and analyzing models.

As a mundane example of a mathematical model, consider the following problem.

> Mary has twice as many apples as John. Mary throws half her apples away, because they are rotten, and John eats one of his. Mary still has twice as many apples as John. How many apples did Mary and John have initially?

Using m and j to denote the numbers of apples that Mary and John have initially, we write formula (0.1) as a mathematical model of the problem.

(0.1) $m = 2 \cdot j$ and $m/2 = 2 \cdot (j - 1)$

Any values of m and j that make (0.1) true could be the numbers of apples that Mary and John had initially. Notice how much more succinct our mathematical model is than the English description of the problem. The mathematical model also has other virtues, as we now see.

Virtues of mathematical models

What is it about mathematical models that makes them so useful? One of their key virtues is the following:

A mathematical model may be more understandable, concise, precise, or rigorous than an informal description written in a natural language.

To illustrate the benefits of rigor (see Historical note 0.1 on page 3), consider an algorithm to compute b, an integer approximation to $\sqrt{n}$ for some integer n. This algorithm is rigorously specified by giving a *precondition*, which can be assumed to hold before execution of the algorithm, and a *postcondition*, which describes what is to be true upon termination. For computing $\sqrt{n}$, the precondition is $0 \leq n$, since the square root of a negative number is undefined if we restrict ourselves to results that are not complex numbers.

Formalizing the postcondition requires us to think carefully about what approximations for $\sqrt{n}$ would be acceptable. Three choices are given below, where variable b contains the approximation to $\sqrt{n}$.

Choice 1: $b^2 \leq n < (b+1)^2$

Choice 2: $abs(b^2 - n) \leq abs((b+1)^2 - n)$ and
$abs(b^2 - n) \leq abs((b-1)^2 - n)$

Choice 3: $(b-1)^2 < n \leq b^2$

Choice 1 corresponds to computing the largest integer that is at most $\sqrt{n}$; choice 2, to computing the integer closest to $\sqrt{n}$; choice 3, to computing the smallest integer that is at least $\sqrt{n}$.

Note that in the informal English specification of the problem, we simply wrote "an approximation to $\sqrt{n}$". In the mathematical formulation, we were forced to be precise in specifying exactly what approximation was acceptable —rigor guided us to a more thorough analysis.

A second important virtue of mathematical models is:

Answers to questions about an object or phenomenon can often be computed directly using a mathematical model of the object or phenomenon.

The discovery of the planet Neptune illustrates this virtue. As early as the seventeenth century, Kepler, Newton, and others formulated mathematical models of planetary motion, based on observing the stars and planets. In the early 1800's, it was discovered that observations of the planet Uranus did not agree with the mathematical models being used. The discrepancies between observations and models received so much attention that, in 1843, the Royal Society of Sciences of Göttingen, Germany, offered a prize for a satisfactory theory of the motions of Uranus. Scientists conjectured that

---

HISTORICAL NOTE 0.1. WEBSTER AND HILBERT ON RIGOR

Calls for more *rigor* in programming and related areas have often been met with glassy-eyed stares, frozen features, and stiffening of backs —as if the listener equated rigor with rigor mortis. In searching for reasons for this reaction, we looked up "rigor" in *Webster's Third International Dictionary*. We found the following meanings: "often harsh inflexibility in opinion, temper, or judgement; the quality of being unyielding or inflexible; an act of severity or harshness; a condition that makes life difficult". No wonder people were unsympathetic to calls for more rigor!

Only in the fourth definition of "rigor" did our intended meaning surface: "strict precision or exactness".

The brilliant and influential mathematician David Hilbert (see Historical note 6.1 on page 111) also called for rigor. Here is what he had to say, in a famous lecture to the Second International Congress of Mathematicians held in Paris in 1900 (see [32]).

> It remains to discuss briefly what general requirements may be justly laid down for the solution of a mathematical problem. I should say first of all this: that it shall be possible to establish the correctness of the solution by means of a finite number of steps based on a finite number of hypotheses which are implied in the statement of the problem and which must be exactly formulated. This requirement of logical deduction by means of a finite number of processes is simply the requirement of rigor in reasoning ....
>
> It is an error to believe that rigor in proof is the enemy of simplicity. On the contrary, we find it confirmed by numerous examples that the rigorous method is at the same time the simpler and the more easily comprehended. The very effort for rigor forces us to discover simpler methods of proof. It also frequently leads the way to methods which are more capable of development than the old methods of less rigor.

Our respect for rigor is the same, and we hope that studying this text will give you a better feeling for rigor and its application.

---

the orbit of Uranus was being affected by an unknown planet. Some two-to-three years of calculation (all by hand!) uncovered the probable position for the unknown planet. Searching that area with telescopes led to the discovery of the planet Neptune in 1846.

There is a third important advantage in using mathematical models:

> Mathematics provides methods for reasoning: for manipulating expressions, for proving properties from and about expressions, and for obtaining new results from known ones. This reasoning can be done without knowing or caring what the symbols being manipulated mean.

That is, there are rules for performing syntactic manipulations, without regard for semantics. We use these rules to learn, in a syntactic fashion, things about the model and the phenomenon it models. Only the initial and final formulations need be interpreted in terms of the original problem.

Here is a simple example of syntactic manipulation. Suppose we want an expression equivalent to Einstein's equation $e = m \cdot c^2$ that shows how to calculate m given e. Without thinking much about it, you will write $m = e/c^2$. In school, you learned rules for manipulating arithmetic expressions and practiced them so much that you can now apply them automatically, often several rules at a time. Below, in some detail, is a calculation of $e/c^2 = m$ from $e = m \cdot c^2$.

$$e = m \cdot c^2$$
$$= \quad \langle \text{Divide both sides by the non-zero } c^2 \rangle$$
$$e/c^2 = (m \cdot c^2)/c^2$$
$$= \quad \langle \text{Associativity} \rangle$$
$$e/c^2 = m \cdot (c^2/c^2)$$
$$= \quad \langle c^2/c^2 = 1 \rangle$$
$$e/c^2 = m \cdot 1$$
$$= \quad \langle m \cdot 1 = m \rangle$$
$$e/c^2 = m$$

In this calculation, between each pair of expressions appears a line with an equals sign and a hint (within brackets $\langle$ and $\rangle$). The equals sign indicates that the expressions are equal, and the hint indicates why. Since equality is *transitive* (which means that from $b = c$ and $c = d$ we can conclude that $b = d$ holds), we conclude that $e = m \cdot c^2$ is equivalent to $e/c^2 = m$.

We can understand each of the above manipulations without knowing what m, e, and c denote, that is, without knowing that the equations being manipulated are models of the relation between energy and matter. We are able to reason syntactically.

We expect that you are accustomed to manipulating arithmetic expressions but probably have had little experience manipulating boolean expressions (as found in programming languages like Pascal). For example, given that p and q together imply r and that r is false, what can be inferred about p and about q? Familiarity with manipulating boolean expressions would allow you to conclude that at least one of p and q is false. As another example, consider the following English statement about an array b.

Every value in array segment $b[1..n]$ that is not in $b[i..j]$ is in $b[i..j]$.

Do you know how to make sense of this apparent gibberish? Try formulating this sentence as a boolean expression, simplifying the boolean expression, and then translating back into English. (We do this in a later chapter.)

Manipulation of boolean expressions is obviously a useful skill for programmers, so our first task in this text is to help you develop that skill. Acquiring skill in manipulation will require a great deal of practice on your part, but the time spent will be worthwhile. This skill will be of service to you in much of the mathematical work you do later, including computer programming. And, of course, we will use such syntactic manipulations throughout this text.

Beyond syntactic manipulation

Although the initial chapters of this text emphasize syntactic manipulation of formulas, the text addresses much more. For example, we hope to convey a sense of taste and style in inventing notation and using formalism. Some notations are ambiguous, while others are suited for only certain tasks. Perhaps you have had experience using the various arithmetic calculators on the market. Keying in $13 + 5 \cdot 6$ on many will produce 108, which is not consistent with the value a mathematician would ascribe to that expression: 43 (i.e. $13 + (5 \cdot 6)$). Other calculators process "reverse polish notation", so one can enter either $13\ 5 + 6\cdot$ or $13\ 5\ 6\cdot +$, depending on whether $(13 + 5) \cdot 6$ or $13 + (5 \cdot 6)$ is desired. Without explicit parentheses or operator-precedence rules, our usual infix notation for arithmetic expressions is ambiguous. And, apparently, infix notation is not the only notation for specifying arithmetic calculations.

This text also introduces you to a number of useful abstractions and their properties. You are no doubt comfortable using integer variables to model integer-valued quantities. Richer, more powerful abstractions are needed to model other phenomena of interest. Mathematicians have invented a collection of such general-purpose abstractions. The *set* allows us to model and reason about collections of objects, like groups of cities in the Northeast or the Sun Belt. The *relation* allows us to model and reason about relationships between objects in sets. For example, relation $<$ on integers characterizes the relative magnitudes of two integers and the "adjacent" relation on cities tells whether it is possible to drive from one city directly to another. We also discuss in this text various types of infinities —abstractions that may not have a counterpart in reality but nevertheless are useful in understanding questions that arise in connection with the foundations of mathematics.

In science and engineering, mastery of a subject is equivalent to being able to reason about the subject. One finds many different notions of what

---

HISTORICAL NOTE 0.2. Starting With Zero

The first chapter of this text is numbered 0. In some situations, it does not matter whether we start with 0 or 1, and we might start with 0 simply for the shock value. In other situations, starting with 0 is really the best choice. More and more programming languages, for example, number the first element of a character string 0, because it makes many manipulations easier. And, memory locations in a computer are numbered starting with 0. Numbering from 0 also makes sense in some non-computer situations. What's the lowest score you can get on a test? How old are you on your first birthday —at your birth?

Too many people write 1st for *first*, 2nd for *second*, 3rd for *third*, and so on. You won't find us doing that, because that would lead to writing 0th when numbering starts at 0, and that makes no sense. *First* means "before all others or anything else in time, rank, order, etc.", so how could anything come before the first? If counting starts at 0, then 0 is the 1st number.

The concept of zero was developed by the Hindus, and the small circle they used to denote it was given the Sanskrit word for *vacant*. The concept and symbol were transliterated into Arabic and then into Latin about 1200 A.D. Before that, the Romans and the western world had no symbol for zero. That may explain partially why starting from 1 has been ingrained in our society for so long.

---

constitutes the embodiment of such reasoning, the "proof". At one end of the spectrum are highly stylized proofs, like those you learned in high-school geometry; at the other end, are informal English language arguments typically found in introductory calculus texts. Underlying all these proof styles is a small number of simple, domain-independent methods. In this text we discuss these methods —mathematical induction, proof by contradiction, the pigeonhole principle, and so on. We also discuss various styles and formats for proofs. Formal logic is the basis for these discussions, because it abstracts the notion of a reasoning system.

Finally, in this text we apply what we advocate in domains of particular interest in computing science. The design of combinational circuits, a key aspect of hardware design, is closely related to the study of boolean expressions. Both involve reasoning about "variables" (called wires in a circuit design) that can take one of two values and expressions obtained by combining those variables with boolean operators (gates in a circuit design). Reasoning about programs is also treated in this text —the importance of this task should be obvious.

Chapter 1

Textual Substitution, Equality, and Assignment

W e introduce *textual substitution* and illustrate its application to reasoning about equality and about the assignment statement in programming languages. We discuss Leibniz's definition of equality and formalize it in terms of textual substitution (see Historical note 1.2). We give a proof format for showing equality of two expressions.

1.1 Preliminaries

Recall the syntax [1] of conventional mathematical expressions. Expressions are constructed from constants, variables, and operators (like $+$, $\cdot$, $<$, and $=$). We can define the syntax of simple expressions as follows:

- A constant (e.g. 231) or variable (e.g. x) is an expression.

- If E is an expression, then (E) is an expression.

- If $\circ$ is a unary prefix operator [2] and E is an expression, then $\circ E$ is an expression, with operand E. For example, the negation symbol $-$ is used as a unary operator, so -5 is an expression.

- If $\star$ is a binary infix operator and D and E are expressions, then $D \star E$ is an expression, with operands D and E. For example, the symbols $+$ (for addition) and $\cdot$ (for multiplication or product) are binary operators, so $1+2$ and $(-5) \cdot (3+x)$ are expressions.

Parentheses are used in expressions to indicate aggregation (to *aggregate* means to bring together). Thus, $2 \cdot (3+5)$ denotes the product of 2 and $3+5$. Precedences are assigned to operators in order to reduce the need

---

[1] *Syntax* refers to the structure of expressions, or the rules for putting symbols together to form an expression. *Semantics* refers to the meaning of expressions, or how they are evaluated.

[2] A unary operator has one operand. A binary operator has two operands. A prefix operator is written before its operands, as in -5. An infix operator is written between its operands, as in $x+2$.

for parentheses. The precedences assigned to all operators used in this text are given in a precedence table on the inside front cover. For example, in $(4+2) \cdot 3$, parentheses indicate that $4+2$ is multiplied by 3, while in $(4 \cdot 2)+3$, the parentheses can be omitted because multiplication, according to the table, has higher precedence than addition.

An expression can contain variables, and evaluating such an expression requires knowing what values to use for these variables. To this end, we introduce the notion of a *state*. A state is simply a list of variables with associated values. For example, in the state consisting of $(x, 5)$ and $(y, 6)$, variable x is associated with the value 5 and variable y with 6.

Evaluation of an expression E in a state is performed by replacing all variables in E by their values in the state and then computing the value of the resulting expression. For example, evaluating $x - y + 2$ in the state just given consists of replacing variables x and y by their values to yield $5 - 6 + 2$ and then evaluating that to yield 1.

1.2   Textual substitution

Let E and R be expressions and let x be a variable. We use the notation

$$E[x := R] \qquad \text{or} \qquad E^x_R$$

to denote an expression that is the same as E but with all occurrences of x replaced by "(R)". The act of replacing all occurrences of x by "(R)" in E is called *textual substitution*. Examples are given below.

Expression	Result	Unnecessary parentheses removed
$x[x := z + 2]$	$(z + 2)$	$z + 2$
$(x + y)[x := z + 2]$	$((z + 2) + y)$	$z + 2 + y$
$(x \cdot y)[x := z + 2]$	$((z + 2) \cdot y)$	$(z + 2) \cdot y$

Observe that the parentheses delimiting R can be deleted if they are not needed. We often combine the steps of textual substitution and removal of unnecessary parentheses, saying, for example, simply that $(x+y)[x := z+2]$ equals $z + 2 + y$.

For x a list $x_1, \ldots, x_n$ of distinct variables and R a list $R_1, \ldots, R_n$ of expressions, the *simultaneous textual substitution $E[x := R]$* denotes the simultaneous replacement in E of the variables of x by the corresponding expressions of R, each expression being enclosed in parentheses. For example, $(z + y)[z, y := 5, 6]$ is $((5) + (6))$, which simplifies to $5 + 6$, and $(z + y)[z, y := y \cdot y, w]$ is $((y \cdot y) + (w))$, which simplifies to $y \cdot y + w$.

HISTORICAL NOTE 1.1. GOTTFRIED WILHELM LEIBNIZ (1646–1716)

Mathematics, law, religion, history, literature, logic, and philosophy all owe debts to Leibniz. A man of tremendous energy, he read, wrote, and thought incessantly. He was refused a law doctorate at Leipzig at the age of 20, basically because the faculty was envious of him. So, he traveled to the University of Altdorf in Nuremberg and submitted an essay on the historical method of teaching law, which he composed during the trip. Altdorf not only awarded him a doctorate but offered him a professorship. He turned it down.

Leibniz spent much of his life as a diplomat, librarian, historian, and genealogist in the service of the nobility —the last 40 years were spent with the Duke of Hanover. Leibniz's work brought him into contact with nobility and their problems. In one essay written for an employer, he urged the European states to work together in the conquest of the non-Christian world in the middle east. He worked actively to reunite the Catholic and Protestant churches and wrote treatises that looked for common ground between them. At one point, he was offered the post of librarian at the Vatican but declined because he did not want to become a Catholic.

As a mathematician, Leibniz is best known, along with Isaac Newton, for the development of calculus —blame your Freshman Calculus course on them. The controversy between these two giants is legendary. Leibniz was also far ahead of his time in dreaming of a "general method in which all truths of the reason would be reduced to a kind of calculation". Its principal utility would be in reasoning performed by operations on symbols —even geometry would be handled this way, without need for diagrams and figures. Thus, Leibniz foresaw symbol manipulation as we know it today.

Textual substitution has a higher precedence than any operator listed in the precedence table on the inside front cover. Consequently, in the first case below, the substitution is performed only on subexpression y. In the second case, parentheses are used to indicate that the substitution is being applied to $z + y$, rather than to y alone.

$$z + y[z, y := 5, 6] \text{ is } z + 6$$
$$(z + y)[z, y := 5, 6] \text{ is } 5 + 6$$

The alternative notation E_R^x is more concise than $E[x := R]$, but, at least to programmers, $E[x := R]$ is more suggestive of the operation that it denotes than is E_R^x. Hence, we tend to use $E[x := R]$ as long as a formula fits easily on one line.

Note that x in $E[x := R]$ must be a list of *distinct* variables —all the variables in x must be different. Also, note that textual substitution is defined only for replacing variables and not for replacing expressions. Further examples of textual substitution appear in Table 1.1.

Textual substitution is left associative, so that $E[x := R][y := Q]$ is defined to be $(E[x := R])[y := Q]$, a copy of E in which every occurrence of x has been replaced by R and then every y has been replaced by Q. This means that, in general, $E[x := R][y := Q]$ is different from $E[x, y := R, Q]$, as illustrated by the following two textual substitutions:

$$(x + 2 \cdot y)[x, y := y, x] \quad \text{and} \quad (x + 2 \cdot y)[x := y][y := x] \quad .$$

Textual substitution and hidden variables

At times, we name an expression and then use its name within another expression. For example, we may give the name Q to an expression using

$$Q : \frac{-b + \sqrt{b^2 - 4 \cdot a \cdot c}}{2 \cdot a} .$$

We can then abbreviate the expression $x = (-b + \sqrt{b^2 - 4 \cdot a \cdot c})/(2 \cdot a)$ by $x = Q$.

However, the expression $x = Q$ then has three hidden variables, a, b, and c, and these must be taken into account when a textual substitution is performed. For example, the textual substitution $(x = Q)[b := 5]$ yields $(x = Q')$, where $Q' = (-5 + \sqrt{5^2 - 4 \cdot a \cdot c})/(2 \cdot a)$.

Inference rule Substitution

Our first use of textual substitution comes in the form of an *inference rule*, which provides a syntactic mechanism for deriving "truths", or *theorems* as we call them. Later we see that theorems correspond to expressions that are true in all states. An inference rule consists of a list of expressions,

---

TABLE 1.1. Examples of Textual Substitution

Substitution for one variable

$35[x := 2] = 35$
$y[x := 2] = y$
$x[x := 2] = 2$
$(x \cdot x + y)[x := c + y] = (c + y) \cdot (c + y) + y$
$(x^2 + y^2 + x^3)[x := x + y] = (x + y)^2 + y^2 + (x + y)^3$

Substitution for several variables

$(x + y + y)[x, y := z, w] = z + w + w$
$(x + y + y)[x, y := 2 \cdot y, x \cdot z] = 2 \cdot y + x \cdot z + x \cdot z$
$(x + 2 \cdot y)[x, y := y, x] = y + 2 \cdot x$
$(x + 2 \cdot y \cdot z)[x, y, z := z, x, y] = z + 2 \cdot x \cdot y$

called its *premises* or *hypotheses*, above a line and an expression, called its *conclusion*, below the line. It asserts that if the premises are theorems, then the conclusion is a theorem.

Inference rule Substitution uses an expression E, a list of variables v, and a corresponding list of expressions F:

(1.1)   **Substitution:** $$\frac{E}{E[v := F]}$$

This rule asserts that if E is a theorem, then so is E with all occurrences of the variables of v replaced by the corresponding expressions of F. For example, if $x + y = y + x$ (this is E) is a theorem, Substitution allows us to conclude that $b + 3 = 3 + b$ (this is $E[x, y := b, 3]$) is also a theorem.

Here is another example. Suppose the expression $2 \cdot x / 2 = x$ is a theorem. By Substitution (1.1), we can conclude that $(2 \cdot x / 2 = x)[x := j]$, i.e. $2 \cdot j / 2 = j$, is also a theorem.

It should be noted that an inference rule like Substitution (1.1) is really a scheme that represents an infinite set of rules —one rule for each combination of an expression E, list of variables v, and list of expressions F. For example, we can instantiate E, v, and F of Substitution (1.1) with $2 \cdot x / 2 = x$, x, and $j + 5$, respectively, to obtain the inference rule

$$\frac{2 \cdot x / 2 = x}{(2 \cdot x / 2 = x)[x := j + 5]} \quad \text{or} \quad \frac{2 \cdot x / 2 = x}{2 \cdot (j + 5) / 2 = j + 5} \ .$$

1.3   Textual substitution and equality

Evaluation of the expression $X = Y$ in a state yields the value *true* if expressions X and Y have the same value and yields *false* if they have different values. This characterization of equality is in terms of expression evaluation. For reasoning about expressions, a more useful characterization would be a set of laws that can be used to show that two expressions are equal, without calculating their values. For example, you know that $x = y$ equals $y = x$, regardless of the values of x and y. A collection of such laws can be regarded as a definition of equality, provided two expressions have the same value in all states iff [3] one expression can be translated into the other according to these laws.

We now give four laws that characterize equality. The first two are expressions that we postulate are theorems (and they are true in every state).

---

[3] Mathematicians use *iff* as an abbreviation for *if and only if*. Thus b iff c holds provided (i) b holds if c holds and (ii) c holds if b holds.

(1.2)   **Reflexivity:** $x = x$

(1.3)   **Symmetry** [4] **:** $(x = y) = (y = x)$

The third law for equality, *transitivity*, is given as an inference rule.

(1.4)   **Transitivity:** $\dfrac{X = Y,\ Y = Z}{X = Z}$

We read this inference rule as: from $X = Y$ and $Y = Z$, conclude $X = Z$. For example, from $x+y = w+1$ and $w+1 = 7$ we conclude, by Transitivity (1.4), $x + y = 7$. As another example, on page 4, we gave a proof that $(e = m \cdot c^2) = (e/c^2 = m)$. It is Transitivity that allows us to conclude that the first expression $e = m \cdot c^2$ equals the third, then equals the fourth, and finally equals the fifth expression, $e/c^2 = m$.

A fourth law of equality was articulated by Gottfried Wilhelm Leibniz, some 350 years ago (see Historical Note 1.2). In modern terminology, we paraphrase Leibniz's rule as follows.

> Two expressions are equal in all states iff replacing one by the other in any expression E does not change the value of E (in any state).

A consequence of this law can be formalized as an inference rule (see also Exercise 1.4):

(1.5)   **Leibniz:** $\dfrac{X = Y}{E[z := X] = E[z := Y]}$

Variable z is used in the conclusion of (1.5) because textual substitution is defined for the replacement of a variable but not for the replacement of an expression. In one copy of E, z is replaced by X, and in the other copy, it is replaced by Y. Effectively, this use of variable z allows replacement of an instance of X in $E[z := X]$ by Y.

Here is an example of the use of Leibniz (1.5). Assume that $b+3 = c+5$ is a theorem. We can conclude that $d + b + 3 = d + c + 5$ is a theorem, by choosing X, Y, and E of Leibniz, as follows.

$$X:\ b+3 \qquad E:\ d+z$$
$$Y:\ c+5 \qquad z:\ z$$

---

[4] A binary operator $\star$ (or function f with two parameters) is called *symmetric*, or *commutative*, if $x \star y = y \star x$ (or $f(x,y) = f(y,x)$) for all arguments x and y. Hence, (1.3) asserts that $=$ is a symmetric operator.

HISTORICAL NOTE 1.2. LEIBNIZ'S DEFINITION OF EQUALITY

Def. 1. Two terms are the same (*eadem*) if one can be substituted for the other without altering the truth of any statement (*salva veritate*). If we have A and B, and A enters into some true proposition, and the substitution of B for A wherever it appears results in a new proposition that is likewise true, and if this can be done for every proposition, then A and B are said to be the *same*; and conversely, if A and B are the same, they can be substituted for one another as I have said. Terms that are the same are also called *coincident* (*coincidentia*); A and A are, of course, said to be the same, but if A and B are the same, they are called *coincident*.

Def. 2. Terms that are not the same, that is, terms that cannot always be substituted for one another, are *different* (*diversa*).

Corollary. Whence also, whatever terms are not different are the same.

Charact. 1. $A \infty B$ signifies that A and B are the same, or *coincident*.

Charact. 1. A non B signifies that A and B are *different*.

(From [29, page 291], which is an English translation of the Latin version of Leibniz's work found in [19]. Note that Leibniz used the sign ∞ for equality.)

1.4   Leibniz's rule and function evaluation

A function is a rule for computing a value v (say) from another value w (say). Value w is called the *argument* and v the corresponding result. For example, consider the function g defined by

$$(1.6) \quad g(z) = 3 \cdot z + 6 \ .$$

Function g has the value of $3 \cdot w + 6$ for any argument w. The argument is designated in a *function application*, which is a form of expression. The conventional notation for the function application that applies g to the argument 5 is $g(5)$; it yields the value of $3 \cdot 5 + 6$. In order to reduce the use of parentheses when writing function definitions and function applications, we use the notation $g.5$ instead of $g(5)$ when the parameter or argument is an identifier or constant. For example, function g of (1.6) might be defined by

$$g.z : 3 \cdot z + 6 \ .$$

We give two examples of evaluation of function applications.

$g.5$		$g(y + 2)$
$=$ ⟨Apply function⟩		$=$ ⟨Apply function⟩
$3 \cdot 5 + 6$		$3 \cdot (y + 2) + 6$
$=$ ⟨Arithmetic⟩		
21		

Function application can be defined in terms of textual substitution: If

(1.7) $g.z : E$

defines function g, then function application $g.X$ for any argument X is defined by $g.X = E[z := X]$. This close correspondence between function application and textual substitution suggests that Leibniz (1.5) links equality and function application, and we can reformulate (1.5) as

(1.8)   **Leibniz:** $\dfrac{X = Y}{g.X = g.Y}$.

This rule indicates that from the equality of X and Y we can deduce the equality of function applications $g.X$ and $g.Y$. This fundamental property of equality and function application holds for any function g and expressions X and Y.

In fact, any expression can (momentarily) be viewed as a function of one or more of its variables. For example, we can consider $x + y$ as a function $gx.x = x + y$ or as a function $gy.y = x + y$. Hence, the two Leibniz rules (1.5) and (1.8) are just two different forms of the same rule.

1.5   Reasoning using Leibniz's rule

Leibniz (1.5) allows us to "substitute equals for equals" in an expression without changing the value of that expression. It therefore gives a method for demonstrating that two expressions are equal. In this method, the format we use to show an application of Leibniz is

$$
\begin{aligned}
&E[z := X] \\
= \quad &\langle X = Y \rangle \\
&E[z := Y] \quad .
\end{aligned}
$$

The first and third lines are the equal expressions of the conclusion in Leibniz; the hint on the middle line is the premise $X = Y$. The hint is indented and delimited by $\langle$ and $\rangle$. Variable z of Leibniz is not mentioned at all.

Here is an illustration of the use of Leibniz from the problem of John, Mary, and the rotten apples on page 1.

$$
\begin{aligned}
&m/2 = 2 \cdot (j - 1) \\
= \quad &\langle m = 2 \cdot j, \text{ by } (0.1) \rangle \\
&2 \cdot j/2 = 2 \cdot (j - 1)
\end{aligned}
$$

Here, E of Leibniz is $z/2 = 2 \cdot (j - 1)$, X is m, and Y is $2 \cdot j$.

Leibniz is often used in conjunction with Substitution (1.1), in the following manner. Suppose we know that the following is a theorem:

(1.9) $2 \cdot x / 2 = x$.

The following calculation uses both Leibniz and Substitution.

$$2 \cdot j / 2 = 2 \cdot (j - 1)$$
$$= \quad \langle (1.9), \text{ with } x := j \rangle$$
$$j = 2 \cdot (j - 1)$$

We are using Leibniz with the premise $2 \cdot j / 2 = j$. We can use this premise only if it is a theorem. It is, because $2 \cdot x / 2 = x$ is a theorem and, therefore, by Substitution, $(2 \cdot x / 2 = x)[x := j]$ is a theorem.

If a use of Substitution is simple enough, as in this case, we may leave off the indication "with $x := j$" and sometimes even the rule number from the hint, writing simply

$$2 \cdot j / 2 = 2 \cdot (j - 1)$$
$$= \quad \langle 2 \cdot x / 2 = x \rangle$$
$$j = 2 \cdot (j - 1)$$

We may also place an explanatory comment in a hint (after a dash —), as in the following hint.

$$\langle 2 \cdot x / 2 = x \text{ —note that } / \text{ is division} \rangle$$

A proof that involves a sequence of applications of Leibniz has the following general form:

$$E0$$
$$= \quad \langle \text{Explanation of why } E0 = E1, \text{ using Leibniz} \rangle$$
$$E1$$
$$= \quad \langle \text{Explanation of why } E1 = E2, \text{ using Leibniz} \rangle$$
$$E2$$
$$= \quad \langle \text{Explanation of why } E2 = E3, \text{ using Leibniz} \rangle$$
$$E3$$

The proof establishes that $E0 = E3$, by Transitivity (1.4) and the individual steps $E0 = E1$, $E1 = E2$, and $E2 = E3$.

Most proofs of equalities in this text use the format just introduced. In it, the expressions Ei are aligned but indented past the column in which $=$ appears, the hints are aligned and indented a bit further, and the hints are delimited by $\langle$ and $\rangle$. Each hint gives the premise $X = Y$ for an application of Leibniz. Parentheses are *never* placed around the expressions

---

HISTORICAL NOTE 1.3. SYMBOLS FOR EQUALITY

The history of the signs for equality is so interesting and involved that Cajori [7] devotes 12 pages to it. In the fifteenth century, a number of different symbols were used infrequently for equality, including the dash. Generally, however, equality was expressed using words like *aequales, esgale, gleich*, and sometimes by the abbreviated form *aeq*.

The use of = for equality was introduced by Robert Recorde in 1577 [31], who wrote,

> And to auoide the tediouse repetition of these woordes: is equalle
> to: I will sette as I doe often in woorke use, a paire of parallels, or
> Gemowe lines of one lengthe, thus: ====, bicause noe .2. thynges,
> can be more equalle.

Recorde viewed = only as an abbreviation and not as a boolean function. The concept of function took over 100 years to develop (Leibniz introduced the term in 1694), and the notion of boolean took another 150 years (George Boole introduced it in about 1850)!

In spite of the appropriateness of Recorde's symbol for equality, it did not appear in print again until sixty-one years later, many authors preferring to use a word rather than a symbol for equality. One problem was that = was in use at the time for at least five different purposes. Also, there were competing symbols for equality, to cite a few: [, |, ⊓, ∞ , and 2|2 (by Hérigone, in 1634, who also used 3|2 and 2|3 for > and <).

In the late seventeenth century, = became the favorite in England for equality. However, = faced real competition on the continent, where Descartes had used the symbol ∞ for Taurus to denote equality in 1637 in [11]. On the continent, most authors used either Descartes' symbol or no symbol at all. But in the eighteenth century, = gradually won out, in large part due to the adoption of = by Newton and Leibniz at the close of the seventeenth century.

Today, Recorde's =, the only symbol he introduced, is universally embraced. Equality is one of our most important concepts, and it deserves a unique symbol. The use of = for assignment in FORTRAN has only caused confusion, as has the use of = for assignment and == for equality in C.

$E0$, $E1$, etc., because the line breaks in the proof take their place. Adhere carefully to this format; the more standard our communication mechanism, the easier time we have understanding each other.

1.6   The assignment statement

In the previous section, we showed how textual substitution was inextricably intertwined with equality. We now show a correspondence between textual substitution and the assignment statement that allows programmers to reason about assignment.

Execution of the assignment statement

(1.10) $x := E$

evaluates expression E and stores the result in variable x.[5] Assignment $x := E$ is read as "x becomes E".[6]

Execution of (1.10) in a state stores in x the value of E in that state, thus changing the state. For example, suppose the state consists of $(v, 5)$, $(w, 4)$, $(x, 8)$ and consider the assignment $v := v + w$. The value of $v + w$ in the state is 9, so executing $v := v + w$ stores 9 in v, changing the state to $(v, 9)$, $(w, 4)$, $(x, 8)$.

Just as important as how to execute an assignment statement is a way to reason about its effect. For example, from a precondition for an assignment,[7] how can we determine a corresponding postcondition? Or, from a postcondition, can we determine a suitable precondition? The conventional way of indicating a precondition and a postcondition for a statement S is

(1.11) $\{P\}\ S\ \{Q\}$,

where P is the precondition and Q is the postcondition, This is known as a *Hoare triple*, after C.A.R. Hoare (see Historical note 1.4), who invented the notation in giving the first definition for a programming language in terms of how programs could be proved correct with respect to their specifications rather than in terms of how they could be executed.

For example,

$$\{x = 0\}\ x := x + 1\ \{x > 0\}$$

is a Hoare triple that is *valid* iff execution of $x := x + 1$ in any state in which x is 0 terminates in a state in which $x > 0$. Here are two other

---

[5] For the time being, we assume that E always has a value and that the value can be stored in x. We treat the more general case later, in Sec. 10.2.

[6] Perhaps because of the use of $=$ for assignment in FORTRAN (see Historical note 1.3), assignment is often read as "x equals E". This causes great confusion. The first author learned to distinguish between $=$ and $:=$ while giving a lecture in Marktoberdorf, Germany, in 1975. At one point, he wrote "$:=$" on the board but pronounced it "equals". Immediately, the voice of Edsger W. Dijkstra boomed from the back of the room: "becomes!". After a disconcerted pause, the first author said, "Thank you; if I make the same mistake again, please let me know.", and went on. Once more during the lecture the mistake was made, followed by a booming "becomes" and a "Thank you". The first author has never made that mistake again! The second author, having received his undergraduate education at Cornell, has never experienced this difficulty.

[7] Recall from Chapter 0 that a precondition of a statement is an assertion about the program variables in a state in which the statement may be executed, and a postcondition is an assertion about the states in which it may terminate.

valid Hoare triples for the assignment statement $x := x + 1$.

$$\{x > 5\} \quad x := x + 1 \quad \{x > 0\}$$
$$\{x + 1 > 0\} \quad x := x + 1 \quad \{x > 0\}$$

The Hoare triple

$$\{x = 5\} \ x := x + 1 \ \{x = 7\}$$

is *not* valid, because execution of $x := x + 1$ in a state in which $x = 5$ does not terminate in a state in which $x = 7$.

Formula (1.12) below schematically defines valid Hoare triples for an assignment $x := E$ in terms of textual substitution: for any postcondition R, a suitable precondition is $R[x := E]$. Thus, the precondition is calculated from the assignment and the postcondition.[8]

(1.12) **Definition of assignment:** $\{R[x := E]\} \ x := E \ \{R\}$

As an example, consider the assignment $x := x + 1$ and postcondition $x > 4$. Thus, in definition (1.12) we would take E to be $x + 1$ and R to be $x > 4$. We conclude that a precondition for a valid triple is $(x > 4)[x := x + 1]$, which is $x + 1 > 4$.

Here are more examples of the use of definition (1.12).

$$\{x + 1 > 5\} \quad x := x + 1 \quad \{x > 5\}$$
$$\{5 \neq 5\} \quad x := 5 \quad \{x \neq 5\}$$
$$\{x^2 > x^2 \cdot y\} \quad x := x^2 \quad \{x > x \cdot y\}$$

Let us see why definition (1.12) is consistent with execution of $x := E$. Call the initial program state s and the final state s'. We will show that $R[x := E]$ has the same value in s as R does in s'. This suffices because then execution begun in a state in which $R[x := E]$ is *true* will terminate in a state in which R is *true*.

Note that R and $R[x := E]$ are exactly the same except that where R has an occurrence of x, $R[x := E]$ has an occurrence of "(E)". Since each variable except x has the same value in states s and s', we need only show that the value of E in state s equals the value of x in state s'. This last fact holds because execution of $x := E$ begun in state s stores into x the value of E in s.

---

[8] The tendency is to expect the postcondition to be calculated from the precondition and to expect the definition to be $\{R\} \ x := E \ \{R[x := E]\}$. Fight this intuition, for it is not consistent with how the assignment is executed. For example, using this incorrect rule, we would obtain $\{x = 0\} \ x := 2 \ \{(x = 0)[x := 2]\}$, which is invalid. This is because when the assignment terminates, the resulting state does not satisfy the postcondition, *false*. As will be seen later, definition (1.12) works well with methodologies for the formal development of programs.

HISTORICAL NOTE 1.4. C.A.R. HOARE (1934–)

C. Anthony (Tony) R. Hoare, a Fellow of the Royal Society and Professor of Computer Science at Oxford University, has made fundamental contributions to programming and programming languages. He is the creator of the sorting algorithm Quicksort. His 1969 axiomatic definition of a programming language gave us a basic manipulative method for reasoning about programs. He has developed major programming constructs for concurrency, e.g. the monitor and CSP (Communicating Sequential Processes), and has made deep and fundamental contributions to the theory of programming languages.

Hoare received the ACM Turing Award in 1980. His Turing lecture, *The Emperor's old clothes*, illustrates well why he has had so much impact in our field. It is original, perceptive, elegant, extremely well-written, and pertinent to the technical and social problems of the field.

Twenty-two of Hoare's most major contributions have been collected in [22]. The last essay, *Envoi*, explains part of Hoare's success. It begins with "I enjoy writing." The fourth paragraph begins with "I enjoy rewriting" and discusses how and how many, many times he rewrites an article before it is suitable to show to friends and colleagues. After that, he says, his article may sit in a drawer for many months as responses accumulate. Finally, it will be entirely rewritten before being submitted for publication. A paper may be rewritten eight or more times before it appears in print. Hoare mentions that throughout the development of an idea, the most important requirement is that it be clearly and convincingly explained. Obviously, Hoare understands that communicating ideas well is just as important as having ideas.

As a final note, the house owned by Hoare's in-laws was the setting for much of the movie *Room with a View*.

In some programming languages, the assignment statement is extended to the *multiple assignment* $x_1, x_2, \ldots, x_n := E_1, E_2, \ldots, E_n$, where the x_i are distinct variables and the E_i are expressions. The multiple assignment is executed as follows. First evaluate all the expressions E_i to yield values v_i (say); then assign v_1 to x_1, v_2 to x_2, ..., and finally v_n to x_n. Note that all expressions are evaluated before any assignments are performed. Thus, the last two examples of multiple assignment in Table 1.2 are equivalent.

TABLE 1.2. EXAMPLES OF MULTIPLE ASSIGNMENTS

$x, y := y, x$	Swap x and y
$x, i := 0, 0$	Store 0 in x and i
$i, x := i + 1, x + i$	Add 1 to i and i to x
$x, i := x + i, i + 1$	Add 1 to i and i to x

Definition (1.12) for assignment actually holds for the multiple assignment, when one considers x in (1.12) to be a list of distinct variables and E to be a list of expressions. Thus, multiple assignment is defined in terms of simultaneous textual substitution. Table 1.3 gives valid Hoare triples for several assignments; in each, the precondition is determined using (1.12).

The preconditions for the last two assignments of Table 1.3 are identical, even though the variables and expressions in the assignments appear in a different order. Note also that in the last two cases, the precondition equals the postcondition (to see this, subtract i from the LHS[9] and RHS of the precondition). When the precondition and the postcondition are equal, we say that the precondition is *maintained* by execution of the assignments, or equivalently, the assignment *maintains* or preserves the precondition.

To see the difference between multiple assignment and a sequence of assignments, consider

$$x, y := x + y, x + y \quad \text{and} \quad x := x + y; \ y := x + y.$$

In initial state $(x, 2)$ and $(y, 3)$, execution of the first sets both x and y to 5, but execution of the second would set x to 5 and then y to 8. So they are different. The preconditions that are constructed using Definition (1.12) for $x, y := E, F$ and for the sequence $x := E; \ y := F$ with postcondition R are given below. In the second case, the definition is first used to find the precondition for the last assignment, which is also the postcondition for the first assignment; then the definition is used again to find the precondition for the first assignment.

$$\{R[x, y := E, F]\} \ x, y := E, F \ \{R\}$$
$$\{R[y := F][x := E]\} \ x := E; \ y := F \ \{R\}$$

---

[9] LHS and RHS stand for the *lefthand side* and *righthand side* of the equation.

TABLE 1.3. Examples of Hoare Triples for Multiple Assignment

$$\{y > x\} \ x, y := y, x \ \{x > y\}$$

$$\{x + i = 1 + 2 + \cdots + (i + 1 - 1)\}$$
$$x, i := x + i, i + 1$$
$$\{x = 1 + 2 + \cdots + (i - 1)\}$$

$$\{x + i = 1 + 2 + \cdots + (i + 1 - 1)\}$$
$$i, x := i + 1, x + i$$
$$\{x = 1 + 2 + \cdots + (i - 1)\}$$

We showed above that $R[x, y := \ldots]$ is, in general, different from $R[x := \ldots][y := \ldots]$, so it should be no surprise that these two assignments have different effects.

It is a shame that the multiple assignment is not included in more programming languages. The programmer is frequently called upon to specify a state change that involves modifying several variables in one step, where the values assigned all depend on the initial state, and the multiple assignment is ideally suited for this task.

Exercises for Chapter 1

1.1 Perform the following textual substitutions. Be careful with parenthesization and remove unnecessary parentheses.

(a) $x[x := b + 2]$
(b) $x + y \cdot x[x := b + 2]$
(c) $(x + y \cdot x)[x := b + 2]$
(d) $(x + x \cdot 2)[x := x \cdot y]$
(e) $(x + x \cdot 2)[y := x \cdot y]$
(f) $(x + x \cdot y + x \cdot y \cdot z)[x := x + y]$

1.2 Perform the following simultaneous textual substitutions. Be careful with parenthesization and remove unnecessary parentheses.

(a) $x[x, y := b + 2, x + 2]$
(b) $x + y \cdot x[x, y := b + 2, x + 2]$
(c) $(x + y \cdot x)[x, y := b + 2, x + 2]$
(d) $(x + x \cdot 2)[x, y := x \cdot y, x \cdot y]$
(e) $(x + y \cdot 2)[y, x := x \cdot y, x \cdot x]$
(f) $(x + x \cdot y + x \cdot y \cdot z)[x, y := y, x]$

1.3 Perform the following textual substitutions. Be careful with parenthesization and remove unnecessary parentheses.

(a) $x[x := y + 2][y := y \cdot x]$
(b) $x + y \cdot x[x := y + 2][y := y \cdot x]$
(c) $(x + y \cdot x)[x := y + 2][y := y \cdot x]$
(d) $(x + x \cdot 2)[x, y := y, x][x := z]$
(e) $(x + x \cdot 2)[x, y := x, z][x := y]$
(f) $(x + x \cdot y + x \cdot y \cdot z)[x, y := y, x][y := 2 \cdot y]$

1.4 Leibniz's definition of equality given just before inference rule Leibniz (1.5) says that $X = Y$ is true in every state iff $E[z := X] = E[z := Y]$ is true in every state. Inference rule Leibniz (1.5), however, gives only the "if" part. Give an argument to show that the "only if" part follows from Leibniz (1.5). That is, suppose $E[z := X] = E[z := Y]$ is true in every state, for every expression E. Show that $X = Y$ is true in every state.

1.5 Let X, Y, and Z be expressions and z a variable. Let E be an expression, which may or may not contain Z. Here is another version of Leibniz.

$$\textbf{Leibniz:} \quad \frac{Z = X, \; Z = Y}{E[z := X] = E[z := Y]} \; .$$

Show that transitivity of $=$ follows from this definition.

1.6 Inference rule Substitution (1.1) stands for an infinite number of inference rules, each of which is constructed by instantiating expression E, list of variables v, and list of expressions F with different expressions and variables. Show three different instantiations of the inference rule, where E is $x < y \lor x \geq y$.

1.7 Inference rule Leibniz (1.5) stands for an infinite number of inference rules, each of which is constructed by instantiating E, X, and Y with different expressions. Below, are a number of instantiations of Leibniz, with parts missing. Fill in the missing parts and write down what expression E is. Do not simplify. The last two exercises have three answers; give them all.

(a) $\dfrac{x = x + 2}{4 \cdot x + y = ?}$

(b) $\dfrac{2 \cdot y + 1 = 5}{x + (2 \cdot y + 1) \cdot w = ?}$

(c) $\dfrac{x + 1 = y}{3 \cdot (x + 1) + 3 \cdot x + 1 = ?}$

(d) $\dfrac{x = y}{x + x = ?}$

(e) $\dfrac{7 = y + 1}{7 \cdot x + 7 \cdot y = ?}$

1.8 The purpose of this exercise is to reinforce your understanding of the use of Leibniz (1.5) along with a hint in proving two expressions equal. For each of the expressions $E[z := X]$ and hints $X = Y$ below, write the resulting expression $E[z := Y]$. There may be more than one correct answer.

	$E[z := X]$	hint $X = Y$
(a)	$x + y + w$	$x = b + c$
(b)	$x + y + w$	$b \cdot c = y + w$
(c)	$x \cdot (x + y)$	$x + y = y + x$
(d)	$(x + y) \cdot w$	$w = x \cdot y$
(e)	$(x + y) \cdot q \cdot (x + y)$	$y + x = x + y$

1.9 The purpose of this exercise is to reinforce your understanding of the use of Leibniz (1.5) along with a hint in proving two expressions equal. For each of the following pair of expressions $E[z := X]$ and $E[z := Y]$, identify a hint $X = Y$ that would show them to be equal and indicate what E is.

	$E[z := X]$	$E[z := Y]$
(a)	$(x+y)\cdot(x+y)$	$(x+y)\cdot(y+x)$
(b)	$(x+y)\cdot(x+y)$	$(y+x)\cdot(y+x)$
(c)	$x+y+w+x$	$x+y\cdot w+x$
(d)	$x\cdot y\cdot x$	$(y+w)\cdot y\cdot x$
(e)	$x\cdot y\cdot x$	$y\cdot x\cdot x$

1.10 In Sec. 1.3, we stated that the four laws Reflexivity (1.2), Symmetry (1.3), Transitivity (1.4), and Leibniz (1.5) characterized equality. This statement is almost true. View $=$ as a function $eq(x, y)$ that yields a value *true* or *false*. There is one other function that, if used in place of eq in the four laws, satisfies all of them. What is it?

1.11 Using Definition (1.12) of the assignment statement on page 18, determine preconditions for the following statements and postconditions.

	Statement	Postcondition
(a)	$x := x+7$	$x+y > 20$
(b)	$x := x-1$	$x^2 + 2\cdot x = 3$
(c)	$x := x-1$	$(x+1)\cdot(x-1) = 0$
(d)	$y := x+y$	$y = x$
(e)	$y := x+y$	$y = x+y$

Chapter 2

Boolean Expressions

We discuss *boolean expressions*, which are named after George Boole (although the 1971 Compressed OED spells *boolean* as *boolian*!). Boolean expressions are used in one form or another in most programming languages today (e.g. Pascal, FORTRAN, C, Scheme, and Lisp), so the material in this chapter will be familiar to practicing programmers. The chapter also discusses how to model English statements as boolean expressions.

2.1 Syntax and evaluation of boolean expressions

Boolean expressions are constructed from the constants *true* and *false*, boolean variables, which can be associated (only) with the values *true* and *false*, and the boolean operators $\equiv$, $\not\equiv$, $\neg$, $\vee$, $\wedge$, $\Rightarrow$, and $\Leftarrow$. The constants *true* and *false* are often called *boolean values*, and a boolean expression is often said to be *of type boolean*.

We begin by describing the unary boolean operators —those with one operand. We do this by enumerating all boolean functions of one boolean argument, i.e. functions that have one argument of type boolean and that yield a boolean value. Since there are exactly two possible argument values and two possible result values for such a function, there are a total of four boolean functions of one boolean argument. These functions are shown in the table below. Each column to the right of the vertical line describes a function, whose name (if it has one) is given above the line. Each value below the line in such a column is the result of applying the function to the argument value appearing in the same row. Such a table is known as a *truth table*.

		id		$\neg$	
Argument	*true*	*true*	*true*	*false*	*false*
	false	*true*	*false*	*true*	*false*

For example, from the table we see that *id.true = true* and *id.false = false*.

HISTORICAL NOTE 2.1. GEORGE BOOLE (1815–1864)

George Boole was the son of a poor shopkeeper in England. In those days, little in the way of formal education was open to such people. Boole, however, was determined to learn enough to rise above the pitiful existence eked out by his father. He learned Latin and Greek on his own before he was 12 and math from his father, who also had gone beyond his schooling. At 16, young Boole got a job teaching in an elementary school to help support his parents. At 20, he opened his own school and, to prepare his pupils in math, began to study in earnest what the great masters were doing. His at-first-unguided efforts led to many contributions. He was so successful that, in 1849, he was appointed Professor of Mathematics at Queens College, Ireland.

Boole's great contribution was an algebraic basis for logic, something Leibniz had dreamed about 200 years earlier (see Historical Note 1.1 on page 9) and that De Morgan, nine years older than Boole and also a great logician, was unable to devise (see Historical Note 3.1 on page 54). In *The Laws of Thought* [6], Boole's aim was to "investigate the fundamental laws ... by which reasoning is performed, ... give expression to them in the language of a Calculus, and upon this foundation ... establish the Science of Logic ... ". Boole's work is the foundation of all mathematical logic. According to Bertrand Russell, "Pure Mathematics was discovered by Boole in ... *The Laws of Thought*.".

The two functions whose result does not depend on the argument are unnamed. Function *id* is the identity function of one argument; applying it to an argument yields the value of the argument. Function symbol $\neg$, read *negation* or *not*, is used as a prefix operator. For example, we write $\neg false$.

The sixteen boolean functions of two boolean arguments are given in the truth table below. In this table, there are two arguments in each row instead of one (as in the previous truth table).

			$\vee$	$\Leftarrow$		$\Rightarrow$		$=$	$\wedge$	nand ($\equiv$)	$\neq$					nor	
t	*t*	*t*	*t*	*t*	*t*	*t*	*t*	*t*	*t*	*f*	*f*	*f*	*f*	*f*	*f*	*f*	*f*
t	*f*	*t*	*t*	*t*	*t*	*f*	*f*	*f*	*f*	*t*	*t*	*t*	*t*	*f*	*f*	*f*	*f*
f	*t*	*t*	*t*	*f*	*f*	*t*	*t*	*f*	*f*	*t*	*t*	*f*	*f*	*t*	*t*	*f*	*f*
f	*f*	*t*	*f*	*t*	*f*	*t*	*f*	*t*	*f*	*t*	*f*	*t*	*f*	*t*	*f*	*t*	*f*

Note: *true* is abbreviated by *t* and *false* by *f*

Eight of the functions in the table are useful enough to be named. Applications of these functions are written in infix form, so they are called *operators*[1]. For example, $b \vee c$ and $x \Leftarrow y$ denote applications of the functions in the second and third columns of the table, respectively. We now discuss the operators, in the order in which they will be discussed later in Chap. 3.

Operator $=$ is conventional equality. Expression $b = c$ is read as "b equals c". For boolean operands only, equality is given a second name, *equivalence*, and a second symbol, $\equiv$. We read $b \equiv c$ as "b equivales c".[2] The operands of $\equiv$ are called *equivalents*. In Sec. 2.2, we explain the use of the two different symbols for equality.

Operator $\neq$ is conventional inequality. Expression $b \neq c$ is read as "b differs from c". Operator $\neq$ satisfies $(b \neq c) = \neg(b = c)$. For boolean operands only, inequality is given a second name, *inequivalence*, and a second symbol, $\not\equiv$. Inequality is sometimes called *xor*, for *exclusive or*, since it is *true* when exactly one of the two operands is *true*.

Operator $\vee$ is called *disjunction* or *or*. Expression $b \vee c$ is read as "b or c", because it is *true* iff b or c (or both) is *true*. Operands b and c of $b \vee c$ are called *disjuncts*.

Operator $\wedge$ is called *conjunction* or *and*. Expression $b \wedge c$ is read as "b and c", because it is *true* only if both operands b and c are *true*. Operands b and c of $b \wedge c$ are called *conjuncts*.

Operator $\Rightarrow$ is called *implication*. Expression $b \Rightarrow c$ is read as "b implies c" or as "if b then c". Operands b and c are called the *antecedent* and *consequent*, respectively. Note that $b \Rightarrow c$ is *true* if b is *false*. This is consistent with the usual English interpretation of a statement like "If Schneider is ten feet tall, then Gries can walk on the ceiling" as being *true* simply because Schneider is not ten feet tall. False implies anything, as the saying goes. We discuss implication in more detail in Sec. 2.4.

Operator $\Leftarrow$ is called *consequence*. Expression $b \Leftarrow c$ is read as "b follows from c". Operands b and c are called the *consequent* and *antecedent*, respectively. Since $b \Rightarrow c$ is equal to $c \Leftarrow b$ (according to the truth table), $\Leftarrow$ might seem superfluous. Later, we see how it can help make some proofs more palatable.

The names of operators *nand* and *nor* stand for "not and" and "not or", respectively. Expression b *nand* c is equal to $\neg(b \wedge c)$, while b *nor* c is equal to $\neg(b \vee c)$. These operators are useful when implementing switching circuits, as discussed in Sec. 5.2.

---

[1] Boolean operators are also called *connectives*.

[2] The *Oxford English Dictionary* defines *equivale* as "to be equivalent to".

USING TRUTH TABLES TO EVALUATE BOOLEAN EXPRESSIONS

In addition to defining boolean operators, truth tables can be used to compute the value of any boolean expression, in every state. The truth table below gives the value of the expression $p \lor (q \land \neg r)$. The first three columns of each row of the table describe a state by giving values for p, q, and r. Together, the eight rows describe all states. In each row, successive columns to the right of the vertical line contain values for subexpressions of $p \lor (q \land \neg r)$, with each being calculated from the values of *its* subexpressions, which appear to the left in some column. The righthand column of each row contains the value of the entire expression $p \lor (q \land \neg r)$ for the values of p, q, and r given in that row. Such a truth table allows us to determine the value of an expression in any state in a systematic fashion.

p	q	r	$\neg r$	$q \land \neg r$	$p \lor (q \land \neg r)$
t	t	t	f	f	t
t	t	f	t	t	t
t	f	t	f	f	t
t	f	f	t	f	t
f	t	t	f	f	f
f	t	f	t	t	t
f	f	t	f	f	f
f	f	f	t	f	f

PRECEDENCE OF BOOLEAN OPERATORS

A table of precedences of operators appears on the inside front cover.

Not all texts assign $\lor$ and $\land$ the same precedence, as we do. Sometimes, $\lor$ and $+$ are given the same precedence, and $\land$ and $\cdot$ are given another, but higher, precedence. One even finds 1 used for *true*, 0 for *false*, $+$ for $\lor$, and $\cdot$ for $\land$. This overloading of boolean and arithmetic operators can lead to misconceptions, because the rules for manipulation of boolean and arithmetic expressions are different. For example, *true* $\lor$ *true* $\equiv$ *true* evaluates to *true* but $1 + 1 = 1$ evaluates to $2 = 1$ and thus to *false*. Also, the first expression below evaluates to *true* but the second does not.

$$x \lor (y \land z) \equiv (x \lor y) \land (x \lor z)$$
$$x + (y \cdot z) = (x + y) \cdot (x + z)$$

2.2 Equality versus equivalence

The boolean expression $b \equiv c$ is evaluated exactly as $b = c$, except that $\equiv$ can be used only when b and c are boolean expressions. We now discuss the reasons for having two infix symbols $=$ and $\equiv$ for the same boolean operation.

First, giving the binary boolean operators lower precedences than the binary arithmetic operators and assigning different precedences for $=$ and $\equiv$, as in the precedence table on the inside front cover, allows us to avoid parentheses in expressions like

$$x \cdot y = 0 \;\; \equiv \;\; x = 0 \vee y = 0 \;\;.$$

Note how the extra space surrounding $\equiv$ serves as a reminder that $\equiv$ has lower precedence. We often use white space in this manner to help indicate aggregation.

The second reason for using both $=$ and $\equiv$ for equality is that one can be *conjunctional* and the other *associative*. All the operators on line (j) of the precedence table on the inside front cover are *conjunctional*. For $\circ$ and $\star$ conjunctional operators, expression $b \circ c \star d$ is an abbreviation for $b \circ c \wedge c \star d$. For example,

$$b = c < d \text{ is an abbreviation for } b = c \wedge c < d \quad,$$
$$b = c = d \text{ is an abbreviation for } b = c \wedge c = d \quad.$$

Hence, $b = c = d$ and $(b = c) = d$ are different; the former uses $=$ conjunctionally, while the latter does not. In the state with $(b, false)$, $(c, false)$, and $(d, true)$, $b = c = d$ is *false* but $(b = c) = d$ is *true*.

Operator $\equiv$ is *associative*, which means that $b \equiv c \equiv d$, $(b \equiv c) \equiv d$, and $b \equiv (c \equiv d)$ are all equivalent.[3] Being associative, $\equiv$ cannot also be conjunctional. On the other hand, $=$ is conjunctional but not associative. Thus, in formulas without parentheses, sequences of $=$ and sequences of $\equiv$ mean different things.

Treat the conjunctional use of $=$ and other operators as *syntactic sugar*, i.e. as an extension to the basic definition of expressions to make writing some expressions easier. Whenever an expression that contains this syntactic sugar is to be evaluated or manipulated, first remove the syntactic sugar. For example, we evaluate *false = false = true* and *false $\equiv$ false $\equiv$ true* :

---

[3] Binary operator $\circ$ is associative iff $((b \circ c) \circ d) = (b \circ (c \circ d))$ for all b, c, d.

HISTORICAL NOTE 2.2. NOTATIONAL SURPRISES

A few examples of surprises in programming languages will show the need for extreme care in understanding and stating notational conventions. In mathematics, $1/bc$ stands for $1/(bc)$, where the juxtaposition of b and c denotes multiplication of b and c. However, in FORTRAN, Algol, and most other imperative programming languages, $1/b * c$ means $(1/b) * c$ (i.e. c/b). This difference has not caused much confusion, perhaps because of the difference in the two notations: the programming languages require an explicit operator for multiplication, while mathematical notation does not.

PL/1, a popular language of the 1960s and 1970's, exhibits an astonishing oddity: the expression $2 < 1 < 1$ has the value *true*! This is because PL/1 does not view $<$ as conjunctional and is quite happy to insert type conversions where possible. In PL/I, the expression $2 < 1 < 1$ is evaluated as follows.

$$2 < 1 < 1$$
$=$ ⟨$2 < 1 \equiv \ '0'B$, PL/1's representation of *false* ⟩
$$'0'B < 1$$
$=$ ⟨The bit $'0'B$ is converted into the integer 0⟩
$$0 < 1$$
$=$ ⟨$0 < 1 \equiv$ *true*, which is represented by $'1'B$ ⟩
$$'1'B$$
$=$ ⟨$'1'B$ is PL/1's representation of *true* ⟩
$$true$$

$false = false = true$
$=$ ⟨$=$ is conjunctional⟩
$(false = false) \wedge (false = true)$
$=$ ⟨Evaluate both $=$⟩
$true \wedge false$
$=$ ⟨Evaluate $\wedge$⟩
$false$

$false \equiv false \equiv true$
$=$ ⟨Evaluate first $\equiv$⟩
$true \equiv true$
$=$ ⟨Evaluate $\equiv$⟩
$true$

As another example, we show below how to change occurrences of $\equiv$ into occurrences of $=$, and vice versa. In these manipulations, we parenthesize operations $b = c$ and $b \equiv c$ before making the replacements because $=$ and $\equiv$ have different precedences and we do not want the different precedences implicitly to change the structure of the expression.

$b \equiv c \equiv d$
$=$ ⟨Parenthesize⟩
$(b \equiv c) \equiv d$
$=$ ⟨Replace operators⟩
$(b = c) = d$

$b = c = d$
$=$ ⟨$=$ is conjunctional⟩
$b = c \wedge c = d$
$=$ ⟨Parenthesize⟩
$(b = c) \wedge (c = d)$
$=$ ⟨Replace operators⟩
$(b \equiv c) \wedge (c \equiv d)$

This interplay between $=$ and $\equiv$ may seem confusing. It is the product of conventions passed down over the years. The conventions are rarely stated, leading to misunderstandings —for example, see Historical note 2.2.

2.3 Satisfiability, validity, and duality

We now define some terms that will be useful later on.

(2.1)   **Definition.** A boolean expression P is *satisfied* in a state if its value is *true* in that state; P is *satisfiable* if there is a state in which it is satisfied; and P is *valid* if it is satisfied in every state. A valid boolean expression is called a *tautology*.

For example, $p \lor q$ is satisfied in any state that contains the pair $(p, true)$, so it is satisfiable. But it is not valid, since it is not satisfied in a state containing $(p, false)$ and $(q, false)$. Expression $p \lor p \equiv p$ is valid.

Being familiar with boolean expressions includes having a familiarity with various simple expressions that are valid —i.e. are *true* in all states. The following definition of duality helps reduce the number of valid expressions one has to remember. Examples of duals are given in Table 2.1.

(2.2)   **Definition.** The *dual* P_D of a boolean expression P is constructed from P by interchanging occurrences of

$$true \text{ and } false,$$
$$\land \text{ and } \lor,$$
$$\equiv \text{ and } \not\equiv,$$
$$\Rightarrow \text{ and } \not\Leftarrow, \text{ and}$$
$$\Leftarrow \text{ and } \not\Rightarrow.$$

We use the notion of duality to state Metatheorem [4] Duality (2.3). We give (2.3a) without proof, because its proof requires techniques that we

---

[4] See the footnote on page 45 for a definition of "metatheorem".

TABLE 2.1. Examples of Duals

P	P_D
$p \lor q$	$p \land q$
$p \Rightarrow q$	$p \not\Leftarrow q$
$p \equiv \neg p$	$p \not\equiv \neg p$
$false \not\equiv true \lor p$	$true \equiv false \land p$
$\neg p \land \neg q \equiv r$	$\neg p \lor \neg q \not\equiv r$

have not yet developed. See Exercises 12.43–12.46.

(2.3) **Metatheorem Duality.**
    (a) P is valid iff $\neg P_D$ is valid.
    (b) $P \equiv Q$ is valid iff $P_D \equiv Q_D$ is valid.

Table 2.2 illustrates Metatheorem (2.3). In Table 2.2, all the expressions on the left are valid; hence, so are the expressions on the right. The two expressions on the last line of this table are called "De Morgan's laws", after Augustus De Morgan (see Historical note 3.1 on page 54). Remembering one of these two valid expressions is enough, because the other one can be obtained using Duality.

2.4 Modeling English propositions

We use the term *proposition* for a statement that can be interpreted as being either *true* or *false*. An example of a proposition is

(2.4) Henry VIII had one son and Cleopatra had two.

We now investigate how a proposition can be translated into a boolean expression. There are at least two reasons for performing such translations. First, English is often ambiguous, and the translation process may force us to identify and resolve the ambiguity. In the same way, lawyers write in a very stylized manner, which has evolved partly to avoid ambiguity (and partly to baffle the uninitiated). A second reason to translate propositions into boolean expressions is that we can then analyze, reason about, manipulate, and simplify the expressions (using rules introduced in the next chapter). As we will see, rules of logic provide an effective alternative to reasoning in English.

TABLE 2.2. Using Duality to Generate Valid Expressions

P (valid)	$\neg P_D$ (also valid)
true	$\neg false$
$p \vee true$	$\neg(p \wedge false)$
$p \vee \neg p$	$\neg(p \wedge \neg p)$

$P \equiv Q$ (valid)	$P_D \equiv Q_D$ (also valid)
$true \equiv true$	$false \equiv false$
$p \vee q \equiv q \vee p$	$p \wedge q \equiv q \wedge p$
$p \equiv q \equiv q \equiv p$	$p \not\equiv q \equiv q \not\equiv p$
$\neg(p \vee q) \equiv \neg p \wedge \neg q$	$\neg(p \wedge q) \equiv \neg p \vee \neg q$

One trivial way to translate a proposition into a boolean expression is simply to create a boolean variable to denote that proposition. For example, we can use variable p to stand for proposition (2.4), with the meaning that p is *true* exactly when (2.4) is:

p : Henry VIII had one son and Cleopatra had two.

A boolean variable that can denote a proposition is sometimes called a *propositional variable*, but we will stick to the term *boolean variable*.

Note that (2.4) contains two subpropositions, "Henry VIII had one son" and "Cleopatra had two (sons)". If we give names to these propositions:

x : Henry VIII had one son,
y : Cleopatra had two (sons),

we can rewrite proposition (2.4) as the English statement "x and y", which we can then translate into the boolean expression $x \land y$. Hence, another translation of (2.4) would be $x \land y$.

Obviously, the translation of a proposition into a boolean expression depends on which of its subpropositions are represented by boolean variables. The smaller the subpropositions so represented, the more logical structure the resulting boolean expression will have.

The process of translating a proposition into a boolean expression can be summarized as follows.

(2.5)   **Translation into a boolean expression.** To translate proposition p into a boolean expression:

1. Introduce boolean variables to denote subpropositions.

2. Replace these subpropositions by their corresponding boolean variables.

3. Translate the result of step 2 into a boolean expression, using "obvious" translations of the English words into operators. Table 2.3 gives examples of translations of English words.

TABLE 2.3. TRANSLATION OF ENGLISH WORDS

and, but	becomes	$\land$
or	becomes	$\lor$
not	becomes	$\neg$
it is not the case that	becomes	$\neg$
if p then q	becomes	$p \Rightarrow q$

In programming, there is a tendency to use long identifiers to convey meaning. This is not advisable here, for long identifiers make expressions unwieldy, and symbolic manipulation then becomes painful. Further, manipulation is generally performed according to given rules, without regard for the meaning of the identifiers, so knowing their meaning is of no benefit. Use short identifiers.

We now give other examples of translating propositions into boolean expressions. First, we introduce two more boolean variables, so that we have

x : Henry VIII had one son,

y : Cleopatra had two (sons),

z : I'll eat my hat,

w : 1 is prime.

We then have the following sentences and their translations.

proposition	translation
Henry VIII had one son or I'll eat my hat.	$x \lor z$.
Henry VIII had one son and 1 is not prime.	$x \land \neg w$.
If 1 is prime and Cleopatra had two sons, I'll eat my hat.	$w \land y \Rightarrow z$.

In the second example, some rearrangement of the sentence was necessary before the translation could be performed. The phrase "1 is not prime" had to be rephrased as "it is not the case that 1 is prime", so that it could be translated into "it is not the case that w" and finally into "$\neg w$".

Due to the subtleties, vagaries, and ambiguities of English, translation from English into boolean expressions is not always easy. English is so flexible that it would be impossible to give rules for translating all English statements. Below, we limit the discussion to some subtle and intricate points in performing the translation.

TRANSLATION OF "OR"

The word "or" in English is sometimes used in an inclusive sense and sometimes in an exclusive sense. The sentence "Wear a blue shirt or blue socks" would be considered inclusive, since you could wear both. On the other hand, "I'll spend my two-day vacation in Florida or Vermont" — i.e. "I'll spend my two-day vacation in Florida or I'll spend my two-day vacation in Vermont"— would be considered exclusive, since one cannot spend the two days in both places simultaneously. The inclusive sense of "b or c" is translated as $b \lor c$. The exclusive sense can be translated into $b \not\equiv c$, since the exclusive or of b and c is *true* exactly when one of them is *true* and the other *false*. The exclusive sense can also be written as $b \equiv \neg c$.

DEALING WITH IMPLICATION

Sentences of the form "If b then c" or "If b, c" are usually translated as $b \Rightarrow c$. For example, consider the sentence "If you don't eat your spinach, I'll spank you". Using variable *es* for "you eat your spinach" and variable *sy* for "I'll spank you", we translate this as $\neg es \Rightarrow sy$. Note that this expression is *true* if you eat your spinach, i.e. if $\neg es$ is *false* then so is $\neg es \Rightarrow sy$. This fact may seem strange at first. But note that the two sentences

> If you don't eat your spinach, I'll spank you
> Eat your spinach or I'll spank you

have the same meaning. Therefore, since the second is *true* if you eat your spinach, the first should be also. This equivalence between $\neg es \Rightarrow sy$ and $es \lor sy$ will be revisited in the next chapter, but it can be deduced using the truth tables on pages 25–26.

Sometimes, an implication is subtly hidden in a proposition. For example, consider the sentence "Every name in the Ithaca telephone directory is in the New York City telephone directory". This can be rewritten to reveal an implication: "If a name is in the Ithaca telephone directory, then it is in the New York City telephone directory".

IMPLICATION VERSUS EQUIVALENCE

Some "If" phrases in English are more accurately regarded as equivalences and not as implications. For example, when we say "If two sides of a triangle are equal, the triangle is isosceles", we might be defining "the triangle is isosceles" to mean "the triangle has two sides equal". Thus, using the propositions

> t :   two sides of the triangle are equal,
> *is* :   the triangle is isosceles,

we would translate this sentence as $t \equiv is$.

Oddly enough, English handles equivalence (i.e. equality) awkwardly. For example, to write "If two sides of a triangle are equal, the triangle is isosceles" unambiguously, we would have to write something like one of the following alternatives

Two sides of a triangle are equal iff the triangle is isosceles.

Two sides of a triangle are equal exactly when the triangle is isosceles.

"Two sides of a triangle are equal" is the same as "the triangle is isosceles".

and all of these are slightly awkward. [5]

The expression $bob2 \equiv bob1 \equiv bob0$ is even more difficult to translate into colloquial English. For example, if $bobj$ stands for "Bob has sight in j of his eyes", we might want to verbalize $bob2 \equiv bob1 \equiv bob0$ as

> Bob has full sight is equivalent to Bob has only partial sight is equivalent to Bob is blind.

This makes little sense in English, even though its translation $bob2 \equiv bob1 \equiv bob0$ is *true*, since exactly one of $bob0$, $bob1$, and $bob2$ is *true*.

NECESSITY AND SUFFICIENCY

When we say,

> To stay dry, it's sufficient to wear a raincoat.

we mean that if you wear a raincoat, then you will stay dry. Introducing variables sd for "stay dry" and wr for "wear a raincoat", we can formalize the above statement as $wr \Rightarrow sd$.

On the other hand,

> To stay dry, it's necessary to wear a raincoat

means that you will stay dry *only* if you wear a raincoat. In other words, staying dry implies wearing a raincoat: $sd \Rightarrow wr$. (This statement is actually *false*, since you could use an umbrella.)

Thus, "x is sufficient for y" means $x \Rightarrow y$, "x is necessary for y" means $y \Rightarrow x$, and "x is necessary and sufficient for y" means $(x \Rightarrow y) \wedge (y \Rightarrow x)$. A shorter way of saying "x is necessary and sufficient for y" is "x if and only if y", or "x iff y", and a shorter translation is $x \equiv y$.

---

[5] An anecdote provides further evidence of lack of familiarity with equivalence. Some electrical engineers were once asked what they would call the negation of the boolean binary operator that is *true* when its two operands differ. Electrical engineers use operator *xor*, or *exclusive or*, for the operator that is *true* when its operands differ, so they decided to call its negation the *exclusive nor*. They did not realize that *xor* is $\not\equiv$, so that its negation is $\equiv$!

STATEMENTS WITH MATHEMATICAL CONSTITUENTS

One way to formalize a statement like " $x > 0$ or $y = 3$ " is to associate boolean variables with the mathematical substatements. For example, if we associate xg with $x > 0$ and $y3$ with $y = 3$, we can then write the expression as $xg \lor y3$. In Chap. 9, we will extend the language of boolean expressions to the predicate calculus, so that it will no longer be necessary to replace such mathematical substatements by boolean variables.

A FINAL EXAMPLE

As a final example of translating English propositions into boolean expressions, consider the following paragraph. [6]

> If Superman were able and willing to prevent evil, he would do so. If Superman were unable to prevent evil, he would be impotent; if he were unwilling to prevent evil, he would be malevolent. Superman does not prevent evil. If Superman exists, he is neither impotent nor malevolent. Therefore, Superman does not exist.

This paragraph consists of assumptions about Superman and one conclusion (Superman does not exist), which is supposed to follow from those assumptions. In order to write this whole paragraph as an expression, we first associate identifiers with the primitive subpropositions:

a : Superman is able to prevent evil.
w : Superman is willing to prevent evil.
i : Superman is impotent.
m : Superman is malevolent.
p : Superman prevents evil.
e : Superman exists.

We then have, in order, the following translations of the first four sentences of the paragraph (we have given a name to each):

$F0: a \land w \Rightarrow p$,
$F1: (\neg a \Rightarrow i) \land (\neg w \Rightarrow m)$,
$F2: \neg p$,
$F3: e \Rightarrow \neg i \land \neg m$.

---

[6] This example, taken from [2], is adapted from an argument about the nonexistence of God in [17].

The paragraph about Superman asserts that its last sentence follows from the first four, so it can be written as the following expression:

$$F0 \wedge F1 \wedge F2 \wedge F3 \;\Rightarrow\; \neg e \;\;.$$

The reason for giving a name (i.e. a boolean variable) to each sentence now becomes clear; had we used the sentences themselves instead of their names, the final expression would have been long and unwieldy. To determine the validity of the Superman paragraph, we have to see whether this expression is *true* in all states. Rather than do this (there are $2^6 = 64$ states to check!), let us wait until we have learned rules and methods for manipulating and simplifying expressions given in the next chapter.

Exercises for Chapter 2

2.1 Each line below contains an expression and two states $S0$ and $S1$ (using t for *true* and f for *false*). Evaluate the expression in both states.

	expression	m	n	p	q	m	n	p	q
		state $S0$				state $S1$			
(a)	$\neg(m \vee n)$	t	f	t	t	f	t	t	t
(b)	$\neg m \vee n$	t	f	t	t	f	t	t	t
(c)	$\neg(m \wedge n)$	t	f	t	t	f	t	t	t
(d)	$\neg m \wedge n$	t	f	t	t	f	t	t	t
(e)	$(m \vee n) \Rightarrow p$	t	f	t	t	t	t	f	t
(f)	$m \vee (n \Rightarrow p)$	t	f	t	t	t	t	f	t
(g)	$(m \equiv n) \wedge (p \equiv q)$	f	f	t	f	t	f	t	f
(h)	$(m \equiv (n \wedge (p \equiv q)))$	f	f	t	f	t	f	t	f
(i)	$(m \equiv (n \wedge p \equiv q)$	f	f	t	f	t	f	t	f
(j)	$(m \equiv n) \wedge (p \Rightarrow q)$	f	t	f	t	t	t	f	f
(k)	$(m \equiv n \wedge p) \Rightarrow q$	f	t	f	t	t	t	f	f
(l)	$(m \Rightarrow n) \Rightarrow (p \Rightarrow q)$	f	f	f	f	t	t	t	t
(m)	$(m \Rightarrow (n \Rightarrow p)) \Rightarrow q$	f	f	f	f	t	t	t	t

2.2 Write truth tables to compute values for the following expressions in all states.

(a) $b \vee c \vee d$ (e) $\neg b \Rightarrow (b \vee c)$

(b) $b \wedge c \wedge d$ (f) $\neg b \equiv (b \vee c)$

(c) $b \wedge (c \vee d)$ (g) $(\neg b \equiv c) \vee b$

(d) $b \vee (c \wedge d)$ (h) $(b \equiv c) \equiv (b \Rightarrow c) \wedge (c \Rightarrow b)$

2.3 Write the duals P_D for each of the following expressions P .

(a) $b \vee c \vee true$ (e) $\neg false \Rightarrow b \vee c$

(b) $b \wedge c \wedge d$ (f) $\neg b \Leftarrow b \vee c$

(c) $b \wedge (c \vee \neg d)$ (g) $(\neg b \equiv true) \vee b$

(d) $b \vee (c \wedge d)$ (h) $(b \equiv c) \equiv (b \Rightarrow c) \wedge (c \Rightarrow b)$

2.4 For each expression $P \equiv Q$ below, write the expression $P_D \equiv Q_D$.

(a) $\quad p \equiv q$ (e) $\quad true \Rightarrow p \equiv p$

(b) $\quad p \wedge p \equiv p$ (f) $\quad false \Rightarrow p \equiv true$

(c) $\quad p \Rightarrow p \equiv true$ (g) $\quad p \wedge (q \vee r) \equiv (p \wedge q) \vee (p \wedge r)$

(d) $\quad p \Rightarrow q \equiv \neg p \vee q$ (h) $\quad p \equiv q \equiv q \equiv p$

2.5 Translate the following English statements into boolean expressions.

(a) Whether or not it's raining, I'm going swimming.

(b) If it's raining I'm not going swimming.

(c) It's raining cats and dogs.

(d) It's raining cats or dogs.

(e) If it rains cats and dogs I'll eat my hat, but I won't go swimming.

(f) If it rains cats and dogs while I am going swimming, I'll eat my hat.

2.6 Translate the following English statements into boolean expressions.

(a) None or both of p and q is $true$.

(b) Exactly one of p and q is $true$.

(c) Zero, two, or four of p, q, r, and s are $true$.

(d) One or three of p, q, r, and s are $true$.

2.7 Give names to the primitive components (e.g. $x < y$ and $x = y$) of the following English sentences and translate the sentences into boolean expressions.

(a) $x < y$ or $x = y$.

(b) Either $x < y$, $x = y$, or $x > y$.

(c) If $x > y$ and $y > z$, then $v = w$.

(d) The following are all $true$: $x < y$, $y < z$, and $v = w$.

(e) At most one of the following is $true$: $x < y$, $y < z$, and $v = w$.

(f) None of the following are $true$: $x < y$, $y < z$, and $v = w$.

(g) The following are not all $true$ at the same time: $x < y$, $y < z$, and $v = w$.

(h) When $x < y$, then $y < z$; when $x \geq y$, then $v = w$.

(i) When $x < y$, then $y < z$ means that $v = w$, but if $x \geq y$ then $y > z$ does not hold; however, if $v = w$ then $x < y$.

(j) If execution of program P is begun with $x < y$, then execution terminates with $y = 2^x$.

(k) Execution of program P begun with $x < 0$ will not terminate.

2.8 Translate the following English statement into a boolean expression. v is in $b[1..10]$ means that if v is in $b[11..20]$ then it is not in $b[11..20]$.

2.9 The Tardy Bus Problem, taken from [1], has three assumptions:

1. If Bill takes the bus, then Bill misses his appointment if the bus is late.

2. Bill shouldn't go home if Bill misses his appointment and Bill feels downcast.

3. If Bill doesn't get the job, he feels downcast and shouldn't go home.

The problem has eight conjectures:

4. If Bill takes the bus, then Bill does get the job if the bus is late.

5. Bill gets the job, if Bill misses his appointment and he should go home.

6. If the bus is late and Bill feels downcast and he goes home, then he shouldn't take the bus.

7. Bill doesn't take the bus if, the bus is late and Bill doesn't get the job.

8. If Bill doesn't miss his appointment, then Bill shouldn't go home and Bill doesn't get the job.

9. Bill feels downcast if the bus is late or Bill misses his appointment.

10. If Bill takes the bus and the bus is late and he goes home, then he gets the job.

11. If Bill takes the bus but doesn't get the job, then either the bus is on time or he shouldn't go home.

Translate the assumptions and conjectures into boolean expressions. Write down a boolean expression that stands for "conjecture (11) follows from the three assumptions".

2.10 Solve the following puzzle. A certain island is inhabited by people who either always tell the truth or always lie and who respond to questions with a yes or a no. A tourist comes to a fork in the road, where one branch leads to a restaurant and the other does not. There is no sign indicating which branch to take, but there is an islander standing at the fork. What single yes/no question can the tourist ask to find the way to the restaurant?

Hint: Let p stand for "the islander at the fork always tells the truth" and let q stand for "the left-hand branch leads to the restaurant". Let E stand for a boolean expression such that, whether the islander tells the truth or lies, the answer to the question "Is E *true*?" will be yes iff the left-hand branch leads to the restaurant. Construct the truth table that E must have, in terms of p and q, and then design an appropriate E according to the truth table.

Chapter 3

Propositional Calculus

T his chapter offers an alternative to the widely-held view that boolean expressions are defined by how they are evaluated. Here, expressions are defined in terms of how they can be manipulated. Our goals are to convey a sense of how one manipulates boolean expressions and to teach heuristics and principles for developing proofs. By working out the exercises, the reader can develop a manipulative skill that will prove valuable in later work.

3.1 Preliminaries

A *calculus* is a method or process of reasoning by calculation with symbols. [1] This chapter presents a *propositional* calculus. It is so named because it is a method of calculating with boolean expressions that involve propositional variables (see page 33). We call our propositional calculus *equational logic* **E**.

One part of **E** is a set of *axioms*, which are certain boolean expressions that define basic manipulative properties of boolean operators. As an example, for operator $\lor$, the axiom $p \lor q \equiv q \lor p$ indicates that $\lor$ is symmetric in its two operands, i.e. the value of a disjunction is unchanged if its operands are swapped.

The other part of our propositional calculus consists of three inference rules: Leibniz (1.5), Transitivity (1.4), and Substitution (1.1). We repeat them here, as a reminder, formulated in terms of identifiers that will typically be used in this chapter: $P, Q, R, \ldots$ for arbitrary boolean expressions and $p, q, r, \ldots$ for boolean variables.

$$\textbf{Leibniz:} \quad \frac{P = Q}{E[r := P] = E[r := Q]}$$

$$\textbf{Transitivity:} \quad \frac{P = Q, \; Q = R}{P = R}$$

$$\textbf{Substitution:} \quad \frac{P}{P[r := Q]}$$

---

[1] From *Webster's Third New International Dictionary*.

A *theorem* of our propositional calculus is either (i) an axiom, (ii) the conclusion of an inference rule whose premises are theorems, or (iii) a boolean expression that, using the inference rules, is proved equal to an axiom or a previously proved theorem. Our proofs will follow the format discussed in Sec. 1.5, although variations will emerge.

Choosing different axioms for our calculus could lead to the same set of theorems, and many texts do use other axioms. Moreover, the sequence in which the boolean operators can be introduced, each being defined in terms of previously defined operators, is not unique. For example, sometimes $\wedge$, $\vee$, and $\neg$ are defined first, then $\Rightarrow$, and finally $\equiv$ and $\not\equiv$. In view of the importance of Leibniz and equivalence in our calculus, we choose the order $\equiv$ (and $=$), $\neg$ and $\not\equiv$ (and $\neq$), $\vee$, $\wedge$, and finally $\Rightarrow$ and $\Leftarrow$.[2]

All theorems of our propositional calculus are valid (see Def. (2.1) on page 31). This fact can be established by (i) checking each axiom with a truth table and (ii) arguing for each inference rule that if its premises are valid then so is its conclusion.

Not only are all theorems valid, but all valid expressions are theorems of our calculus (although we do not prove this fact). Theoremhood and validity are one and the same. Hence, Metatheorem Duality (2.3a) —which says that the negation of the dual of a valid expression is itself valid— can be used to discover theorems. However, in this chapter we do not use Duality to derive theorems. The goal of this chapter is not simply to discover theorems but to acquire a skill in manipulation and in developing proofs.

Helpful hints

One goal of this chapter is to present some heuristics for deriving proofs. To become proficient at using these heuristics requires practice, and the reader would do well to record in a loose-leaf notebook proofs of *all* theorems stated in this chapter. All theorems either are proved herein or are exercises, sometimes accompanied by hints.

When reading this chapter, avoid the temptation to evaluate the boolean expressions being discussed. Simply derive theorems. The skill of manipulating formulas, without regard for their meaning, is extremely useful in all of mathematics, and studying this chapter will help you acquire this skill.

Do not be discouraged by the number of theorems. You do not have to memorize them all. It will suffice to become familiar with them and how they are organized, so that you can find the ones you need when developing a proof. The more practice you have using the theorems, the more they will

---

[2] Remember that $=$ and $\equiv$ are interchangeable in formulas, without special mention (subject to the caveats mentioned in Sec. 2.2).

become your formal friends, who serve you in your mathematical work.

One final point. Just as you will struggle to develop nice short proofs, so did we. The proofs you read here are our final versions. Many of them were reworked two, three, or more times. Although we discuss heuristics for proof development, do not be deceived and think that, once these heuristics are mastered, all proofs will be easy to develop. Like good prose, good proofs are the result of thinking, analysis, writing, and revision. Practice, of course, makes the task easier. Tools have to be kept clean and oiled to be of use.

3.2   Equivalence and true

Equivalence is associative. This property is formalized as a manipulative property by the following axiom.

(3.1)   **Axiom, Associativity of** $\equiv$**:** $((p \equiv q) \equiv r) \cdot \equiv (p \equiv (q \equiv r))$

Associativity allows us to be informal and insert or delete pairs of parentheses in sequences of equivalences, just as we do with sequences of additions (e.g. $w + x + y + z$ is equivalent to $w + (x + y) + z$). Hence, we can write

$$p \equiv q \equiv r \quad \text{instead of} \quad p \equiv (q \equiv r) \quad \text{or} \quad (p \equiv q) \equiv r \quad .$$

Keeping axiom (3.1) in mind, we express the second axiom, symmetry, without parentheses.

(3.2)   **Axiom, Symmetry of** $\equiv$**:** $p \equiv q \equiv q \equiv p$

You can see why this axiom is called *symmetry* by imagining parentheses as follows: $(p \equiv q) \equiv (q \equiv p)$.

We now give our first proof, of the following theorem:

$$p \equiv p \ \equiv \ q \equiv q \quad .$$

Remember that the axiom of associativity allows us to parenthesize an expression such as (3.2) in several ways. In the following proof, we parenthesize (3.2) as $(p \equiv q \equiv q) \equiv p$, so that, using Leibniz, we can replace $p \equiv q \equiv q$ in an expression by p.

$$
\begin{aligned}
& p \ \equiv \ p \equiv q \equiv q \\
= \quad & \langle \text{Symmetry of } \equiv (3.2) \text{ —replace } p \equiv q \equiv q \text{ by } p \rangle \\
& p \ \equiv \ p \\
= \quad & \langle \text{Symmetry of } \equiv (3.2) \text{ —replace first } p \text{ by } p \equiv q \equiv q \rangle \\
& p \equiv q \equiv q \ \equiv \ p
\end{aligned}
$$

Since the final expression is axiom (3.2), and since, by the definition of theorem on page 42, any expression that is proved equal to an axiom is a theorem, the first expression has been proved to be a theorem.

The final axiom of this section introduces the constant symbol *true* as an abbreviation for $q \equiv q$ —using a constant symbol is reasonable because the value of $q \equiv q$ does not depend on the value of q.

(3.3)   **Axiom, Identity of $\equiv$:** $true \equiv q \equiv q$

We call *true* the *identity* of $\equiv$ because, as can be seen from the axiom of symmetry and (3.3), $p = (true \equiv p)$ and $(p \equiv true) = p$.[3] We can now prove the following two theorems.

Two theorems

(3.4)   *true*

(3.5)   **Reflexivity of $\equiv$:** $p \equiv p$

The proof of the second theorem is left to the reader (Exercise 3.3). To show that *true* is a theorem, we show that it equivales axiom (3.3):

$$true$$
$$= \quad \langle \text{Identity of } \equiv (3.3), \text{ with } q := true \,\rangle$$
$$true \equiv true$$
$$= \quad \langle \text{Identity of } \equiv (3.3) \text{ —replace the second } true \,\rangle$$
$$true \equiv q \equiv q \qquad \text{—Identity of } \equiv (3.3)$$

Axioms Identity (3.3) and Symmetry (3.2) imply that occurrences of "$\equiv true$" (or "$true \equiv$") in an expression are redundant. Thus, $Q \equiv true$ may be replaced by Q in any expression without changing the value of the expression. Therefore, we usually eliminate such occurrences unless something (e.g. symmetry) encourages us to leave them in.

For theorems of the form $P \equiv Q$, another proof method is available: transform P to Q as shown to the left below.

$$true$$
$$= \quad \langle (3.3), \ true \equiv q \equiv q \,\rangle$$

P		$P \equiv P$
$=$	$\langle \text{Hint 0} \rangle$	$= \quad \langle \text{Hint 0} \rangle$
R		$P \equiv R$
$=$	$\langle \text{Hint 1} \rangle$	$= \quad \langle \text{Hint 1} \rangle$
$\ldots$		$\ldots$
$=$	$\langle \text{Hint 2} \rangle$	$= \quad \langle \text{Hint 2} \rangle$
Q		$P \equiv Q$

---

[3] U is the identity of operation $\circ$ iff $b = b \circ U = U \circ b$ for all b. U is a left identity if $b = U \circ b$ for all b. U is a right identity if $b = b \circ U$ for all b.

In order to justify this technique, we show how such a proof can be converted mechanically into a proof that transforms *true* to $P \equiv Q$. The conversion is shown to the right above. It is obtained from the proof on the left by adding "$P \equiv$" to the beginning of each formula and adding a step at the beginning of the proof. For all steps (except the additional one), the hints are the same.

We summarize this new proof method:

(3.6)   **Proof method.** To prove that $P \equiv Q$ is a theorem, transform P to Q or Q to P using Leibniz.

We end this section with the following metatheorem [4], whose proof is left as an exercise.

(3.7)   **Metatheorem.** Any two theorems are equivalent.

3.3   Negation, inequivalence, and false

We introduce three axioms. The first defines *false*; the first and second together define negation, $\neg$; and the third defines inequivalence, $\not\equiv$.

(3.8)   **Axiom, Definition of** *false* **:** $false \equiv \neg true$

(3.9)   **Axiom, Distributivity of** $\neg$ **over** $\equiv$ **:** $\neg(p \equiv q) \equiv \neg p \equiv q$

(3.10) **Axiom, Definition of** $\not\equiv$ **:** $(p \not\equiv q) \equiv \neg(p \equiv q)$

Theorems (3.11)–(3.19) below can now be proved. Double negation (3.12) asserts that negation is its own inverse [5]. Double negation is used in English occasionally. For example, one might say "That was not done unintentionally" instead of "That was done intentionally".

Mutual associativity of $\equiv$ and $\not\equiv$, (3.18), allows us to omit parentheses in mixed sequences of $\not\equiv$ and $\equiv$, as, for example, in theorem (3.19). Mutual interchangeability is startling at first; it allows the exchange of adjacent occurrences of $\equiv$ and $\not\equiv$.

---

[4] A *theorem* in our technical sense is a boolean expression that is proved equal to an axiom. A *metatheorem* is a general statement about our logic that we prove to be *true*.

[5] Function g is the *inverse* of function f if $g(f.x) = x$ for all x.

Theorems relating $\equiv$, $\not\equiv$, $\neg$, and *false*

(3.11) $\neg p \equiv q \equiv p \equiv \neg q$

(3.12) **Double negation:** $\neg\neg p \equiv p$

(3.13) **Negation of *false*:** $\neg false \equiv true$

(3.14) $(p \not\equiv q) \equiv \neg p \equiv q$

(3.15) $\neg p \equiv p \equiv false$

(3.16) **Symmetry of $\not\equiv$:** $(p \not\equiv q) \equiv (q \not\equiv p)$

(3.17) **Associativity of $\not\equiv$:** $((p \not\equiv q) \not\equiv r) \equiv (p \not\equiv (q \not\equiv r))$

(3.18) **Mutual associativity:** $((p \not\equiv q) \equiv r) \equiv (p \not\equiv (q \equiv r))$

(3.19) **Mutual interchangeability:** $p \not\equiv q \equiv r \equiv p \equiv q \not\equiv r$

At this point, we note an interesting and useful fact about sequences of equivalences. The boolean expression

(3.20) $P0 \equiv P1 \equiv \cdots \equiv Pn$

is *true* exactly when an even number of the Pi are *false*. Why? By Identity of $\equiv$ (3.3), each subexpression *false* $\equiv$ *false* can be replaced by *true* until either one or zero *false* equivalents remain, in which case the sequence is *false* or *true*. For example, we can determine without any additional formal manipulation that *false* $\equiv$ *false* $\equiv$ *false* $\equiv$ *true* is *false*, because three (an odd number) of its equivalents are *false*.

We can use this fact about sequences of equivalences in formalizing certain English statements. Below, the second and fourth examples rely on the fact that "not an even number are *true*" equivales "an odd number are *true*".

> None or both of p and q is *true*: $p \equiv q$.
> Exactly one of p and q is *true*: $\neg(p \equiv q)$, or $p \not\equiv q$.
> Zero, two, or four of p, q, r, and s are *true*: $p \equiv q \equiv r \equiv s$.
> One or three of p, q, r, and s are *true*: $\neg(p \equiv q \equiv r \equiv s)$.

Proof heuristics and principles

We now discuss some heuristics for developing proofs. The first heuristic is illustrated by a proof of (3.11), $\neg p \equiv q \equiv p \equiv \neg q$. To prove (3.11), we try to transform it to $true$. This requires finding a theorem to use with Leibniz.[6] We match patterns in order to determine which theorems are applicable. Axiom (3.1) does not match at all; (3.2) can be used, but basically only to swap two operands, which does not seem useful; (3.3) can be used only to add an equivalent $true$; (3.4) is of no use; (3.5) can be used only to replace something by itself; and (3.8) and (3.10) do not match at all. This leaves (3.9) —its RHS, $\neg p \equiv q$, appears in the theorem to be proved.

This reasoning uses an important heuristic for developing proofs:

(3.21) **Heuristic.** Identify applicable theorems by matching the structure of expressions or subexpressions. The operators that appear in a boolean expression and the shape of its subexpressions can focus the choice of theorems to be used in manipulating it.

Obviously, the more theorems you know by heart and the more practice you have in pattern matching, the easier it will be to develop proofs.

We proceed with the proof of (3.11). It is given below. Note that Symmetry of $\equiv$ is used in the second step of the proof, without explicit mention: the substitution used is $p \equiv \neg q \;\equiv\; \neg(p \equiv q)$ but the hint is $\neg(q \equiv p) \equiv \neg q \equiv p$. These two expressions are the same, up to symmetry of equivalence. To shorten proofs, Symmetry and Associativity axioms for all binary operators are often used without mention. Finally, in two of the hints we mention the substitution used to effect the transformation. Later, we omit such hints when they are obvious.

$$\neg p \equiv q \equiv p \equiv \neg q$$
$$= \quad \langle (3.9), \ \neg(p \equiv q) \ \equiv \ \neg p \equiv q \rangle$$
$$\neg(p \equiv q) \equiv p \equiv \neg q$$
$$= \quad \langle (3.9), \text{ with } p,q := q,p \ \text{—i.e. } \neg(q \equiv p) \ \equiv \ \neg q \equiv p \rangle$$
$$\neg(p \equiv q) \equiv \neg(p \equiv q) \quad \text{—Reflexivity of } \equiv \ (3.5)$$

Theorem (3.11) can be proved in other ways, as well. For example, we could use proof method (3.6) and transform $\neg p \equiv q$ into $p \equiv \neg q$. We could also transform $\neg p \equiv p$ into $\neg q \equiv q$. Or, we could begin with $\neg p$ and transform it into $q \equiv p \equiv \neg q$.

---

[6] Only a previously proved theorem or an axiom may used, so its number should be less than the number of the theorem being proved. The only theorems available for use are (3.1)–(3.5) and (3.8)–(3.10).

Theorem (3.11) can be proved in so many ways, even though we have only a few axioms and theorems to use, because its structure affords many possibilities for manipulation. This capability is both a blessing, because it gives flexibility, and a curse, because we may have to investigate several options in developing a proof. As a rule, then, do not be satisfied with your first proof, for it may not be the shortest or simplest.

Now consider the following two proofs of (3.15), $\neg p \equiv p \equiv false$.

$$\neg p \equiv p \equiv false$$
$$= \quad \langle (3.9),\ \neg(p \equiv q) \ \equiv\ \neg p \equiv q\,,\ \text{with}\ q := p \rangle$$
$$\neg(p \equiv p) \equiv false$$
$$= \quad \langle \text{Identity of } \equiv (3.3),\ \text{with}\ q := p \rangle$$
$$\neg true \equiv false \quad \text{—theorem (3.8)}$$

$$\neg p \equiv p$$
$$= \quad \langle (3.9),\ \neg(p \equiv q) \ \equiv\ \neg p \equiv q\,,\ \text{with}\ q := p \rangle$$
$$\neg(p \equiv p)$$
$$= \quad \langle \text{Identity of } \equiv (3.3),\ \text{with}\ q := p \rangle$$
$$\neg true$$
$$= \quad \langle (3.8) \rangle$$
$$false$$

Which proof do you prefer? The first proof has fewer steps, but it requires copying "$\equiv false$" on every line. Here, the difference is slim, but if the part to be copied were longer, the whole proof would look longer and more complicated. There is also more danger of making mistakes in copying a part many times. For the sake of brevity, ease of reading, and avoidance of mistakes, adhere to the following principle.

(3.22) **Principle:** Structure proofs to avoid repeating the same subexpression on many lines.

We end this section with one final heuristic. It describes an oft-used pattern for proving some property of an operator that is defined in terms of another:

(3.23) **Heuristic of Definition Elimination:** To prove a theorem concerning an operator ∘ that is defined in terms of another, say •, expand the definition of ∘ to arrive at a formula that contains • ; exploit properties of • to manipulate the formula; and then (possibly) reintroduce ∘ using its definition.

To illustrate the use of this heuristic, we prove (3.16), $(p \not\equiv q) \ \equiv\ (q \not\equiv p)$. Here, ∘ is $\not\equiv$ and • is $\equiv$.

$$p \not\equiv q$$
$$=\quad \langle \text{Def. of } \not\equiv \ (3.10) \rangle$$
$$\neg(p \equiv q)$$
$$=\quad \langle \text{Symmetry of } \equiv \ (3.2) \rangle$$
$$\neg(q \equiv p)$$
$$=\quad \langle \text{Def. of } \not\equiv \ (3.10), \text{ with } p,q := q,p \rangle$$
$$q \not\equiv p$$

3.4 Disjunction

The disjunction operator $\vee$ is defined by the following five axioms.

(3.24) **Axiom, Symmetry of** $\vee$: $p \vee q \equiv q \vee p$

(3.25) **Axiom, Associativity of** $\vee$: $(p \vee q) \vee r \equiv p \vee (q \vee r)$

(3.26) **Axiom, Idempotency [7] of** $\vee$: $p \vee p \equiv p$

(3.27) **Axiom, Distributivity of** $\vee$ **over** $\equiv$:
$$p \vee (q \equiv r) \equiv p \vee q \equiv p \vee r$$

(3.28) **Axiom, Excluded Middle:** $p \vee \neg p$

Distributivity (3.27) can be viewed in two ways, much like distributivity of $\cdot$ over $+$. Replacing the LHS of (3.27) by the RHS could be called "multiplying out"; replacing the RHS by the LHS, "factoring".

Axiom Excluded Middle can be interpreted to mean that in any state either p or $\neg p$ is *true*; there is no middle ground.

With the five axioms for $\vee$, we can prove the following theorems.

Theorems concerning $\vee$

(3.29) **Zero [8] of** $\vee$: $p \vee true \equiv true$

(3.30) **Identity of** $\vee$: $p \vee false \equiv p$

(3.31) **Distributivity of** $\vee$ **over** $\vee$: $p \vee (q \vee r) \equiv (p \vee q) \vee (p \vee r)$

(3.32) $p \vee q \equiv p \vee \neg q \equiv p$

---

[7] A binary operator $\circ$ is *idempotent* if $x \circ x = x$ for all x. Multiplication $\cdot$ and addition $+$ of integers are not idempotent, but $\vee$ and $\wedge$ are.

MORE PROOF HEURISTICS AND PRINCIPLES

By Proof method (3.6), we can prove $P \equiv Q$ by transforming either P to Q or Q to P. The following heuristic helps in deciding which to try.

(3.33) **Heuristic:** To prove $P \equiv Q$, transform the expression with the most structure (either P or Q) into the other.

To illustrate the use of this heuristic, we develop a proof of theorem (3.29), $p \vee true \equiv true$. Its LHS has the most structure, so we begin with it. Its structure suggests which theorems may be used in applying Leibniz. The only theorems available about $\vee$ are axioms (3.24)–(3.27), and the only likely way of using them is to introduce an equivalence into the LHS, so that, perhaps, distributivity (3.27) can be used. Accordingly, we use axiom identity of equivalence, (3.3), to replace $true$, yielding $p \vee (p \equiv p)$. This is the germ of the following proof.

$$p \vee true$$
$$=\quad \langle \text{Identity of } \equiv (3.3) \rangle$$
$$p \vee (p \equiv p)$$
$$=\quad \langle \text{Distributivity of } \vee \text{ over } \equiv (3.27) \rangle$$
$$p \vee p \equiv p \vee p$$
$$=\quad \langle \text{Identity of } \equiv (3.3) \rangle$$
$$true$$

Suppose we had tried instead to transform the RHS of $p \vee true \equiv true$ to its LHS. The structure of its RHS, $true$, gives absolutely no insight into where to begin! So the heuristic of beginning with the side with the most structure makes sense here.

We could reverse the above proof, as shown below. This is a bad proof to present, because readers have little motivation for the beginning of the proof and hence are not able to visualize developing the proofs themselves. The first step is a rabbit pulled out of a hat.

$$true$$
$$=\quad \langle \text{Identity of } \equiv (3.3) \rangle$$
$$p \vee p \equiv p \vee p$$
$$=\quad \langle \text{Distributivity of } \vee \text{ over } \equiv (3.27) \rangle$$
$$p \vee (p \equiv p)$$
$$=\quad \langle \text{Identity of } \equiv (3.3) \rangle$$
$$p \vee true$$

---

[8] Z is a *zero* of a binary operation $\circ$ if $x \circ Z = Z \circ x = Z$, for all x. Z is a *left zero* if $Z \circ x = Z$, for all x. Z is a *right zero* if $x \circ Z = Z$, for all x. The term *zero* comes from the fact that 0 is the zero of $\cdot$.

A general principle, then, is the following.

(3.34) **Principle:** Structure proofs to minimize the number of rabbits pulled out of a hat —make each step seem obvious, based on the structure of the expression and the goal of the manipulation.

3.5 Conjunction

We define conjunction $\wedge$ with a single axiom, called the *Golden rule*. Convince yourself that (3.35) is valid by constructing a truth table for it.

(3.35) **Axiom, Golden rule** : $p \wedge q \equiv p \equiv q \equiv p \vee q$

We chose the Golden rule to define $\wedge$ because it is amazingly versatile, given the associativity and symmetry of $\equiv$. For example, one view is that it defines $p \wedge q$ as $p \equiv q \equiv p \vee q$, but it can also be rewritten as

$$(p \equiv q) \equiv (p \wedge q \equiv p \vee q) \ ,$$

which indicates that p and q are equal iff their conjunction and disjunction are equal.

With the Golden rule, we can prove a host of theorems that relate $\wedge$ to the already-defined operators. We now give these theorems, offering comments as appropriate.

The first theorems state that $\wedge$ is symmetric, associative, and idempotent and relate $\wedge$ to constants *true* and *false*.

Basic properties of $\wedge$

(3.36) **Symmetry of** $\wedge$: $p \wedge q \equiv q \wedge p$

(3.37) **Associativity of** $\wedge$: $(p \wedge q) \wedge r \equiv p \wedge (q \wedge r)$

(3.38) **Idempotency of** $\wedge$: $p \wedge p \equiv p$

(3.39) **Identity of** $\wedge$: $p \wedge true \equiv p$

(3.40) **Zero of** $\wedge$: $p \wedge false \equiv false$

(3.41) **Distributivity of** $\wedge$ **over** $\wedge$:
$$p \wedge (q \wedge r) \equiv (p \wedge q) \wedge (p \wedge r)$$

(3.42) **Contradiction:** $p \wedge \neg p \equiv false$

Theorems (3.43) below are called *absorption*, because subexpression q is absorbed into p. Theorems (3.44) are similar, with $\neg p$ being absorbed. The laws of distributivity that follow relate $\lor$ and $\land$. The Laws of De Morgan are named after their discoverer —see Historical note 3.1.

Theorems relating $\land$ and $\lor$

(3.43) **Absorption:** (a) $p \land (p \lor q) \equiv p$

(b) $p \lor (p \land q) \equiv p$

(3.44) **Absorption:** (a) $p \land (\neg p \lor q) \equiv p \land q$

(b) $p \lor (\neg p \land q) \equiv p \lor q$

(3.45) **Distributivity of $\lor$ over $\land$:**

$$p \lor (q \land r) \equiv (p \lor q) \land (p \lor r)$$

(3.46) **Distributivity of $\land$ over $\lor$:**

$$p \land (q \lor r) \equiv (p \land q) \lor (p \land r)$$

(3.47) **De Morgan:** (a) $\neg(p \land q) \equiv \neg p \lor \neg q$

(b) $\neg(p \lor q) \equiv \neg p \land \neg q$

The next group of theorems relate conjunction and equivalence. Theorem (3.48) is similar to (3.32), $p \lor q \equiv p \lor \neg q \equiv p$. Theorem (3.49) shows how $\land$ distributes over $\equiv$. Study (3.49) carefully, because it is too easy to miss or forget the rather odd last equivalent, p. Theorem (3.50) is obtained by replacing r by p in (3.49) and simplifying.

Theorems relating conjunction and equivalence

(3.48) $p \land q \equiv p \land \neg q \equiv \neg p$

(3.49) $p \land (q \equiv r) \equiv p \land q \equiv p \land r \equiv p$

(3.50) $p \land (q \equiv p) \equiv p \land q$

(3.51) **Replacement:** $(p \equiv q) \land (r \equiv p) \equiv (p \equiv q) \land (r \equiv q)$

In most propositional calculi, (3.52) and (3.53) are used to define $\equiv$ and $\not\equiv$. The first theorem indicates that $p \equiv q$ holds exactly when p and q are both *true* or both *false*. The second theorem indicates that $p \not\equiv q$ holds exactly when one of them is *true* and the other is *false*.

Alternative definitions of $\equiv$ and $\not\equiv$

(3.52) **Definition of** $\equiv$: $p \equiv q \equiv (p \wedge q) \vee (\neg p \wedge \neg q)$

(3.53) **Exclusive or:** $p \not\equiv q \equiv (\neg p \wedge q) \vee (p \wedge \neg q)$

STRUCTURING PROOFS USING LEMMAS

When a proof becomes long or complicated, it sometimes helps to impose structure by separating the proof into lemmas [9] . This process may bring to light interesting facts that might otherwise have remained hidden. It can also shorten the proof, if a lemma is used more than once. The same sort of advantages accrue from the judicious use of procedures in programming.

(3.54) **Principle:** Lemmas can provide structure, bring to light interesting facts, and ultimately shorten a proof.

We illustrate this principle by developing a proof of associativity of $\wedge$, (3.37). We can begin with the LHS, $(p \wedge q) \wedge r$, and attempt to transform it into the RHS. The *only* thing we can do at first is to replace the conjunctions using the Golden rule (using heuristic (3.23) of definition elimination), and, after this, we decide to distribute $\vee$ through $\equiv$ as much as possible:

$$
\begin{aligned}
& (p \wedge q) \wedge r \\
= \quad & \langle \text{Golden rule (3.35)} \rangle \\
& (p \equiv q \equiv p \vee q) \wedge r \\
= \quad & \langle \text{Golden rule (3.35), with } p, q := (p \equiv q \equiv p \vee q), r \rangle \\
& p \equiv q \equiv p \vee q \equiv r \equiv (p \equiv q \equiv p \vee q) \vee r \\
= \quad & \langle \text{Distributivity of } \vee \text{ over } \equiv \text{ (3.27)} \rangle \\
& p \equiv q \equiv p \vee q \equiv r \equiv p \vee r \equiv q \vee r \equiv p \vee q \vee r \\
= \quad & \langle \text{Symmetry and associativity of } \equiv \text{ and } \vee \rangle \\
& p \equiv q \equiv r \equiv p \vee q \equiv q \vee r \equiv r \vee p \equiv p \vee q \vee r
\end{aligned}
$$

We have shown that $(p \wedge q) \wedge r$ equivales the equivalence of all possible nonempty unique disjunctions of p, q, and r:

---

[9] The *lemma* is the lower of the two bracts enclosing the flower in the spikelet of grasses; also called *flowering glume*. Well, a *lemma* is also an auxiliary theorem used in a proof of some other theorem. The difference between "lemma" and "theorem" is in the eye of the beholder. The theorem is the thing we are interested in; the lemma, just a small theorem needed in its proof. (These definitions are taken from *Webster's Third New International Dictionary*.)

HISTORICAL NOTE 3.1. AUGUSTUS DE MORGAN (1806–1871)

De Morgan was born in India and educated at Trinity College, Cambridge. He spent most of his career as a professor of mathematics at the University of London (beginning at age 22!). De Morgan and George Boole are responsible for the great renaissance of logic in the 19th century. De Morgan was a founder and the first president of the London Mathematical Society and a founder of the British Association for the Advancement of Science. He was also deeply religious, but he abhorred any suspicion of sectarianism and turned down a fellowship at Cambridge to avoid such suspicion.

De Morgan rarely socialized. In reply to a friend who thought he worked too hard, De Morgan once wrote, "I have never been *hard* working, but I have been very *continuously* at work. I have never *sought* relaxation. And why? Because it would have killed me. Amusement is real hard work to me."

We might expect such a person to be dull and ponderous. Not De Morgan. He had a real sense of humor, as can be seen in his witty *A Budget of Paradoxes*, which exposed the writings of people who tried to do impossible things like squaring the circle. ('Budget' meant 'collection', 'stock', 'supply'; 'paradox', a tenet or proposition contrary to received opinion.) De Morgan's *Budget* contains all sorts of digressions —including anagrams of "Augustus De Morgan", like "Great gun! Do us a sum!" If you cannot find a copy of *Budget*, then do obtain U. Dudley's interesting *A Budget of Trisections* (1987), which was inspired by De Morgan's book.

(3.55) $(p \wedge q) \wedge r \equiv$

$\qquad p \equiv q \equiv r \equiv p \vee q \equiv q \vee r \equiv r \vee p \equiv p \vee q \vee r$

This equivalence is interesting enough by itself to leave as a lemma. Moreover, since $\equiv$ and $\vee$ are both associative and symmetric, we would hope that the RHS of (3.37) would also equivale the RHS of lemma (3.55), and then (3.55) can be used more than once. So we construct our proof of (3.37) afresh, starting with its RHS and trying to use (3.55). Here is the result.

$\qquad p \wedge (q \wedge r)$
$= \qquad \langle \text{Symmetry of } \wedge \ (3.36) \rangle$
$\qquad (q \wedge r) \wedge p$
$= \qquad \langle (3.55), \text{ with } p, q, r := q, r, p \rangle$
$\qquad q \equiv r \equiv p \equiv q \vee r \equiv r \vee p \equiv p \vee q \equiv q \vee r \vee p$
$= \qquad \langle \text{Symmetry and associativity of } \equiv \text{ and } \vee \rangle$
$\qquad p \equiv q \equiv r \equiv p \vee q \equiv q \vee r \equiv r \vee p \equiv p \vee q \vee r$
$= \qquad \langle (3.55) \rangle$
$\qquad (p \wedge q) \wedge r$

USING THE GOLDEN RULE

Since the Golden rule is the definition of $\wedge$, it can be used with the heuristic of Definition Elimination, (3.23) on page 48. In fact, the Golden rule is used to eliminate the conjunction in the first step of the proofs of almost all the theorems listed in this section on conjunction. In most of these theorems, an equivalent with a conjunction has the most structure, and heuristic (3.33) on page 50 suggests beginning with that equivalent.

As an example, we prove theorem (3.44a), $p \wedge (\neg p \vee q) \equiv p \wedge q$.

$$p \wedge (\neg p \vee q)$$
$$= \quad \langle \text{Golden rule (3.35), with } q := \neg p \vee q \rangle$$
$$p \equiv \neg p \vee q \equiv p \vee \neg p \vee q$$
$$= \quad \langle \text{Excluded middle (3.28)} \rangle$$
$$p \equiv \neg p \vee q \equiv true \vee q$$
$$= \quad \langle (3.29), \; true \vee p \equiv true \rangle$$
$$p \equiv \neg p \vee q \equiv true$$
$$= \quad \langle \text{Identity of } \equiv (3.3), \text{ with } q := p \equiv \neg p \vee q \rangle$$
$$p \equiv \neg p \vee q$$
$$= \quad \langle (3.32), \; p \vee q \equiv p \vee \neg q \equiv p,$$
$$\qquad \text{with } p, q := q, p \; \text{—to eliminate operator } \neg \rangle$$
$$p \equiv p \vee q \equiv q$$
$$= \quad \langle \text{Golden rule (3.35)} \rangle$$
$$p \wedge q$$

The Golden rule has four equivalents. Therefore, it can be used to replace one equivalent by three, two equivalents by two, or three equivalents by one. The idea of replacing more than one equivalent takes getting used to, so here are some examples. First, we prove (3.39), $p \wedge true \equiv p$, by showing that it equivales a previously proved theorem.

$$p \wedge true \equiv p$$
$$= \quad \langle \text{Golden rule (3.35) —replace two equivalents} \rangle$$
$$p \vee true \equiv true \quad \text{—Zero of } \vee \; (3.29)$$

We now prove theorem (3.49), $p \wedge (q \equiv r) \equiv p \wedge q \equiv p \wedge r \equiv p$. Read the hints carefully, because they describe in detail how the Golden rule is being used. It is used three times, and each time *two* equivalents are replaced. The proof begins with first and last equivalents of (3.49) and ends with the two middle equivalents —we prove something of the form $w \equiv x \equiv y \equiv z$ by transforming $w \equiv z$ to $x \equiv y$.

$$p \wedge (q \equiv r) \equiv p$$
$$= \quad \langle \text{Golden rule, with } q := q \equiv r \; \text{—replace two equivalents} \rangle$$
$$p \vee (q \equiv r) \equiv q \equiv r$$

$$= \quad \langle \text{Distributivity of } \vee \text{ over } \equiv (3.27) \rangle$$
$$p \vee q \equiv p \vee r \equiv q \equiv r$$
$$= \quad \langle \text{Symmetry of } \equiv (3.2) \rangle$$
$$p \vee q \equiv q \quad \equiv \quad p \vee r \equiv r$$
$$= \quad \langle \text{Golden rule, twice}$$
$$\qquad \text{—replace } p \vee q \equiv q \text{ and } p \vee r \equiv r \rangle$$
$$p \wedge q \equiv p \quad \equiv \quad p \wedge r \equiv p$$
$$= \quad \langle \text{Symmetry of } \equiv (3.2),\ p \equiv q \equiv q \equiv p,$$
$$\qquad \text{with } q := p \wedge q \equiv p \wedge r \rangle$$
$$p \wedge q \equiv p \wedge r$$

In summary, we have the following heuristic.

(3.56) **Heuristic:** Exploit the ability to parse theorems like the Golden rule in many different ways.

3.6 Implication

We now define and investigate two final operators, implication $\Rightarrow$ and consequence $\Leftarrow$.

(3.57) **Axiom, Definition of Implication:** $p \Rightarrow q \equiv p \vee q \equiv q$

(3.58) **Axiom, Consequence:** $p \Leftarrow q \equiv q \Rightarrow p$

Because of the similarity of $\Rightarrow$ and $\Leftarrow$, we give only theorems that involve $\Rightarrow$; corresponding ones for $\Leftarrow$ follow immediately from (3.58).

The first thing to note about implication is that it can be written in many ways. Besides the next three theorems, other ways of rewriting implication are given in Exercises 3.44–3.46. Theorem (3.59) or (3.60) is sometimes used as the definition of implication.

Rewriting implication

(3.59) **Definition of Implication:** $p \Rightarrow q \equiv \neg p \vee q$

(3.60) **Definition of Implication:** $p \Rightarrow q \equiv p \wedge q \equiv p$

(3.61) **Contrapositive:** $p \Rightarrow q \equiv \neg q \Rightarrow \neg p$

Theorems (3.62) and (3.63) show how to eliminate $\equiv$ as the consequent, while (3.65) shows how to shunt a conjunct from the antecedent to the consequent.

Miscellaneous theorems about implication

(3.62) $p \Rightarrow (q \equiv r) \equiv p \wedge q \equiv p \wedge r$

(3.63) **Distributivity of $\Rightarrow$ over $\equiv$:**
$$p \Rightarrow (q \equiv r) \equiv p \Rightarrow q \equiv p \Rightarrow r$$

(3.64) $p \Rightarrow (q \Rightarrow r) \equiv (p \Rightarrow q) \Rightarrow (p \Rightarrow r)$

(3.65) **Shunting:** $p \wedge q \Rightarrow r \equiv p \Rightarrow (q \Rightarrow r)$

(3.66) $p \wedge (p \Rightarrow q) \equiv p \wedge q$

(3.67) $p \wedge (q \Rightarrow p) \equiv p$

(3.68) $p \vee (p \Rightarrow q) \equiv true$

(3.69) $p \vee (q \Rightarrow p) \equiv q \Rightarrow p$

(3.70) $p \vee q \Rightarrow p \wedge q \equiv p \equiv q$

The next five theorems relate $\Rightarrow$ and the boolean constants. Theorem (3.71) asserts that $\Rightarrow$ is reflexive; the others give the value of an implication that has a constant as an operand. Theorems (3.72) and (3.73) indicate that implication is not symmetric; this is why a non-symmetric symbol, $\Rightarrow$, is chosen for it. From three of the theorems, " $\equiv true$ " could have been omitted; we leave it in for uniformity with the other two theorems.

Implication and boolean constants

(3.71) **Reflexivity of $\Rightarrow$:** $p \Rightarrow p \equiv true$

(3.72) **Right zero of $\Rightarrow$:** $p \Rightarrow true \equiv true$

(3.73) **Left identity of $\Rightarrow$:** $true \Rightarrow p \equiv p$

(3.74) $p \Rightarrow false \equiv \neg p$

(3.75) $false \Rightarrow p \equiv true$

Theorem (3.76a) below is obtained from (3.71), $p \Rightarrow p \equiv true$, by deleting the redundant " $\equiv true$ " and replacing consequent p by $p \vee q$; this leaves the implication valid. Each of the theorems (3.76a)–(3.76e) is called *weakening* or *strengthening*[10], depending on whether it is used to transform the antecedent into the consequent, thus weakening it, or to transform the consequent into the antecedent, thus strengthening it.

Weakening, strengthening, and Modus ponens

(3.76)   **Weakening/strengthening:** (a) $p \Rightarrow p \vee q$

(b) $p \wedge q \Rightarrow p$

(c) $p \wedge q \Rightarrow p \vee q$

(d) $p \vee (q \wedge r) \Rightarrow p \vee q$

(e) $p \wedge q \Rightarrow p \wedge (q \vee r)$

(3.77)   **Modus ponens:** $p \wedge (p \Rightarrow q) \Rightarrow q$.

Modus ponens (see (3.77)) is Latin for *Method of the bridge*. In many propositional calculi, a form of Modus ponens is one of the major inference rules —this is discussed in more detail on Sec. 6.2. Modus ponens takes a back seat in our calculus because of our emphasis on equational reasoning. Nevertheless, it is extremely useful at times.

The next two theorems embody case analysis. The first indicates that proving $p \vee q \Rightarrow r$ can be done by proving separately the cases $p \Rightarrow r$ and $q \Rightarrow r$. Similarly, the second indicates how a proof of r can be broken into two cases. Such proofs are often done informally in English. We return to case analysis on page 73 and more formally on page 115.

Forms of case analysis

(3.78) $(p \Rightarrow r) \wedge (q \Rightarrow r) \equiv (p \vee q \Rightarrow r)$

(3.79) $(p \Rightarrow r) \wedge (\neg p \Rightarrow r) \equiv r$

In most propositional calculi, equivalence is the last operator to be defined and is defined as "mutual implication". Thus, (3.80) below typically is made an axiom. We down-play implication in our calculus because, as an unsymmetric operator, it is harder to manipulate. Indeed, we can often progress most easily in a proof by eliminating implication from the expression at hand (using the heuristic of Definition elimination, (3.23)).

---

[10] Suppose $P \Rightarrow Q$. Then we say that P is *stronger than* Q and Q is *weaker than* P. This is because Q is *true* in more (or at least the same) states than P. That is, P imposes more restrictions on a state. The strongest formula is *false* and the weakest is *true* .

Theorem (3.81) is a direct corollary of mutual implication. Theorem (3.82a) is the usual definition of transitivity of implication. The two theorems following it are akin to transitivity, and that is why they are placed here. All three transitivity theorems are used in Sec. 4.1 to justify a proof format that allows shorter proofs.

Mutual implication and transitivity

(3.80) **Mutual implication:** $(p \Rightarrow q) \wedge (q \Rightarrow p) \equiv p \equiv q$

(3.81) **Antisymmetry** [11] : $(p \Rightarrow q) \wedge (q \Rightarrow p) \Rightarrow (p \equiv q)$

(3.82) **Transitivity:** (a) $(p \Rightarrow q) \wedge (q \Rightarrow r) \Rightarrow (p \Rightarrow r)$

(b) $(p \equiv q) \wedge (q \Rightarrow r) \Rightarrow (p \Rightarrow r)$

(c) $(p \Rightarrow q) \wedge (q \equiv r) \Rightarrow (p \Rightarrow r)$

PROVING THEOREMS CONCERNING IMPLICATION

Many of theorems (3.59)–(3.82) can be proved quite simply using the principles and heuristics outlined in previous parts of this chapter, so we relegate their proofs to the exercises. We limit our discussion here to some general remarks and prove a few of the more difficult theorems.

The heuristic of definition elimination, (3.23) on page 48, is useful in dealing with implication. For this purpose, look upon theorems (3.59)–(3.61) as well as axiom (3.57) as being definitions of implication. The shape of the goal of the manipulation should provide insight into which definition to choose. To illustrate, we prove (3.62), $p \Rightarrow (q \equiv r) \equiv p \wedge q \equiv p \wedge r$. Because (3.62) contains conjunctions, theorem (3.60) seems promising, since it shows how to replace an implication by introducing a conjunction:

$$p \Rightarrow (q \equiv r)$$
$$= \quad \langle \text{Definition of implication (3.60)} \rangle$$
$$p \wedge (q \equiv r) \equiv p$$
$$= \quad \langle (3.49),\ p \wedge (q \equiv r) \equiv p \wedge q \equiv p \wedge r \equiv p \rangle$$
$$p \wedge q \equiv p \wedge r$$

We made several attempts at proving mutual implication (3.80). The first one began by replacing each of the conjuncts of $(p \Rightarrow q) \wedge (q \Rightarrow p)$ using (3.59), $p \Rightarrow q \equiv \neg p \vee q$, to arrive at $(\neg p \vee q) \wedge (\neg q \vee p)$ as the LHS. Then,

---

[11] A binary relation $\circ$ is *antisymmetric* if $x \circ y \wedge y \circ x \Rightarrow x = y$ holds for all x and y. For example, $\leq$ and $\geq$ are antisymmetric.

because of the need to introduce equivalence at some point (since it is in the RHS of (3.80)), we used the Golden rule. The proof was complicated, so we threw it away and tried again —and again. Our final proof eliminates implication and then heads for a form in which the alternative definition of equivalence can be used:

$$(p \Rightarrow q) \land (q \Rightarrow p)$$
$= \quad$ ⟨Definition of implication (3.59), twice⟩
$$(\neg p \lor q) \land (\neg q \lor p)$$
$= \quad$ ⟨Distributivity of $\land$ over $\lor$ (3.46), thrice⟩
$$(\neg p \land \neg q) \lor (\neg p \land p) \lor (q \land \neg q) \lor (q \land p)$$
$= \quad$ ⟨Contradiction (3.42), twice; Identity of $\lor$ (3.30), twice⟩
$$(\neg p \land \neg q) \lor (q \land p)$$
$= \quad$ ⟨Alternative definition of $\equiv$ (3.52)⟩
$$p \equiv q$$

Here is a short proof of (3.82a), transitivity of $\Rightarrow$:

$$(p \Rightarrow q) \land (q \Rightarrow r) \Rightarrow (p \Rightarrow r)$$
$= \quad$ ⟨Shunting (3.65), with $p, q := (p \Rightarrow q) \land (q \Rightarrow r), p$
$\quad\quad$ —to shunt the p in the consequent to the antecedent⟩
$$p \land (p \Rightarrow q) \land (q \Rightarrow r) \Rightarrow r$$
$= \quad$ ⟨(3.66), $p \land (p \Rightarrow q) \equiv p \land q$
$\quad\quad$ —replace first two conjuncts⟩
$$p \land q \land (q \Rightarrow r) \Rightarrow r$$
$= \quad$ ⟨(3.66) —again, to replace second and third conjuncts—
$\quad\quad$ with $p, q := q, r$ ⟩
$$p \land q \land r \Rightarrow r \quad \text{—Strengthening (3.76b)}$$

LEIBNIZ'S RULE AS AN AXIOM

On page 12, we introduced Leibniz (1.5):

$$\frac{X = Y}{E[z := X] = E[z := Y]} \quad \text{or} \quad \frac{X = Y}{E_X^z = E_Y^z}$$

Now that we have introduced operator $\Rightarrow$, we can give a version of Leibniz as an axiom scheme:

(3.83) **Axiom, Leibniz:** $(e = f) \Rightarrow (E_e^z = E_f^z)$ (E any expression)

Inference rule Leibniz says, "if $X = Y$ is valid, i.e. *true* in all states, then so is $E[z := X] = E[z := Y]$." Axiom (3.83), on the other hand, says, "if $e = f$ is *true* in a state, then $E[z := e] = E[z := f]$ is *true* in that state." Thus, the inference rule and the axiom are not quite the same.

We show why the implication of Axiom (3.83) does not hold in the other direction. Let E be $false \land z$, e be $true$, and f be $false$. Then $E[z := e] = E[z := f]$ is $true$ but $e = f$ is $false$.

The following rules of substitution follow directly from axiom (3.83).

Rules of substitution

(3.84) **Substitution:** (a) $(e = f) \land E_e^z \;\equiv\; (e = f) \land E_f^z$

(b) $(e = f) \Rightarrow E_e^z \;\equiv\; (e = f) \Rightarrow E_f^z$

(c) $q \land (e = f) \Rightarrow E_e^z \;\equiv\; q \land (e = f) \Rightarrow E_f^z$

The first theorem below indicates that any occurrence of the antecedent of an implication in the consequent may be replaced by $true$; the second extends the first to the case that the antecedent is a conjunction. The third and fourth theorems provide for a similar replacement of (disjuncts of) the consequent in the antecedent. Theorem (3.89), attributed to Claude Shannon (see Historical note 5.1 on page 93), provides for a case analysis based on the possible values $true$ and $false$ of p.

Replacing variables by boolean constants

(3.85) **Replace by** $true$: (a) $p \Rightarrow E_p^z \;\equiv\; p \Rightarrow E_{true}^z$

(b) $q \land p \Rightarrow E_p^z \;\equiv\; q \land p \Rightarrow E_{true}^z$

(3.86) **Replace by** $false$: (a) $E_p^z \Rightarrow p \;\equiv\; E_{false}^z \Rightarrow p$

(b) $E_p^z \Rightarrow p \lor q \;\equiv\; E_{false}^z \Rightarrow p \lor q$

(3.87) **Replace by** $true$: $p \land E_p^z \;\equiv\; p \land E_{true}^z$

(3.88) **Replace by** $false$: $p \lor E_p^z \;\equiv\; p \lor E_{false}^z$

(3.89) **Shannon:** $E_p^z \;\equiv\; (p \land E_{true}^z) \lor (\neg p \land E_{false}^z)$

We illustrate the use of these theorems in proving $p \land q \Rightarrow (p \equiv q)$.

$$p \land q \Rightarrow (p \equiv q)$$
$$= \quad \langle \text{Replace by } true \ (3.85b) \rangle$$
$$p \land q \Rightarrow (true \equiv q)$$
$$= \quad \langle \text{Replace by } true \ (3.85b) \rangle$$
$$p \land q \Rightarrow (true \equiv true)$$
$$= \quad \langle \text{Identity of } \equiv \ (3.3) \rangle$$
$$p \land q \Rightarrow true \quad \text{—theorem (3.72)}$$

Exercises for Chapter 3

3.1 We have defined $\equiv$ using three axioms. Assuming that the symbol *true* is identified with the symbol *true* of the previous chapter on boolean expressions, do the axioms uniquely identify operator $\equiv$? Answer this question by seeing which of the 16 possible binary operators $\circ$ (say) given in the truth table on page 26 satisfy $((p \circ q) \circ r) \circ (p \circ (q \circ r))$, $p \circ q \circ q \circ p$, and $true \circ q \circ q$. (For example, the operator given by the last column does not satisfy $true \circ q \circ q$, since the operator always yields f.)

3.2 Use truth tables to show that axioms (3.1), (3.2), and (3.3) are valid (*true* in every state).

3.3 Prove Reflexivity of $\equiv$ (3.5), $p \equiv p$.

3.4 Prove the following metatheorem. $Q \equiv true$ is a theorem iff Q is a theorem.

3.5 Prove the following metatheorem. Any two theorems are equivalent.

3.6 Assume that operator $\equiv$ is identified with operator $\equiv$ of Sec. 2.1 (see Exercise 3.1) and *true* is identified with the symbol *true* of Sec. 2.1. Prove that axioms (3.8) and (3.9) uniquely define operator $\neg$. That is, determine which of the four prefix operators $\circ$ defined in the truth table on page 26 satisfy $false \equiv \circ true$ and $\circ(p \equiv q) \equiv \circ p \equiv q$.

Exercises on negation, inequivalence, and false

3.7 Prove theorem (3.11) in three different ways: start with $\neg p \equiv q$ and transform it to $p \equiv \neg q$, start with $\neg p \equiv p$ and transform it into $q \equiv \neg q$, and start with $\neg p$ and transform it into $q \equiv p \equiv \neg q$. Compare these three proofs and the one given on page 47. Which is simpler or shorter?

3.8 Prove Double negation (3.12), $\neg \neg p \equiv p$.

3.9 Prove Negation of *false* (3.13), $\neg false \equiv true$.

3.10 Prove theorem (3.14), $(p \not\equiv q) \equiv \neg p \equiv q$.

3.11 Prove theorem (3.15) by transforming $\neg p \equiv p \equiv false$ to *true* using (3.11). The proof should require only two uses of Leibniz.

3.12 Prove Associativity of $\not\equiv$ (3.17), $((p \not\equiv q) \not\equiv r) \equiv (p \not\equiv (q \not\equiv r))$, using the heuristic of Definition elimination (3.23) —by eliminating $\not\equiv$, using a property of $\equiv$, and reintroducing $\not\equiv$.

3.13 Prove Mutual associativity (3.18), $((p \not\equiv q) \equiv r) \equiv (p \not\equiv (q \equiv r))$, using the heuristic of Definition elimination (3.23) —by eliminating $\not\equiv$, using a property of $\equiv$, and reintroducing $\not\equiv$.

3.14 Prove Mutual interchangeability (3.19), $p \not\equiv q \equiv r \equiv p \equiv q \not\equiv r$, using the heuristic of Definition elimination (3.23) —by eliminating $\not\equiv$, using a property of $\equiv$, and reintroducing $\not\equiv$.

Exercises on disjunction

3.15 Assume that $\equiv$, $\neg$, *true*, and *false* have meanings as given in Sec. 2.1. Show that axioms (3.24)–(3.28) uniquely determine operator $\vee$ —only one of the operators of the truth table for binary operators on page 26 can be assigned to it.

3.16 Prove that the zero of a binary operator $\oplus$ is unique. (An object is unique if, when we assume that two of them B and C exist, we can prove $B = C$.)

3.17 Prove Identity of $\vee$ (3.30), $p \vee false \equiv p$, by transforming its more structured side into its simpler side. Theorem (3.15) may be a suitable way to introduce an equivalence.

3.18 Prove Distributivity of $\vee$ over $\vee$ (3.31), $p \vee (q \vee r) \equiv (p \vee q) \vee (p \vee r)$. The proof requires only the symmetry, associativity, and idempotency of $\vee$.

3.19 Prove theorem (3.32), $p \vee q \equiv p \vee \neg q \equiv p$. Note that the pattern $p \vee q \equiv p \vee \neg q$ matches the RHS of distributivity axiom (3.27), with $r := \neg q$, so consider transforming $p \vee q \equiv p \vee \neg q$ to p.

Exercises on conjunction

3.20 Show the validity of the Golden rule, (3.35), by constructing a truth table for it.

3.21 Prove that the only distinct formulas (up to interchanging p and q) involving variables p, q, $\equiv$, and $\vee$ are: p, $p \equiv p$, $p \equiv q$, $p \vee q$, $p \vee q \equiv q$, and $p \equiv q \equiv p \vee q$.

3.22 Prove Symmetry of $\wedge$ (3.36), $p \wedge q \equiv q \wedge p$, using the heuristic of Definition elimination (3.23) —eliminate $\wedge$ (using its definition, the Golden rule), manipulate, and then reintroduce $\wedge$.

3.23 Prove Idempotency of $\wedge$ (3.38), $p \wedge p \equiv p$, using the heuristic of Definition elimination (3.23) —eliminate $\wedge$ (using its definition, the Golden rule) and manipulate.

3.24 Prove Zero of $\wedge$ (3.40), $p \wedge false \equiv false$, using the heuristic of Definition elimination (3.23) —eliminate $\wedge$ (using its definition, the Golden rule) and manipulate.

3.25 Prove Distributivity of $\wedge$ over $\wedge$ (3.41), $p \wedge (q \wedge r) \equiv (p \wedge q) \wedge (p \wedge r)$.

3.26 Prove Contradiction (3.42), $p \wedge \neg p \equiv false$, using the heuristic of Definition elimination (3.23) —eliminate $\wedge$ (using its definition, the Golden rule) and manipulate.

3.27 Prove Absorption (3.43a), $p \wedge (p \vee q) \equiv p$, using the heuristic of Definition elimination (3.23) —eliminate $\wedge$ (using its definition, the Golden rule) and manipulate.

3.28 Prove Absorption (3.43b), $p \vee (p \wedge q) \equiv p$. Use the Golden rule.

3.29 Prove Absorption (3.44b), $p \lor (\neg p \land q) \equiv p \lor q$. Use the Golden rule and manipulate.

3.30 Prove Distributivity of $\lor$ over $\land$ (3.45), $p \lor (q \land r) \equiv (p \lor q) \land (p \lor r)$, using the heuristic of Definition elimination (3.23) —eliminate $\land$ (using its definition, the Golden rule), manipulate, and reintroduce $\land$ using the Golden rule again.

3.31 Prove Distributivity of $\land$ over $\lor$ (3.46). It cannot be proved in the same manner as Distributivity of $\lor$ over $\land$ (3.45) because $\land$ does not distribute over $\equiv$ so nicely. Instead, prove it using (3.45) and Absorption.

3.32 Prove De Morgan (3.47a), $\neg(p \land q) \equiv \neg p \lor \neg q$. Start by using the Golden rule; (3.32) should come in handy.

3.33 Prove De Morgan (3.47b), $\neg(p \lor q) \equiv \neg p \land \neg q$, beginning with the LHS and using the Golden rule.

3.34 Prove $(p \land q) \lor (p \land \neg q) \equiv p$.

3.35 Prove (3.48), $p \land q \equiv p \land \neg q \equiv \neg p$. Theorem (3.32) should come in handy.

3.36 Prove (3.50), $p \land (q \equiv p) \equiv p \land q$, using (3.49) with the instantiation $r := p$.

3.37 Prove Replacement (3.51), $(p \equiv q) \land (r \equiv p) \equiv (p \equiv q) \land (r \equiv q)$, by proving that the LHS and the RHS each equivale $p \equiv q \equiv r \equiv p \lor q \equiv q \lor r \equiv r \lor p$. The transformation of the LHS (or the RHS) to this expression can be done by applying (3.27) three times.

3.38 Prove Replacement (3.51), $(p \equiv q) \land (r \equiv p) \equiv (p \equiv q) \land (r \equiv q)$, by making immediate use of Distributivity of $\land$ over $\equiv$ (3.49) to replace both equivalents.

3.39 Prove Definition of $\equiv$ (3.52), $p \equiv q \equiv (p \land q) \lor (\neg p \land \neg q)$. Hint: Apply theorem (3.32), $p \lor q \equiv p \lor \neg q \equiv p$, to the RHS.

3.40 Prove Exclusive or (3.53), $p \not\equiv q \equiv (\neg p \land q) \lor (p \land \neg q)$. Hint: Try to apply Definition of $\equiv$ (3.52).

Exercises on implication

3.41 Prove Implication (3.59), $p \Rightarrow q \equiv \neg p \lor q$. At one point of the proof, you may find theorem (3.32) useful.

3.42 Prove Implication (3.60), $p \Rightarrow q \equiv p \land q \equiv p$.

3.43 Prove Contrapositive (3.61), $p \Rightarrow q \equiv \neg q \Rightarrow \neg p$.

3.44 Prove $p \Rightarrow q \equiv \neg(p \land \neg q)$. Axiom (3.57) may not be the best choice to eliminate the implication.

3.45 Prove $p \Rightarrow q \equiv \neg p \lor \neg q \equiv \neg p$.

3.46 Prove $p \Rightarrow q \equiv \neg p \wedge \neg q \equiv \neg q$.

3.47 Prove Distributivity of $\Rightarrow$ over $\equiv$ (3.63), $p \Rightarrow (q \equiv r) \equiv p \Rightarrow q \equiv p \Rightarrow r$.

3.48 Prove theorem (3.64), $p \Rightarrow (q \Rightarrow r) \equiv (p \Rightarrow q) \Rightarrow (p \Rightarrow r)$.

3.49 Prove Shunting (3.65), $p \wedge q \Rightarrow r \equiv p \Rightarrow (q \Rightarrow r)$. Use the heuristic of Definition Elimination, (3.23), on page 48. Use of one of (3.59)–(3.61) instead of (3.57) to remove the implication will be more fruitful.

3.50 Prove theorem (3.66), $p \wedge (p \Rightarrow q) \equiv p \wedge q$. Hint: Try to eliminate the implication in a manner that allows an Absorption law to be used.

3.51 Prove theorem (3.67), $p \wedge (q \Rightarrow p) \equiv p$. Hint: Try to eliminate the implication in a manner that allows an Absorption law to be used.

3.52 Prove theorem (3.68), $p \vee (p \Rightarrow q) \equiv true$. Hint: Use (3.59) to eliminate the implication.

3.53 Prove theorem (3.69), $p \vee (q \Rightarrow p) \equiv q \Rightarrow p$. Hint: use (3.59) to eliminate the implication.

3.54 Prove theorem (3.70), $p \vee q \Rightarrow p \wedge q \equiv p \equiv q$. Hint: Start with $p \vee q \Rightarrow p \wedge q$ and remove the implication. Head toward a use of the alternative definition of $\equiv$.

3.55 Prove Reflexivity of $\Rightarrow$ (3.71), $p \Rightarrow p \equiv true$.

3.56 Prove Right zero of $\Rightarrow$ (3.72), $p \Rightarrow true \equiv true$.

3.57 Prove Left identity of $\Rightarrow$ (3.73), $true \Rightarrow p \equiv p$.

3.58 Prove theorem (3.74), $p \Rightarrow false \equiv \neg p$.

3.59 Prove theorem (3.75), $false \Rightarrow p \equiv true$.

3.60 Prove Weakening/strengthening (3.76a), $p \Rightarrow p \vee q$. After eliminating the implication (in a suitable manner), you may find it helpful to use a law of Absorption.

3.61 Prove Weakening/strengthening (3.76b), $p \wedge q \Rightarrow p$. The hint of the preceding exercise applies here also.

3.62 Prove Weakening/strengthening (3.76c), $p \wedge q \Rightarrow p \vee q$. The hint of the preceding exercise applies here also.

3.63 Prove Weakening/strengthening (3.76d), $p \vee (q \wedge r) \Rightarrow p \vee q$. Since the main operator in this expression is $\vee$, one idea is to remove the implication using (3.57). Alternatively, it can be proved in *one* step.

3.64 Prove Weakening/strengthening (3.76e), $p \wedge q \Rightarrow p \wedge (q \vee r)$. Since the main operator in this expression is $\wedge$, one idea is to remove the implication using (3.60). Alternatively, it can be proved in *one* step.

3.65 Prove Modus ponens, (3.77), $p \wedge (p \Rightarrow q) \Rightarrow q$. Hint: Use theorem (3.66).

3.66 Prove theorem (3.78), $(p \Rightarrow r) \wedge (q \Rightarrow r) \equiv (p \vee q \Rightarrow r)$.

3.67 Prove theorem (3.79), $(p \Rightarrow r) \wedge (\neg p \Rightarrow r) \equiv r$.

3.68 Prove Mutual implication (3.80). Begin by replacing each conjunct in the LHS using (3.59) and then use the Golden rule.

3.69 Prove Antisymmetry (3.81) in two steps (use Mutual implication (3.80)).

3.70 Prove Transitivity of implication (3.82a). Start with the whole expression, transform each of the four implications in it using (3.59), and then massage.

3.71 Prove Transitivity (3.82b), $(p \equiv q) \wedge (q \Rightarrow r) \Rightarrow (p \Rightarrow r)$. Use Mutual implication (3.80), Transitivity (3.82a), and Shunting (3.65).

3.72 Prove Transitivity (3.82c), $(p \Rightarrow q) \wedge (q \equiv r) \Rightarrow (p \Rightarrow r)$. Use Mutual implication (3.80), Transitivity (3.82a), and Shunting (3.65).

Exercises on Leibniz's rule as an axiom

3.73 Prove Substitution (3.84a), $(e = f) \wedge E_e^z \equiv (e = f) \wedge E_f^z$. Begin with Leibniz (3.83) and replace the implication.

3.74 Prove Substitution (3.84b), $(e = f) \Rightarrow E[z := e] \equiv (e = f) \Rightarrow E[z := f]$.

3.75 Prove Substitution (3.84c), $q \wedge (e = f) \Rightarrow E_e^z \equiv q \wedge (e = f) \Rightarrow E_f^z$. Use Shunting (3.65).

3.76 Prove Replace by *true* (3.85a), $p \Rightarrow E[z := p] \equiv p \Rightarrow E[z := true]$. In order to be able to use (3.84b), introduce the equivalent *true* into the antecedent.

3.77 Prove Replace by *true* (3.85b), $q \wedge p \Rightarrow E[z := p] \equiv q \wedge p \Rightarrow E[z := true]$.

3.78 Prove Replace by *false* (3.86a), $E[z := p] \Rightarrow p \equiv E[z := false] \Rightarrow p$.

3.79 Prove Replace by *false* (3.86b), $E[z := p] \Rightarrow p \vee q \equiv E[z := false] \Rightarrow p \vee q$.

3.80 Prove Replace by *true* (3.87), $p \wedge E[z := p] \equiv p \wedge E[z := true]$.

3.81 Prove Replace by *false* (3.88), $p \vee E[z := p] \equiv p \vee E[z := false]$.

3.82 Prove $p \Rightarrow (q \Rightarrow p)$ using theorem (3.85a) in the first step.

3.83 Prove Shannon (3.89), $E_p^z \equiv (p \wedge E_{true}^z) \vee (\neg p \wedge E_{false}^z)$.

3.84 Prove Weakening/strengthening (3.76e), $p \wedge q \Rightarrow p \wedge (q \vee r)$, using Replace by *true* (3.85b).

Exercises on duals

3.85 Consider any expression P of the form *true*, $q \wedge r$, $q \equiv r$, or $q \Rightarrow r$, and consider its dual P_D (see Def. (2.2) on page 31). Prove that $P \equiv \neg P_D$ for expressions of the form given above, provided it holds for their subexpressions. Hint: By the definition of the dual, for an operation like $\wedge$, $(q \wedge r)_D \equiv q_D \vee r_D$.

3.86 Look through the equivalences that are theorems in this chapter (only up through theorem (3.53)) and put them in pairs $P \equiv Q$ and $P_D \equiv Q_D$. (For example, Symmetry of $\equiv$ (3.2) and Symmetry of $\not\equiv$ (3.16) form such a pair.)

3.87 Make a list of all theorems P in this chapter (only up through theorem (3.53)) that are equivalences for which $\neg P_D$ is not listed as a theorem in this chapter (see the previous exercise).

Exercises on normal forms

A boolean expression is in *conjunctive normal form* if it has the form

$$E_0 \wedge E_1 \wedge \ldots \wedge E_{n-1}$$

where each E_i is a disjunction of variables and negations of variables. For example, the following expression is in conjunctive normal form.

$$(a \vee \neg b) \wedge (a \vee b \vee c) \wedge (\neg a)$$

An expression is in *disjunctive normal form* if it has the form

$$E_0 \vee E_1 \vee \ldots \ldots E_{n-1}$$

where each E_i is a conjunction of variables and negations of variables. For example, the following expression is in disjunctive normal form.

$$(a \wedge \neg b) \vee (a \wedge b \wedge c) \vee (\neg a)$$

In electrical engineering, where conjunctive and disjunctive normal forms are used in dealing with circuits, an expression of the form $V_0 \vee \cdots \vee V_n$, where each V_i is a variable, is called a *maxterm*, for the following reason. If one considers *false* < *true*, then $x \vee y$ is the maximum of x and y, so the maxterm is the maximum of its operands. Similarly, an expression of the form $V_0 \wedge \cdots \wedge V_n$ is called a *minterm*.

3.88 The following truth table defines a set of states of variables a, b, c, d. Give a boolean expression in disjunctive normal form that is *true* in exactly the states defined by the truth table. Based on this example, outline a procedure that translates any such truth table into an equivalent boolean expression in disjunctive normal form.

a	b	c	d
t	t	t	f
t	f	t	f
f	t	t	f

Since every boolean expression can be described by such a truth table, every boolean expression can be transformed to disjunctive normal form.

3.89 The following truth table defines a set of states of variables a, b, c, d. Give a boolean expression in conjunctive normal form that is *true* in exactly the states defined by the truth table. Based on this example, describe a procedure that translates any such truth table into an equivalent boolean expression in conjunctive normal form.

a	b	c	d
t	t	t	f
t	t	t	t
t	f	t	f
f	t	t	f

Since every boolean expression can be described by such a truth table, every boolean expression can be transformed to conjunctive normal form.

Chapter 4

Relaxing the Proof Style

$\mathbf{I}$ n the previous chapter, we defined the propositional calculus, discussed proof strategies and heuristics, and proved many theorems. In this chapter, we provide some flexibility in the use of the propositional calculus. First, we introduce an extension of our proof format in order to shorten some proofs of implications. Second, we show how to present proofs in a less formal style. In doing so, we relate classical proof methods to proofs in the propositional calculus.

4.1 An abbreviation for proving implications

Step away from propositional calculus for a moment and consider arithmetic relations. Suppose we have concluded that $b = d - 1$ holds. Since $d - 1 < d$, we infer $b < d$. We are proving $b < d$ using a law of transitivity, $x = y \ \wedge \ y < z \ \Rightarrow \ x < z$.

We can extend our notion of proofs of equality in Sec. 1.5 and give this proof of $b < d$ as shown below. In this proof, we are making implicit use of the law $x = y \ \wedge \ y < z \ \Rightarrow \ x < z$.

$$
\begin{array}{ll}
& b \\
= & \langle \text{Some hint} \rangle \\
& d - 1 \\
< & \langle \text{Definition of} < \rangle \\
& d
\end{array}
$$

A similar proof format can be used whenever we have a relation $\circ$ (say) that satisfies transitivity laws like $x=y \wedge y \circ z \Rightarrow x \circ z$ and $b \circ c \wedge c \circ d \Rightarrow b \circ d$. (We already have transitivity of equality.) In particular, we can extend the proof format for our propositional calculus in this fashion because of theorems (3.82a)–(3.82c). Given $p \equiv q$ and $q \Rightarrow r$, we would demonstrate that $p \Rightarrow r$ holds using the following proof.

$$
\begin{array}{ll}
& p \\
= & \langle \text{Why } p \equiv q \rangle \\
& q \\
\Rightarrow & \langle \text{Why } q \Rightarrow r \rangle \\
& r
\end{array}
$$

Formally, in order to accept proofs in this format, we have to show that we can translate this proof into a proof of $p \Rightarrow r$ that does not use the extension. Here is such a proof, which uses the same two theorems.

$$(p \equiv q) \wedge (q \Rightarrow r) \Rightarrow (p \Rightarrow r) \qquad \text{—Transitivity (3.82b)}$$
$$= \quad \langle \text{Why } p \equiv q \equiv true \rangle$$
$$true \wedge (q \Rightarrow r) \Rightarrow (p \Rightarrow r)$$
$$= \quad \langle \text{Why } q \Rightarrow r \equiv true \rangle$$
$$true \wedge true \Rightarrow (p \Rightarrow r)$$
$$= \quad \langle \text{Idempotency of } \wedge \text{ (3.38); Left identity of } \Rightarrow \text{ (3.73)} \rangle$$
$$p \Rightarrow r$$

Generalizing, we allow any number of $\equiv$ steps and $\Rightarrow$ steps to be used in the proof format. Similarly, from a sequence of $\Leftarrow$ and $\equiv$ steps we conclude that the first expression is a consequence of the last.

The following theorems can be proved quite simply using the new format.

Additional theorems concerning implication

(4.1) $p \Rightarrow (q \Rightarrow p)$

(4.2)   **Monotonicity [1] of** $\vee$: $(p \Rightarrow q) \Rightarrow (p \vee r \Rightarrow q \vee r)$

(4.3)   **Monotonicity of** $\wedge$: $(p \Rightarrow q) \Rightarrow (p \wedge r \Rightarrow q \wedge r)$

We develop a proof of (4.2) in order to illustrate the use of our abbreviation for proofs by implication. We begin with the consequent, since it has more structure, and transform it into the antecedent, keeping in mind the goal, antecedent $p \Rightarrow q$. The first step is to eliminate the implication. Any of the four "definitions" of implication (3.57), (3.59), (3.60), and (3.61) could be used for this. Here, we use (3.57) so that all the operators on both sides of the resulting equivalence are disjunctions. For the step of weakening or strengthening (which puts $\Rightarrow$ or $\Leftarrow$ as the operator in the left column), (3.76a), (3.76b), and (3.76c) are often useful.

$$p \vee r \Rightarrow q \vee r$$
$$= \quad \langle (3.57), \ p \Rightarrow q \equiv p \vee q \equiv q \rangle$$
$$p \vee r \vee q \vee r \equiv q \vee r$$
$$= \quad \langle \text{Idempotency of } \vee \text{ (3.26)} \rangle$$
$$p \vee q \vee r \equiv q \vee r$$
$$= \quad \langle \text{Distributivity of } \vee \text{ over } \equiv \text{ (3.27)},$$
$$\text{with } p, q, r := r, p \vee q, q \rangle$$

---

[1] A boolean function f is *monotonic* if $(x \Rightarrow y) \Rightarrow (f.x \Rightarrow f.y)$.

$$(p \lor q \equiv q) \lor r$$
$$\Leftarrow \quad \langle \text{Weakening (3.76a)} \rangle$$
$$p \lor q \equiv q$$
$$= \quad \langle (3.57) \text{ —again} \rangle$$
$$p \Rightarrow q$$

Note that by starting with the consequent, we were forced to use $\Leftarrow$. Starting with the antecedent, as shown below, allows us to use $\Rightarrow$ instead.

$$p \Rightarrow q$$
$$= \quad \langle (3.57), \; p \Rightarrow q \equiv p \lor q \equiv q \rangle$$
$$p \lor q \equiv q$$
$$\Rightarrow \quad \langle \text{Weakening (3.76a)} \rangle$$
$$(p \lor q \equiv q) \lor r$$
$$= \quad \langle \text{Distributivity of } \lor \text{ over } \equiv \text{ (3.27)} \rangle$$
$$p \lor q \lor r \equiv q \lor r$$
$$= \quad \langle \text{Idempotency of } \lor \text{ (3.26)} \rangle$$
$$p \lor r \lor q \lor r \equiv q \lor r$$
$$= \quad \langle (3.57) \text{ —again} \rangle$$
$$p \lor r \Rightarrow q \lor r$$

However, a rabbit is pulled out of the hat in this second proof, contradicting principle (3.34) on page 51: in the second step, disjunct r is introduced without any motivation. This example, again, illustrates that the direction a proof takes may determine whether it appears simple and "opportunity driven", i.e. whether the shapes of the expressions guide each step in a straightforward manner.

4.2 Additional proof techniques

When dealing with proofs of boolean expressions, our equational logic suffices. When dealing with other domains of interest (e.g. integers, sequences, or trees), where we use inductively defined objects, partial functions and the like, a few additional proof techniques become useful. In this section, we introduce these techniques. In doing so, we can begin looking at the relation between formal and informal proofs.

ASSUMING THE ANTECEDENT

A common practice in mathematics is to prove an implication $P \Rightarrow Q$ by assuming the antecedent P and proving the consequent Q. By "assuming the antecedent" we mean thinking of it, momentarily, as an axiom and thus equivalent to *true*. In the proof of consequent Q, each variable in the new axiom P is treated as a constant, so that Substitution (1.1) cannot be used to replace the variable. Later, we discuss the need for this restriction.

We justify this method of proof with the following metatheorem.

(4.4)   **(Extended) Deduction Theorem.** Suppose adding P_1, ... , P_n as axioms to propositional logic **E**, with the variables of the P_i considered to be constants, allows Q to be proved. Then $P_1 \wedge$ $... \wedge P_n \Rightarrow Q$ is a theorem.

The proof of this metatheorem involves showing how a proof of Q using P_1, ... , P_n as additional axioms can be mechanically transformed into a proof of $P_1 \wedge ... \wedge P_n \Rightarrow Q$. The description of the transformation is long and tedious, and we do not give it here.

Below, we give a proof of $p \wedge q \Rightarrow (p \equiv q)$ using metatheorem (4.4). The proof illustrates how we say in English that the conjuncts of the antecedent are "assumed", or added as axioms to the logic.

Proof. To prove $p \wedge q \Rightarrow (p \equiv q)$, we assume the conjuncts of its antecedent and prove its consequent:

$$
\begin{array}{ll}
& p \\
= & \langle \text{Assumption } p \rangle \\
& true \\
= & \langle \text{Assumption } q \rangle \\
& q \qquad\qquad\qquad\qquad\qquad\qquad\qquad\qquad\qquad\qquad\qquad \Box
\end{array}
$$

If a proof is long, it may be difficult to remember the assumptions. In this case, we place the assumptions at the beginning of the proof, as in the following example. The first line alerts the reader that a proof is being conducted by assuming the conjuncts of the antecedent and proving the consequent.

$$
\begin{array}{ll}
\textbf{Assume } p, q \\
& p \\
= & \langle \text{Assumption } p \rangle \\
& true \\
= & \langle \text{Assumption } q \rangle \\
& q \qquad\qquad\qquad\qquad\qquad\qquad\qquad\qquad\qquad\qquad\qquad \Box
\end{array}
$$

Metatheorem (4.4) requires that all variables in the assumed expression be viewed as constants throughout the proof of Q, so that Substitution (1.1) cannot be used to replace them. The following incorrect proof of $(b \equiv c) \Rightarrow (d \equiv c)$ (which is not valid) shows why this is necessary. The proof is incorrect because b in the assumption is replaced using the rule of Substitution.

Assume $b \equiv c$ (proof incorrect)
    d

$=$ $\langle$Assumption $b \equiv c$, with $b := d$ —i.e. $d \equiv c \rangle$
    c

Proofs by assumption can be hierarchical. For example, we prove

$$(p \Rightarrow p') \Rightarrow ((q \Rightarrow q') \Rightarrow (p \wedge q \Rightarrow p' \wedge q')) \ .$$

Our proof assumes first $p \Rightarrow p'$ and then $q \Rightarrow q'$. However, $p \Rightarrow p'$ is not in a suitable form for use in this proof; by (3.60), it is equivalent to $p \wedge p' \equiv p$, and this formula is needed in the proof. Rather than write and prove $p \wedge p' \equiv p$ separately, as a lemma, we simply say that it holds and rely on the reader's experience to fill in details if deemed necessary. Here, then, is the proof.

Assume $p \Rightarrow p'$ (which is equivalent to $p \wedge p' \equiv p$)
  Assume $q \Rightarrow q'$ (which is equivalent to $q \wedge q' \equiv q$)
    $p \wedge q$
  $=$ $\langle$Assumption $p \wedge p' \equiv p \rangle$
    $p \wedge p' \wedge q$
  $=$ $\langle$Assumption $q \wedge q' \equiv q \rangle$
    $p \wedge p' \wedge q \wedge q'$
  $\Rightarrow$ $\langle$Weakening (3.76b)$\rangle$
    $p' \wedge q'$

PROOF BY CASE ANALYSIS

A proof of P (say) by case analysis proceeds as follows. Find *cases* (boolean expressions) Q and R (say) such that $Q \vee R$ holds. Then show that P holds in each case: $Q \Rightarrow P$ and $R \Rightarrow P$. One could have a 3-case analysis, or a 4-case analysis, and so on; the disjunction of all the cases must be *true* and each case must imply P.

It is usually best to avoid case analysis. A single thread of reasoning is usually easier to comprehend than several. A proof by case analysis can be much longer than a proof that avoids it, simply because each case needs a separate proof and because one must ensure with an additional proof that all possibilities are enumerated by the cases. This situation occurs with our proof of the law of contradiction, given below. Further, use of nested case analysis can lead to an explosion in the number of cases to be considered —much like the use of nested conditional statements in programs, which we all know becomes unwieldy. However, case analysis cannot always be avoided, and we need good methods for handling it.

Our first formalization of case analysis depends on Shannon (3.89):

$$E_p^z \;\equiv\; (p \wedge E_{true}^z) \vee (\neg p \wedge E_{false}^z) \quad .$$

Using (3.89), we can justify the following metatheorem, which indicates that we can prove a theorem by considering two cases. In the first case, one of its variables is replaced by *true*, and in the second case, the same variable is replaced by *false*.

(4.5)   **Metatheorem Case analysis.** If $E[z := true]$ and $E[z := false]$ are theorems, then so is $E[z := p]$.

We prove (4.5). Under the hypotheses of the theorem, we have,

$$E[z := p]$$
$=$ ⟨Shannon (3.89)⟩
$$(p \wedge E[z := true]) \vee (\neg p \wedge E[z := false])$$
$=$ ⟨Hypotheses of (4.5) together with Exercise 3.4⟩
$$(p \wedge true) \vee (\neg p \wedge true)$$
$=$ ⟨Identity of $\wedge$ (3.39), twice⟩
$$p \vee \neg p \quad \text{—Excluded Middle (3.28)}$$

We illustrate this kind of case analysis with two proofs of Contradiction (3.42), $p \wedge \neg p \equiv false$; they should be compared to the equational proof requested in Exercise 3.26. The first proof is in English:

Proof. If p is *true*, then the LHS of the formula is $true \wedge \neg true$, which, by Identity of $\wedge$ (3.39)) and the Definition of *false* (3.8) is equivalent to *false*. If p is *false*, then the LHS of the formula is $false \wedge \neg false$, which, by Zero of $\wedge$ (3.40) is equivalent to *false*. Hence, in both cases, the LHS is equivalent to *false* and the formula is *true*. Therefore, by metatheorem (4.5), the formula is *true*.   □

The second proof illustrates a stylized form of proof by case analysis that makes the structure of the proof clearer.

> **Prove:** $p \wedge \neg p \equiv false$
> **By Shannon**
> **Case** $(p \wedge \neg p \equiv false)[p := true]$
> $=$ ⟨Textual substitution⟩
> $true \wedge \neg true \equiv false$
> $=$ ⟨Identity of $\wedge$ (3.39); Definition of *false* (3.8)⟩
> $false \equiv false$ —which is Reflexivity of $\equiv$ (3.5)
> **Case** $(p \wedge \neg p \equiv false)[p := false]$
> $=$ ⟨Textual substitution⟩
> $false \wedge \neg false \equiv false$
> $=$ ⟨Zero of $\wedge$ (3.40)⟩
> $false \equiv false$ —which is Reflexivity of $\equiv$ (3.5)

In addition to its use in proofs, the case analysis embodied in (4.5) is applicable as a sort of partial evaluation, to check quickly if a formula could be a theorem. Choose a suitable variable and see what the value of the formula is when the variable is *true* and when it is *false*. If in either case the value is *false*, the formula is not a theorem. Such a check does not work so easily for other domains, like the integers, because there are too many different values of the variable to check.

We now turn attention to a more general kind of case analysis, which is based on the following theorem (see Exercise 4.12).

$$(4.6) \quad (p \vee q \vee r) \wedge (p \Rightarrow s) \wedge (q \Rightarrow s) \wedge (r \Rightarrow s) \ \Rightarrow \ s$$

This theorem justifies a three-case analysis; the antecedent indicates that at least one of the cases p, q, and r is true in each state and that each case implies s. It should be clear that the same kind of theorem, as well as the results of this subsection, will hold for any number of cases. Here, we treat only the three-case analysis.

A format for a three-case analysis is given in Fig. 4.1. Using a three-case analysis, we can prove S by splitting the state-space into three parts P, Q, and R (which may overlap) and then proving that in each case S holds. For example, suppose we define the Fibonacci numbers $f.i$ for i a natural number by

$$f.i = \begin{cases} 0 & \text{if } i = 0 \\ 1 & \text{if } i = 1 \\ f(i-1) + f(i-2) & \text{if } i > 1 \ . \end{cases}$$

A proof of some property of f is then likely to use the three-case analysis suggested by this definition, looking separately at the cases $i = 0$, $i = 1$, and $i > 1$. Such a proof is almost forced by the three-part definition of f —although by noticing that $f.i = i$ for $0 \leq i \leq 1$, a two-case analysis might suffice. In general, reducing the number of cases used in defining an object can reduce the work necessary for proving its properties.

FIGURE 4.1. STYLIZED PROOF BY CASE ANALYSIS

Prove: S
  By cases: P, Q, R
    (proof of $P \vee Q \vee R$ —omitted if obvious)
    Case P : (proof of $P \Rightarrow S$)
    Case Q : (proof of $Q \Rightarrow S$)
    Case R : (proof of $R \Rightarrow S$)

PROOF BY MUTUAL IMPLICATION

A proof by mutual implication of an equivalence $P \equiv Q$ is performed as follows:

(4.7) **Proof method.** To prove $P \equiv Q$, prove $P \Rightarrow Q$ and $Q \Rightarrow P$.

Such a proof rests on theorem (3.80), which we repeat here:

$$(p \Rightarrow q) \wedge (q \Rightarrow p) \equiv (p \equiv q) \quad .$$

Certain forms of equational proof involve mutual implication in disguise. Consider a proof of $P \equiv Q$ of the form:

$$
\begin{array}{cl}
P & \\
= & \langle \text{Hint} \rangle \\
& \cdots \\
= & \langle \text{Hint} \rangle \\
P \vee Q & \\
= & \langle \text{Hint} \rangle \\
& \cdots \\
= & \langle \text{Hint} \rangle \\
Q &
\end{array}
$$

This proof establishes $(P \equiv P \vee Q)$ and $(P \vee Q \equiv Q)$. Since $P \equiv P \vee Q$ equivales $Q \Rightarrow P$ and $P \vee Q \equiv Q$ equivales $P \Rightarrow Q$, the proof establishes

$$(Q \Rightarrow P) \wedge (P \Rightarrow Q) \quad .$$

But this formula is the LHS of (3.80). Hence, the proof is really just a proof by mutual implication of $P \equiv Q$.

In writing this section of the text, we searched for a good example of proof by mutual implication. Several texts on discrete mathematics used mutual implication to prove the following theorem. Let $even.i$ stand for "i is a multiple of 2", i.e. $i = 2 \cdot k$ for some natural number k.

(4.8) **Theorem.** For any natural number i, $even.i \equiv even(i^2)$.

One proof by mutual implication in the literature proved $LHS \Rightarrow RHS$ and the contrapositive of $RHS \Rightarrow LHS$, $odd.i \Rightarrow odd(i^2)$. Both of these proofs were essentially in English. We made these proofs calculational and polished them until they were as clear as we could make them. We then realized that our proof of $LHS \Rightarrow RHS$, given below, had become a proof of $LHS \equiv RHS$, so that a proof by mutual implication was not needed! Half the proof was thrown away! This story illustrates how formalizing can shorten and simplify an argument.

$even.i$

$=$ $\langle$Definition of $even\,\rangle$
       $i = 2 \cdot k$ (for some natural number k)

$=$ $\langle\, x = y \;\equiv\; x^2 = y^2$ (for natural numbers $x, y\,\rangle)$
       $i^2 = (2 \cdot k)^2$ (for some natural number k)

$=$ $\langle$Arithmetic$\rangle$
       $i^2 = 2 \cdot (2 \cdot k^2)$ (for some natural number k)

$=$ $\langle$Definition of $even\,\rangle$
       $even(i^2)$

In Chap. 3, we used mutual implication to prove an equivalence only once, even though we proved over 60 theorems (counting the exercises). Just like case analysis, a proof by mutual implication is generally going to be longer than a direct proof that avoids it, and we suggest eschewing mutual implication where possible. However, there are situations where mutual implication *must* be used to prove an equivalence $P \equiv Q$. This occurs when the proofs of $P \Rightarrow Q$ and $Q \Rightarrow P$ rely on different properties. See, for example, the proof of Theorem (12.26) in Chap. 12.

PROOF BY CONTRADICTION

Another common practice in mathematics for proving a theorem P is to assume P is *false* and derive a contradiction (that is, derive *false* or something equivalent to *false*). The formal basis for such a proof is theorem (3.74), $p \Rightarrow false \;\equiv\; \neg p$. With the substitution $p := \neg p$, and using double negation (3.12), we derive the theorem

(4.9)   **Proof by contradiction:** $\neg p \Rightarrow false \equiv p$.

Hence, having proved that $\neg P \Rightarrow false$ is a theorem, we can conclude that P is a theorem as well.

Formula $\neg P \Rightarrow false$ is usually proved using the method of the previous subsection: assume $\neg P$ and prove *false*. A shortcut is often taken: instead of proving *false* directly, prove something that is obviously equivalent to *false*, like $Q \wedge \neg Q$.

This proof method is overused —many proofs by contradiction can be more simply written using a direct method. Often, this overuse arises from trying to do too much of the proof in English. As an example, consider the following theorem and its (informal) proof by contradiction.

(4.10)   **Theorem.** Let u be a left identity and v be a right identity of operator $\circ$, i.e. $u \circ x = x$ and $x \circ v = x$ for all x. Then $u = v$.

Proof. We assume $u \neq v$ and prove a contradiction. Consider the expression $u \circ v$. Since u is a left identity, this expression equals v; since v is a right identity, this expression equals u; hence, $u = v$, but this contradicts the assumption $u \neq v$. Hence the assumption is *false*, and $u = v$.   □

Here is a much simpler, straightforward equational proof.

$$u$$
$$= \quad \langle v \text{ is a right identity} \rangle$$
$$u \circ v$$
$$= \quad \langle u \text{ is a left identity} \rangle$$
$$v$$

That the formal proof is much simpler is no accident. Using formal tools, and not even letting contorted English sentences come into one's thoughts, can often lead to simpler arguments. Let the formal tools do the work.

Here is a case where proof by contradiction is sensible. Consider writing a function *Halt* that would test whether execution of an input-free program (or any imperative statement) halts. (By "input-free" we mean that the program does not read from files or refer to global variables.) The first line, a comment, is a specification for *Halt*; it indicates that a function application $Halt(P)$ equivales the value of the statement "P halts".

$\{Halt(P) \equiv P \text{ halts}\}$
function $Halt(P : string) : bool$;
begin ... **end**

Parameter P is a string of characters. Presumably, *Halt* analyzes P much the way a compiler does, but the compiler generates a program in some machine language while *Halt* just determines whether P halts.

Function *Halt* would be very useful. However, Alan Turing proved in the 1930's (see Historical note 4.1) that it cannot be written.

(4.11) **Theorem.** Function *Halt* does not exist.

Proof. Assume *Halt* exists and consider the following procedure.

procedure B;
begin while $Halt(\text{"call } B\text{"})$ **do** *skip* **end**

Note that the argument of the call on *Halt* in the body of B is a call on B itself. We observe the following.

"**call** B" halts
$= \quad \langle \text{inspection of } B\text{'s procedure body} \rangle$
$\neg Halt(\text{"call } B\text{"})$
$= \quad \langle \text{Definition of } Halt \text{ —see comment on function } Halt \rangle$
$\neg (\text{"call } B\text{" halts})$

HISTORICAL NOTE 4.1. ALAN M. TURING (1912–1954)

Alan Turing is the legendary figure after whom the ACM's Annual *Turing Award* is named —computer science's equivalent of the Nobel Prize. His stature comes from work he did when he was 23 while on a student fellowship at Cambridge University, work that was fundamental in a field that did not exist yet: computer science.

Turing was taken by Hilbert's claim that mathematics would be decidable (see Historical note 6.1), i.e. in principle, there would be a mechanical procedure for determining whether any statement was true or false. Turing developed an abstract form of computer (before computers existed) to carry out mechanical procedures. This mathematical computer, which now bears the name *Turing machine*, is still of great interest today. Turing gave convincing evidence that the Turing machine was *universal*: any "computable function" could be written as a Turing machine. Using Turing machines, Turing then proved that decidability was out of the question. For example, the halting problem discussed on page 78 is undecidable; there is no procedure for determining in a finite time whether an arbitrary program will halt.

Turing was also a key player on the team at Bletchley that deciphered German messages during World War II. He was a prime developer of both the electronics and the architecture of the British computer ACE (starting in 1945) and was the first to recognize the full potential of a stored-program computer that could create its own instructions. A paper of his in 1949 is viewed as the first instance of a program-correctness proof.

The last two years of Turing's life are a sad commentary on the times. In 1952, Turing was charged with 12 counts of "committing an act of gross indecency" with another male. He was a homosexual. Both men pleaded guilty, but Turing felt no guilt and lived through the proceedings in a seemingly detached manner. His punishment was a year of probation, during which he had to take the female hormone estrogen to reduce his sexual libido. His intellectual life went on as before. However, in June 1954, with no warning and no note of explanation, he committed suicide by taking cyanide. (See the excellent biography [23].)

We have derived a contradiction, so we have disproved the assumption that *Halt* exists.   □

PROOF BY CONTRAPOSITIVE

An implication $P \Rightarrow Q$ is sometimes proved as follows. First assume P; then prove Q by contradiction, i.e. assume $\neg Q$ and prove *false*. Such a proof is not as clear as we might hope, and there is a better way:

(4.12) **Proof method:** Prove $P \Rightarrow Q$ by proving its contrapositive $\neg Q \Rightarrow \neg P$ (see (3.61)).

Here is an example: we prove $x + y \geq 2 \Rightarrow x \geq 1 \lor y \geq 1$. By Contrapositive (3.61), De Morgan, and arithmetic, this formula is equivalent to $x < 1 \land y < 1 \Rightarrow x + y < 2$, and we prove the latter formula by assuming the antecedent and proving the consequent.

$$
\begin{array}{ll}
& x + y \\
< & \langle \text{Assumptions } x < 1 \text{ and } y < 1 \rangle \\
& 1 + 1 \\
= & \langle \text{Arithmetic} \rangle \\
& 2
\end{array}
$$

Exercises for Chapter 4

Exercises on an abbreviation for implications

4.1 Prove theorem (4.1), $p \Rightarrow (q \Rightarrow p)$, using the method of Sec. 4.1.

4.2 Prove Monotonicity of $\land$ (4.3), $(p \Rightarrow q) \Rightarrow (p \land r \Rightarrow q \land r)$, using the method of Sec. 4.1. Start with the consequent, since it has more structure.

4.3 Prove Weakening/strengthening (3.76d), $p \lor (q \land r) \Rightarrow p \lor q$, using the method of Sec. 4.1. Start with the antecedent, since it has more structure, and distribute.

4.4 Prove $(p \Rightarrow q) \land (r \Rightarrow s) \Rightarrow (p \lor r \Rightarrow q \lor s)$, using the proof format of Sec. 4.1. You may first want to remove the implications in the antecedent, distribute as much as possible, and then use theorem (3.76d) and an absorption theorem.

4.5 Prove $(p \Rightarrow q) \land (r \Rightarrow s) \Rightarrow (p \land r \Rightarrow q \land s)$, using the proof format of Sec. 4.1. Before using the proof format, you may first want to using Shunting (3.65) to move $p \land r$ into the antecedent.

Exercises on additional proof techniques

4.6 Prove $p \Rightarrow (q \Rightarrow p)$ by the method of assuming the antecedent.

4.7 Prove $(\neg p \Rightarrow q) \Rightarrow ((p \Rightarrow q) \Rightarrow q)$ by the method of assuming the antecedent.

4.8 Prove $p \land q \Rightarrow (p \equiv q)$ by the method of assuming the antecedent.

4.9 Prove $(p \Rightarrow p') \land (q \Rightarrow q') \Rightarrow (p \lor q \Rightarrow p' \lor q')$ by the method of assuming the antecedent.

4.10 Prove Modus ponens (3.77), $p \land (p \Rightarrow q) \Rightarrow q$, by the method of assuming the antecedent.

4.11 Prove the following theorem using Metatheorem Case analysis (4.5):

$$(p \vee q) \wedge r \; \equiv \; (p \wedge r) \vee (q \wedge r) \quad .$$

4.12 Prove theorem (4.6), $(p \vee q \vee r) \wedge (p \Rightarrow s) \wedge (q \Rightarrow s) \wedge (r \Rightarrow s) \; \Rightarrow \; s$.

4.13 Let $x \downarrow y$ be the minimum of integers x and y, defined by $x \downarrow y = ($**if** $x \leq y$ **then** x **else** $y)$. Prove that $\downarrow$ is symmetric, i.e. $b \downarrow c = c \downarrow b$. How many cases do you have to consider? You may use the necessary rules of integer arithmetic, for example, that $b \leq c \; \equiv \; b = c \vee b < c$ and that $b < c \; \equiv \; c > b$.

4.14 Prove by case analysis that $\downarrow$ is associative, i.e. that $b \downarrow (c \downarrow d) = (b \downarrow c) \downarrow d$ (see the previous exercise). How many cases do you have to consider, based on the definition of $\downarrow$?

4.15 Consider the discussion on page 76 that shows how a proof of $P \equiv Q$ with $P \vee Q$ as an intermediate step can be viewed as a proof by mutual implication. Write a similar discussion to show how a proof of $P \equiv Q$ with $P \wedge Q$ as an intermediate step can be viewed as a proof by mutual implication.

Chapter 5

Applications of Propositional Calculus

W e look at two applications of propositional calculus. The first is its use in solving various "word problems", such as the superman story on page 37. Formalizing such problems in propositional calculus allows us to solve them more easily than we could using English alone. As a second application, we show how propositional calculus can be used in the design of combinational digital circuits.

5.1 Solving word problems

We can reason about English statements by formalizing them as boolean expressions and manipulating the formalization. This technique has at least two uses. First, we can check an English argument by formalizing it as a boolean expression and then proving the expression to be a theorem. Of course, the expression may not be a theorem (which means that the English argument from which it was derived is unsound). In this case, our attempt at proving the expression may lead us to a counterexample —a state in which the expression is *false* .

Second, we can use propositional logic to help solve word problems and puzzles. The challenging puzzles in this chapter (and its exercises) concerning Portia, Superman, the maid and the butler, the island of Marr, and knights and knaves were taken from Backhouse [2], Smullyan [37], and Wickelgren [45]. If, after studying this chapter, you want to try additional recreational puzzles, get Smullyan's book, which contains 270 of them. Smullyan's other books [38, 39] are also recommended.

CHECKING ENGLISH ARGUMENTS

We can check an argument given in English by formalizing it and proving the formalization to be a theorem. Consider argument (5.1) below. It starts with two English sentences, each of which states a fact that is asserted to be *true* . These are followed by a conclusion, which is supposed to be supported by the facts. The conclusion is introduced by the word "hence".

(5.1)   If Joe fails to submit a project in course CS414, then he fails the
course. If Joe fails CS414, then he cannot graduate. Hence, if Joe
graduates, he must have submitted a project.

Most of the arguments considered in this section have this form, although
the number of facts will vary, different words like "thus" and "therefore"
will introduce the conclusion, and the facts and conclusion may be hidden
by obtuse wording.

Let us call the facts of (5.1) $F0$ and $F1$ and call the conclusion C.
Then, from the truth of $F0$ and $F1$, C is to be derived. That is, we have
to prove $F0 \land F1 \Rightarrow C$.

We now translate these facts and conclusion into propositional calculus.
We associate identifiers with the primitive propositions:

s :   Joe submits a project in CS414.
f :   Joe fails CS414.
g :   Joe graduates.

$F0$ is formalized as $\neg s \Rightarrow f$, $F1$ as $f \Rightarrow \neg g$, and C as $g \Rightarrow s$. To
check the soundness of (5.1), we prove $F0 \land F1 \Rightarrow C$:

$$(\neg s \Rightarrow f) \land (f \Rightarrow \neg g) \Rightarrow (g \Rightarrow s) .$$

We prove this theorem by transforming its antecedent into its consequent:

$$
\begin{aligned}
&(\neg s \Rightarrow f) \land (f \Rightarrow \neg g) \\
\Rightarrow \quad &\langle \text{Transitivity of} \Rightarrow (3.82a) \rangle \\
&\neg s \Rightarrow \neg g \\
= \quad &\langle \text{Contrapositive } (3.61) \rangle \\
&g \Rightarrow s
\end{aligned}
$$

Actually, you should question whether English statement (5.1) really is
an argument. An argument is a coherent set of facts and reasons that gives
evidence of the truth of some statement. But (5.1) does not give any reasons
at all, it simply states the theorem to be proved, $F0 \land F1 \Rightarrow C$! It is up
to the reader, without help from the writer, to prove the theorem because
none of the steps of the proof are provided. Perhaps this is why so many
arguments that are couched in English are difficult to understand.

Constructing a Counterexample

When an English argument is not sound, attempting to formalize and prove
it can lead to a counterexample —an assignment of values to its variables

or primitive propositions that makes the argument *false*. Consider the
following argument:

(5.2)   If X is greater than zero, then if Y is zero then Z is zero.
Variable Y is zero. Hence, either X is greater than zero or Z is
zero.

This argument consists of two facts and a conclusion drawn from them. We
begin formalizing the argument by associating identifiers with its primitive
propositions.

$$x: \quad X \text{ is greater than zero.}$$
$$y: \quad Y \text{ is zero.}$$
$$z: \quad Z \text{ is zero.}$$

We can then formalize (5.2) as

(5.3) $(x \Rightarrow (y \Rightarrow z)) \wedge y \Rightarrow x \vee z$.

The antecedent has the most structure, so we manipulate it.

$$
\begin{array}{ll}
& (x \Rightarrow (y \Rightarrow z)) \wedge y \\
= & \langle \text{Shunting (3.65), twice} \rangle \\
& (y \Rightarrow (x \Rightarrow z)) \wedge y \\
= & \langle (3.66),\ p \wedge (p \Rightarrow q) \equiv p \wedge q \rangle \\
& (x \Rightarrow z) \wedge y
\end{array}
$$

Compare the last form of the antecedent, $(x \Rightarrow z) \wedge y$, with consequent
$x \vee z$. Variable y has nothing to do with the consequent, and $x \Rightarrow$
z (i.e. $(\neg x \vee z)$) does not imply $x \vee z$. Hence, we should suspect that
(5.3) is not valid and that argument (5.2) is not sound. So we look for a
counterexample.

How can we find a counterexample? Based on the form of an expression,
we can determine what values of its operands make the expression *false*,
by using Table 5.1. This table, then, helps in constructing counterexamples.

Expression (5.3) is an implication, so, based on Table 5.1, for it to be
false its consequent must be *false*, and this requires $x = z = false$. Then

---

TABLE 5.1. COUNTEREXAMPLES FOR EXPRESSIONS

expression	counterexample 1	counterexample 2
$p \wedge q$	$p = false$	$q = false$
$p \vee q$	$p = q = false$	
$p \equiv q$	$p = true,\ q = false$	$p = false,\ q = true$
$p \not\equiv q$	$p = q = true$	$p = q = false$
$p \Rightarrow q$	$p = true,\ q = false$	

y must be chosen to make the antecedent *true* , which requires $y = true$. Hence, the counterexample, is $x = z = false$ and $y = true$.

Whether one starts the search for a counterexample of an implication by attempting to make the consequent *false* or by attempting to make the antecedent *true* depends on their shape and content. Working with the one with the fewest different variables is usually easier.

MAKING SENSE OF AN ENGLISH SENTENCE

We can use propositional logic to understand English sentences better. Consider the following English statement, which seems preposterous.

> Value v is in $b[1..10]$ means that if v is in $b[11..20]$ then it is not in $b[11..20]$.

We associate boolean variables with primitives of the sentence.

$x :$ v is in $b[1..10]$.
$y :$ v is in $b[11..20]$.

Then the sentence is formalized as $x \equiv y \Rightarrow \neg y$. We simplify it.

$$
\begin{aligned}
& x \equiv y \Rightarrow \neg y \\
= \quad & \langle \text{Rewrite implication (3.59)} \rangle \\
& x \equiv \neg y \lor \neg y \\
= \quad & \langle \text{Idempotency of } \lor \ (3.26) \rangle \\
& x \equiv \neg y
\end{aligned}
$$

Translating back into English, we see that the sentence has the meaning "v is in $b[1..10]$ means that it is not in $b[11..20]$" —any value in the first half of b is not in the second half. In this case, propositional logic helped us clarify a seemingly gibberish sentence.

SOLVING PUZZLES: PORTIA'S SUITOR'S DILEMMA

Consider the following, which is a simplification of a situation in Shakespeare's *Merchant of Venice*. Portia has a gold casket and a silver casket and has placed a picture of herself in one of them. On the caskets, she has written the following inscriptions:

> Gold: The portrait is not in here.
> Silver: Exactly one of these inscriptions is true.

Portia explains to her suitor that each inscription may be *true* or *false* , but that she has placed her portrait in one of the caskets in a manner that is consistent with this truth or falsity of the inscriptions. If he can choose the casket with her portrait, she will marry him —in those days, that's what suitors wanted. The problem for the suitor is to use the inscriptions (although they could be *true* or *false*) to determine which casket contains her portrait.

To begin solving the problem, we formalize it. We introduce four variables to stand for primitive propositions:

gc : The portrait is in the gold casket.

sc : The portrait is in the silver casket.

g : The portrait is not in the gold casket.
(This the inscription on the gold casket.)

s : Exactly one of g and s is *true* .
(This the inscription on the silver casket.)

Using these propositions, we proceed as follows. First, the fact that the portrait is in exactly one place can be written as [1]

$$F0 : gc \equiv \neg sc \ .$$

Next, inscription g on the gold casket is the negation of gc .

$$F1 : g \equiv \neg gc \ .$$

Taking a cue from $F0$, we see that inscription s on the silver casket is equivalent to $s \equiv \neg g$. We do not want to claim that $s \equiv \neg g$ is a fact, since we do not know whether this inscription is *true* ; we only want to claim that inscription s equivales $s \equiv \neg g$. Hence, we arrive at $F2$:

$$F2 : s \equiv (s \equiv \neg g) \ .$$

Expressions $F0$, $F1$, and $F2$ formalize the problem. We now determine whether we can derive either gc or sc from them. $F2$, which has the most structure, looks the most promising for manipulation:

$$
\begin{aligned}
& s \equiv s \equiv \neg g \\
= \quad & \langle \text{Symmetry of } \equiv (3.2) \text{ —so } \neg g \equiv s \equiv s \equiv \neg g \ \rangle \\
& \neg g \\
= \quad & \langle \ F1 \,; \text{ Double negation } (3.12) \rangle \\
& gc
\end{aligned}
$$

---

[1] Those not facile with equivalence will write this as $(gc \wedge \neg sc) \vee (\neg gc \wedge sc)$ or as $(gc \vee sc) \wedge \neg(gc \wedge sc)$. But $gc \equiv \neg sc$ is shorter and, because it is an equivalence, easier to handle.

Hence, from $F1$ and $F2$ ($F0$ is not needed) we conclude gc. The portrait is in the gold casket.

We should make sure that $F0$, $F1$, and $F2$ are not contradictory, i.e. that there is at least one assignment of values to g, s, gc, and sc that makes all three *true*. If $F0 \wedge F1 \wedge F2$ were *false* in every state, then the propositional logic together with $F0$, $F1$, and $F2$ would be inconsistent (see Def. (7.1)), and anything could be proved. Were that the case, the assumption could not be satisfied, so we would conclude that the problem had no solution.

With $gc = true$, the additional assignments $sc = false$, $g = false$, and $s = false$ (or $s = true$!) satisfy $F0$, $F1$, and $F2$. Note that it does not matter whether the inscription on the silver casket is *true* or *false*.

This example illustrates how effective the calculational style of proof can be. Through a rather simple formalization and calculation, we have solved what seemed to be a complicated problem.

MORE ON INCONSISTENCIES

Formalizing the previous puzzle did not lead to an inconsistency. We now analyze a similar puzzle whose formalization is inconsistent. Consider again Portia's suitor's problem, and suppose that Portia writes a different inscription on the silver casket:

s' : This inscription is false.

A formalization of this inscription is

$F2'$: $s' \equiv \neg s'$.

But $F2'$ is *true* in no state; it is equivalent to *false*. Adding $F2'$ as an axiom of propositional logic, then, would be taking *false* as an axiom, and from *false*, anything can be proved (3.75). With the addition of $F2'$, our logic becomes inconsistent and thus useless. $F2'$ is absurd and cannot be part of any mathematical model of reality, and we conclude that this puzzle has no solution.

An inconsistency can also arise from an interplay between axioms. For example, suppose $F0$ is already an axiom. If we now add the axiom $F5$: $gc \equiv sc$, then $F0 \wedge F5 \equiv false$, so the system is inconsistent.

ANOTHER PUZZLE: DOES SUPERMAN EXIST?

Page 37 contains an English argument that Superman does not exist:

If Superman were able and willing to prevent evil, he would do so. If Superman were unable to prevent evil, he would be impotent; if he were unwilling to prevent evil, he would be malevolent. Superman does not prevent evil. If Superman exists, he is neither impotent nor malevolent. Therefore, Superman does not exist.

We want to use the propositional calculus to determine whether this argument is sound —whether the conclusion "Superman does not exist" follows from the previous sentences. As on page 37, we associate variables with the primitive propositions:

a : Superman is able to prevent evil.
w : Superman is willing to prevent evil.
i : Superman is impotent.
m : Superman is malevolent.
p : Superman prevents evil.
e : Superman exists.

The first four sentences can be formalized as

$F0 : a \land w \Rightarrow p$

$F1 : (\neg a \Rightarrow i) \land (\neg w \Rightarrow m)$

$F2 : \neg p$

$F3 : e \Rightarrow \neg i \land \neg m$

and the Superman argument is equivalent to the boolean expression

$(5.4) \quad F0 \land F1 \land F2 \land F3 \Rightarrow \neg e$.

One way to prove (5.4) is to assume the four conjuncts of the antecedent and prove the consequent. That is, we begin by manipulating the consequent $\neg e$. Beginning with $\neg e$, we see only *one* way to proceed. The only assumption in which e appears is $F3$. If we translate $F3$ into its contrapositive $\neg(\neg i \land \neg m) \Rightarrow \neg e$, $\neg e$ emerges. (See (3.61) for the contrapositive of an implication).

Assume $F0$, $F1$, $F2$, $F3$
    $\neg e$
  $\Leftarrow$ ⟨Contrapositive $\neg(\neg i \land \neg m) \Rightarrow \neg e$ of $F3$
        —the only other place e appears⟩
    $\neg(\neg i \land \neg m)$
  $=$ ⟨De Morgan (3.47a); Double negation (3.12), twice⟩
    $i \lor m$
  $\Leftarrow$ ⟨First conjunct of $F1$ and Monotonicity (4.2)⟩
    $\neg a \lor m$

$$\Leftarrow \quad \langle \text{Second conjunct of } F1 \text{ and Monotonicity } (4.2) \rangle$$
$$\neg a \lor \neg w$$
$$= \quad \langle \text{De Morgan } (3.47a) \rangle$$
$$\neg(a \land w)$$
$$\Leftarrow \quad \langle \text{Contrapositive } \neg p \Rightarrow \neg(a \land w) \text{ of } F0 \rangle$$
$$\neg p \quad \text{—this is } F2$$

We conclude that (5.4) is a theorem, so the argument of the Superman paragraph is sound.

This calculation illustrates an important point. We started with the consequent $\neg e$ and worked "backward" toward the assumptions. In this case, working backwards was a real help, for at each step there was essentially no choice about what to do next! The only choice was in the order in which to use the conjuncts of $F1$, and this choice was immaterial to the proof development. Proofs in which there is no choice at each step are particularly nice, because the reader can see that each step is directed by a formula's structure and is not a rabbit pulled out of a hat.

5.2   Combinational digital circuits

Digital circuits are electronic circuits whose inputs and outputs denote the boolean constants *false* and *true*. In one common scheme, each input and output is a wire; low signal voltages represent *false* and high voltages represent *true*.

Digital circuits can be designed to perform arithmetic operations, to process text, and even to execute programs. This is because numbers, characters, and program operations all can be represented by sequences of bits 0 and 1 (see Sec. 15.5), and a bit can be implemented as a boolean constant. Conventionally, 0 is represented by *false* and 1 by *true*.

A *combinational digital circuit* is a digital circuit whose outputs at any time are determined solely by the values of its inputs at that time — previous inputs and outputs have no effect on the current output. Such circuits cannot implement components, such as a memory, whose operation depends on past inputs and outputs. Still, a significant portion of most digital circuitry is combinational, and many circuits in computers are entirely combinational. For example, the arithmetic-logical unit (ALU) and memory-addressing circuitry of most computers are combinational.

A circuit can be described by a *circuit diagram*, which describes a collection of *gates* and their interconnections. Each gate is a component whose output is a boolean function of its inputs. Circuit-diagram symbols for three representative gates, which compute conjunctions, disjunctions, and negations, are given in Fig. 5.1. The gate for negation is called an *inverter* Input wires of a gate usually enter from the left and top; output wires emerge from the right and bottom.

A circuit appears on the right in Fig. 5.2. Its wires are labeled. Wires that always have the same signal may be given the same name. A black dot where wires cross represents a connection that allows a signal to be directed to several places. For example, input wire a in Fig. 5.2 is directed to three places: two and-gates and an inverter.

In a combinational circuit, the output of no gate is connected (either directly or through a series of wires and gates) to its inputs. This topological restriction ensures that the input of a gate is not influenced by a past output of the gate, the hallmark of a combinational circuit.

On the left in Fig. 5.2 is a "black-box" symbol for the circuit, giving it a name *HA* and showing the relative positions of inputs a and b and outputs c and s. This symbol is similar to a procedure heading, which names a procedure and describes its parameters, in a programming language. A black-box symbol (without the parameter names) can be used to denote the circuit when it is used as a component in a larger circuit. Such a use is similar to a procedure call. See the right side of Fig. 5.4 on page 100 for an example.

CIRCUIT DIAGRAMS AND BOOLEAN EXPRESSIONS

In a combinational circuit, each output is determined solely by current inputs. Thus, a circuit with n inputs x_1, x_2, ..., x_n and m outputs z_1, z_2, ..., z_m implements, for each output z_i, a boolean function of n arguments. This idea was first observed and exploited by Claude Shannon in his Masters thesis some 55 years ago (see Historical note 5.1).

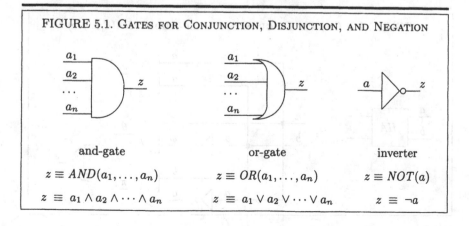

FIGURE 5.1. GATES FOR CONJUNCTION, DISJUNCTION, AND NEGATION

and-gate

$z \equiv AND(a_1, \ldots, a_n)$

$z \equiv a_1 \wedge a_2 \wedge \cdots \wedge a_n$

or-gate

$z \equiv OR(a_1, \ldots, a_n)$

$z \equiv a_1 \vee a_2 \vee \cdots \vee a_n$

inverter

$z \equiv NOT(a)$

$z \equiv \neg a$

A boolean expression that is equivalent to a given circuit can be constructed as follows. First, be sure that each wire of the circuit is labeled. Then, for each gate, write a conjunct that embodies the relationship that the gate implements between its inputs and outputs. Whenever different names x and y are given to connected wires, include the conjunct $x = y$ in the boolean expression.

Here is the boolean expression constructed from circuit HA of Fig. 5.2. Greek letters are used for the internal wires to distinguish them from the inputs and outputs.

$$(5.5) \quad c = (a \wedge b) \wedge$$
$$\phi = \neg b \ \wedge \ \theta = (a \wedge \phi) \wedge$$
$$\omega = \neg a \ \wedge \ \pi = (b \wedge \omega) \wedge$$
$$s = (\theta \vee \pi)$$

Let C be the boolean expression constructed from a circuit $\widehat{C}$. From the construction, we can see that an assignment of boolean constants to the names of the wires is possible for $\widehat{C}$ iff the assignment satisfies C. For example, the assignment

$$a = f, \ b = t, c = f, \phi = f, \ \theta = f, \ \omega = t, \ \pi = t, \ s = t$$

is possible for the circuit of Fig. 5.2 and also satisfies (5.5). (We abbreviate *false* as f and *true* as t.) On the other hand, an assignment with $a = f$, $b = f$, $c = t$ is not possible because the uppermost and-gate ensures that $c \equiv a \wedge b$ holds.

Thus, $\widehat{C}$ and C are two different representations of the same object, and we can use them interchangeably.

Above, we showed how to construct a boolean expression from a circuit. We now show how to construct a circuit (in terms of the gates of Fig. 5.1)

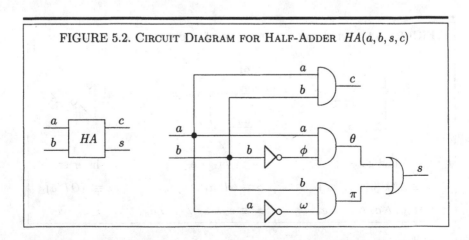

FIGURE 5.2. Circuit Diagram for Half-Adder $HA(a, b, s, c)$

HISTORICAL NOTE 5.1. CLAUDE SHANNON (1916-)

As an undergrad at MIT, Shannon majored in both math and electrical engineering, and this combination led him to write one of the most important Master's theses of all time [35]. Shannon showed how Boole's ideas on logic could be used in the design of electronic circuits, thus revolutionizing the field.

Some ten years later, while working for AT&T Bell Labs, Shannon started the field now called *information theory* and in [36] established the framework for the efficient transmission of electronic data. This framework is the basis for all systems that store, process, and transmit data in digital form, including your modem, fax machine, and compact disk. So important is [36] that it has been called the Magna Carta of the communications age.

Shannon defined the binary unit to be the basic unit of information —John Tukey then abbreviated "binary unit" to "bit", the term we use today.

Shannon has many interests, one of which is juggling —he was known for riding a unicycle through Bell Labs while juggling four balls. He loves gadgets and has built a juggling manikin that looks like comedian W.C. Fields, a mechanical mouse that finds its way through a maze, and a computer that calculates in Roman numerals. He has stated that "I've always pursued my interests without much regard for financial value or value to the world; I've spent lots of time on totally useless things."

This material was gleaned from the profile [25] of Claude Shannon.

---

from a boolean expression C, provided C has the following form:

$$(5.6) \quad z_1 = E_1(x) \ \wedge \ z_2 = E_2(x) \ \wedge \ \ldots \ \wedge \ z_m = E_m(x)$$

where

- the z_i are outputs,

- x is the vector of inputs, and

- each E_i is a boolean expression that involves only x, *true*, *false*, and the operators $\wedge$, $\vee$, and $\neg$. (A boolean constant is implemented by a wire connected to a constant voltage source.)

Each E_i of (5.6) can be written in terms of the three boolean functions

$$(5.7) \quad \begin{array}{ll} NOT(a): & \neg a \\ AND(a_1, \ldots, a_n): & a_1 \wedge \cdots \wedge a_n \quad \text{(for } n \geq 2) \\ OR(a_1, \ldots, a_n): & a_1 \vee \cdots \vee a_n \quad \text{(for } n \geq 2) \ . \end{array}$$

A boolean expression $z \equiv F(A_1, A_2, \ldots A_n)$ is implemented by a circuit whose output z comes from an F-gate whose inputs are connected to the outputs of circuits that compute $A_1, \ldots, A_n$.

To illustrate, consider the boolean expression $s \equiv (\neg a \wedge b) \vee (a \wedge \neg b)$. We have

$$s \equiv (\neg a \wedge b) \vee (a \wedge \neg b)$$
$$= \quad \langle \text{Definition of } OR \text{ gate (5.7)} \rangle$$
$$s \equiv OR(\neg a \wedge b, a \wedge \neg b)$$
$$= \quad \langle \text{Definition of } AND \text{ gate (5.7), twice} \rangle$$
$$s \equiv OR(AND(\neg a, b), AND(a, \neg b))$$
$$= \quad \langle \text{Definition of an inverter (5.7), twice} \rangle$$
$$s \equiv OR(AND(NOT(a), b), AND(a, NOT(b)))$$

Thus, we conclude that output s is the output of a 2-input or-gate. One of the inputs of this or-gate is connected to the output of a circuit that computes $\neg a \wedge b$; the other input, to a circuit that computes $a \wedge \neg b$.

The circuit for $\neg a \wedge b$ (the first input to the or-gate) is a 2-input and-gate, with one input coming from a circuit for $\neg a$ and the other input being wire b.

The circuit for $b \wedge \neg a$ (the second input to the or-gate) is a 2-input and-gate, with one input being wire a and the other input coming from a circuit for $\neg b$.

This yields the part of the circuit of Fig. 5.2 that computes s, but without names for internal wires.

In summary, for every combinational circuit $\widehat{C}$ we have a boolean expression C, and for every boolean expression of form (5.6), we can construct a circuit. Note, however, that the boolean expression constructed from a circuit references internal-wire names, while the circuit constructed from a boolean expression does not have names on its internal wires. This difference is revisited in the next section.

FROM SPECIFICATION TO IMPLEMENTATION

A boolean expression, or equivalently a truth table, can serve as a *specification* of a combinational circuit, in which case we say that the circuit *implements* the specification. So, we can use the propositional calculus to manipulate and analyze combinational circuits. To do so, we have to investigate the notions of specification and implementation.

We are interested in the input-output behavior of a circuit $\widehat{C}$. A specification of $\widehat{C}$ should indicate for each set of inputs what the corresponding outputs should be. We use the name *behavior* to denote this assignment of input and output values. For example, one behavior for the circuit of Fig. 5.2 is $a = t$, $b = f$, $c = f$, $s = t$. Thus, each behavior is a state that assigns boolean values to variables modeling the circuit's inputs and outputs (but not to internal wires).

Consider a specification S, given as a boolean expression, and a circuit $\widehat{C}$. Remember that boolean expression C and circuit $\widehat{C}$ are equivalent representations. If $\widehat{C}$ implements S, then every behavior of $\widehat{C}$ should satisfy S. In other words, suppose

$\widehat{C}$ produces values $Z_1, \ldots, Z_m$ on output wires $z_1, \ldots, z_m$ when given input values $X_1, \ldots, X_n$ on input lines $x_1, \ldots, x_n$.

Then for $\widehat{C}$ to implement S, any state that contains all the associations $x_i = X_i$ and $z_j = Z_j$ should satisfy specification S. That is, if C is *true* in a state, then S should be *true* in that state as well, or, equivalently, $C \Rightarrow S$ should be valid. Hence, we define an *implementation* as follows.

(5.8) **Definition.** Circuit $\widehat{C}$ *implements* specification S exactly when $C \Rightarrow S$ is valid.

We have recast the question of whether a circuit implements a specification as a question about validity of a boolean expression.

As an example, we prove that the circuit of Fig. 5.2 implements specification (5.10). The proof below uses Substitution (3.84a) and Weakening (3.76b) to eliminate the names of the internal wires.

$$
\begin{aligned}
&HA \\
= \quad &\langle (5.5) \text{ is } HA \rangle \\
&c = (a \wedge b) \ \wedge \ \phi = \neg b \ \wedge \ \theta = (a \wedge \phi) \ \wedge \\
&\omega = \neg a \ \wedge \ \pi = (b \wedge \omega) \ \wedge \ s = (\theta \vee \pi) \\
= \quad &\langle \text{Substitution (3.84a) —for } \phi \text{ and } \omega \rangle \\
&c = (a \wedge b) \ \wedge \ \phi = \neg b \ \wedge \ \theta = (a \wedge \neg b) \ \wedge \\
&\omega = \neg a \ \wedge \ \pi = (b \wedge \neg a) \ \wedge \ s = (\theta \vee \pi) \\
\Rightarrow \quad &\langle \text{Weakening (3.76b) —eliminate } \phi \text{ and } \omega \rangle \\
&c = (a \wedge b) \ \wedge \ \theta = (a \wedge \neg b) \ \wedge \ \pi = (b \wedge \neg a) \ \wedge \ s = (\theta \vee \pi) \\
= \quad &\langle \text{Substitution (3.84a) —for } \theta \text{ and } \pi \rangle \\
&c = (a \wedge b) \ \wedge \ \theta = (a \wedge \neg b) \ \wedge \ \pi = (b \wedge \neg a) \ \wedge \\
&s = ((a \wedge \neg b) \vee (b \wedge \neg a)) \\
\Rightarrow \quad &\langle \text{Weakening (3.76b) —eliminate } \theta \text{ and } \pi \rangle \\
&c = (a \wedge b) \ \wedge \ s = ((a \wedge \neg b) \vee (b \wedge \neg a)) \\
= \quad &\langle \text{Exclusive or (3.53)} \rangle \\
&c = (a \wedge b) \ \wedge \ s = (a \not\equiv b)
\end{aligned}
$$

Note that, according to Definition (5.8), a specification that is equivalent to *true* is satisfied by every implementation. This is because $C \Rightarrow true$ is a theorem no matter what C is. Having every circuit implement *true* is reasonable, since *true* imposes no requirements on implementations — *true* is satisfied by every state and thus imposes no restrictions on outputs for any input configuration. Similarly, an implementation that is equivalent to *false* satisfies every specification, since *false* $\Rightarrow S$ is a theorem for every S. This is not upsetting once we realize that a circuit characterized by *false* does not exist. A circuit that produces outputs cannot be specified by *false* (a specification that prohibits outputs). [2]

Besides using boolean expressions, we can use truth tables to specify combinational circuits. The truth table would have one column for each input and each output and one row for each possible combination of input values.

For example, suppose we want to specify a circuit that adds two bits a and b to yield a sum bit s and carry c:

$$\begin{array}{c} a \\ + \, b \\ \hline c \ s \end{array}$$

This addition can be defined by giving the values of s and c in four cases:

$$\begin{array}{cccc} 1 & 1 & 0 & 0 \\ + \, 1 & + \, 0 & + \, 1 & + \, 0 \\ \hline 1 \ 0 & 0 \ 1 & 0 \ 1 & 0 \ 0 \end{array}$$

Using the standard representation of *false* for 0 and *true* for 1, we rewrite this definition of s and c as a truth table.

(5.9)

a	b	s	c
t	t	f	t
t	f	t	f
f	t	t	f
f	f	f	f

This truth table can be expressed more succinctly as

(5.10) $HA(a,b,s,c): \quad s = (a \not\equiv b) \ \wedge \ c = (a \wedge b)$

or, using Exclusive or (3.53), as

$$s = ((\neg a \wedge b) \vee (a \wedge \neg b)) \ \wedge \ c = (a \wedge b)$$

This specification for a one-bit adder is implemented by the circuit of Fig. 5.2, as we proved above. The circuit is called a *half-adder* because, as we see later, two such half-adders are required to build a circuit to add a column of two n-digit binary numbers.

A specification should characterize the desired behavior of the circuit and nothing more. Eschewing superfluous restrictions and details gives freedom to the implementor, who ultimately must design a circuit to satisfy the specification. In this sense, (5.10) is a better specification of a half-adder than specification (5.5) of the circuit of Fig. 5.2, because (5.5) unnecessarily

---

[2] Even the circuit with one output z that is always *false* is characterized by a non-*false* expression: $\neg z$.

refers to internal wires (e.g. ω) and imposes other irrelevant structure on the implementation (e.g. two inverters, three and-gates, and an or-gate).

DON'T CARE CONDITIONS

Definition (5.8) can be used to guide the design of a circuit. Given a specification S, we need only manipulate S to yield a boolean expression C that satisfies $C \Rightarrow S$ and that has the form described in (5.6). Then, from C, we construct the circuit as described earlier.

We now illustrate this approach to circuit design. The truth tables of Table 5.2 specify a circuit that yields *false* if the number of *true* input wires is less than 2 and *true* if the number of *true* input wires equals 2. In the left truth table, a row for the case when all three input wires are *true* has been omitted, presumably because, in the context in which the implementation is to be used, it does not matter what output is produced for that input. The right truth table uses a convention of electrical engineers to indicate this "Don't care" condition: the value D in the top row means that either t or f is an acceptable result. By having this Don't-care condition, the author of the specification has given the implementor some freedom, so that there is more opportunity for an efficient and simple implementation.

Our first task in implementing this specification is to write an equivalent boolean expression. Each row of the left truth table of Table 5.2, e.g.

a	b	c	z
...			
t	t	f	t
...			

indicates that the given inputs on the wires imply the given output. Hence,

---

TABLE 5.2. TRUTH TABLE FOR PARTIAL MAJORITY CIRCUIT

a	b	c	z		a	b	c	z
					t	t	t	D
t	t	f	t		t	t	f	t
t	f	t	t		t	f	t	t
t	f	f	f		t	f	f	f
f	t	t	t	or	f	t	t	t
f	t	f	f		f	t	f	f
f	f	t	f		f	f	t	f
f	f	f	f		f	f	f	f

Don't-care case implicit   Don't-care case explicit

each row contributes an implication to the boolean expression. For example, the row above contributes

$$a \wedge b \wedge \neg c \Rightarrow z \quad .$$

In this manner, we can construct the following boolean expression, which is equivalent to the truth tables of Table 5.2. [3]

(5.11) $(a \wedge b \wedge \neg c \Rightarrow z) \wedge$
$(a \wedge \neg b \wedge c \Rightarrow z) \wedge$
$(a \wedge \neg b \wedge \neg c \Rightarrow \neg z) \wedge$
$(\neg a \wedge b \wedge c \Rightarrow z) \wedge$
$(\neg a \wedge b \wedge \neg c \Rightarrow \neg z) \wedge$
$(\neg a \wedge \neg b \wedge c \Rightarrow \neg z) \wedge$
$(\neg a \wedge \neg b \wedge \neg c \Rightarrow \neg z)$

We can use theorem (3.78), which is

$$p \vee q \Rightarrow r \ \equiv \ (p \Rightarrow r) \wedge (q \Rightarrow r) \quad ,$$

to aggregate antecedents in (5.11) and then simplify, to yield

(5.12) $((\neg a \wedge \neg b) \vee (\neg b \wedge \neg c) \vee (\neg a \wedge \neg c) \Rightarrow \neg z) \wedge$
$((\neg a \wedge b \wedge c) \vee (a \wedge \neg b \wedge c) \vee (a \wedge b \wedge \neg c) \Rightarrow z)$

This expression does not have form (5.6) (see page 93), from which a circuit could be derived. It also contains more operations than we would like —remember, each operation is implemented by a gate, and it makes sense to try to minimize the number of gates used. Perhaps we can resolve the Don't-care condition in Table 5.2 in a way that allows us to simplify the boolean expression even further. There are two possibilities: replacing D by *true* adds the implication $a \wedge b \wedge c \Rightarrow z$ as a conjunct to (5.12); replacing D by *false* adds $a \wedge b \wedge c \Rightarrow \neg z$. We can investigate the consequence of adding each conjunct. Here, we investigate only the conjunct $a \wedge b \wedge c \Rightarrow z$, since its introduction does result in a simplification. Exercise 5.14 concerns proving that $a \wedge b \wedge c \Rightarrow z$ together with (5.12) is equivalent to

(5.13) $z \equiv (a \wedge b) \vee (b \wedge c) \vee (c \wedge a) \quad .$

---

[3] In the construction, a don't-care row like

$$t \ \ t \ \ f \ \mid \ D$$

contributes the conjunct $a \wedge b \wedge \neg c \Rightarrow z \vee \neg z$, since D signifies that the result can be either *true* or *false*. Because the consequent is equivalent to *true*, the implication is itself *true* and can be omitted entirely.

Since we added a conjunct to (5.12) and then manipulated using only Leibniz to arrive at (5.13), we have (5.13) $\Rightarrow$ (5.12), so, by definition, a circuit for (5.13) implements (5.12). Further, (5.13) is in form (5.6), and it can easily be turned into a circuit. The resulting circuit is given in Fig. 5.3.

Expression (5.13) is *false* when fewer than two of a, b, and c are *true* and *true* when at least two are *true*. Thus, it specifies a circuit whose output is *true* exactly when a majority of its three inputs are *true*.

USING HIGHER-LEVEL BUILDING BLOCKS

We can use Definition (5.8) to derive implementations of circuit specifications in terms of higher-level building blocks. This is now illustrated with the design of a binary adder. Starting with a specification that characterizes addition of two unsigned binary numbers, we obtain an expression that (i) implies the specification and (ii) is in terms of boolean expressions for half-adders (circuits now at hand), conjunction, disjunction, and negation.

Adding unsigned binary numbers is similar to adding unsigned decimal numbers; the sole difference is that only the two bits 0 and 1 are available instead of the ten digits $0, \ldots, 9$. Below, we give an example, show how the addends are encoded using boolean constants, and then give the general case:

$$
\begin{array}{ccc}
\begin{array}{r} 1\ 0\ 0\ 1 \\ +\ 1\ 1\ 0\ 1 \\ \hline 1\ 0\ 1\ 1\ 0 \end{array}
&
\begin{array}{r} t\ f\ f\ t \\ +\ t\ t\ f\ t \\ \hline t\ f\ t\ t\ f \end{array}
&
\begin{array}{r} a_{n-1} \ldots a_1\ a_0 \\ +\ b_{n-1} \ldots b_1\ b_0 \\ \hline d_n\ d_{n-1} \ldots d_1\ d_0 \end{array}
\end{array}
$$

In the general case, each d_i is the least significant bit of the sum of a_i, b_i, and the carry from the previous column, except that (i) the carry into

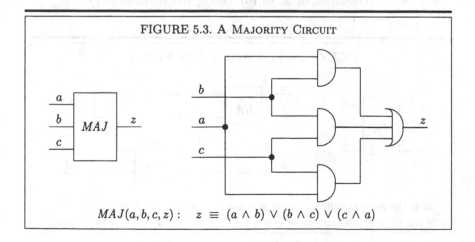

FIGURE 5.3. A MAJORITY CIRCUIT

$$MAJ(a, b, c, z): \quad z \equiv (a \land b) \lor (b \land c) \lor (c \land a)$$

column 0 is 0 and (ii) the carry from column $n-1$ is d_n. We define the addition as follows. Let c_i be the carry from column i. Then, the result of adding $a_{n-1} \ldots a_0$ and $b_{n-1} \ldots b_0$ is given by

- $c_{-1} = 0$. In the boolean representation, $c_{-1} = f$.

- Each carry c_i, $0 \leq i < n$, is 1 iff at least two of a_i, b_i, and c_{i-1}, are 1. As a boolean expression, c_i is defined by

$$c_i \equiv (a_i \wedge b_i) \vee (b_i \wedge c_{i-1}) \vee (c_{i-1} \wedge a_i) \ .$$

- Each d_i, $0 \leq i < n$, is the least significant bit of the 2-bit result of adding bits a_i, b_i, and c_{i-1}. Investigation shows that it is 1 iff an odd number of a_i, b_i, and c_{i-1} are 1. As a boolean expression, d_i is defined by

$$d_i \equiv (a_i \equiv b_i \equiv c_{i-1}) \ .$$

- d_n is the carry from position $n-1$, i.e. $d_n = c_n$.

The key building block for an n-bit adder is a *full adder*, a circuit that calculates sum d and carry e that result from adding bits a, b, c:

(5.14) $d = (a \equiv b \equiv c) \ \wedge \ e = ((a \wedge b) \vee (b \wedge c) \vee (c \wedge a)) \ .$

We can manipulate this expression so that it can be implemented using half-adders. In each step of the manipulation, we seek to rewrite the current line into one involving subexpressions of the form $A \not\equiv B$ and $A \wedge B$, since these are the outputs of a half-adder (see (5.10)).

$$d = (a \equiv b \equiv c)$$
$$= \quad \langle \text{Definition of } \not\equiv \ (3.10), \text{ twice, and Associativity} \rangle$$

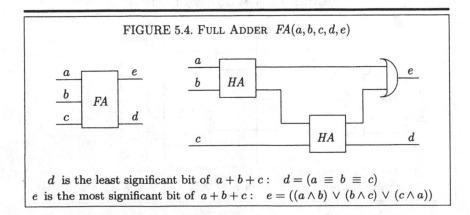

FIGURE 5.4. Full Adder $FA(a, b, c, d, e)$

d is the least significant bit of $a+b+c$: $d = (a \equiv b \equiv c)$
e is the most significant bit of $a+b+c$: $e = ((a \wedge b) \vee (b \wedge c) \vee (c \wedge a))$

$$d = (a \not\equiv b \not\equiv c)$$
$$= \quad \langle \text{Associativity —insert parentheses} \rangle$$
$$d = ((a \not\equiv b) \not\equiv c)$$

Now, $a \not\equiv b$ is output argument u of $HA(a, b, u, \_)$ and d is then an output argument of $HA(u, c, d, \_)$. Hence, a circuit that computes d results simply from connecting a and b to the inputs of one half-adder and connecting its output, along with c, to the inputs of another half-adder, as shown in Fig. 5.4.

We now manipulate the definition of e; in doing so, we try to use the the two half-adders used in implementing d.

$$e \equiv (a \wedge b) \vee (b \wedge c) \vee (c \wedge a)$$
$$= \quad \langle \text{Distributivity of } \wedge \text{ over } \vee \text{ (3.46)} \rangle$$
$$e \equiv (a \wedge b) \vee ((a \vee b) \wedge c)$$
$$= \quad \langle \text{Absorption (3.44b), with } p, q := a \wedge b, (a \vee b) \wedge c \rangle$$
$$e \equiv (a \wedge b) \vee (\neg(a \wedge b) \wedge (a \vee b) \wedge c)$$
$$= \quad \langle \text{Propositional calculus —see Exercise 5.17} \rangle$$
$$e \equiv (a \wedge b) \vee ((a \not\equiv b) \wedge c)$$
$$= \quad \langle \text{Definition of gate } OR \rangle$$
$$e \equiv OR(a \wedge b, (a \not\equiv b) \wedge c)$$

Now, $a \wedge b$ is the fourth argument p (say) of $HA(a, b, u, p)$. Further, $(a \not\equiv b) \wedge c$ is output s of $HA(u, c, d, s)$. Hence, we can implement the full-adder as in Fig. 5.4 and use $FA(a, b, c, d, e)$ to compute d and e.

It is now a simple matter to connect a series of full adders to compute the sum of $a_{n-1} \ldots a_0$ and $b_{n-1} \ldots b_0$. We give such an adder in Fig. 5.5, for the case $n = 3$.

NAND AND OTHER BUILDING BLOCKS

The collection of gates in Fig. 5.1 is a natural set of building blocks, because any boolean expression can be written using conjunction, disjunction, and negation (operations of the form $P \equiv Q$ can be replaced by $(P \wedge Q) \vee (\neg P \wedge \neg Q)$, and $P \Rightarrow Q$ can be replaced by $\neg P \vee Q$). In fact, a set of 2-input and-gates, 2-input or-gates and inverters suffices. This is because an n-input and-gate can be implemented using a cascaded network of 2-input and-gates, as the following manipulation shows.

$$AND(a_1, a_2, a_3, \ldots, a_n)$$
$$= \quad \langle \text{Definition of } AND \rangle$$
$$a_1 \wedge a_2 \wedge a_3 \wedge \cdots \wedge a_n$$
$$= \quad \langle \text{Associativity of } \wedge \text{ (3.37); Definition of } AND \rangle$$

$AND(a_1, a_2) \wedge a_3 \wedge \cdots \wedge a_n$

$=$ ⟨Associativity of $\wedge$ (3.37); Definition of AND⟩

$AND(AND(a_1, a_2), a_3) \wedge \cdots \wedge a_n$

$=$ $\cdots$

A similar argument demonstrates that an n-input or-gate can be implemented by a network of 2-input or-gates.

Other collections of gates can also suffice as a *universal* set of building blocks. For example, surprising though it may seem, any boolean function can be implemented using a single type of gate: the nand-gate. An n-input nand-gate is defined by

$$NAND(a_1, a_2, \cdots, a_n) : \quad \neg(a_1 \wedge a_2 \wedge \cdots \wedge a_n) \quad .$$

By the argument just given for implementing n-input and-gates (or-gates) using 2-input and-gates (or-gates), we can establish the universality of $NAND$. We show how to implement an inverter, a 2-input and-gate, and a 2-input or-gate in terms of 2-input nand-gates. We give these implementations in Fig. 5.6.

$\neg a$

$=$ ⟨Identity of $\vee$ (3.30)⟩

$\neg a \vee false$

$=$ ⟨Definition of *false* (3.8); De Morgan (3.47a)⟩

$\neg(a \wedge true)$

$=$ ⟨Definition of $NAND$⟩

$NAND(a, true)$

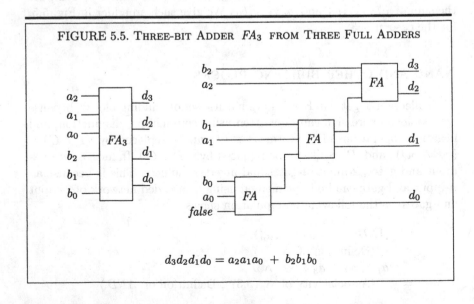

FIGURE 5.5. THREE-BIT ADDER FA_3 FROM THREE FULL ADDERS

$$d_3 d_2 d_1 d_0 = a_2 a_1 a_0 + b_2 b_1 b_0$$

$a \wedge b$

= ⟨Double negation (3.12), twice; De Morgan (3.47a)⟩

$\neg(\neg a \vee \neg b)$

= ⟨Zero of $\wedge$ (3.40)⟩

$\neg((\neg a \vee \neg b) \wedge true)$

= ⟨De Morgan (3.47b)⟩

$\neg(\neg(a \wedge b) \wedge true)$

= ⟨Definition of $NAND$, twice⟩

$NAND(NAND(a, b), true)$

$a \vee b$

= ⟨Double negation (3.12), twice; De Morgan (3.47b)⟩

$\neg(\neg a \wedge \neg b)$

= ⟨Zero of $\wedge$ (3.40)⟩

$\neg(\neg(a \wedge true) \wedge \neg(b \wedge true))$

= ⟨Definition of $NAND$, thrice⟩

$NAND(NAND(a, true), NAND(b, true))$

Thus, the nand-gate is universal. The exercises ask you to establish that the nor-gate is also universal, where a two-input nor-gate is specified by

$NOR(a, b) : \neg(a \vee b)$.

An obvious question is to identify the merits of different universal sets of building blocks. Unquestionably, the 3-gate collection of Fig. 5.1 is the most natural for the neophyte circuit designer. As a practical matter, however, it is best when the number of distinct gates in a (universal) collection is small. Construction of any object is easier if it involves fewer kinds of building

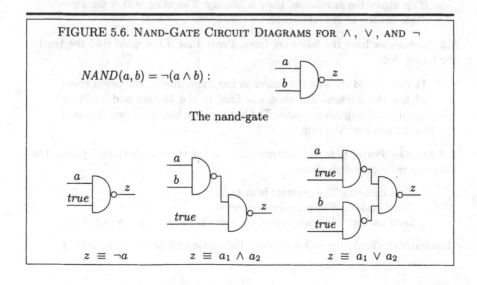

FIGURE 5.6. NAND-GATE CIRCUIT DIAGRAMS FOR $\wedge$, $\vee$, AND $\neg$

$NAND(a, b) = \neg(a \wedge b)$:

The nand-gate

$z \equiv \neg a$ $z \equiv a_1 \wedge a_2$ $z \equiv a_1 \vee a_2$

blocks. A second practical concern is the complexity of each building block; obviously, we seek simpler building blocks, since each is then cheaper to design and produce. Fortuitously, nand-gates and nor-gates are easier to implement with transistors than are and-gates and or-gates. Since the nand-gate by itself and the nor-gate by itself are each universal, rarely is the 3-gate set of Fig. 5.1 used in actual hardware; combinational circuits are, in fact, most often constructed using nand-gates.

Exercises for Chapter 5

5.1 Formalize the following arguments and either prove that they are valid or find a counterexample.

(a) Either the program does not terminate or n eventually becomes 0. If n becomes 0, m will eventually be 0. The program terminates. Therefore, m will eventually be 0.

(b) If the initialization is correct and if the loop terminates, then P is *true* in the final state. P is *true* in the final state. Therefore, if the initialization is correct, the loop terminates.

(c) If there is a man on the moon, the moon is made of cheese, and if the moon is made of cheese then I am a monkey. Either no man is on the moon or the moon is not made of cheese. Therefore either the moon is not made of cheese or I am a monkey.

(d) If Joe loves Mary, then either mom is mad or father is sad. Father is sad. Therefore, if mom is mad then Joe doesn't love Mary.

5.2 Prove that the following argument is valid if the "or" in it is considered to be inclusive and invalid if it is considered exclusive.

If an algorithm is reliable, then it is okay. Therefore, either an algorithm is okay or it is unreliable.

5.3 Suppose we have the following facts. Prove that if the maid told the truth, the butler lied.

The maid said she saw the butler in the living room. The living room adjoins the kitchen. The shot was fired in the kitchen and could be heard in all adjoining rooms. The butler, who had good hearing, said he did not hear the shot.

5.4 Suppose Portia puts her picture into one of three caskets and places the following inscriptions on them:

Gold casket: The portrait is in here.
Silver casket: The portrait is in here.
Lead casket: At least two of the caskets have a false inscription.

Which casket should the suitor choose? Formalize and calculate an answer.

5.5 Suppose Portia puts a dagger in one of three caskets and places the following inscriptions on the caskets:

> Gold casket: The dagger is in this casket.
> Silver casket: The dagger is not in this casket.
> Lead casket: At most one of the caskets has a true inscription.

Portia tells her suitor to pick a casket that does not contain the dagger. Which casket should the suitor choose? Formalize and calculate an answer.

5.6 This set of questions concerns an island of knights and knaves. Knights always tell the truth and knaves always lie. In formalizing these questions, associate identifiers as follows:

> b : B is a knight.
> c : C is a knight.
> d : D is a knight.

If B says a statement " X ", this gives rise to the expression $b \equiv X$, since if b , then B is a knight and tells the truth, and if $\neg b$, B is a knave and lies.

(a) Someone asks B "are you a knight?" He replies, "If I am a knight, I'll eat my hat." Prove that B has to eat his hat.

(b) Inhabitant B says of inhabitant C , "If C is a knight, then I am a knave." What are B and C ?

(c) It is rumored that gold is buried on the island. You ask B whether there is gold on the island. He replies, "There is gold on the island if and only if I am a knight." Can it be determined whether B is a knight or a knave? Can it be determined whether there is gold on the island?

(d) Three inhabitants are standing together in the garden. A non-inhabitant passes by and asks B , "Are you a knight or a knave?" B answers, but so indistinctly that the stranger cannot understand. The stranger then asks C , "What did B say?" C replies, " B said that he is a knave." At this point, the third man, D , says, "Don't believe C ; he's lying!" What are C and D ?

Hint: Only C 's and D 's statements are relevant to the problem. Also, D 's remark that C is lying is equivalent to saying that C is a knave.

(e) B , C , and D are sitting together. C says, "There is one knight among us." D says, "You're lying." What can you tell about the knighthood or knavehood of the three?

Here is a hint. One can describe the fact that 1 or 3 of them are knights by the rather nice expression $b \equiv c \equiv d$, since this expression is *true* when the number of *false* operands is even. Restricting it further to 1 knight requires only the additional conjunct $\neg(b \wedge c \wedge d)$. See the discussion beginning on page 46.

(f) A non-inhabitant meets three inhabitants, B , C , and D , and asks a question, and B replies indistinctly. So the stranger asks C , "What did he say?" C replies, " B said that there was one knight among us." Then D says, "Don't believe C ; he's lying." What are C and D ? Hint: See the hint on the previous problem.

(g) In the group of three inhabitants, B says that all three are knaves and C says that exactly one of the three is a knight. What are B , C , and D ?

Hint: See the hint on the previous problem.

5.7 The country of Marr is inhabited by two types of people: liars always lie and truars always tell the truth —sounds like a knight-knave problem, eh? At a cocktail party, the newly appointed United States ambassador to Marr talked to three inhabitants. Joan remarked that Shawn and Peter were liars. Shawn denied he was a liar, but Peter said that Shawn was indeed a liar. From this information, can you tell how many of the three are liars and how many are truars?

5.8 In Marr (see the previous exercise), the Nelsons, who are truars, were leaving their four children with a new babysitter, Nancy, for the evening. Before they left, they told Nancy that three of their children were consistent liars but that one of them was a truar. While she was preparing dinner, one of the children broke a vase in the next room. Nancy rushed into the room and asked who broke the vase. The children's answers were:

> *Betty*: Steve broke the vase,
> *Steve*: John broke it,
> *Laura*: I didn't break it,
> *John*: Steve lied when he said he broke it.

Nancy quickly determined who broke the vase. Who was it? Solve the problem by formalizing and calculating.

Here is a hint. Let b, s, l, and j stand for Betty, Steve, Laura, and John are truars, respectively. That 1 or 3 children are liars can be written as

$$F1 : b \equiv s \equiv l \not\equiv j \quad ,$$

since this expression is *true* exactly when one or three of the operands are *false*. This is a nice expression to work with. Restricting it to three liars (i.e. one truar) requires the additional conjuncts $\neg(x \wedge y)$ for x and y different identifiers drawn from b, s, l, and j. The same kind of expressions can be used to deal with the fact that exactly one child broke the vase.

5.9 Exercise 2.9 on page 39 gives three assumptions and 8 conjectures about the Tardy Bus Problem. Translate each of the assumptions and conjectures into the propositional calculus and determine formally which of the 8 conjectures follow from the three assumptions and which do not.

Exercises on combinational digital circuits

5.10 (a) Construct boolean expressions for the following circuits. (b) Remove the internal names from your answers to (a), thus arriving at specifications for the circuits. (c) Construct truth tables for the expressions of part (b).

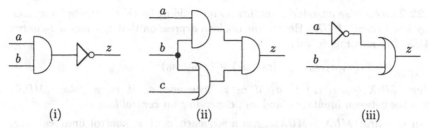

(i)   (ii)   (iii)

5.11 (a) Construct boolean expression for the following circuits. (b) Remove the internal names from your answers to (a), thus arriving at specifications for the circuits. (c) Construct truth table for the boolean expressions of (b).

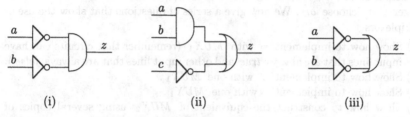

(i)   (ii)   (iii)

5.12 How many different truth tables are there for combinational circuits having 2 inputs and 1 output? Two inputs and 2 outputs?

5.13 Prove $(5.11) \equiv (5.12)$ in more detail than done in the text.

5.14 Prove $(5.13) \Rightarrow (5.12)$ by proving that $(a \wedge b \wedge c \Rightarrow z) \wedge (5.12) \equiv (5.13)$.

5.15 Draw a circuit for each expression, where a, b, and c are the inputs and z and s the outputs. Use only inverters, 2-and gates, and 2-or gates,

(a) $z \equiv a \wedge b \wedge c$
(b) $z \equiv (a \vee b) \wedge a$
(c) $z \equiv \neg a \wedge (a \vee b)$
(d) $(z \equiv a \wedge \neg b) \wedge (s \equiv a \wedge b)$
(e) $(z \equiv a \wedge b) \wedge (s \equiv a \wedge b \wedge c)$

5.16 Suppose you are given a boolean expression that is implemented by a circuit, with the input variables and output variables identified. Explain how to figure out what the "don't care" states are.

5.17 Prove theorem $\neg(a \wedge b) \wedge (a \vee b) \equiv (a \not\equiv b)$.

5.18 (a) The implementation of $AND(a_1, \ldots, a_n)$ in terms of 2-input and-gates can have an input traverse n levels of gates. A faster circuit is possible, where no input signal has to traverse more than $log_2(n)$ levels of gates. Derive this circuit for the case where n is a power of 2.
(b) Suppose n is not a power of 2. Derive an implementation that still gives reasonable, logarithmic performance.

5.19 Derive implementations of $\neg$, $\vee$, and $\wedge$ in terms of nor-gates.

5.20 Derive an implementation of a nand-gate in terms of nor-gates.

5.21 Derive an implementation of a nor-gate in terms of nand-gates.

5.22 This exercise introduces another useful building block, the *multiplexor*, usually abbreviated MUX. Here is the boolean expression that specifies a *1-control* MUX (see also Fig. 5.7(a)):

$$MUX_1(a_0, a_1, c_0): \quad (c_0 \wedge a_1) \vee (\neg c_0 \wedge a_0)$$

Thus, $MUX_1(a_0, a_1, c_0)$ is a_1 if c_0 is *true* and a_0 if c_0 is *false*. MUX_1 switches between inputs a_0 and a_1, depending on *control line* c_0.

An *n-control MUX*, MUX_n, has a sequence c of n control lines c_0, ..., c_{n-1} and a sequence a of 2^n other inputs $a_0, ..., a_{2^n-1}$ (see Fig. 5.7(b)). Here is the best way to think of MUX_n. Consider the n control lines $c_0, ..., c_{n-1}$ as the binary representation of a natural number C in the range $0..2^n - 1$. Then the output of $MUX_n(c, a)$ is a_C —i.e. $MUX_n(c, a)$ uses the control lines as an integer C to choose a_C. We now give a series of questions that show the use of multiplexors.

(a) Show how to implement $\neg$ with MUX_1 (remember that circuits can have input lines that are always *true* and other input lines that are always *false*).
(b) Show how to implement $\vee$ with one MUX_1.
(c) Show how to implement $\wedge$ with one MUX_1.
(d) Show how to construct the equivalent of MUX_2 using several copies of MUX_1.
(e) Let $n > 0$, where $2^{m-1} \leq n < 2^m$ for some m. Show how to implement any circuit having n input variables using MUX_m.
(f) Show how to implement a truth table with n input variables and one output variable using one n-control multiplexor.

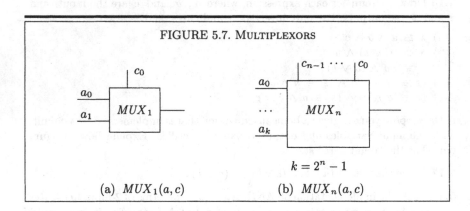

FIGURE 5.7. MULTIPLEXORS

(a) $MUX_1(a, c)$ (b) $MUX_n(a, c)$

Chapter 6

Hilbert-style Proofs

We present hierarchical, *Hilbert-style proofs* (see Historical note 6.1) as an alternative to the equational proof system **E** of Chap. 3. The Hilbert style of proof is used often in teaching geometry in high school. To illustrate a propositional logic in the Hilbert style, we give a *natural deduction* logic, **ND**. Using this logic, we formalize informal proof methods used in mathematics and discuss various proof styles and their advantages.

6.1 Hilbert-style proofs

Our Hilbert style of proof consists of (i) the theorem to be proved, on one line, followed by (ii) a sequence of numbered lines, each of which contains a theorem and an explanation of why it is a theorem. The last numbered line must be the theorem being proved.

A Hilbert-style proof is given below. Line 1 is a theorem by virtue of Leibniz with premise (3.8). Line 2 is theorem (3.12), with $p := true$. Line 3 follows from Transitivity of equality (1.4) with the two premises appearing on lines 1 and 2. Note that line 3 is the theorem being proved.

$$\neg false \equiv true$$

1	$\neg false \equiv \neg\neg true$	Leibniz, (3.8) $false \equiv \neg true$
2	$\neg\neg true \equiv true$	Substitution, (3.12) $\neg\neg p \equiv p$
3	$\neg false \equiv true$	Transitivity (1.4), 1, 2

Here is a corresponding equational proof.

$$\neg false$$
$$=\quad \langle (3.8),\ false \equiv \neg true \rangle$$
$$\neg\neg true$$
$$=\quad \langle (3.12),\ \neg\neg p \equiv p \rangle$$
$$true$$

The explanation given for a line of a Hilbert-style proof justifies the validity of the line. Therefore, it must be one of the following.

- The number of an axiom or a previously proved theorem, if the formula on the line is that axiom or theorem.

- The name of an inference rule that is used to derive the theorem. Suppose line k has the form

$$k \quad R \quad inf\text{-}rule, i, j, \ldots$$

where R is the theorem being proved. Then

$$\frac{\text{formula on line } i \,, \text{formula on line } j \,, \ldots}{R}$$

must be an instance of rule $inf\text{-}rule$. Instead of being a line number, i and j may be references to axioms or theorems previously proved elsewhere.

As in equational proofs, applications of symmetry and associativity can be made without mention.

It should be clear from the example above that any equational proof can be transformed mechanically into a Hilbert-style proof, and vice versa. But there are significant differences in the styles. The structure of an equational proof allows implicit use of inference rules Leibniz, Transitivity of equality, and Substitution. In the Hilbert style, the structure is no help in this regard, so all uses of inference rules must be mentioned explicitly. With only the three inference rules available at this point, the equational style is preferable. However, the Hilbert style has other advantages. Additional inference rules may be used, and the Hilbert style can be easily extended to incorporate subproofs within a proof, as we now show.

SUBPROOFS AND SCOPE

Consider the following Hilbert-style proof.

$$p \wedge \neg p \equiv p \vee \neg p \equiv \mathit{false}$$
1 $\; p \wedge \neg p \equiv p \vee \neg p \equiv p \equiv \neg p$ Substitution, Golden r. (3.35)
2 $\; p \equiv \neg p \equiv \mathit{false}$ (3.15)
3 $\; p \wedge \neg p \equiv p \vee \neg p \equiv \mathit{false}$ Transitivity (1.4), 1, 2

Now, suppose that Theorem (3.15) had not yet been proved. We could prove it as a lemma, but, for locality, it is sometimes preferable to include a proof as a subproof, as shown on the next page. Note how the subproof is indented and numbered.

To understand such a proof, focus your attention on one level at a time. For example, in the proof below, first look at the outer proof —only lines

HISTORICAL NOTE 6.1. DAVID HILBERT (1862–1943)

David Hilbert's broad interests and originality made him a pioneer in many different fields, including number theory, geometry, formal foundations of mathematics, and physics. So enormous were his accomplishments that, at the age of 43, he was the runner-up for the first Bolyai prize, given by the Hungarian Academy of Sciences to the mathematician who had most contributed to mathematics in the past 25 years; later, he received the second Bolyai prize to be awarded. Hilbert set the tone and direction of mathematics. For example, in 1900, at an international mathematical congress, he discussed the future of mathematics, posing what are now known as *Hilbert's problems*. (There were 10 in the talk but 23 in the accompanying paper.) Some are still unsolved.

Hilbert spent most of his working life at the University of Göttingen, Germany, which, under the influence of Klein, Hilbert, Minkowski, and Runge, flourished in mathematics as no other place has. Mathematicians came to study from all over the world. Hilbert himself advised 69 Ph.D. students, many with names that resound throughout mathematics. A sad note was the decline of this great institution in the 1930's, as Jewish scientists were forced out. In 1933, the Nazi minister of education asked Hilbert how mathematics was, now that it had been freed of Jewish influence. "Mathematics in Göttingen?", replied Hilbert, "There is really none any more."

Leibniz dreamt of having a general method for reducing mathematics to calculation. Boole and De Morgan provided a basis for it. And Hilbert worked to make it a practical reality. When close to 60, he proposed to formalize all of mathematics as an axiomatic system in which theorems would be proved purely by symbol manipulation. Hilbert felt that mathematics should be *complete* (all truths should be provable), *consistent* (nothing false should be provable), and *decidable* (there should be a mechanical procedure for deciding whether any assertion is true or false). Hilbert felt that this formalization could solve foundational arguments concerning classical versus constructive mathematics rampant at the time (see Sec. 7.2). Gödel shattered Hilbert's dreams (see Historical note 7.1), and yet Hilbert's program had a profound effect on the field. See Historical note 0.1 and Reid's biography [32].

1, 2, and 3— and check that it is indeed a proof of the theorem on the first line. In doing this checking, study the justifications for lines substantiated by inference rules, but do not check subproofs. Next, check the subproofs.

$p \wedge \neg p \equiv p \vee \neg p \equiv false$
1 $p \wedge \neg p \equiv p \vee \neg p \equiv p \equiv \neg p$ Substitution, Golden r. (3.35)
2 $p \equiv \neg p \equiv false$
   2.1 $p \equiv \neg p \equiv \neg(p \equiv p)$ Substitution, (3.9)
                                          —with $q := p$
   2.2 $\neg(p \equiv p) \equiv \neg true$ Leibniz, Identity of $\equiv$ (3.3)
   2.3 $p \equiv \neg p \equiv \neg true$ Transitivity (1.4), 2.1, 2.2

2.4 $\neg true \equiv false$ Definition of $false$ (3.8)
2.5 $p \equiv \neg p \equiv false$ Transitivity (1.4), 2.3, 2.4
3 $p \wedge \neg p \equiv p \vee \neg p \equiv false$ Transitivity (1.4), 1, 2

Note how the indentation of a subproof helps one understand the proof structure, in the same way that indenting substatements of a Pascal or C program helps one see program structure. It takes practice to read such a proof efficiently, partly because of its hierarchical nature and partly because there may not be any direct connection between formulas on adjacent lines. Each formula is being staged for a later use.

Actually, we could present a subproof in-line with its outer proof, as shown below. However, the proof then becomes more difficult to understand because its structure is hidden. The subproof mechanism allows us to engineer proofs for ease of reading.

$p \wedge \neg p \equiv p \vee \neg p \equiv false$
1 $p \wedge \neg p \equiv p \vee \neg p \equiv p \equiv \neg p$ Substitution, Golden r. (3.35)
2 $p \equiv \neg p \equiv \neg(p \equiv p)$ Substitution, (3.9)
                                                   —with $q := p$
3 $\neg(p \equiv p) \equiv \neg true$ Leibniz, Identity of (3.3)
4 $p \equiv \neg p \equiv \neg true$ Transitivity (1.4), 2, 3
5 $\neg true \equiv false$ Definition of $false$ (3.8)
6 $p \equiv \neg p \equiv false$ Transitivity (1.4), 4, 5
7 $p \wedge \neg p \equiv p \vee \neg p \equiv false$ Transitivity (1.4), 1, 6

Finally, a subproof can be given in the equational style:

$p \wedge \neg p \equiv p \vee \neg p \equiv false$
1 $p \wedge \neg p \equiv p \vee \neg p \equiv p \equiv \neg p$ Substitution, Golden r. (3.35)
2 $p \equiv \neg p \equiv false$
    $p \equiv \neg p$
= $\langle$Distributivity of $\neg$ over $\equiv$ (3.9)$\rangle$
    $\neg(p \equiv p)$
= $\langle$Identity of $\equiv$ (3.3)$\rangle$
    $\neg true$
= $\langle$Def. of $false$ (3.8)$\rangle$
    $false$
3 $p \wedge \neg p \equiv p \vee \neg p \equiv false$ Transitivity (1.4), 1, 2

With the introduction of subproofs, we have to be careful about referencing theorems on previous lines of a proof. A line may refer to previous lines of the proof (or subproof) in which it directly occurs. A line of a subproof may also reference previous lines of surrounding proofs. Such *global* references are similar to global references in imperative programming languages in which a procedure may contain references to variables declared in its surrounding scope.

We state this scope rule more formally as an inductive definition.

(6.1)   **Scope rule.** For k an integer, line k of a proof may contain references to lines 1 through $k-1$. A line numbered $\ldots j.k$ (say) can reference lines $\ldots j.1$ through $\ldots j.(k-1)$ as well as whatever can be referenced by line $\ldots j$.

According to the scope rule, the following is not a proof because it has two invalid references.

$\ldots$

$\begin{array}{lll} 4 \ p \Rightarrow (p \Rightarrow \neg p) & & \text{INVALID PROOF!} \\ \quad 4.1 \ p & 5 & \text{(invalid forward reference)} \\ \quad 4.2 \ p \Rightarrow (p \Rightarrow \neg p) & 4 & \text{(invalid reference)} \\ \quad 4.3 \ p \Rightarrow \neg p & & \\ \quad\quad p \Rightarrow \neg p & & \\ \quad\quad = \quad \langle \text{Left identity of} \Rightarrow (3.73) \rangle & & \\ \quad\quad p \quad \text{—line 4.1} & & \\ 5 \ p & & \end{array}$

6.2 Natural deduction

The inference rules used thus far —Leibniz, Transitivity of equality, and Substitution— are not particularly suited to the Hilbert style. Different inference rules will put the Hilbert style in a better light. In this section, we present an entirely different propositional logic, **ND**, called a *natural deduction* logic, which uses inference rules that are more attuned to the Hilbert style. Natural deduction is due to Gerhard Gentzen —see Historical note 6.2.

Table 6.1 presents the inference rules for **ND**. There are two inference rules for each operator and each constant: one rule shows how to introduce the symbol into a theorem and the other rule shows how to eliminate it. For each operator or constant $\star$, the rules are named $\star$-I and $\star$-E. For example, the introduction and elimination rules for $\wedge$ are $\wedge$-I and $\wedge$-E. As before, each inference rule is a schema, and substituting boolean expressions for the variables P, Q, and R in it yields an inference rule.

Natural deduction is noteworthy for several reasons. First, **ND** has no axioms. Actually, in any logic, one can view an inference rule that has no premise as an axiom. But in **ND**, all inference rules have premises, so there really are no axioms.

Second, theorem Modus ponens (3.77), $p \wedge (p \Rightarrow q) \Rightarrow q$, is an inference rule in **ND** (rule $\Rightarrow$-E).

Third, Deduction theorem (4.4), a metatheorem of the equational calculus, is an inference rule in **ND** (rule $\Rightarrow$-I). While $\Rightarrow$-E allows an implication to be eliminated from a formula, $\Rightarrow$-I allows an implication to be introduced. Rules $\Rightarrow$-I and $\Rightarrow$-E are a hallmark of natural deduction.

Rule $\Rightarrow$-I has the *sequent* $P_1, \ldots, P_n \vdash Q$ as its premise. We now explain sequents. In logic **E**, we postulated axioms (call them $A_0, \ldots, A_n$ for now) and then proved theorems using them. We did not prove each theorem Q (say) in isolation; instead, we proved that Q *follows from* some formulas. Logicians express this relationship between a theorem and the formulas assumed for its proof as the sequent

$$A_0, \ldots, A_n \vdash Q \quad \text{or} \quad \vdash_L Q \ ,$$

where L is the name of the logic with axioms $A_0, \ldots, A_n$.

Symbol $\vdash$ is called the "turnstile", and the A_i are called the *premises* of the sequent. The sequent $A_0, \ldots, A_n \vdash Q$ is read as "Q is provable from $A_0, \ldots, A_n$." (The order of the A_i is immaterial.) The sequent $\vdash_L Q$ is read as "Q is provable in logic L" —i.e. using the axioms of L. Often, when the logic is unambiguous from the context, the subscript L is omitted. Thus, $\vdash Q$ means that Q is a theorem in the logic at hand. In Chap. 3, we could have placed the turnstile before each theorem.

Note the difference between the sequent $A_0, \ldots, A_n \vdash Q$ and the formula

---

TABLE 6.1. INFERENCE RULES FOR **ND**

Introduction rules	Elimination rules
$\wedge$-I : $\dfrac{P,\ Q}{P \wedge Q}$	$\wedge$-E : $\dfrac{P \wedge Q}{P}, \quad \dfrac{P \wedge Q}{Q}$
$\vee$-I : $\dfrac{P}{P \vee Q}, \quad \dfrac{P}{Q \vee P}$	$\vee$-E : $\dfrac{P \vee Q,\ P \Rightarrow R,\ Q \Rightarrow R}{R}$
$\Rightarrow$-I : $\dfrac{P_1, \ldots, P_n \vdash Q}{P_1 \wedge \cdots \wedge P_n \Rightarrow Q}$	$\Rightarrow$-E : $\dfrac{P,\ P \Rightarrow Q}{Q}$
$\equiv$-I : $\dfrac{P \Rightarrow Q,\ Q \Rightarrow P}{P \equiv Q}$	$\equiv$-E : $\dfrac{P \equiv Q}{P \Rightarrow Q}, \quad \dfrac{P \equiv Q}{Q \Rightarrow P}$
$\neg$-I : $\dfrac{P \vdash Q \wedge \neg Q}{\neg P}$	$\neg$-E : $\dfrac{\neg P \vdash Q \wedge \neg Q}{P}$
true-I : $\dfrac{P \equiv P}{true}$	*true*-E : $\dfrac{true}{P \equiv P}$
false-I : $\dfrac{\neg true}{false}$	*false*-E : $\dfrac{\neg false}{true}$

$A_0 \wedge \cdots \wedge A_n \Rightarrow Q$. The sequent is not a boolean expression. The sequent asserts that Q can be proved from $A_0, \ldots, A_n$. Formula $A_0 \wedge \cdots \wedge A_n \Rightarrow Q$, on the other hand, is a boolean expression (but it need not be a theorem). However, in **ND**, the sequent and formula are related by inference rule $\Rightarrow$-I. [1]

Since a sequent $P \vdash Q$ can be a premise of an inference rule, subproofs arise naturally in **ND**-proofs. To conclude $P \Rightarrow Q$ using $\Rightarrow$-I, one needs a proof of $P \vdash Q$, and this proof can be presented as a subproof.

We now illustrate the use of the inference rules of logic **ND**, in the order in which they appear in Table 6.1. We begin with a proof that from $p \wedge q$ we can conclude $q \wedge p$. The proof is in the Hilbert style, with one addition. In an explanation, to refer to a premise of a sequent we use "pr a" for the first premise, "pr b" for the second, and so on. Table 6.2 summarizes the forms of references to a premise.

$$p \wedge q \vdash q \wedge p$$

1	$p \wedge q$	pr a
2	p	$\wedge$-E, 1
3	q	$\wedge$-E, 1
4	$q \wedge p$	$\wedge$-I, 3, 2

As another example, we prove that from $p \wedge q$ we can infer $p \wedge (q \vee r)$. Here, we use rule $\vee$-I. In this proof, to save space, we do not write the premise on a separate line of the proof but just refer to it in the justification of an inference rule.

$$p \wedge q \vdash p \wedge (q \vee r)$$

1	p	$\wedge$-E, pr a
2	q	$\wedge$-E, pr a
3	$q \vee r$	$\vee$-I, 2
4	$p \wedge (q \vee r)$	$\wedge$-I, 1, 3

The two proofs given so far illustrate the nature of many proofs in natural deduction; expressions are picked apart to get at their constituents and then new expressions are built up from the constituents.

We give another proof, which uses rule $\vee$-E. This rule embodies case analysis. If each of P and Q imply R, and if at least one of P and Q holds, we can conclude R.

---

[1] In logic **E**, the sequent and the formula are related by Deduction theorem (4.4), which can now be rephrased using sequents as "If $P_0, \ldots, P_n \vdash Q$, then $\vdash P_0 \wedge \ldots \wedge P_n \Rightarrow Q$."

HISTORICAL NOTE 6.2. GERHARD GENTZEN (1909–1945)

Gerhard Gentzen was born in Pomerania, which at the time was in north-eastern Germany but is now part of Poland. Even as a young boy, he had declared that mathematics was the only subject he would ever be able to study. He received his doctorate in 1933 at the age of 23 at Göttingen. In 1934, he became David Hilbert's assistant at Göttingen, and this profoundly affected his later work. For example, his work on axiomatic methods stems in part from his concerns with the aims of Hilbert's program for providing firm foundations for mathematics (see Historical note 6.1), and his natural-deduction system of logic was developed "to set up a formal system that comes as close as possible to natural reasoning."

Gentzen's tragic end illustrates how stupid war is. Conscripted into the German armed forces at the outbreak of World War II, he became seriously ill, was placed in a military hospital for three months, and was then freed from military duty. He returned to Göttingen, but in 1943 was requested to move to the German University of Prague. In 1945, all the professors at Prague were jailed by the local authorities. Amidst all the turmoil of that time, after several months of physical hardship, Gentzen died in his cell of malnutrition.

After reading about natural-deduction proof systems, the reader may want to tackle Gentzen's original paper [18] on natural deduction, which appears (in English) in the volume [42] of his collected papers.

$(p \lor (q \land r)), (p \Rightarrow s), (q \land r \Rightarrow s) \vdash s \lor p$

1 s $\lor$-E, pr a, pr b, pr c

2 $s \lor p$ $\lor$-I, 1

SUBPROOFS

To prove $P \Rightarrow Q$ using rule $\Rightarrow$-I, we must have a proof of $P \vdash Q$. The proof of $P \vdash Q$ can be embedded as a subproof within the proof of $P \Rightarrow Q$. The following illustrates this. Note the explanation "pr $1.a$" on line 1.1 below, which refers to the first premise of line 1. See Table 6.2 for the forms of references to premises of a sequent.

TABLE 6.2. FORMS OF REFERENCES TO A PREMISE

The general form is "pr *line-number* . *letter*", where *line-number* gives the line number of a sequent with that premise and *letter* refers to the premise (a is the first, b is the second, etc.). For example, "pr $1.2.b$" refers to the second premise of line 1.2.

If "*line-number*." is missing, the reference is to the sequent on the unnumbered first line, which contains the theorem to be proved.

$\vdash p \Rightarrow p \vee q$
1   $p \vdash p \vee q$
   1.1 $p \vee q$ $\vee$-I, pr 1.a
2   $p \Rightarrow p \vee q$ $\Rightarrow$-I, 1

The next two theorems will be used later to reprove shunting theorem (3.62). On line 1.3 of the first proof below, we see our first use of rule $\Rightarrow$-E, Modus ponens. The second proof shows a proof within a proof within a proof.

(6.2) $p \Rightarrow (q \Rightarrow r) \vdash p \wedge q \Rightarrow r$
   1   $p \wedge q \vdash r$
      1.1 p $\wedge$-E, pr 1.a
      1.2 $q \Rightarrow r$ $\Rightarrow$-E, 1.1, pr a
      1.3 q $\wedge$-E, pr 1.a
      1.4 r $\Rightarrow$-E, 1.3, 1.2
   2   $p \wedge q \Rightarrow r$ $\Rightarrow$-I, 1

(6.3) $p \wedge q \Rightarrow r \vdash p \Rightarrow (q \Rightarrow r)$
   1   $p \vdash q \Rightarrow r$
      1.1 $q \vdash r$
         1.1.1 $p \wedge q$ $\wedge$-I, pr 1.a, pr 1.1.a
         1.1.2 r $\Rightarrow$-E, 1.1.1, pr a
      1.2 $q \Rightarrow r$ $\Rightarrow$-I, 1.1
   2   $p \Rightarrow (q \Rightarrow r)$ $\Rightarrow$-I, 1

ON THE USE OF EQUIVALENCE

Rules $\equiv$-I and $\equiv$-E of **ND** indicate that equivalence $P \equiv Q$ is viewed as an abbreviation of $(P \Rightarrow Q) \wedge (Q \Rightarrow P)$. This means that equivalence is proved in **ND** only by mutual implication. As an example of such a proof, we reprove Shunting (3.65).

$\vdash p \wedge q \Rightarrow r \equiv p \Rightarrow (q \Rightarrow r)$
1   $(p \wedge q \Rightarrow r) \Rightarrow (p \Rightarrow (q \Rightarrow r))$ $\Rightarrow$-I, proof (6.3)
2   $(p \Rightarrow (q \Rightarrow r)) \Rightarrow (p \wedge q \Rightarrow r)$ $\Rightarrow$-I, proof (6.2)
3   $p \wedge q \Rightarrow r \equiv p \Rightarrow (q \Rightarrow r)$ $\equiv$-I, 1, 2

PROOF BY CONTRADICTION

Proof by contradiction (see page 77) in **ND** is embodied in rules $\neg$-I and $\neg$-E. These rules are similar, and one might think that one would suffice.

However, if one of them is omitted, several valid formulas are no longer theorems, as explained in Sec. 7.2.

To illustrate proof by contradiction, we prove the law of Double Negation, (3.12). The proof is longer than its proof in the equational system because **ND** relegates equivalence to a minor role.

$$(6.4) \quad \vdash p \equiv \neg\neg p$$

$$
\begin{array}{lll}
1 & p \vdash \neg\neg p & \\
& 1.1 \quad \neg p \vdash p \wedge \neg p & \\
& \quad 1.1.1 \quad p \wedge \neg p & \wedge\text{-I, pr } 1.a, \text{ pr } 1.1.a \\
& 1.2 \quad \neg\neg p & \neg\text{-I, } 1.1 \\
2 & p \Rightarrow \neg\neg p & \Rightarrow\text{-I, } 1 \\
3 & \neg\neg p \vdash p & \\
& 3.1 \quad \neg p \vdash \neg p \wedge \neg\neg p & \\
& \quad 3.1.1 \quad \neg p \wedge \neg\neg p & \wedge\text{-I, pr } 3.1.a, \text{ pr } 3.a \\
& 3.2 \quad p & \neg\text{-E, } 3.1 \\
4 & \neg\neg p \Rightarrow p & \Rightarrow\text{-I, } 3 \\
5 & p \equiv \neg\neg p & \equiv\text{-I, } 2, 4
\end{array}
$$

We now show that from p and $\neg p$ one can prove anything. This makes sense; from a contradiction anything can be proved.

$$
\begin{array}{lll}
& p, \neg p \vdash q & \\
1 & \neg q \vdash p \wedge \neg p & \\
& 1.1 \quad p \wedge \neg p & \wedge\text{-I, pr a, pr b} \\
2 & q & \neg\text{-E, } 1
\end{array}
$$

THE CONSTANTS TRUE AND FALSE

According to the **ND** inference rules for introducing and eliminating constants, the constant *true* is equivalent to $P \equiv P$ and *false* is equivalent to $\neg(P \equiv P)$. Hence, one can have a propositional logic that does not use the two constants, but instead uses the equivalent expressions $P \equiv P$ and $\neg(P \equiv P)$. Strictly speaking, the constants are not necessary, which should not be surprising. In the equational logic, we also introduced *true* in axiom (3.3) to abbreviate $p \equiv p$, and *false* to abbreviate $\neg true$.

6.3   Additional proof formats

We now look at reformatting proofs in **ND** to make them more readable. The benefits can be seen in proofs that use rule $\Rightarrow$-E. Such a proof has the following structure.

```
 ...
1  P                    ... explanation
 ...
3  P ⇒ Q                ... explanation
4  Q                    ⇒-E , 1, 3
5  ...
```

This structure does not make clear the intent as one reads from beginning to end, especially if the proofs of P and $P \Rightarrow Q$ require several lines. Why is P being proved? $P \Rightarrow Q$? For this frequently used rule, we introduce a special format that makes some proofs easier to read:

```
 ...
1  Q                 by Modus ponens
   1.1  P               ... explanation
   1.2  P ⇒ Q           ... explanation
2  ...
```

In this format, the theorem to be proved, Q, comes first, along with the explanation "by Modus ponens". This alerts the reader that the indented subproof will consist of (at least) two lines containing some theorem P and the implication $P \Rightarrow Q$ —there may be other lines if they are needed to prove these two.

The general idea, then, is to announce the shape of the proof early rather than keep the reader in suspense.

Table 6.3 contains some additional proof formats, each of which corresponds to an inference rule of **ND**. (In proofs, other lines may be needed besides the ones shown in the table in order to prove the ones shown.) These inference rules and proof formats correspond to frequently used methods of proof, like proof by contradiction, proof by case analysis, and proof by mutual implication. Natural deduction does indeed formalize many of the informal ideas that are used in proving theorems.

Compare the following proof with the proof of the same (except for one application of shunting) theorem given on page 73. The natural-deduction style of picking formulas apart and then building up new ones works well with formulas involving many implications. Note that the proof does not use the special format for Modus ponens, even though Modus ponens is used twice. Proving the premises of Modus ponens was not long and difficult, and the special format was not needed.

$$(p \Rightarrow p') \land (q \Rightarrow q') \ \Rightarrow \ (p \land q \ \Rightarrow \ p' \land q') \qquad \text{by Deduction}$$
```
1  p ⇒ p'              Assumption
2  q ⇒ q'              Assumption
3  p ∧ q ⇒ p' ∧ q'     by Deduction    (continued on next page)
```

3.1	p	$\wedge$-E , Assumption 3
3.2	p'	$\Rightarrow$-E , 3.1, 1
3.3	q	$\wedge$-E , Assumption 3
3.4	q'	$\Rightarrow$-E , 3.3, 2
3.5	$p' \wedge q'$	$\wedge$-I , 3.2, 3.4

6.4   Styles of reasoning

The point of a proof is to provide convincing evidence of the correctness of some statement. Almost every statement to be proved, no matter what the domain, contains propositional elements —implication, disjunction, etc. Thus, propositional logic, in one form or another, is the glue that binds rea-

---

TABLE 6.3. Additional Proof Formats

Inference rule	Proof format	
$\Rightarrow$-E : $\dfrac{P,\ P \Rightarrow Q}{Q}$	Q 1 P 2 $P \Rightarrow Q$	by Modus ponens ... *explanation* ... *explanation*
$\Rightarrow$-I : $\dfrac{P1, P2 \vdash Q}{P1 \wedge P2 \Rightarrow Q}$	$P1 \wedge P2 \Rightarrow Q$ 1 $P1$ 2 $P2$ ... i Q	by Deduction Assumption Assumption ... *explanation*
$\neg$-E : $\dfrac{\neg P \vdash Q \wedge \neg Q}{P}$	P 1 $\neg P$ ... i $Q \wedge \neg Q$	by Contradiction Assumption ... *explanation*
$\neg$-I : $\dfrac{P \vdash Q \wedge \neg Q}{\neg P}$	$\neg P$ 1 P ... i $Q \wedge \neg Q$	by Contradiction Assumption ... *explanation*
$\vee$-E : $\dfrac{P \vee Q,\ P \Rightarrow R,\ Q \Rightarrow R}{R}$	R 1 $P \vee Q$ 2 $P \Rightarrow R$ 3 $Q \Rightarrow R$	by Case analysis ... *explanation* ... *explanation* ... *explanation*
$\equiv$-I : $\dfrac{P \Rightarrow Q,\ Q \Rightarrow P}{P \equiv Q}$	$P \equiv Q$ 1 $P \Rightarrow Q$ 2 $Q \Rightarrow P$	by Mutual implication ... *explanation* ... *explanation*

soning together. Studying different styles of propositional proof and gaining experience with them increases understanding of reasoning methods.

We have given two styles of formal, propositional proof. Logic **E** uses an equational style; logic **ND**, a hierarchical Hilbert style. Proofs in these styles constitute convincing evidence because each step of a proof can be checked, independently of the others, either by a human or by a computer.

An informal proof written in English might seem easier to read (but not always!), although often it will omit steps or contain subtle errors, since English is rich but ambiguous. Thus, one could argue that no proof in a natural language could be construed as convincing evidence. However, a good informal proof can be viewed as an outline or set of instructions for constructing a formal proof in some specified formal logical system. Such an informal proof could be considered convincing evidence. Thus, informal proofs are legitimate reasoning tools when they serve as descriptions or outlines of formal proofs.

One purpose of mathematics and logic is to prevent complexity from overwhelming. One can take refuge in formalism when informal reasoning becomes too intricate. Formal proofs are often simpler than their informal counterparts in English, because inferences can be based purely on well-defined syntactic manipulation. Finally, formal proof styles provide insight into strategies for *developing* proofs. Chap. 3 discussed a number of such strategies.

For everyday use, we prefer the equational style. The equational style is based on equivalence and the rules of equality. It allows us to calculate with logical formulas in the same way we learned to calculate with arithmetic formulas in elementary school and high school. The equational style does not mimic the usual way of thinking in English —in fact, English, which is more attuned to implication, does very poorly with equivalence. For example, there is no simple way to say $P \equiv Q \equiv R$ in English, while $P \Rightarrow Q \Rightarrow R$ can be read as "If P, then if Q, then R". The equational style is often an effective alternative to reasoning in English, because it allows a concise and precise argument for something that would be complicated and contorted in English. Portia's suitor's problem on page 86 is a good example of this. Further, when dealing with the development of programs and their proofs, we have found the equational style to be far more useful than natural deduction, and we use it most of the time.

Why was the natural deduction style invented? Gentzen felt that the pure Hilbert style of proof (as a sequence of theorems, with no hierarchical structure) was "rather far removed from the forms of deduction used in practice in mathematical proofs", and he wanted "to set up a formal system that comes as close as possible to actual reasoning" [42, p. 68]. And, in general, natural deduction seems to mirror English arguments well. For example, rule $\wedge$-I can be read in English as "If P and Q are true, then so is $P \wedge Q$." However, mirroring English arguments is not necessarily a good thing. Indeed, English arguments are not always effective, since they tend to be long, unwieldy, and contorted, with many case analyses. Why should formalizing them be an improvement? Furthermore, natural deduction proofs appear to require far more steps than proofs in the equational style. Even for some of Gentzen's proofs in [42], it has been said that simplicity and elegance of procedure are sacrificed to the demands of 'naturalness'.

The natural deduction style is heavily used in mechanical proof-checking systems, perhaps because of its modus operandi of picking an expression to pieces and then building up the desired expression from the pieces. And its method of nesting proofs is useful for structuring large proofs. A mechanical system based on natural deduction can be useful, especially when one is not interested in reading a proof but just in knowing that a theorem has been proved.

If you have mastered both the equational style and the natural-deduction style of proof, you will always be able to use the one that best fits the context in which you are working. Formal tools are supposed to help. Sometimes just the formal ideas used informally are of benefit. Sometimes, a blend of formal tools is better. Thus, we can have the best of all worlds; on any particular problem and in any particular context, use the style that bests suits it.

Exercises for Chapter 6

6.1 Using the example on page 109 of an equational proof and its Hilbert-style counterpart, give a procedure to transform any equational proof into a Hilbert-style proof.

In the following exercises, use the proof format of Table 6.3 in proving theorems in logic **ND**.

6.2 Prove $p \Rightarrow (q \Rightarrow p)$.

6.3 Prove $p \Rightarrow (q \Rightarrow p \wedge q)$.

6.4 Prove $(p \wedge q) \wedge r \Rightarrow p \wedge (q \wedge r)$.

6.5 Prove $(p \Rightarrow q) \Rightarrow (p \wedge r \Rightarrow q \wedge r)$.

6.6 Prove $(p \Rightarrow (q \Rightarrow r)) \Rightarrow (q \Rightarrow (p \Rightarrow r))$.

6.7 Prove $(p \Rightarrow q) \wedge (p \Rightarrow r) \Rightarrow (p \Rightarrow q \wedge r)$.

6.8 Prove $(p \Rightarrow q) \wedge (q \Rightarrow r) \Rightarrow (p \Rightarrow r)$.

6.9 Prove $(q \Rightarrow r) \Rightarrow ((p \Rightarrow q) \Rightarrow (p \Rightarrow r))$.

6.10 Prove $(p \Rightarrow (q \Rightarrow r)) \Rightarrow ((p \Rightarrow q) \Rightarrow (p \Rightarrow r))$.

6.11 Prove $(p \Rightarrow \neg p) \Rightarrow \neg p$.

6.12 Prove $(p \wedge \neg q) \Rightarrow \neg(p \Rightarrow q)$.

6.13 Prove $(p \Rightarrow q) \wedge (p \Rightarrow \neg q) \Rightarrow \neg p$.

6.14 Proof of $p \vee \neg p$.

6.15 Prove $\neg(p \wedge \neg p)$.

6.16 Prove $p \Rightarrow q \equiv \neg q \Rightarrow \neg p$.

6.17 Prove $p \Rightarrow q \equiv \neg p \vee q$.

Chapter 7

Formal Logic

W e study the general notion of a formal logical system and its *interpretations*. Thus, we discuss both syntax (proof theory) and semantics (model theory) for logics. We also study *constructive* logic in a propositional setting.

7.1   Formal logical systems

PROOF THEORY

A *formal logical system*, or *logic*, is a set of rules defined in terms of

- a set of *symbols*,

- a set of *formulas* constructed from the symbols,

- a set of distinguished formulas called *axioms*, and

- a set of *inference rules*.

The set of formulas is called the *language* of the logic. The language is defined syntactically; there is no notion of meaning or semantics in a logic per se.

Inference rules allow formulas to be derived from other formulas. Inference rules have the form

$$\frac{H_1, H_2, \ldots, H_n}{C}$$

where formulas $H_1, H_2, \ldots, H_n$ are the *premises* (or *hypotheses*) of the inference rule and formula C is its *conclusion*. A formula is a *theorem* of the logic if it is an axiom or if it can be generated from the axioms and already proved theorems using the inference rules. A *proof* that a formula is a theorem is an argument that shows how the inference rules are used to generate the formula.

For equational logic $\mathbf{E}$ of Chap. 3, the symbols are $(\,,\,)\,,\,=\,,\,\neq\,,\,\equiv\,,\,\not\equiv\,,$ $\neg\,,\,\vee\,,\,\wedge\,,\,\Rightarrow\,,\,\Leftarrow\,,$ the constants *true* and *false* , and boolean variables

p, q , etc. The formulas are the boolean expressions constructed using these symbols. **E** has 15 axioms, starting with Associativity of $\equiv$, (3.1). Its inference rules are Leibniz (1.5), Transitivity of equality (1.4), and Substitution (1.1). Its theorems are the formulas that can be shown to be equal to an axiom using these inference rules.

A logic can only be useful if it makes some distinction between formulas:

(7.1)   **Definition.** A logic is *consistent* if at least one of its formulas is a theorem and at least one is not; otherwise, the logic is *inconsistent*.

For example, logic **E** is consistent, because *true* is a theorem and *false* is not. Adding *false* $\equiv$ *true* as an axiom to **E** would make it inconsistent.

Table 7.1 presents another logic, PQ-L, due to Hofstadter [24]. PQ-L is slightly perplexing because we do not say what the formulas, axioms, and inference rules mean. PQ-L forcefully illustrates the view that a logic is a system for manipulating symbols, independent of meaning.

Below are three formulas of PQ-L.

$$- - - P - Q - -$$
$$P \; Q \; -$$
$$- P - Q - -$$

PQ-L uses the Hilbert style of proof. Here is a proof of theoremhood of $- - - - P - - - Q - - - - - - - -$. This theorem, together with the fact that $- P - Q -$ is not a theorem, tells us that PQ-L is consistent.

$$- - - - P - - - Q - - - - - - - -$$

1. $- P - Q - -$		Axiom 0
2. $- - P - Q - - -$		Axiom 1
3. $- - - P - - Q - - - - -$		Inf. rule, 1, 2
4. $- - - - P - - - Q - - - - - - -$		Inf. rule, 1, 3

TABLE 7.1. LOGIC PQ–L

Symbols: P, Q, –

Formulas: Sequences of the form $a \, P \, b \, Q \, c$, where a , b , and c denote finite sequences of zero or more dashes –.

Axioms: $0: \; - P - Q - -$
                  $1: \; - - P - Q - - -$

Inference Rule: $$\frac{a \, P \, b \, Q \, c, \quad d \, P \, e \, Q \, f}{a \, d \, P \, b \, e \, Q \, c \, f}$$

MODEL THEORY

Typically, the formulas of a logic are intended to be statements about some *domain of discourse*, that is, some area of interest. We give the formulas a meaning with respect to this domain by defining which formulas are true statements and which are false statements about the domain.

An *interpretation* assigns meaning to the operators, constants, and variables of a logic. For example, we can give formulas of PQ–L meaning by providing the following interpretation.

(7.2) **Addition-equality Interpretation**. A formula a P b Q c is mapped to $\#a + \#b = \#c$, where $\#x$ denotes the number of dashes in sequence x.

For example, formulas $-$ P Q $-$ and $-$ P $- -$ Q $- - -$ are mapped to $1 + 0 = 1$ and $1 + 2 = 3$, which are *true*, and $-$ P $-$ Q $-$ is mapped to $1 + 1 = 1$, which is *false*. Also, axiom $-$ P $-$ Q $- -$ of PQ-L is interpreted as $1 + 1 = 2$ and axiom $- -$ P $-$ Q $- - -$ as $2 + 1 = 3$.

Because a logic is purely a syntactic object, it may have more than one interpretation. For example, here is a second interpretation for PQ–L.

(7.3) **Addition-inequality Interpretation**. A formula a P b Q c is mapped to *true* iff $\#a + \#b \leq \#c$, where $\#x$ denotes the number of dashes in sequence x.

Interpretations (7.2) and (7.3) are different. The first maps $-$ P $-$ Q $- - -$ to *false*, since $1 + 1 = 3$ is *false*. The second maps this formula to *true*, since $1 + 1 \leq 3$ is *true*.

In a logic in which formulas have variables, an interpretation associates a value with each variable. Each interpretation gives the meaning of formulas with a different variable-value association, so the complete meaning of a formula is given by a set of interpretations. Conventionally, we split such an interpretation into two parts: one gives a fixed meaning to the operators and constants; the other supplies values for variables, i.e. denotes a state.

For example, consider logic **E**, with its standard notion of evaluation: an interpretation of an expression gives the value of the expression in some state. We can formalize this notion as follows.

(7.4) **Standard interpretation of expressions of E.** For an expression P without variables, let *eval*.'P' be the value of P, as explained in Sec. 2.1. (Note that *eval* gives "meaning" to the operators and constants of **E**.) Let Q be any expression, and let s be a state that gives values to all the variables of Q. Define $s.$'Q' to be a copy of Q in which all its variables are replaced by their corresponding values in state s. Then function f given by $f.$'Q' $= eval(s.$'Q') is an *interpretation* for Q.

We can give other, less conventional, interpretations to **E** as well. For example, one interpretation maps every expression to *false*. Such an interpretation provides no connection between the logic, which tells us which formulas are theorems and which are not, and the domain of discourse.

On page 31, we defined satisfiability and validity of boolean expressions. We now extend the definition to cover satisfiability and validity of a formula with respect to any logic and interpretation.

(7.5)   **Definition.** Let S be a set of interpretations for a logic and F be a formula of the logic. F is *satisfiable* (under S) iff at least one interpretation of S maps F to *true*. F is *valid* (under S) iff every interpretation in S maps F to *true*.

In terms of **E**, a boolean expression F is satisfiable iff F evaluates to *true* in at least one state. F is valid iff F evaluates to *true* in all states.

An interpretation is a model for a logic iff every theorem is mapped to *true* by the interpretation.

The next definition gives terminology for describing the relationship between a logic and a set of interpretations for it.

(7.6)   **Definition.** A logic is *sound* iff every theorem is valid. A logic is *complete* iff every valid formula is a theorem.

Soundness means that the theorems are true statements about the domain of discourse. Completeness means that every valid formula can be proved.

E is sound and complete with respect to standard interpretation (7.4). Adding the axiom $p \land \neg p$ would make **E** unsound, because $p \land \neg p$ is unsatisfiable. Logic PQ-L is sound with respect to interpretation (7.2) but not complete, because the valid formula $-$ PQ $-$ is not a theorem of PQ-L.

A sound and complete logic allows exactly the valid formulas to be proved. Failure to prove that a formula is a theorem in such a logic cannot be attributed to weakness of the logic. Unfortunately, many domains of discourse of concern to us —arithmetic truths, program behavior, and so on— do not have sound and complete axiomatizations. This is a consequence of Gödel's incompleteness theorem (see Historical note 7.1), which states that no formal logical system that axiomatizes arithmetic can be both sound and complete. Fortunately, this incompleteness is not a problem in practice.

In order to isolate sources of incompleteness in a logic, the logic can be defined in a hierarchical fashion. A logic L^+ is an *extension* of logic L if the symbols, formulas, axioms, and inference rules of L are included in L^+. For example, we obtain a *predicate logic* by extending a propositional logic with variables that may be associated with other types of variables (e.g. the integers) and by introducing *predicates* on those variables, i.e. boolean functions of those variables (e.g. $less(x, y)$, or $x < y$).

HISTORICAL NOTE 7.1. KURT GÖDEL (1906–1978)

In 1929, in his 35-page Ph.D. thesis, 23-year old Gödel proved that a logical system similar to our predicate calculus of Chap. 9 was complete. (The predicate calculus is an extension of the propositional calculus.)

Just over a year later, he wrote paper [20], which has been called the greatest single piece of work in the whole history of mathematical logic. Hilbert had proposed a program to completely formalize mathematics and show that it was complete, consistent, and decidable —all facts could be proved, nothing false could be proved, and any statement could be proved to be either true or false (see Historical note 6.1). In [20], Gödel showed that Hilbert's program could not be realized. He showed that any formal system that included arithmetic was either incomplete or inconsistent, and he exhibited arithmetic statements that were not decidable. But [20] did much more. As Wang says in [44], it "pulled together, consolidated, and raised previous work to a much higher level in nearly all directions, proving surprising central results, making old concepts precise, introducing new concepts, and opening up wholly new horizons."

Gödel was born and raised in Czechoslovakia and did his Ph.D. work at the University of Vienna. After several professional visits to the Institute for Advanced Study at Princeton, he left Austria in 1940 (with his wife) to spend the rest of his life in Princeton. There, he and Einstein were good friends. Most of his work on mathematical logic was done before 1940, and he spent much of his later working life on philosophy.

Because of bad health, Gödel never visited Europe after 1940, and he did surprisingly little travel within the U.S. (He spent some time in the 1930's in sanatoriums for nervous depression, was relatively frail, and had various physical problems.) He could not even attend the conference organized to celebrate his sixtieth birthday at Ohio State. Unfortunately, relatively few people saw this great man in action.

A DECISION PROCEDURE FOR PROPOSITIONAL LOGIC

In **E**, or any equivalent propositional logic, there is a simple way to determine whether a formula is a theorem: just check its validity.

(7.7)   **Decision Procedure for logic E.** Compute the interpretation of a formula F in every possible state of the state space defined by the boolean variables in F. F is a theorem of **E** iff it is mapped to *true* in every state.

Determining whether a boolean expression involving n boolean variables is valid requires checking 2^n cases, since that is the size of the state space defined by n boolean variables. This decision procedure is time-consuming for formulas involving a large number of variables.

Not all logics have decision procedures. That is, for some logics, there is no algorithm to tell whether an arbitrary formula is a theorem or not. In fact, most logics that deal with interesting domains of discourse, like the integers, do not have decision procedures.

7.2 Constructive logics

Let P be any mathematical statement whose truth is not known. For example, we could take as P the statement "there are an infinite number of twin primes" (twin primes are prime numbers that differ by 2; 11 and 13 are twin primes). Given such a P, we can define x (say) as follows.

$$x = \begin{cases} 0 \text{ if } P \text{ is } \textit{true} \\ 1 \text{ if } P \text{ is } \textit{false} \end{cases}$$

This definition defines x unambiguously. And yet, since we do not know whether P holds, we cannot compute the value of x! We have given a *non-constructive* definition of x —a *constructive* definition would tell us how to calculate the value of x.

We can also prove things in a non-constructive way. We give an example that deals with real numbers. A real number is *rational* if it can be written in the form b/c for two integers b and c ($c \neq 0$); otherwise it is *irrational*. The number $1/3$ is rational, while $\sqrt{2}$ and the number π (the ratio of the circumference of a circle to its diameter) are irrational.

(7.8) **Theorem.** There exist two irrational numbers b and c such that b^c is rational.

Proof. The proof is by case analysis: $(\sqrt{2})^{\sqrt{2}}$ is either rational or irrational.

Case $(\sqrt{2})^{\sqrt{2}}$ is rational. Choose $b = c = \sqrt{2}$.

Case $(\sqrt{2})^{\sqrt{2}}$ is irrational. Choose $b = (\sqrt{2})^{\sqrt{2}}$ and $c = \sqrt{2}$. Since 2 is rational, we can show that b^c is rational:

$$\begin{aligned} & ((\sqrt{2})^{\sqrt{2}})^{\sqrt{2}} \quad \text{—this is } b^c \\ = \ & \langle \text{Arithmetic} \rangle \\ & (\sqrt{2})^{\sqrt{2}\sqrt{2}} \\ = \ & \langle \text{Arithmetic} \rangle \\ & 2 \end{aligned}$$

$\square$

This proof of the existence of rational b^c does not show us how to construct b^c, since we do not know whether $(\sqrt{2})^{\sqrt{2}}$ is rational. It is a non-constructive proof.

Constructive mathematics is the branch of mathematics in which each definition or proof of existence of an object provides an algorithm for com-

---

HISTORICAL NOTE 7.2. CONSTRUCTIVE MATHEMATICS

One of the first proponents of the constructive style was Leopold Kronecker (see Historical note 7.3), who did not believe in the existence of π because he did not know how to construct it. Kronecker railed against non-constructive mathematics, such as Cantor's theory of infinite sets (see Chap. 20), as a dangerous mathematical insanity. Others followed in Kronecker's footsteps — Poincaré, Borel, Lebesgue, Brouwer, Heyting, and Weyl, to name a few— although they were often more liberal in their views than Kronecker. Brouwer's work is called *intuitionistic mathematics*, since it is based on the thesis that mathematics is based on primitive intuitions. Intuitionistic mathematics rejects the law of the excluded middle.

Two modern-day events brought constructive mathematics into the limelight. The first was E. Bishop's work on constructive mathematics in the 1960's —see his book [4] and also [5]. Bishop's book has "an ultimate goal: to hasten the day when constructive mathematics will be the accepted norm". The second event was the development of the computer, because a computer system can extract an algorithm from a constructive proof and then run the algorithm. Nuprl [8] is perhaps the first software system to mechanize constructive logic and extract programs from proofs.

For more on constructive mathematics and logic, turn to Bishop's book or [43]. See also Historical notes on Hilbert (page 111), Gentzen (page 116), and Kronecker (page 132).

---

puting it. (In some versions of constructive mathematics, and there are several, it is enough to provide an algorithm to construct as close an approximation to an object as we desire, even if the object cannot be computed exactly. This kind of constructive mathematics would allow as objects the irrational numbers, while the stricter form of constructive mathematics, which is usually called *finitistic* mathematics, would not.)

Constructive mathematics has increased in popularity with the advent of computers. Embedded in a constructive proof that an object exists is an algorithm for constructing the object. A computer program can analyze such a proof and extract an algorithm from it, so the programming task can be replaced by the task of proving theorems. (However, this interesting idea has not yet been brought to fruition in a practical way.)

Of course, the algorithm extracted from a proof may not be as efficient as we want, so we may want to develop another proof whose embedded algorithm is more efficient. Mathematicians often develop different proofs in their search for brief, concise, and simple ones. The development of constructive mathematics provides a new basis for comparing proofs: how fast are the algorithms that can be extracted from them?

We now introduce a constructive propositional logic. This logic is based on the following principles, which tell us how a constructive mathematical

---

HISTORICAL NOTE 7.3. Leopold Kronecker (1823–1891)

Kronecker received his doctorate in mathematics from the University of Berlin when he was 22. For the next 8 years, he managed his family businesses, pursuing mathematics as a recreation, until he was financially able to retire at the age of 30. Such is life.

At the age of 38, he began lecturing at the University of Berlin, receiving a professorship when he was 59. He made major contributions to the theories of elliptic functions, algebraic equations, and algebraic numbers. But his philosophy of mathematics and his polemics against those with different opinions are what some people remember him for. Kronecker really liked people. His house was open to his pupils, and he was known for his generous hospitality. But he was also quite vociferous about his views, most notably about constructive mathematics. If a thing could not be constructed, it did not exist and should not be talked about, and he railed against others, like Weierstrass, Hilbert, and Cantor, who thought otherwise. "God made the integers," said Kronecker, "all the rest is the work of man".

For more on this story of constructive versus classical mathematics, see Historical notes 7.2 on constructive mathematics (page 131), 6.1 on Hilbert (page 111), and 20.1 on Cantor (page 464).

---

proof of an expression should be built from proofs of its constituents. These principles are called the *BHK*-interpretation of constructive mathematics, after the people who were involved in their formulation: Brouwer, Heyting, and Kolmogorov.

1. A proof of $P \wedge Q$ is given by presenting a proof of P and a proof of Q.

2. A proof of $P \vee Q$ is given by presenting either a proof of P or a proof of Q (and indicating which it is).

3. A proof of $P \Rightarrow Q$ is a procedure that permits us to transform a proof of P into a proof of Q.

4. The constant *false*, which is a contradiction, has no proof.

5. A proof of $\neg P$ is a procedure that transforms any hypothetical proof of P into a proof of a contradiction.

The fifth principle can be explained as follows. The constant *false* is not true, so there exists no proof for it. Hence, if we show that *false* follows from a hypothetical proof of P, then P itself is *false*. We regard the proof of $P \vdash false$ as a proof of $\neg P$.

In constructive logic, the law of the Excluded Middle, $P \vee \neg P$, is not a theorem. For if it were, then, for any expression P, we could construct a proof either of P or of $\neg P$ (see principle 2 of the *BHK*-interpretation, given above). Since there are many statements in mathematics that no one has been able to prove or disprove, for example whether there are an infinite number of twin primes, we cannot accept the law of the Excluded Middle in a constructive system.

The inference rules for a constructive propositional logic are given in Table 7.2. Each premise of each inference rule is a sequent, which we interpret as a proof that its conclusion follows from its premise. Similarly, we make each conclusion a sequent. Each rule mirrors a corresponding principle in the *BHK*-interpretation given above. Note that we have given different names to the two rules for $\vee$-introduction, so that the rule's name indicates which of the two operands is being used as evidence that $P \vee Q$ holds.

Consider rule $\Rightarrow$-I. Instantiating P_1, Q, n with $P, false, 1$ yields

$$\frac{P \vdash false}{P \Rightarrow false} \quad,$$

which, since $\neg P$ denotes $P \Rightarrow false$, can be written as

$$(7.9) \quad \frac{P \vdash false}{\neg P}.$$

TABLE 7.2. RULES FOR CONSTRUCTIVE NATURAL DEDUCTION

Introduction rules	Elimination rules
$\wedge$-I: $\dfrac{\vdash P, \ \vdash Q}{\vdash P \wedge Q}$	$\wedge$-E$_l$: $\dfrac{\vdash P \wedge Q}{\vdash P}$, $\wedge$-E$_r$: $\dfrac{\vdash P \wedge Q}{\vdash Q}$
$\vee$-I$_l$: $\dfrac{\vdash P}{\vdash P \vee Q}$, $\vee$-I$_r$: $\dfrac{\vdash Q}{\vdash P \vee Q}$	$\vee$-E: $\dfrac{\vdash P \vee Q, \ P \vdash R, \ Q \vdash R}{\vdash R}$
$\Rightarrow$-I: $\dfrac{P_1, \ldots, P_n \vdash Q}{\vdash P_1 \wedge \cdots \wedge P_n \Rightarrow Q}$	$\Rightarrow$-E: $\dfrac{\vdash P, \ \vdash P \Rightarrow Q}{\vdash Q}$
false-I: (none)	*false*-E: $\dfrac{\vdash false}{\vdash P}$

$$
\begin{array}{lll}
P \equiv Q & \text{denotes} & (P \Rightarrow Q) \wedge (Q \Rightarrow P) \\
\neg P & \text{denotes} & P \Rightarrow false \\
true & \text{denotes} & \neg false
\end{array}
$$

However, there is no corresponding inference rule to eliminate $\neg$;

$$(7.10) \quad \frac{\neg P \vdash false}{P}$$

is *not* an inference rule of constructive propositional logic. The absence of (7.10) has drastic repercussions on what can be proved. For example, $\neg\neg p \Rightarrow p$ is not a theorem! But $p \Rightarrow \neg\neg p$ is. It is interesting to compare the following constructive proof of $p \Rightarrow \neg\neg p$ (i.e. $p \Rightarrow ((p \Rightarrow false) \Rightarrow false)$) with the non-constructive proof in **ND** given on page 118 (as a subproof of the proof of $p \equiv \neg\neg p$).

$$\vdash p \Rightarrow ((p \Rightarrow false) \Rightarrow false)$$

1	$p \vdash (p \Rightarrow false) \Rightarrow false$	
	1.1 $p \Rightarrow false \vdash false$	
	1.1.1 *false*	$\Rightarrow$-E, pr 1.a, pr 1.1.a
	1.2 $(p \Rightarrow false) \Rightarrow false$	$\Rightarrow$-I, 1.1
2	$p \Rightarrow ((p \Rightarrow false) \Rightarrow false)$	$\Rightarrow$-I, 1

Since $\neg\neg p \Rightarrow p$ is not a theorem, neither is $\neg\neg p \equiv p$. The law of Double Negation does not hold in constructive propositional logic. If we add inference rule (7.10) to the logic, we leave the realm of constructive logic and have a non-constructive propositional logic, which has the same theorems as logics **E** and **ND**.

Formula $p \vee \neg p$ is not a theorem of constructive logic, but $\neg\neg(p \vee \neg p)$ is. To prove it, we first have to rewrite it to eliminate $\neg$:

(7.11)	$\vdash (p \vee (p \Rightarrow false) \Rightarrow false) \Rightarrow false$	
	1 $\quad p \vee (p \Rightarrow false) \Rightarrow false \vdash false$	
	1.1 $p \vdash false$	
	1.1.1 $p \vee (p \Rightarrow false)$	$\vee$-I$_l$, pr 1.1.a
	1.1.2 *false*	$\Rightarrow$-E, 1.1.1, pr 1.a
	1.2 $p \Rightarrow false$	$\Rightarrow$-I, 1.1
	1.3 $p \vee (p \Rightarrow false)$	$\vee$-I$_r$, 1.2
	1.4 *false*	$\Rightarrow$-E, 1.3, pr 1.a
	2 $\quad (p \vee (p \Rightarrow false) \Rightarrow false) \Rightarrow false$	$\Rightarrow$-I, 1

We have barely touched the surface of constructive logic. Do not make a judgement for or against constructive logic until you thoroughly understand it and have seen it being used.

Exercises for Chapter 7

7.1 Give a finite set of axioms that can be added to PQ-L to make it sound and complete under Addition-Equality Interpretation (7.2).

7.2 Consider the logic defined as follows, where a, b, and c denote finite sequences of zero or more o's.

Symbols:	M, I, o
Formulas:	Sequences of the form a M b I c
Axiom:	ooMooIoooo

Inference Rule R1: $\dfrac{a \text{ M } b \text{ I } c}{a\,a \text{ M } b \text{ I } c\,c}$

Inference Rule R2: $\dfrac{a \text{ M } b\,b \text{ I } c\,c}{a\,a \text{ M } b \text{ I } c\,c}$

(a) Give 5 formulas of this logic.

(b) State and prove five theorems of the logic.

(c) Give an interpretation of the logic that makes multiplication of integers a model.

(d) Give a formula that is *true* according to your interpretation but is not a theorem. Argue (informally) why it cannot be a theorem.

7.3 Two possible definitions for soundness of an inference rule are:

> **Theorem-Soundness.** An inference rule is considered *sound* if a formula derived using it is valid whenever the premises used in the inference are theorems.

> **Model-Soundness.** An inference rule is considered *sound* if a formula derived using it is valid whenever the premises used in the inference are valid.

What are the advantages/disadvantages of considering axiomatizations in which all inference rules satisfy Theorem-Soundness versus Model-Soundness?

7.4 Recall from Exercise 3.88 (page 67) that a formula is in conjunctive normal form if it is the conjunction of propositional logic formulas, each of which is a disjunction of boolean variables. For such a formula P, denote by $\#P$ the minimum number of boolean variables that, if all were *false*, would cause P to be *false*. Describe a procedure to calculate $\#P$ for P a formula in conjunctive normal form in which no variable is negated.

(Solving this problem turns out to be useful in determining the fault-tolerance of a distributed system. Each variable denotes whether a give processor is faulty, and the formula is *true* iff the system has sufficient resources to work.)

7.5 This exercise concerns a new calculus, the 01◇-calculus. The symbols of the calculus are

- Variables x, y, z, ...
- The three constant symbols 0, 1, and ◇.
- The binary infix predicate symbol $=$.
- Parentheses.

Formulas of the calculus have the form $\alpha = \beta$, where α and β are sequences of one or more constants, possibly with balanced parentheses as usual to indicate aggregation. Examples of expressions are

$$0 \diamond = 1 \quad \text{and} \quad (\diamond x)(\diamond \diamond) = 10 \ .$$

By definition, if parentheses are omitted, left association is assumed, so that 0100 is shorthand for $((01)0)0$.

The 01 $\diamond$ -calculus has inference rules Leibniz, Substitution, and Transitivity of equality $=$. *Note that symmetry and reflexivity of $=$ are not axioms*, so be extremely careful in applying inference rules. There are four axioms:

Left zero: $0 \diamond = 1$
Zero: $x0 \diamond = x1$
Left one: $1 \diamond = 10$
One: $x1 \diamond = x \diamond 0$

A *theorem* of the calculus is either an axiom or an expression $X = Y$ such that X can be transformed into Y using the inference rules. As an example, we prove that $011 \diamond 0 = 1000$ is a theorem.

$$011 \diamond 0$$
$$= \quad \langle \text{Axiom One, with } x := 01 \rangle$$
$$01 \diamond 00$$
$$= \quad \langle \text{Axiom One, with } x := 0 \rangle$$
$$0 \diamond 000$$
$$= \quad \langle \text{Axiom Left zero} \rangle$$
$$1000$$

We now give meaning to the 01 $\diamond$ -calculus by defining interpretations for it. A *state* assigns to each variable a natural number or $\diamond$. For example, the state $\{(x, 19), (y, \diamond)\}$ assigns 19 to z and $\diamond$ to y . Expressions are evaluated in a state by first replacing each variable by its value in the state and then applying the following rules. (In the rules, $x \succ y$ means that x evaluates to y).

$$m\,n \succ (2 \cdot m + n) \qquad \text{(for integers } m \text{ and } n)$$
$$m \diamond \succ (m + 1) \qquad \text{(for integer } m)$$
$$\diamond n \succ (2 + n) \qquad \text{(for integer } n)$$
$$\diamond \diamond \succ 2$$
$$(x = x) \succ \textit{true} \qquad \text{(for } x \text{ an integer)}$$
$$(x = y) \succ \textit{false} \qquad \text{(for } x \text{ and } y \text{ different integers)}$$

Perform the following exercises.

(a) Prove: $0 \diamond \diamond = 10$.
(b) Prove: $0 \diamond \diamond \diamond = 11$.
(c) Prove: $0 \diamond \diamond \diamond \diamond = 100$.
(d) Prove: $0 \diamond \diamond \diamond \diamond \diamond = 101$.
(e) Why does the following metatheorem concerning the 01 $\diamond$ -calculus hold: Every theorem of the 01 $\diamond$ -calculus contains a $\diamond$.
(f) Does it follow from the previous question that the 01 $\diamond$ -calculus is consistent?
(g) Evaluate the expression 1011 in the state $\{(x, 19), (y, \diamond)\}$. In doing these and other evaluations, fully parenthesize the expression in order to be able to distinguish characters 0 and 1 from integers. Thus, evaluate the expression $(((1)0)1)1$.

(h) Evaluate the expression $1011\diamond$ in the state $\{(x, 19), (y, \diamond)\}$.

(i) Evaluate the expression $1011 \diamond \diamond$ in the state $\{(x, 19), (y, \diamond)\}$.

(j) Evaluate the expression $1x0y\diamond$ in the state $\{(x, 19), (y, \diamond)\}$.

(k) Evaluate the expression $0(\diamond y) = 10$ in the state $\{(x, 19), (y, \diamond)\}$.

(l) Find a *model* and a *counterexample* for the expression $x \diamond 0 = 10$.

(m) Show that the $01\diamond$-calculus with the interpretation given above is sound by checking that all four axioms are valid and that all three inference rules preserve validity.

(n) Show that the $01\diamond$-calculus is incomplete by finding a valid expression that is not a theorem.

(o) Show that the expression $x1 = \diamond\diamond$ is unsatisfiable.

Chapter 8

Quantification

W̶e introduce *quantification* for any symmetric and associative operator. Summing a set of values (using addition $+$) and "oring" together a set of values (using disjunction $\vee$) can be expressed using quantification. Quantification is important in the predicate calculus of the next chapter, and it is used in most later chapters.

8.1 On types

In programming languages, a *type* denotes the (nonempty) set of values that can be associated with a variable. Thus far, we have been dealing mainly with type *bool*, or $\mathbb{B}$ as we write it from now on. It is the set of values *true* and *false*. We now begin dealing with other types as well —see Table 8.1. The introduction of types causes us to refine our notion of an expression. To be an expression, not only must a sequence of symbols satisfy the normal rules of syntax concerning balanced parentheses, etc., it must also be *type correct*. Thus, some expressions that earlier seemed okay will no longer be called expressions because they do not satisfy the typing rules.

Every expression E has a type t (say), which we can declare by writing $E{:}t$. For example, since the constant 1 has type $\mathbb{Z}$ and *true* has type $\mathbb{B}$, we may write $1{:}\mathbb{Z}$ and *true*$:\mathbb{B}$. Similarly, every variable has a type. Sometimes, the type of a variable is mentioned in the text accompanying an

---

TABLE 8.1. SOME BASIC TYPES

Name	Symbol	Type (set of values)
integer	$\mathbb{Z}$	integers: $\ldots, -3, -2, -1, 0, 1, 2, 3, \ldots$
nat	$\mathbb{N}$	natural numbers: $0, 1, 2, \ldots$
positive	$\mathbb{Z}^+$	positive integers: $1, 2, 3, \ldots$
negative	$\mathbb{Z}^-$	negative integers: $-1, -2, -3, \ldots$
rational	$\mathbb{Q}$	rational numbers i/j for i, j integers, $j \neq 0$
reals	$\mathbb{R}$	real numbers
positive reals	$\mathbb{R}^+$	positive real numbers
bool	$\mathbb{B}$	booleans: *true*, *false*

expression that uses the variable, and sometimes it is given in some sort of a declaration, much like a programming-language declaration **var** x:*integer* . However, when the type of a variable is not important to the discussion, in the interest of brevity and clarity we may omit it.

We may want to declare the type of a subexpression of an expression, in order to make the expression absolutely clear to the reader. For example, we might write 1^n as

$$(1\!:\!\mathbb{Z})^{n:\mathbb{N}}$$

to indicate that 1 is an integer and n a natural number. This convention may be useful, for example, in a context where 1 could denote an identity matrix as well as an integer and where we want to make clear that n is nonnegative. Any subexpression of an expression may be annotated with its type. Here is a fully typed expression: $((x\!:\!\mathbb{N}+y\!:\!\mathbb{N})\cdot x\!:\!\mathbb{N})\!:\!\mathbb{N}$.

Besides constants and variables, the only other kind of expression we have encountered thus far is function application. [1] Each function has a type, which describes the types of its parameters and the type of its result. If the parameters $p_1, \ldots, p_n$ of function f have types $t_1, \ldots, t_n$ and the result of the function has type r , then f has type $t_1 \times \ldots \times t_n \to r$. We indicate this by writing

(8.1) $f : t_1 \times \cdots \times t_n \to r$.

(The reason for this strange-looking syntax will become clear in Chap. 14.) Here are some examples of functions and their types.

function	type	typical function application
plus	$\mathbb{Z} \times \mathbb{Z} \to \mathbb{Z}$	$plus(1,3)$ or $1+3$
not	$\mathbb{B} \to \mathbb{B}$	$not.true$ or $\neg true$
less	$\mathbb{Z} \times \mathbb{Z} \to \mathbb{B}$	$less(5,3)$ or $5 < 3$

For function f with type as shown in (8.1), we define function application $f(a_1, \ldots, a_n)$ to be an expression iff each argument a_i has type t_i . The type of the function application is then r . In this way, "expressionhood", as well as the type of the expression, is determined from the types of its operands.

It is important to recognize that type and type correctness, as we have defined them, are syntactic notions. Type correctness depends only on the sequence of symbols in the proposed expression, and not on evaluation of the expression (in a state). For example, $(1/(x\!:\!\mathbb{Z}))\!:\!\mathbb{R}$ is an expression, even though its evaluation is undefined if $x = 0$.

---

[1] Operations like $x+y$ and $-x$ are simply convenient ways of writing function applications $plus(x,y)$ and $minus.x$.

For any type (or set) t and expression E, we define the expression $E \in t$ to be the value of "E is in t". For example, we might write

$$(8.2) \quad i \in \mathbb{N} \;\Rightarrow\; -i \le 0 \quad .$$

Thus, $E \in t$ is an expression, just like $x < y$, which is evaluated when the expression in which it appears is evaluated, while $x{:}t$ is simply the expression x annotated with its type. The connection between the syntactic annotation $E{:}t$ and the expression $E \in t$ can be expressed as follows:

> If E has type t, i.e. $E{:}t$, then $E \in t$ evaluates to *true* in all states in which E is well defined.

A language with syntactic rules that assign a type to each expression is called *strongly typed*. Pascal, Ada, and ML are strongly typed programming languages. Strong typing provides a measure of syntactic control, in two ways. First, as we will see, it frees us from having to place expressions $E \in t$ in various places within expressions. If the syntax indicates $E{:}t$, then $E \in t$ necessarily holds (if E is defined). Second, when the language is implemented, strong typing allows some errors to be detected early by a compiler, editor, or other software tool.

A language without syntactic typing rules is called *untyped*. Lisp, Scheme, and Prolog are untyped. In an untyped language, $-true$ is an expression, and the mistake in it is considered to be a semantic error, which is detected (if at all) only when the expression is evaluated. Many texts on logic deal only with untyped logics, in which case the only type available is the type consisting of all possible values. Since there is only one type, it is unnamed. In this case, heavy use of expressions $E \in t$ (for different sets t) is made.

With the notion of type, some restrictions are needed to ensure type correctness during manipulations:

(8.3)   In a textual substitution $E[x := F]$, x and F must have the same type.

(8.4)   Equality $b = c$ is defined only if b and c have the same types. That is, equality $=$ has type $t \times t \to \mathbb{B}$, for any type t.

Restriction (8.3) ensures that making a textual substitution does not produce a non-expression. Restriction (8.4) ensures that application of Leibniz or Substitution does not violate restriction (8.3).

A number of issues have been glossed over in this brief introduction to types. For example, the natural numbers $\mathbb{N}$ are a subset of the integers $\mathbb{Z}$, so $1{:}\mathbb{Z}$ and $1{:}\mathbb{N}$ are both suitable declarations. We obviously need a notion of subtypes, as well as a notion of *overloading* of both constants and operators, so that the same constants and operators can be used in more than one way. We also need a notion of *polymorphism*. As an example, function $= : t \times t \to bool$ is polymorphic because it is defined for any type t. We shall not delve into these issues, because that would detract from our current task, the study of quantification.

8.2   Syntax and interpretation of quantification

The reader is probably familiar with the following notation (in which expression e may refer to i).

(8.5) $\Sigma_{i=1}^{n} e$

Formula (8.5) stands for $e_1^i + e_2^i + \cdots + e_n^i$ —in words, for the sum of the values $e[i := v]$ for integers v in the range $1..n$. Here is an example.

$$\Sigma_{i=1}^{3} i^2 = 1^2 + 2^2 + 3^2$$

Henceforth, we use the linear notation

$$(\Sigma i \mid 1 \leq i \leq n : e) \quad \text{or} \quad (+i \mid 1 \leq i \leq n : e)$$

instead of (8.5), for several reasons:

- The parentheses in the linear notation make explicit the *scope* of the *dummy* or *quantified variable* i: the places where i can be referenced. This scope comprises the expressions within the parentheses. Note that Σi (or $+i$) acts as a declaration, introducing dummy i. This dummy is not a variable in the usual sense, for it does not obtain a value from the state in which the expression is evaluated.

- The linear notation makes it easier to write more general ranges for i. We can write *any* boolean expression to describe the values of i for which e should be summed. For example, using *even.i* for "i is even" and *odd.i* for "i is odd", we have

$$
\begin{aligned}
(+i \mid 1 \leq i \leq 7 \wedge \text{even}.i : i) &= 2 + 4 + 6 \quad, \\
(+i \mid 1 \leq i \leq 7 \wedge \text{odd}.i : 2{\cdot}i) &= 2{\cdot}1 + 2{\cdot}3 + 2{\cdot}5 + 2{\cdot}7 \quad.
\end{aligned}
$$

- The linear notation extends more easily to allow more than one dummy, as shown in the following example. In determining which values i^j are being summed, we choose all combinations of i and j that satisfy the *range* $1 \leq i \leq 2 \wedge 3 \leq j \leq 4$:

$$(+i, j \mid 1 \leq i \leq 2 \wedge 3 \leq j \leq 4 : i^j) = 1^3 + 1^4 + 2^3 + 2^4 \quad.$$

In the summations above, our intent is that dummy i ranges over integer values, rather than real values (say). However, the formulas do not tell us this. To make the type explicit, we can write $(+i{:}\mathbb{Z} \mid 1 \leq i \leq 2 : e)$.

What has been said about summation generalizes to other operators. Let $\star$ be any binary operator that is symmetric, is associative, and has an identity u (say):[2]

> **Symmetry:** $b \star c = c \star b$
>
> **Associativity:** $(b \star c) \star d = b \star (c \star d)$
>
> **Identity** $u:$ $u \star b = b = b \star u$

For example, for $\star$ and u, we could choose $+$ and 0, $\cdot$ and 1, $\wedge$ and $true$, or $\vee$ and $false$. The general form of a *quantification* over $\star$ is exemplified by

$$(8.6) \quad (\star x{:}t1, y{:}t2 \mid R : P)$$

where:

- Variables x and y are distinct. They are called the *bound variables* or *dummies* of the quantification. There may be one or more dummies.

- $t1$ and $t2$ are the types of dummies x and y. If $t1$ and $t2$ are the same type, we may write $(\star x, y{:}t1 \mid R : P)$. In the interest of brevity, we usually omit the type when it is obvious from the context, writing simply $(\star x, y \mid R : P)$.

- R, a boolean expression, is the *range* of the quantification —values assumed by x and y satisfy R. R may refer to dummies x and y. If the range is omitted, as in $(\star x \mid : P)$, then the range $true$ is meant.

- P, an expression, is the *body* of the quantification. P may refer to dummies x and y.

- The type of the result of the quantification is the type of P.

Expression $(\star x{:}X \mid R : P)$ denotes the application of operator $\star$ to the values P for all x in X for which range R is true[3].

---

[2] A set of values together with an operator $\star$ that is associative and has an identity is called a *monoid*. It is an *abelian* monoid, after Niels Henrik Abel (see Historical note 8.1), if $\star$ is also symmetric. Abelian monoids occur often in mathematics. The integers with operator $+$ and identity 0 forms an abelian monoid, as do the reals with operator $\cdot$ and identity 1.

[3] Later, we define this notation more formally by stating its properties as axioms, just as we did for the boolean operators $\equiv$, $\neg$, etc.

HISTORICAL NOTE 8.1. NIELS HENRIK ABEL (1802–1829)

Scientifically, we see Abel in this text only through abelian monoids. But he contributed enormously to mathematics, in spite of poverty and neglect by other mathematicians. Abel's father, a pastor, died when Abel was 18. Thereafter, he had to care for his mother and six siblings. He had help from a few who recognized his mathematical genius, but Norway was experiencing severe poverty, and life was difficult for all.

When about 21, Abel solved a problem that had confounded mathematicians for centuries: how to find the roots of $ax^5 + bx^4 + cx^3 + dx^2 + ex + f$ with a finite number of additions, subtractions, multiplications, divisions, and extractions of roots. Abel proved that the task was, in general, impossible! He used what little money he had to print the result himself.

Abel then received a small grant to travel in Europe. He hoped that talking to the great mathematicians would gain him entree into mathematical circles and provide him with a good position, but he was not well received. Gauss, for example, refused to read Abel's paper on the impossibility of solving quintic equations, believing Abel to be just another amateur.

On his two-year trip, Abel did have the good fortune to meet A.L. Crelle, who perceived Abel's greatness. Shortly thereafter, Crelle began publishing the first periodical in the world devoted exclusively to mathematical research, *Journal für die reine und angewandte Mathematik* (Journal for pure and applied mathematics). The first three volumes contained 22 of Abel's papers. Crelle showed Abel off and tried to get him a professorship in the University of Berlin, but to no avail, and Abel remained an outsider. His famous paper on transcendental functions, presented to the Paris Academy of Sciences when he was 24, was misplaced by Cauchy, almost lost, and only published 17 years later. Jacobi called it the most important mathematical discovery of the century.

Abel returned home from his trip, poor and sick with tuberculosis but still doing mathematics. In 1829, at the age of 26, he died. Two days later, a letter arrived from Crelle saying that Berlin was offering him a professorship after all. A year later, the Paris Academy of Sciences made some amends by awarding Abel the Grand Prize in Mathematics.

Here are examples of quantifications, assuming, as we do throughout this chapter, that i has type $\mathbb{Z}$.

$$
\begin{aligned}
(+i \mid 0 \leq i < 4 : i\cdot 8) &= 0\cdot 8 + 1\cdot 8 + 2\cdot 8 + 3\cdot 8 \\
(\cdot i \mid 0 \leq i < 3 : i + (i+1)) &= (0+1) \cdot (1+2) \cdot (2+3) \\
(\wedge i \mid 0 \leq i < 2 : i\cdot d \neq 6) &\equiv 0\cdot d \neq 6 \ \wedge\ 1\cdot d \neq 6 \\
(\vee i \mid 0 \leq i < 21 : b[i] = 0) &\equiv b[0] = 0 \ \vee\ \cdots\ \vee\ b[20] = 0
\end{aligned}
$$

Many notations are used for quantification. Different ways are used to express range R and body P, and, for operators $\vee$ and $\wedge$, the range is not given as a separate entity. For example, one sees the following.

$$\Sigma_{i=1}^n x_i \qquad\qquad \text{for} \quad (+i \mid 1 \le i \le n : x_i)$$
$$\forall i.1 \le i \ \Rightarrow\ x_i = 0 \qquad \text{for} \quad (\wedge i \mid 1 \le i : x_i = 0)$$
$$(\forall i)1 \le i \ \Rightarrow\ x_i = 0 \qquad \text{for} \quad (\wedge i \mid 1 \le i : x_i = 0)$$
$$\exists i . 1 \le i \wedge x_i = 0 \qquad \text{for} \quad (\vee i \mid 1 \le i : x_i = 0)$$

We use the linear notation $(\star x \mid R : P)$ throughout, for all quantifications, but we will bow to convention and use a different symbol for $\star$ in certain cases. In particular, in Chaps. 9 and 15 we write

$$
\begin{array}{lll}
(+x \mid R : P) & \text{as} & (\Sigma x \mid R : P) \\
(\cdot x \mid R : P) & \text{as} & (\Pi x \mid R : P) \\
(\vee x \mid R : P) & \text{as} & (\exists x \mid R : P) \\
(\wedge x \mid R : P) & \text{as} & (\forall x \mid R : P)
\end{array}
$$

SCOPE

The expression

(8.7) $(\wedge i \mid : x \cdot i = 0)$

asserts that x multiplied by any integer equals 0. This fact is true only if $x = 0$, so (8.7) is equivalent to the expression $x = 0$. Thus, the value of (8.7) in a state depends on the value of x in the state but *not* on the value of i. Further, it should be clear that the meaning of (8.7) does not change when dummy i is renamed:

$$(\wedge j \mid : x \cdot j = 0) \ = \ (\wedge i \mid : x \cdot i = 0) \ .$$

We introduce terminology to help distinguish the different roles played by i and x in (8.7). Occurrences of x in (8.7) are said to be *free*. The *scope* of dummy i, i.e. the places in which it can appear, is the range and the body of (8.7). All occurrences of i in the scope of dummy i are said to be *bound* to dummy i.

According to these definitions, all occurrences of a variable in an expression without quantifications are free. In Chap. 3, for example, every use of "variable" could be replaced by "free variable".

Now consider the expression

(8.8) $i > 0 \ \vee \ (\wedge i \mid 0 \le i : x \cdot i = 0) \ .$

The leftmost occurrence of i (i.e. the occurrence in $i > 0$) is free, and, during evaluation, it is replaced by the value of i in the state; the other occurrences of i are bound. Variable i is being used in two different ways in (8.8), each with different meaning. The first (i.e. free) occurrence of i refers to a different variable than do the other (i.e. bound) occurrences of

i within the quantification. The use of i with two different meanings in this fashion can be confusing, so we avoid it by renaming dummies.

The scope rules for a dummy are similar to the scope rules for a local variable of a procedure in an Algol-like (Pascal-like) language.[4] Consider the following Pascal procedure.

> **procedure** p (**var** x:*integer*);
>   **var** i:*integer*;
>   **begin** $i := x \cdot x$; $x := 2 \cdot i$ **end**

The scope of local variable i is the procedure body —the text between **begin** and **end**. Any occurrence of i outside the procedure body refers to an entirely different entity, which happens to have the same name. In the same way, the scope of i in $(\star i \mid R : P)$ is R and P. An occurrence of i outside this expression refers to an entirely different entity.

We now define *free* and *bound* occurrences of variables. Remember that it is an *occurrence* of a variable that is free or bound, not the variable itself. Remember also that infix and prefix operators are just forms of function application.

(8.9) **Definition.** The occurrence of i in the expression i is free.

> Suppose an occurrence of i in expression E is free. Then that same occurrence of i is free in (E), in function application $f(\ldots, E, \ldots)$, and in $(\star x \mid E : F)$ and $(\star x \mid F : E)$ provided i is not one of the dummies in list x.

> Define $occurs('v', 'e')$ to mean that at least one variable in the list v of variables occurs free in at least one expression in expression list e.

(8.10) **Definition.** Let an occurrence of i be free in an expression E. That occurrence of i is *bound* (to dummy i) in the expression $(\star x \mid E : F)$ or $(\star x \mid F : E)$ if i is one of the dummies in list x.

> Suppose an occurrence of i is bound in expression E. Then it is also bound (to the same dummy) in (E), $f(\ldots, E, \ldots)$, $(\star x \mid E : F)$ and $(\star x \mid F : E)$.

As an example, consider the expression

$$i + j + (\Sigma i \mid 1 \leq i \leq 10 : b[i]^j) +$$
$$(\Sigma i \mid 1 \leq i \leq 10 : (\Sigma j \mid 1 \leq j \leq 10 : c[i,j])) .$$

---

[4] Algol 60 was the first programming language to make full use of the scope rules defined here, but such scope rules were used in logic long before 1960.

We have the following:

- The leftmost occurrence of i is free in this expression.

- The leftmost occurrence of j and the occurrence of j as an exponent in the first summation are both free.

- All other occurrences of i and j are bound. There are two different dummies i.

TEXTUAL SUBSTITUTION REVISITED

Textual substitution $E[x := F]$ was defined in Sec. 1.2 for E a constant, variable, or function application. We now extend this definition to cover quantification:

(8.11) Provided $\neg occurs(`y`, `x, F`)$,

$$(\star y \mid R : P)[x := F] = (\star y \mid R[x := F] : P[x := F]) \quad .$$

The caveat in (8.11) means that a dummy of list y will have to be replaced by a fresh variable [5] if that dummy occurs free in x or F.

Here are some examples of textual substitution in quantifications.

$$(+x \mid 1 \le x \le 2 : y)[y := y + z] = (+x \mid 1 \le x \le 2 : y + z)$$
$$(+i \mid 0 \le i < n : b[i] = n)[n := m] = (+i \mid 0 \le i < m : b[i] = m)$$
$$(+y \mid 0 \le y < n : b[y] = n)[n := y] = (+j \mid 0 \le j < y : b[j] = y)$$
$$(+y \mid 0 \le y < n : b[y] = n)[y := m] = (+j \mid 0 \le j < n : b[j] = n)$$

In the last two examples, dummy y was first replaced by fresh variable j, as required by the caveat. Changing the dummy ensures that a free occurrence of y in the textual substitution $x := F$ does not become bound.

8.3   Rules about quantification

Consider a language of expressions that includes the operator $\star$. Assume that $\star$ is symmetric and associative and has an identity u. We introduce two inference rules and several axioms that can be used, along with infer-

---

[5] A *fresh variable* is a variable that does not occur in the expressions under consideration.

ence rules Leibniz, Transitivity of equality, and Substitution, to manipulate quantifications over $\star$.[6]

Inference rule Leibniz, (1.5), is supposed to enable substitution of equals for equals in expressions, and for expressions without quantification it is fine. However, for substitutions in quantifications, it is inadequate. For example, since $x + x = 2 \cdot x$ holds, we would expect to be able to prove

$$(+x \mid 0 \leq x < 9 : x + x) = (+x \mid 0 \leq x < 9 : 2 \cdot x) \quad .$$

However, the instance of Leibniz (1.5) that we want to use in this case,

$$\frac{x + x = 2 \cdot x}{(+x \mid 0 \leq x < 9 : z)^z_{x+x} = (+x \mid 0 \leq x < 9 : z)^z_{2 \cdot x}} \quad ,$$

does not work. This is because variable x in the replacing expression $x + x$ is (deliberately) the same as the dummy, so $(+x \mid 0 \leq x < 9 : z)[z := x + x]$ equals $(+y \mid 0 \leq y < 9 : x + x)$ and not $(+x \mid 0 \leq x < 9 : x + x)$, due to the caveat in (8.11).

Two additional inferences rules allow substitution of equals for equals in the range and body of a quantification.

(8.12) **Leibniz:**
$$\frac{P = Q}{(\star x \mid E[z := P] : S) = (\star x \mid E[z := Q] : S)}$$

$$\frac{R \;\Rightarrow\; P = Q}{(\star x \mid R : E[z := P]) = (\star x \mid R : E[z := Q])}$$

As with Leibniz (1.5), we use these inference rules implicitly in substituting equals for equals.

Our first two axioms concern the introduction and elimination of quantifiers. The notation $occurs(\text{`}x\text{'}, \text{`}E\text{'})$ is explained in Def. (8.9).

(8.13) **Axiom, Empty range:** $(\star x \mid false : P) = u$ (the identity of $\star$)

(8.14) **Axiom, One-point rule:** Provided $\neg occurs(\text{`}x\text{'}, \text{`}E\text{'})$,
$$(\star x \mid x = E : P) = P[x := E]$$

---

[6] Actually, most of the axioms require only symmetry and associativity. An identity is required only when an empty range comes into play. For example, if $\star$ does not have an identity, then One-point rule (8.14) does not hold, and axiom Range split (8.16) does not hold if R or S is *false*. Operator $\downarrow$, where $x \downarrow y$ is the minimum of x and y, is an example of an operator for which we can still use quantification, even though $\downarrow$ has no identity.

As an example of the One-point rule, we have

$$(+x \mid x = 3 : x^2) \;=\; 3^2 \quad .$$

We explain the need for the restriction in the One-point rule that x not occur free in E. The LHS of the One-point rule is not dependent on x (in the state in which it is evaluated), since all occurrences of x are bound. Hence, for the equivalence to hold, the RHS also cannot depend on x, and this requires (in general) that x not occur free in E.

(8.15) **Axiom, Distributivity:** Provided each quantification is defined, [7]
$$(\star x \mid R : P) \star (\star x \mid R : Q) \;=\; (\star x \mid R : P \star Q) \quad .$$

Note that the dummies are the same and the ranges are the same in all three quantifications of (8.15). Distributivity holds because operator $\star$ is symmetric and associative, so that the order in which the operands are accumulated has no bearing on the result. As an example, for dummy i of type integer, we have

$$(+i \mid i^2 < 9 : i^2) + (+i \mid i^2 < 9 : i^3) = (+i \mid i^2 < 9 : i^2 + i^3) \quad .$$

The next axiom is called *range split*, because the range $R \vee S$ in its LHS is split into the two ranges R and S in its RHS.

(8.16) **Axiom, Range split:** Provided $R \wedge S \equiv \textit{false}$ and
each quantification is defined,
$$(\star x \mid R \vee S : P) \;=\; (\star x \mid R : P) \star (\star x \mid S : P)$$

Axiom (8.16) may be understood using the following analogy. Suppose one has a bag of Red numbers and $Silver$ numbers to sum. They can be summed in any order, as the LHS of (8.16) implies. The RHS simply specifies a bit about the ordering of summation: sum the Red ones, sum the $Silver$ ones, and add the two sums.

---

[7] The sum $(+i \mid 0 < i : i) = 1 + 2 + 3 + \ldots$ is not defined. Using $0 = i + (-i)$, we have the following instance of Axiom (8.15). Its LHS is 0 but its RHS is undefined.

$$(+i \mid 0 < i : 0) = (+i \mid 0 < i : i) + (+i \mid 0 < i : -i)$$

This is the reason for the caveat on some of the axioms.

The sum $(+i \mid 0 < i : 1/i^2) = 1/1 + 1/4 + 1/8 + 1/16 + \ldots$ is defined to equal $\pi^2/6$, even though it is an infinite sum, because $(+i \mid 0 < i \le n : 1/i^2)$ "converges" to $\pi^2/6$ as n gets larger. Similarly, $(\equiv i \mid 0 < i : \textit{true})$ is defined to equal $\textit{true}$, since $(\equiv i \mid 0 < i \le n : \textit{true}) \equiv \textit{true}$ for all n. But $(\equiv i \mid 0 < i : \textit{false})$ is undefined (why?).

A complete discussion of when a quantification is defined is outside the scope of this text. Entire books are written on the subject of convergence of summations. However, quantifications with finite ranges are always defined, and quantifications using operator $\wedge$ and $\vee$ are always defined.

The restriction that $R \wedge S \equiv \textit{false}$ in the above axiom ensures that an operand is not accumulated twice in the RHS —once because a value x satisfies R and once because the same value x satisfies S. Axiom (8.17) eliminates this restriction by adding to the LHS the accumulation $(\star x \mid R \wedge S : P)$; thus, the values of P that are accumulated twice because values for the dummies satisfy both R and S are accumulated twice on both sides of the equation.

(8.17) **Axiom, Range split:** Provided each quantification is defined,

$$(\star x \mid R \vee S : P) \star (\star x \mid R \wedge S : P) = (\star x \mid R : P) \star (\star x \mid S : P)$$

On the other hand, if operator $\star$ is idempotent —so that $e \star e = e$ for all e, then it does not matter how many times e is accumulated. Hence, we have the theorem

(8.18) **Axiom, Range split for idempotent $\star$:** Provided each quantification is defined,

$$(\star x \mid R \vee S : P) = (\star x \mid R : P) \star (\star x \mid S : P)$$

The next three axioms concern dummies. The first indicates that nested quantifications with the same operator can be interchanged. The second indicates how a single quantification over a list of dummies can be viewed as a nested quantification. The third shows that a dummy can be replaced (in a consistent fashion) by any fresh dummy.

(8.19) **Axiom, Interchange of dummies:** Provided each quantification is defined, $\neg occurs(\text{'}y\text{'}, \text{'}R\text{'})$ and $\neg occurs(\text{'}x\text{'}, \text{'}Q\text{'})$,

$$(\star x \mid R : (\star y \mid Q : P)) = (\star y \mid Q : (\star x \mid R : P))$$

(8.20) **Axiom, Nesting:** Provided $\neg occurs(\text{'}y\text{'}, \text{'}R\text{'})$,

$$(\star x, y \mid R \wedge Q : P) = (\star x \mid R : (\star y \mid Q : P))$$

(8.21) **Axiom, Dummy renaming:** Provided $\neg occurs(\text{'}y\text{'}, \text{'}R, P\text{'})$,

$$(\star x \mid R : P) = (\star y \mid R[x := y] : P[x := y])$$

The "occurs" restrictions on these laws ensure that an expression that contains an occurrence of a dummy is not moved outside (or inside) the scope of that dummy.

We now generalize axiom Dummy renaming (8.21). We motivate this generalization as follows. Consider the expression

$$(+i \mid 2 \leq i \leq 10 : i^2) \quad .$$

Rewriting this expression so that the range starts at 0 instead of 2 yields the following expression.

$$(+k \mid 0 \leq k \leq 8 : (k+2)^2)$$

Here, note that the relationship between i and k is $i = k+2$, or $k = i-2$.

The equality of the two summations above is an instance of the following general theorem, which holds for *any* symmetric and associative binary operator $\star$. Let f be a function that has an inverse f^{-1}, so that $x = f.y \equiv y = f^{-1}.x$. Then

(8.22) **Change of dummy:** Provided $\neg occurs('y', 'R, P')$ and

  f has an inverse,

$$(\star x \mid R : P) = (\star y \mid R[x := f.y] : P[x := f.y])$$

The proof of this theorem illustrates the use of several of the axioms given above. The proof starts with the RHS of (8.22), because it has more structure.

$$(\star y \mid R[x := f.y] : P[x := f.y])$$
$=$ ⟨One-point rule (8.14)
        —Quantification over x has to be introduced. The One-point rule is the *only* rule that can be applied at first.⟩
$$(\star y \mid R[x := f.y] : (\star x \mid x = f.y : P))$$
$=$ ⟨Nesting (8.20) —Moving dummy x to the outside gets us closer to the final form.⟩
$$(\star x, y \mid R[x := f.y] \wedge x = f.y : P)$$
$=$ ⟨Substitution (3.84a) — $R[x := f.y]$ must be removed at some point. This substitution makes it possible.⟩
$$(\star x, y \mid R[x := x] \wedge x = f.y : P)$$
$=$ ⟨ $R[x := x] \equiv R$; Nesting (8.20), $\neg occurs('y', 'R')$
        —Now we can get a quantification in x alone.⟩
$$(\star x \mid R : (\star y \mid x = f.y : P))$$
$=$ ⟨ $x = f.y \equiv y = f^{-1}.x$ —This step prepares for the elimination of y using the One-point rule.⟩
$$(\star x \mid R : (\star y \mid y = f^{-1}.x : P))$$
$=$ ⟨One-point rule (8.14)⟩
$$(\star x \mid R : P[y := f^{-1}.x])$$
$=$ ⟨Textual substitution — $\neg occurs('y', 'P')$ ⟩
$$(\star x \mid R : P)$$

Discovering this proof is not as difficult as it may appear at first, because each step is almost *forced* by the shape of the expression at that point and the shape of the final goal —in fact, in several of the steps there *is* only one choice. The proof changes the side with the most structure into the side with the least structure, as per heuristic (3.33).

8.4 Manipulating ranges

We now illustrate the manipulation of ranges in quantifications, in order to show the application of the axioms introduced thus far and to prepare for later application in proving theorems by induction and in proving properties of programs.

In dealing with quantifications with ranges like $0 \le i < n$ and $0 \le i \le n$, we often want to split the quantification into two quantifications, using Range split (8.16). Two useful cases of this splitting are given in the following theorem.

(8.23) **Theorem Split off term.** For $n : \mathbb{N}$ and dummies $i : \mathbb{N}$,

$$(\star i \mid 0 \le i < n+1 : P) = (\star i \mid 0 \le i < n : P) \star P[i := n]$$
$$(\star i \mid 0 \le i < n+1 : P) = P[i := 0] \star (\star i \mid 0 < i < n+1 : P).$$

Proof. We prove the first formula and leave the second to the reader.

$$
\begin{aligned}
& (\star i \mid 0 \le i < n+1 : P) \\
= \quad & \langle\, 0 \le i < n+1 \;\equiv\; 0 \le i < n \;\vee\; i = n \,\rangle \\
& (\star i \mid 0 \le i < n \;\vee\; i = n : P) \\
= \quad & \langle \text{Range split (8.16)} - 0 \le i < n \;\wedge\; i = n \;\equiv\; \textit{false} \,\rangle \\
& (\star i \mid 0 \le i < n : P) \star (\star i \mid i = n : P) \\
= \quad & \langle \text{One-point rule (8.14)} \rangle \\
& (\star i \mid 0 \le i < n : P) \star P[i := n] \qquad\qquad\qquad \square
\end{aligned}
$$

Here are some examples of the use of Split off term (8.23). In the third example, the range has been written as $0 \le i \le n$ instead of $0 \le i < n+1$. And, in the fourth example, we use the obvious extension of the theorem to a lower bound other than 0.

$$(\Sigma i \mid 0 \le i < n+1 : b[i]) = (\Sigma i \mid 0 \le i < n : b[i]) + b[n]$$

$$(\Pi i \mid 0 \le i < n+1 : b[i]) = b[0] \cdot (\Pi i \mid 0 < i < n : b[i])$$

$$(\forall i \mid 0 \le i \le n : b[i] = 0) = (\forall i \mid 0 \le i < n : b[i] = 0) \wedge b[n] = 0$$

$$(\Pi i \mid 5 \le i \le 10 : i^2) = 5^2 \cdot (\Pi i \mid 5 < i \le 10 : i^2)$$

In splitting a range into two, we are actually making use of the following theorem. Its proof awaits the introduction of axioms for arithmetic, in Chap. 15 (see Exercise 15.39).

(8.24) $b \le c \le d \;\Rightarrow\; (b \le i < d \;\equiv\; b \le i < c \;\vee\; c \le i < d)$

We will use this theorem in the following, more complex, example, which concerns the sum of a certain set of elements of a two-dimensional array

$c[0..n, 0..n]$:

(8.25) $(\Sigma i, j \mid 0 \leq i \leq j < n+1 : c[i,j])$.

We want to prove that this expression is equivalent to

$$(\Sigma i, j \mid 0 \leq i \leq j < n : c[i,j]) + (\Sigma i \mid 0 \leq i \leq n : c[i,n])$$.

The proof requires splitting the range of the quantification, and to do this we rewrite the range $0 \leq i \leq j < n+1$ as a disjoint disjunction. The way to deal with this rewriting is to remember that the range uses an abbreviation:

$$
\begin{aligned}
& 0 \leq i \leq j < n+1 \\
= \quad & \langle \text{Remove abbreviation} \rangle \\
& 0 \leq i \leq j \wedge j < n+1 \\
= \quad & \langle j < n+1 \equiv j < n \vee j = n \rangle \\
& 0 \leq i \leq j \wedge (j < n \vee j = n) \\
= \quad & \langle \text{Distributivity of } \wedge \text{ over } \vee \text{ (3.46)} \rangle \\
& (0 \leq i \leq j \wedge j < n) \vee (0 \leq i \leq j \wedge j = n) \\
= \quad & \langle \text{Reintroduce abbreviation} \rangle \\
& 0 \leq i \leq j < n \vee (0 \leq i \leq j \wedge j = n)
\end{aligned}
$$

Using the last formula, we can now manipulate (8.25) as follows.

$$
\begin{aligned}
& (\Sigma i, j \mid 0 \leq i \leq j < n+1 : c[i,j]) \\
= \quad & \langle \text{Above proof} \rangle \\
& (\Sigma i, j \mid 0 \leq i \leq j < n \vee (0 \leq i \leq j \wedge j = n) : c[i,j]) \\
= \quad & \langle \text{Range split (8.16)} \rangle \\
& (\Sigma i, j \mid 0 \leq i \leq j < n : c[i,j]) \\
& + (\Sigma i, j \mid 0 \leq i \leq j \wedge j = n : c[i,j])
\end{aligned}
$$

Now, the One-point rule and the conjunct $j = n$ lead us to believe that dummy j can be removed from the second summation. We continue the manipulation:

$$
\begin{aligned}
= \quad & \langle \text{Nesting (8.20)} \rangle \\
& (\Sigma i, j \mid 0 \leq i \leq j < n : c[i,j]) \\
& + (\Sigma j \mid j = n : (\Sigma i \mid 0 \leq i \leq j : c[i,j])) \\
= \quad & \langle \text{One-point rule (8.14)} \rangle \\
& (\Sigma i, j \mid 0 \leq i \leq j < n : c[i,j]) + (\Sigma i \mid 0 \leq i \leq n : c[i,n])
\end{aligned}
$$

The manipulation to show that (8.25) equals the expression following it seems rather torturous. This is because we wanted to show you every detail. With some experience, you will be able to perform this manipulation is one step:

$$(\Sigma\, i, j \mid 0 \le i \le j < n + 1 : c[i, j])$$
$$=\quad \langle \text{Range split (8.16); One-point rule (8.14)} \rangle$$
$$(\Sigma\, i, j \mid 0 \le i \le j < n : c[i, j]) \; + \; (\Sigma\, i \mid 0 \le i \le n : c[i, n])$$

Describing ranges

Consider ways to formalize a range that denotes the values $2, \ldots, 15$ of a dummy i. Here are four possibilities.

(a) $2 \le i \le 15$
(b) $2 \le i < 16$
(c) $1 < i \le 15$
(d) $1 < i < 16$

Which is best? Well, that depends on the kinds of manipulations being performed on ranges. One nice point about (b) and (c) is that the number of values of i in the range is equal to the upper bound minus the lower bound: $16 - 2$ or $15 - 1$. With (a) and (d), the number of elements is not so easily calculated. This should bias us towards (b) or (c).

Another operation that is sometimes performed on ranges is to collapse adjacent ranges (or split a range into two adjacent ones):

(a) $2 \le i \le 15 \ \lor \ 16 \le i \le 20 \ \equiv \ 2 \le i \le 20$
(b) $2 \le i < 16 \ \lor \ 16 \le i < 21 \ \equiv \ 2 \le i < 21$
(c) $1 < i \le 15 \ \lor \ 15 < i \le 20 \ \equiv \ 1 < i \le 20$
(d) $1 < i < 16 \ \lor \ 15 < i < 21 \ \equiv \ 1 < i < 21$

Again, (b) and (c) seem easiest to manipulate, because the upper bound of the lower adjacent range equals the lower bound of the upper adjacent range. Collapsing or splitting such ranges is likely to be done with less chance of a mistake. So, all other things being equal, we usually try to use (b) or (c) to describe ranges of integers.

Which of (b) and (c) should we prefer? We often want to describe a range consisting of the first n natural numbers. Using (b), this is easily done: $0 \le i < n$, and the upper bound n is the number of values in the range. Using (c), we are forced to use the unnatural number -1 and to write the range as $-1 < i \le n - 1$. Hence, (b) would appear to be the better choice.

Experiments performed at Xerox PARC concerning the number of errors programmers made using the four forms of range (a)–(d) found that programmers made fewer errors with form (b).

Exercises for Chapter 8

8.1 Given are functions a, b, c, d, and e with types as follows.

$a : A \to B$

$b : B \to C$

$c : C \to A$

$d : A \times C \to D$

$e : B \times B \to E$

State whether each expression below is type correct. If not, explain why. Assume $u{:}A$, $w{:}B$, $x{:}C$, $y{:}D$, and $z{:}E$.

(a) $e(a.u, w)$

(b) $b.x$

(c) $e(a(c.x), a.u)$

(d) $a(c(b(a.y)))$

(e) $d(c.x, c.x)$

8.2 Consider the expression $e \in t$ where t has type $set(\mathbb{Z})$, i.e. set of integers. Give a reasonable type for function $\in$ and for function application $e \in t$.

8.3 Expand the following textual substitutions. If necessary, change the dummy, according to Dummy Renaming (8.21).

(a) $(\star x \mid 0 \le x + r < n : x + v)[v := 3]$

(b) $(\star x \mid 0 \le x + r < n : x + v)[x := 3]$

(c) $(\star x \mid 0 \le x + r < n : x + v)[n := n + x]$

(d) $(\star x \mid 0 \le x < r : (\star y \mid 0 \le y : x + y + n))[n := x + y]$

(e) $(\star x \mid 0 \le x < r : (\star y \mid 0 \le y : x + y + n))[r := y]$

8.4 Give a definition of $E[x := e]$ for all expressions E, including quantifications. The definition should be in terms of the different kinds of expression E, just as the notions of *free* and *bound* were defined. Treat expressions that are constants, variables, parenthesized expressions, unary operations, binary operations, function applications, and quantifications.

8.5 Prove the following theorems. Provided $0 \le n$,

(a) $(\Sigma i \mid 0 \le i < n + 1 : b[i]) = b[0] + (\Sigma i \mid 1 \le i < n + 1 : b[i])$

(b) $(\Sigma i \mid 0 \le i \le n : b[i]) = (\Sigma i \mid 0 \le i < n : b[i]) + b[n]$

(c) $(\Sigma i \mid 0 \le i \le n : b[i]) = b[0] + (\Sigma i \mid 1 \le i \le n : b[i])$

8.6 Prove the following theorems. Provided $0 \le n$,

(a) $(\vee i \mid 0 \le i < n + 1 : b[i] = 0) \equiv$
$(\vee i \mid 0 \le i < n : b[i] = 0) \vee b[n] = 0$

(b) $(\wedge i \mid 0 \le i < n + 1 : b[i] = 0) \equiv$
$(\wedge i \mid 0 \le i < n : b[i] = 0) \wedge b[n] = 0$

(c) $(\vee i \mid 0 \le i < n + 1 : b[i] = 0) \equiv$
$b[0] = 0 \vee (\vee i \mid 0 < i < n + 1 : b[i] = 0)$

(d) $(\wedge i \mid 0 \le i < n + 1 : b[i] = 0) \equiv$
$b[0] = 0 \wedge (\wedge i \mid 0 < i < n + 1 : b[i] = 0)$

8.7 Prove the following theorems:

(a) $(+i \mid 0 \leq i \leq n : i) = (+i \mid 0 \leq i \leq n \land even.i : i) + (+i \mid 0 \leq i \leq n \land odd.i : i)$

(b) $(+i \mid 0 \leq i \leq 10 : 0) = 0$

Chapter 9

Predicate Calculus

$\mathbb{W}$e introduce *predicate logic*, an extension of propositional logic that allows the use of variables of types other than $\mathbb{B}$. This extension leads to a logic with enhanced expressive and deductive power.

PREDICATES AND PREDICATE CALCULUS

Propositional calculus permits reasoning about formulas constructed from boolean variables and boolean operators. Therefore, the expressiveness of the logic is restricted to sentences that can be modeled using boolean expressions. Predicate calculus permits reasoning about a more expressive class of formulas. A predicate-calculus formula is a boolean expression in which some boolean variables may have been replaced by:

- *Predicates*, which are applications of boolean functions whose arguments may be of types other than $\mathbb{B}$. Examples of predicates are $equal(x, x - z + z)$ and $less(x, y + z)$. The function names (e.g. *equal*, *less*) are called *predicate symbols*. Infix notation is sometimes used for predicates, as in $x < y$.

  The arguments of predicates can be expressions having types other than $\mathbb{B}$ (e.g. the integers $\mathbb{Z}$), so arguments may contain variables and constants of these other types. These arguments are called *terms*. Examples of terms are: $x + y$, $max(a, b)$, and $-b + \sqrt{b^2 - 4 \cdot a \cdot c}$.

- Universal and existential quantification, as discussed in this chapter.

Here is a formula of the predicate calculus: $x < y \land x = z \Rightarrow q(x, z + x)$. It contains three predicates: $x < y$, $x = z$, and $q(x, z + x)$. The terms used in this formula are x, y, z, and $z + x$.

The pure predicate calculus includes the axioms of propositional calculus, together with axioms for quantifications $(\land x \mid R : P)$ and $(\lor x \mid R : P)$, which are introduced in the next two sections. The inference rules of the predicate calculus are Substitution (1.1), Transitivity (1.4), Leibniz (1.5), and Leibniz for quantification (8.12). Substitution may be used to replace a variable of any type by any expression of that type.

In the pure predicate calculus, the function symbols are *uninterpreted* (except for equality $=$), so the logic provides no specific rules for manipulating them. With these symbols uninterpreted, we can develop general rules for manipulation that are sound no matter what meanings we ascribe to the function symbols. Thus, the pure predicate calculus is sound in all domains that may be of interest.

We get a *theory* by adding axioms that give meanings to some of the (uninterpreted) function symbols. For example, the *theory of integers* consists of the pure predicate calculus together with axioms for manipulating the operators (i.e. functions) $+$, $-$, $\cdot$, $<$, $\leq$, etc. Thus, the axioms say that $\cdot$ is symmetric and associative and has the zero 0. And, the *theory of sets* provides axioms for manipulating expressions containing operators like $\in$ (membership), $\cup$ (union), and $\cap$ (intersection). We can also form a joint theory of sets and integers, allowing us to reason about expressions that contain both.

The core of all these theories, however, is the pure predicate calculus; it provides the basic machinery for reasoning about, or providing proofs about, all other domains of interest.

9.1   Universal quantification

Conjunction $\wedge$ is symmetric and associative and has the identity *true*. Therefore, it is an instance of $\star$ of the previous chapter. The quantification $(\wedge x \mid R : P)$ is conventionally written as

(9.1) $(\forall x \mid R : P)$.

The symbol $\forall$, which is read as "for all", is called the *universal* quantifier. Expression (9.1) is called a *universal quantification* and is read as "for all x such that R holds, P holds."

General axioms (8.13)–(8.21) hold for $(\forall x \mid R : P)$ and are not repeated here. Note that $\wedge$ is idempotent, so that universal quantification satisfies range-split axiom (8.18). We now introduce additional axioms and theorems for universal quantification.

TRADING WITH UNIVERSAL QUANTIFICATION

Axiom (9.2) allows a range to be moved into the body.

(9.2)   **Axiom, Trading:** $(\forall x \mid R : P) \equiv (\forall x \mid: R \Rightarrow P)$

This axiom allows us to prove several theorems for universal quantification.

Trading theorems for ∀

(9.3)   **Trading:** (a) $(\forall x \mid R : P) \equiv (\forall x \mid: \neg R \vee P)$

                 (b) $(\forall x \mid R : P) \equiv (\forall x \mid: R \wedge P \equiv R)$

                 (c) $(\forall x \mid R : P) \equiv (\forall x \mid: R \vee P \equiv P)$

(9.4)   **Trading:** (a) $(\forall x \mid Q \wedge R : P) \equiv (\forall x \mid Q : R \Rightarrow P)$

                 (b) $(\forall x \mid Q \wedge R : P) \equiv (\forall x \mid Q : \neg R \vee P)$

                 (c) $(\forall x \mid Q \wedge R : P) \equiv (\forall x \mid Q : R \wedge P \equiv R)$

                 (d) $(\forall x \mid Q \wedge R : P) \equiv (\forall x \mid Q : R \vee P \equiv P)$

We prove (9.4a).

$$(\forall x \mid Q \wedge R : P)$$
$$= \quad \langle \text{Trading (9.2)} \rangle$$
$$(\forall x \mid: Q \wedge R \Rightarrow P)$$
$$= \quad \langle \text{Shunting (3.65)} \rangle$$
$$(\forall x \mid: Q \Rightarrow (R \Rightarrow P))$$
$$= \quad \langle \text{Trading (9.2)} \rangle$$
$$(\forall x \mid Q : R \Rightarrow P)$$

DISTRIBUTIVITY WITH UNIVERSAL QUANTIFICATION

The following axiom shows how ∨ distributes over ∀.

(9.5)   **Axiom, Distributivity of ∨ over ∀:**
    Provided $\neg occurs(\text{`}x\text{'}, \text{`}P\text{'})$,

$$P \vee (\forall x \mid R : Q) \equiv (\forall x \mid R : P \vee Q)$$

In the axiom, the expression P that is being moved out of the scope (or into it, depending on your point of view) cannot contain x as a free variable. This restriction ensures that the LHS and the RHS of the axiom refer to the same free variables —otherwise, the LHS and RHS would, in general, not be equivalent.

Axiom (9.5) allows us to prove the following theorems.

Additional theorems for ∀

(9.6)   Provided $\neg occurs(\text{`}x\text{'}, \text{`}P\text{'})$,

    $(\forall x \mid R : P) \equiv P \vee (\forall x \mid: \neg R)$

Additional theorems for ∀ (continued)

(9.7) **Distributivity of ∧ over ∀:** Provided ¬*occurs*('*x*', '*P*'),

$$\neg(\forall x \mid: \neg R) \Rightarrow ((\forall x \mid R : P \land Q) \equiv P \land (\forall x \mid R : Q))$$

(9.8) $(\forall x \mid R : true) \equiv true$

(9.9) $(\forall x \mid R : P \equiv Q) \Rightarrow ((\forall x \mid R : P) \equiv (\forall x \mid R : Q))$

Be careful when using theorem (9.7). A conjunct can be moved outside the scope of the quantification only if the range R is not everywhere *false* (as prescribed by the antecedent ¬(∀*x* |: ¬*R*)). The proof of (9.7) uses the technique of assuming the antecedent (see page 71). We assume the antecedent ¬(∀*x* |: ¬*R*) and prove the consequent:

$(\forall x \mid R : P \land Q)$
= ⟨Distributivity of ∀ over ∧ (8.15)⟩
$(\forall x \mid R : P) \land (\forall x \mid R : Q)$
= ⟨(9.6) —since ¬*occurs*('*x*', '*P*')⟩
$(P \lor (\forall x \mid: \neg R)) \land (\forall x \mid R : Q)$
= ⟨Assumption ¬(∀*x* |: ¬*R*), i.e. (∀*x* |: ¬*R*) ≡ *false*⟩
$(P \lor false) \land (\forall x \mid R : Q)$
= ⟨Identity of ∨ (3.30)⟩
$P \land (\forall x \mid R : Q)$

MANIPULATING THE RANGE AND BODY WITH UNIVERSAL QUANTIFICATION

Theorems (9.10) and (9.11) have counterparts (3.76a) and (3.76b), with similar names, in propositional calculus; indeed, these two theorems are proved using their counterparts.

Weakening, strengthening, and monotonicity for ∀

(9.10) **Range weakening/strengthening:**
$$(\forall x \mid Q \lor R : P) \Rightarrow (\forall x \mid Q : P)$$

(9.11) **Body weakening/strengthening:**
$$(\forall x \mid R : P \land Q) \Rightarrow (\forall x \mid R : P)$$

(9.12) **Monotonicity of ∀:**
$$(\forall x \mid R : Q \Rightarrow P) \Rightarrow ((\forall x \mid R : Q) \Rightarrow (\forall x \mid R : P))$$

Instantiation with Universal Quantification

In many predicate logics, the following law of Instantiation (9.13) is written as an inference rule, and One-point rule (8.14), particularized for universal quantification as $(\forall x \mid x = E : P) = P[x := E]$, does not appear. However, the One-point rule is sharper than Instantiation —it is sharper to replace an expression by an equivalent one than by one that it implies, just as it is sharper to replace an integer expression by one equal to it rather than by one that is greater than or equal to it.

Instantiation for $\forall$

(9.13) **Instantiation:** $(\forall x \mid: P) \Rightarrow P[x := E]$

Nevertheless, there are many situations where Instantiation is useful, and, like symmetry and associativity, it is often used implicitly. For example, suppose we want to prove $B \lor even(x + y) \equiv B \lor even((x + y)^2)$ for integer expression $x + y$. Assuming that

(9.14) $(\forall i : \mathbb{Z} \mid: even.i \equiv even(i^2))$

holds, we would first use Instantiation (9.13) with (9.14) to infer $even(x + y) \equiv even((x + y)^2)$. Then we would give the following proof:

$$
\begin{array}{ll}
& B \lor even(x + y) \\
= & \langle\, even(x + y) \equiv even((x + y)^2) \; \text{—(9.14) instantiated} \\
& \quad \text{with } i := x + y \,\rangle \\
& B \lor even((x + y)^2)
\end{array}
$$

However, we typically take a short cut and simply write

$$
\begin{array}{ll}
& B \lor even(x + y) \\
= & \langle(9.14)\rangle \\
& B \lor even((x + y)^2)
\end{array}
$$

The implicit use of Instantiation is even more concealed if universal quantification itself is not written formally. For example, conventionally,

$$(\forall a, b : \mathbb{Z} \mid: a + b = b + a)$$

may be written as

(9.15) $a + b = b + a$ (for all integers a, b) .

In this form, because universal quantification is a side comment and not part of the formula, it is easy to forget that producing, say, $x \cdot y + z = z + x \cdot y$ from (9.15) requires not Substitution (1.1) but Instantiation (9.13).

ON THEOREMS AND UNIVERSAL QUANTIFICATION

A boolean expression that has free occurrences of variables, like $b \lor x < y$, is called *open*, and its value may differ from state to state. The expression becomes *closed* if we universally quantify over all of its free variables, as in $(\forall b, x, y \,|: b \lor x < y)$. The value of a closed expression does not depend on the state in which it is evaluated, since it has no free variables. Therefore, a closed expression is equivalent either to *true* or to *false*. The following metatheorem characterizes (at least partially) when quantifying over a variable does not change the value of a boolean expression.

(9.16) **Metatheorem.** P is a theorem iff $(\forall x \,|: P)$ is a theorem.

Proof. The proof is by mutual implication.

LHS $\Rightarrow$ RHS. Assume P is a theorem. Then there is a proof of it that transforms P to *true*, using Leibniz, Transitivity of equals, and Substitution:

$$\begin{array}{ll} & P \\ = & \langle \text{Hint } 1 \rangle \\ & \cdots \\ \cdots \\ & \cdots \\ = & \langle \text{Hint } n \rangle \\ & true \end{array}$$

Leibniz (8.12) allows us to turn this proof into a proof of $(\forall x \,|: P)$:

$$\begin{array}{ll} & (\forall x \,|: P) \\ = & \langle \text{Hint } 1 \rangle \\ & \cdots \\ \cdots \\ & \cdots \\ = & \langle \text{Hint } n \rangle \\ & (\forall x \,|: true) \\ = & \langle (9.8),\ (\forall x \mid R : true) \equiv true \rangle \\ & true \end{array}$$

RHS $\Rightarrow$ LHS. Assume $(\forall x \,|: P)$ is a theorem. Using Instantiation (9.13) with x for E, we conclude that P is a theorem. $\square$

Here are some applications of Metatheorem (9.16). Since $p \lor q \equiv q \lor p$ is a theorem, then so are $(\forall p \,|: p \lor q \equiv q \lor p)$, $(\forall q \,|: p \lor q \equiv q \lor p)$, and $(\forall p, q \,|: p \lor q \equiv q \lor p)$.

A standard terminology is often used for proving a universal quantification using Metatheorem (9.16). We say:

To prove $(\forall x \mid R : P)$, we prove P for arbitrary x in range R.

Thus, $(\forall x \mid R : P)$ can be proved by proving $R \Rightarrow P$ (and $R \Rightarrow P$ itself can be proved by assuming antecedent R and proving consequent P). This method of proof is often used informally in mathematics. In fact, we used it on page 76 in proving theorem (4.8), where the universal quantification was expressed in English as "for any natural number i".

9.2   Existential quantification

Disjunction $\vee$ is symmetric and associative and has the identity *false*. Therefore, it is an instance of $\star$ of Sec. 8.2. The quantification $(\vee x \mid R : P)$ is typically written as

$(\exists x \mid R : P)$.

The symbol $\exists$, which is read as "there exists", is called the *existential* quantifier. The expression is called an *existential quantification* and is read as "there exists an x in the range R such that P holds". A value $\hat{x}$ for which $(R \wedge P)[x := \hat{x}]$ is valid is called a *witness* for x in $(\exists x \mid R : P)$.

General axioms (8.13)–(8.21) hold for $(\vee x \mid R : P)$ and are not repeated here. Note that $\vee$ is idempotent, so that existential quantification satisfies Range split (8.18) as well.

We now give additional theorems for existential quantification. We begin with axiom (9.17) below, which relates existential quantification to universal quantification. We call this axiom Generalized De Morgan, and later we will abbreviate it as De Morgan, since it is a generalization of De Morgan's law (3.47a), $\neg(p \wedge q) \equiv \neg p \vee \neg q$. We can get the idea behind this generalization with an example:

$(\vee i \mid 0 \le i < 4 : P)$
$=$ ⟨Eliminate quantification⟩
    $P_0^i \vee P_1^i \vee P_2^i \vee P_3^i$
$=$ ⟨Double negation (3.12); De Morgan (3.47a)⟩
    $\neg(\neg P_0^i \wedge \neg P_1^i \wedge \neg P_2^i \wedge \neg P_3^i)$
$=$ ⟨Introduce quantification⟩
    $\neg(\wedge i \mid 0 \le i < 4 : \neg P)$

Axiom (9.17) can be viewed as a definition of $\exists$, in the sense that it can be used along with the strategy of definition elimination, (3.23), to prove all theorems concerning existential quantification.

(9.17) **Axiom, Generalized De Morgan:**
$(\exists x \mid R : P) \equiv \neg(\forall x \mid R : \neg P)$

Using Double negation and De Morgan's laws, we immediately derive three similar forms of Generalized De Morgan.

Generalized De Morgan

(9.18) **Generalized De Morgan:** (a) $\neg(\exists x \mid R : \neg P) \equiv (\forall x \mid R : P)$

(b) $\neg(\exists x \mid R : P) \equiv (\forall x \mid R : \neg P)$

(c) $(\exists x \mid R : \neg P) \equiv \neg(\forall x \mid R : P)$

The range is the same on both sides of the various Generalized De Morgan theorems. Often, the body of a quantification will be manipulated while the range remains the same; our syntax for quantification facilitates this by keeping this non-changing part out of the way.

On page 31, we introduced the concept of the dual of a boolean expression. To complete the definition of the dual for all boolean expressions, we define the dual of $(\forall x \mid R : P)$ to be $(\exists x \mid R : \neg P)$ (and the dual of $(\exists x \mid R : P)$ to be $(\forall x \mid R : \neg P)$). This definition is consistent with Definition (2.2) given on page 31. That is, Metatheorem (2.3a) still holds: if P is a theorem, then so is $\neg P_D$.

TRADING WITH EXISTENTIAL QUANTIFICATION

The trading theorems for existential quantification are surprisingly different from their counterparts for universal quantification, with a conjunction instead of an implication in the RHS. To understand Trading (9.19), recall our meaning of $\exists$. The LHS of (9.19) states that "there exists a value x in the range R for which P is *true*.". This means that there is a value x for which both R and P are *true*. And that is exactly what the RHS says.

Trading theorems for $\exists$

(9.19) **Trading:** $(\exists x \mid R : P) \equiv (\exists x \mid : R \wedge P)$

(9.20) **Trading:** $(\exists x \mid Q \wedge R : P) \equiv (\exists x \mid Q : R \wedge P)$

Distributivity with existential quantification

The new theorems concerning $\exists$ parallel those of $\forall$.

Additional theorems for $\exists$

(9.21) **Distributivity of $\wedge$ over $\exists$:** Provided $\neg occurs(\text{'}x\text{'}, \text{'}P\text{'})$,
$$P \wedge (\exists x \mid R : Q) \equiv (\exists x \mid R : P \wedge Q)$$

(9.22) Provided $\neg occurs(\text{'}x\text{'}, \text{'}P\text{'})$,
$$(\exists x \mid R : P) \equiv P \wedge (\exists x \mid : R)$$

(9.23) **Distributivity of $\vee$ over $\exists$:** Provided $\neg occurs(\text{'}x\text{'}, \text{'}P\text{'})$,
$$(\exists x \mid : R) \Rightarrow ((\exists x \mid R : P \vee Q) \equiv P \vee (\exists x \mid R : Q))$$

(9.24) $(\exists x \mid R : false) \equiv false$

Manipulating the range and body with existential quantification

The theorems for manipulating the range and term of $\forall$ have counterparts for $\exists$.

Weakening, strengthening, and monotonicity for $\exists$

(9.25) **Range weakening/strengthening:**
$$(\exists x \mid R : P) \Rightarrow (\exists x \mid Q \vee R : P)$$

(9.26) **Body weakening/strengthening:**
$$(\exists x \mid R : P) \Rightarrow (\exists x \mid R : P \vee Q)$$

(9.27) **Monotonicity of $\exists$:**
$$(\forall x \mid R : Q \Rightarrow P) \Rightarrow ((\exists x \mid R : Q) \Rightarrow (\exists x \mid R : P))$$

Introduction of exists and interchange

We have two final theorems for manipulating quantifier $\exists$.

> ### Introduction and interchange for ∃
>
> (9.28) **∃-Introduction:** $P[x := E] \Rightarrow (\exists x \mid : P)$
>
> (9.29) **Interchange of quantifications:**
> Provided $\neg occurs('y', 'R')$ and $\neg occurs('x', 'Q')$,
> $$(\exists x \mid R : (\forall y \mid Q : P)) \Rightarrow (\forall y \mid Q : (\exists x \mid R : P))$$

Theorem (9.28) shows how to introduce an existential quantifier using an implication. One-point rule (8.14) is a sharper way to introduce or eliminate existential quantification.

Theorem (9.29) permits the interchange of $\forall$ and $\exists$. It is an implication, and not an equivalence. The implication does not hold in the other direction for the following reason. For the antecedent of (9.29) to be *true*, there must exist a single value x such that P holds for all y. For the consequent to be *true*, no such single value of x is required; for each y, a different value of x may satisfy P.

We give the proof of (9.29) because it illustrates well how a proof can be "opportunity driven". In our proof, there is a reason for taking each step, although one does not know at that step exactly how the rest of the proof will go. To start, in isolation, neither the antecedent nor the consequent presents much invitation for manipulation, so we take them together. (We give the proof with ranges implicit to make it easier to read; they can be filled in by the reader).

$$(\exists x \mid : (\forall y \mid : P)) \Rightarrow (\forall y \mid : (\exists x \mid : P))$$

$=$ ⟨Implication (3.57), $p \Rightarrow q \equiv p \lor q \equiv q$,
to eliminate the problematic $\Rightarrow$⟩

$$(\exists x \mid : (\forall y \mid : P)) \lor (\forall y \mid : (\exists x \mid : P)) \equiv (\forall y \mid : (\exists x \mid : P))$$

$=$ ⟨Distributivity of $\lor$ over $\forall$ (9.5) —so that the LHS
and RHS have the same outer quantification⟩

$$(\forall y \mid : (\exists x \mid : (\forall y \mid : P)) \lor (\exists x \mid : P)) \equiv (\forall y \mid : (\exists x \mid : P))$$

$=$ ⟨Distributivity (8.15) —so that the LHS
and RHS have the same two outer quantifications⟩

$$(\forall y \mid : (\exists x \mid : (\forall y \mid : P) \lor P)) \equiv (\forall y \mid : (\exists x \mid : P))$$

$=$ ⟨Instantiation (9.13) says $(\forall y \mid : P) \Rightarrow P$,
which by (3.57) is equivalent to $(\forall y \mid : P) \lor P \equiv P$⟩

$$(\forall y \mid : (\exists x \mid : P)) \equiv (\forall y \mid : (\exists x \mid : P))$$ —Reflexivity of equality

WITNESSES

On page 162, we mentioned that free variables in a theorem are implicitly universally quantified. For example, $b \lor c \equiv c \lor b$ is a theorem iff $(\forall b, c \mid b \lor c \equiv c \lor b)$ is a theorem. This allows us to manipulate the simpler, unquantified formulas, instead of their more complex quantified counterparts. A similar technique for existential quantification is embodied in the following theorem.

(9.30) **Metatheorem Witness.** Suppose $\neg occurs(`\hat{x}`, `P, Q, R`)$. Then

$$(\exists x \mid R : P) \Rightarrow Q \text{ is a theorem iff}$$
$$(R \land P)[x := \hat{x}] \Rightarrow Q \text{ is a theorem.}$$

Identifier $\hat{x}$ is called a *witness* for the existential quantification. [1]

Proof. $(\exists x \mid R : P) \Rightarrow Q$
$=$ $\langle \text{Trading (9.19)} \rangle$
           $(\exists x \mid : R \land P) \Rightarrow Q$
$=$ $\langle \text{Implication (3.59); De Morgan (9.18b)} \rangle$
           $(\forall x \mid : \neg(R \land P)) \lor Q$
$=$ $\langle \text{Dummy renaming (8.21), } - \neg occurs(`\hat{x}`, `P, R`) \rangle$
           $(\forall \hat{x} \mid : \neg(R \land P)[x := \hat{x}]) \lor Q$
$=$ $\langle \text{Distributivity of } \lor \text{ over } \forall \text{ (9.5) } - \neg occurs(`\hat{x}`, `Q`) \rangle$
           $(\forall \hat{x} \mid : \neg(R \land P)[x := \hat{x}] \lor Q)$
$=$ $\langle \text{Implication (3.59)} \rangle$
           $(\forall \hat{x} \mid : (R \land P)[x := \hat{x}] \Rightarrow Q)$

By Metatheorem (9.16), the last line is a theorem iff $(R \land P)[x := \hat{x}] \Rightarrow Q$ is a theorem.   □

Metatheorem Witness is often used in the case that $(\exists x \mid R : P)$ is a known theorem (or axiom) and Q is to be proved. In such cases, the proof often proceeds by assuming $(R \land P)[x := \hat{x}]$ and proving Q. We illustrate this technique in proving $a + b = a + c \Rightarrow b = c$. In the proof, we make use of Additive inverse (15.6), which appears later in Chap. 15, specialized to the integers:

(9.31) $(\exists x : \mathbb{Z} \mid : x + a = 0)$.

This axiom says that, for any integer a, there exists another integer x such that $x + a = 0$. We use the assumption $(x + a = 0)[x := \hat{a}]$ with witness $\hat{a}$, i.e. we use the assumption $\hat{a} + a = 0$.

To prove $a + b = a + c \Rightarrow b = c$, we assume the antecedent $a + b = a + c$ (in addition to $\hat{a} + a = 0$) and prove the consequent:

---

[1] Identifier x itself can be used for $\hat{x}$ if x does not occur free Q.

$$b$$
$$= \quad \langle \text{Additive identity (15.3)}, \ 0+b=b \rangle$$
$$0+b$$
$$= \quad \langle (9.31), \ (\exists x{:}\mathbb{Z}\,|{:}\ x+a=0)\,, \ \text{with witness} \ \hat{a} \rangle$$
$$\hat{a}+a+b$$
$$= \quad \langle \text{Assumption} \ a+b=a+c \rangle$$
$$\hat{a}+a+c$$
$$= \quad \langle \, \hat{a}+a=0 \ \text{—again} \rangle$$
$$0+c$$
$$= \quad \langle \text{Additive identity (15.3) —again} \rangle$$
$$c$$

Deduction Theorem (4.4) on page 72 requires that, in proving $(R \wedge P)[x := \hat{x}] \ \Rightarrow \ Q$ by assuming $(R \wedge P)[x := \hat{x}]$ and proving Q, the variables of $(R \wedge P)[x := \hat{x}]$ be considered to be constants. The discussion following Metatheorem (4.4) explains the reason for this restriction.

One more point concerning Metatheorem Witness needs to be emphasized. When two (or more) witnesses are used in a proof, they must be distinct, for the following reason. Suppose we want to prove $(\exists x \,|{:}\ P) \wedge (\exists x \,|{:}\ R) \ \Rightarrow \ Q$. Use Shunting (3.65) to write this expression as

$$(\exists x \,|{:}\ P) \ \Rightarrow \ ((\exists x \,|{:}\ R) \ \Rightarrow \ Q) \quad .$$

Two applications of Metatheorem Witness indicates that this expression is equivalent to

$$P[x := \hat{x}] \ \Rightarrow \ (R[x := x'] \ \Rightarrow \ Q) \quad .$$

Here, $\hat{x}$ and x' must be distinct because of the requirement that $\hat{x}$ not occur free in the consequent $R[x := x'] \ \Rightarrow \ Q$ of this implication.

We illustrate the inconsistency that may arise if two witnesses are given the same name. Consider again (9.31). Use Substitution (1.1) to replace a by $a+5$, yielding theorem

$$(\exists x{:}\mathbb{Z}\,|{:}\ x+a+5=0) \quad .$$

From this theorem, derive using Witness (9.30) the assumption $\hat{a}+a+5 = 0$, where $\hat{a}$ is the witness. From theorem (9.31), derive the assumption $\hat{a}+a = 0$, where the mistake is made of using the same witness $\hat{a}$. Together, these two assumptions yield the contradiction $5 = 0$.

9.3   English to predicate logic

Formalizing a statement in terms of propositional logic does not always provide the opportunity to reason formally about the constituents of the statement. For example, consider the statement "some integer between 80 and n is a multiple of x". We could simply assign a propositional variable S (say) to this statement and use S everywhere this statement might appear, but this formalization does not give us the ability to reason about the statement.

Using predicate calculus, we can write a formalization that offers more chance of manipulation:

$$(\exists i : \mathbb{Z} \mid 80 \leq i \leq n : mult(n, x)) \quad ,$$

where $mult(n, x)$ denotes "n is a multiple of x". And, we could formalize $mult(n, x)$ as well:

$$(\exists m : \mathbb{Z} \mid : n = m \cdot x) \quad .$$

With axioms for the integers, we could then prove various theorems, for example,

$$even.x \ \wedge \ (\exists m : \mathbb{Z} \mid : n = m \cdot x) \ \Rightarrow \ (\exists m : \mathbb{Z} \mid : n = m \cdot x / 2) \quad .$$

Formalizing English (or a mixture of English and mathematics) in predicate logic can help in at least two ways. First, it may expose ambiguities and force precision. As an example, does the phrase "between 80 and n" include 80 and n or not? A formalization in predicate logic must answer this question. Second, having the formalization allows us to use the inference rules of predicate logic to reason formally about objects under consideration. Later, we give an example of such reasoning.

Just as Table 2.3 on page 33 gives a correspondence between English words and boolean operations, so there is a correspondence between English and the additional symbols of predicate logic. We read the universal quantification symbol $\forall$ as "for all", so it is not surprising that appearances of the phrases

every, all, for all, for each, and *any*

signal that a universal quantification is at hand. Here are two examples.

All even integers are multiples of 2 : $(\forall x : \mathbb{Z} \mid even.x : mult(x, 2))$
Every chapter has at least 3 pages: $(\forall c \mid c \in Chap : size.c \geq 3)$

Not all universal quantifications are signaled by explicit phrases. Sometimes, the universal quantification is implicit, and the appearance of an indefinite article may be a clue. The following two sentences illustrate this point.

Even integers are multiples of 2 : $\quad (\forall x : \mathbb{Z} \mid even.x : mult(x, 2))$
An even integer is a multiple of 2 : $(\forall x : \mathbb{Z} \mid even.x : mult(x, 2))$

This is consistent with our convention that free variables in a theorem of the predicate calculus are implicitly universally quantified (see subsection "On theorems and universal quantification" on page 162). When we write

$$x^2 > 0 \quad ,$$

if there is no particular state implied by the discussion, the meaning is that every value x satisfies $x^2 > 0$, that is, $(\forall x \mid : x^2 > 0)$.

Existential quantifications are also signaled by a host of English words:

exists, some, there are, there is, at least one, and *for some*.

Here are two examples.

> Some even integer is divisible by 3:
> $$(\exists x : \mathbb{N} \mid even.x : divisible(x, 3))$$
> There is a chapter with an even number of pages:
> $$(\exists c \mid c \in Chap : even(size.c))$$

It is instructive to contrast the roles of negation and quantification in natural language. Suppose we are asked to negate "All integers are even". "All integers are not even" is incorrect. The negation of the sentence is "Not all integers are even", which we read as "Not (all integers are even)". This phrase is equivalent to "Some integer is not even", as we now show.

> Not (all integers are even)
> $=$ ⟨Formalize in predicate calculus⟩
> $\neg(\forall z : \mathbb{Z} \mid : even.z)$
> $=$ ⟨De Morgan (9.18c)⟩
> $(\exists z : \mathbb{Z} \mid : \neg even.z)$
> $=$ ⟨Return to English⟩
> Some integer is not even

Once again, we see that arguments couched in English are easy to get wrong. Formalizing the English in predicate logic makes it easier to derive consequences systematically, as we have just done.

Formalizing an English statement in the propositional calculus requires associating boolean variables with the subpropositions of the statement. Formalizing an English statement in predicate logic may require defining predicate symbols and other functions to allow us to capture relationships between variables. For example, $mult(x, 2)$, $even.z$, $size.c$ (for c a chapter) all made it possible to formalize the statements above.

Here is another example of formalization. Consider translating

(9.32) Every senior took one mathematics class and passed one programming class.

We introduce the following predicates:

$taken(s, c)$:   Student s completed class c.
$passed(s, c)$: Student s received a passing grade in c.
$senior(s)$:   Student s is a senior.
$math(c)$:   Class c is a mathematics class.
$prog(c)$:   Class c is a programming class.

Then a translation of (9.32) is

(9.33) $(\forall s \mid senior.s : (\exists c, c' \mid: math.c \land taken(s, c) \land$
$$prog.c' \land passed(s, c')))$$

Note the consequence of interchanging the quantification. The formula

$(\exists c, c' \mid: (\forall s \mid senior.s : math.c \land taken(s, c) \land$
$$prog.c' \land passed(s, c')))$$

says that all students took the same math class and passed the same programming class. Some people could claim that this is what is meant by (9.32). The English is ambiguous.

In dealing formally with the domain of students and classes, we would have to develop axioms that capture the properties of student transcripts. For example, given (9.33), proving that every senior took a programming class would probably require an axiom like

$$passed(s, c) \Rightarrow taken(s, c) \ .$$

Developing a useful theory of student transcripts, or any other domain, takes time and effort. In later chapters, we develop theories of sets, sequences, relations, and integers.

ARGUMENTS IN MATHEMATICS

In high-school algebra and calculus you wrote proofs. We now see that they were not really formal. They were in English and never explicitly cited inference rules from a formal logic. If your proofs were correct, then they were informal descriptions that could be translated into formal proofs. And, we now know enough predicate logic actually to construct such proofs formally.

In high school, for example, you might have been asked to determine whether there is some real x for which

$$1/(x^2 + 1) > 1 \ .$$

Formalized in predicate logic, the question is whether

$$(\exists x : \mathbb{R} \mid: 1/(x^2 + 1) > 1)$$

is valid. We can investigate this question as follows.

$$(\exists x \mid : 1/(x^2 + 1) > 1)$$
$$= \quad \langle \text{Arithmetic} \rangle$$
$$(\exists x \mid : 1 > x^2 + 1)$$
$$= \quad \langle \text{Arithmetic} \rangle$$
$$(\exists x \mid : 0 > x^2)$$
$$= \quad \langle \, x^2 \geq 0 \text{ —from the theory of reals} \rangle$$
$$(\exists x \mid : false)$$
$$= \quad \langle (9.24) \rangle$$
$$false$$

Here, we have formalized the part of the manipulation that deals with quantification. The part that deals with algebraic manipulation has been left informal, because we have not yet studied theories of arithmetic.

$f.x$ |

We now look to the domain of functions for another example. Informally, a function is *continuous at a point* c if it doesn't "jump" at c. The function illustrated in this paragraph is continuous at 1 but not at 2.

$f.c$ | $f.x$ |

Continuity of f at a point c is defined as follows. Choose any distance $\epsilon > 0$. Suppose for any such distance ϵ that another distance $\delta > 0$ can be found such that for all points x within distance δ from c, $f.x$ is within distance ϵ from $f.c$. Then f is said to be continuous at c. This definition outlaws functions that "jump", as illustrated in the previous paragraph.

Using $|z|$ to denote the absolute value of z, we formalize the notion of continuity at a point as follows. Function f is continuous at c iff

(9.34) $(\forall \epsilon \mid \epsilon > 0 : (\exists \delta \mid \delta > 0 : (\forall x \mid : |x - c| < \delta \Rightarrow |f.x - f.c| < \epsilon)))$.

Theorem. $f(x) = 3 \cdot x + 15$ is continuous at all points.

Proof. For arbitrary c, we begin by manipulating the consequent of $|x - c| < \delta \Rightarrow |f.x - f.c| < \epsilon$ of the body of (9.34), for the given function f.

$$|3 \cdot x + 15 - (3 \cdot c + 15)| < \epsilon$$
$$= \quad \langle \text{Arithmetic} \rangle$$
$$|3 \cdot (x - c)| < \epsilon$$
$$= \quad \langle \text{Property of } |\ldots| \rangle$$
$$3 \cdot |x - c| < \epsilon$$
$$= \quad \langle \text{Arithmetic} \rangle$$
$$|x - c| < \epsilon/3$$

We have proved $|x - c| < \epsilon/3 \Rightarrow |f.x - f.c| < \epsilon$ for arbitrary c.

Therefore, it appears that we have $\epsilon/3$ as a witness for δ of (9.34). We can now construct the desired quantified formula.

$$|x - c| < \epsilon/3 \;\Rightarrow\; |f.x - f.c| < \epsilon$$
$$=\quad \langle \text{The above is a theorem; use Metatheorem (9.16)} \rangle$$
$$(\forall x \mid : |x - c| < \epsilon/3 \;\Rightarrow\; |f.x - f.c| < \epsilon)$$
$$=\quad \langle \text{One-point rule (8.14)} \rangle$$
$$(\exists \delta \mid \delta = \epsilon/3 : (\forall x \mid : |x - c| < \delta \;\Rightarrow\; |f.x - f.c| < \epsilon))$$
$$\Rightarrow\quad \langle \text{Range weakening (9.25) —since } \epsilon > 0 \,\rangle$$
$$(\exists \delta \mid \delta > 0 : (\forall x \mid : |x - c| < \delta \;\Rightarrow\; |f.x - f.c| < \epsilon))$$
$$=\quad \langle \text{The above is a theorem; use Metatheorem (9.16) twice} \rangle$$
$$(\forall c \mid : (\forall \epsilon \mid \epsilon > 0 : (\exists \delta \mid \delta > 0 :$$
$$(\forall x \mid : |x - c| < \delta \;\Rightarrow\; |f.x - f.c| < \epsilon))))$$
$$=\quad \langle \text{Definition of continuous at } c \rangle$$
$$(\forall c \mid : f \text{ is continuous at } c) \qquad\qquad \square$$

We end this section with predicate-calculus formalizations of three other statements concerning functions. A function is *one-to-one* if for different arguments it yields different values. We can state this as follows.

$$f \text{ is one-to-one}: \quad (\forall x, y \mid x \neq y : f.x \neq f.y)$$

or

$$f \text{ is one-to-one}: \quad (\forall x, y \mid f.x = f.y : x = y)$$

Function g is the *inverse* of function f if g maps $f.x$ back into x. In other words,

$$g \text{ is the inverse of } f : \quad (\forall x \mid : x = g(f.x))$$

A two-argument function is *symmetric* if interchanging the arguments of a function application does not change its value:

$$f \text{ is symmetric} \;\equiv\; (\forall x, y \mid : f(x, y) = f(y, x))$$

Exercises for Chapter 9

9.1 Prove that Distributivity of $\lor$ over $\forall$ (9.5), $P \lor (\forall x \mid R : Q) \equiv (\forall x \mid R : P \lor Q)$ (provided x does not occur free in P), follows from a similar expression with all ranges *true*: $P \lor (\forall x \mid : Q) \equiv (\forall x \mid : P \lor Q)$ (provided x does not occur free in P). This means we could have used a simpler axiom.

9.2 Prove that $(\forall x \mid R : P) \land (\forall x \mid R : Q) \equiv (\forall x \mid R : P \land Q)$ follows from a similar expression with all ranges *true*: $(\forall x \mid : P) \land (\forall x \mid : Q) \equiv (\forall x \mid : P \land Q)$.

9.3 Prove theorem (9.6), $(\forall x \mid R : P) \equiv P \vee (\forall x \mid: \neg R)$ (provided x does not occur free in P). Beginning with the LHS and trading seems appropriate, since the RHS has *true* as a range.

9.4 Prove theorem (9.8), $(\forall x \mid R : true) \equiv true$. Trading with the LHS will yield a formula to which some form of distributivity may be applied. Or, use (9.6).

9.5 Prove theorem (9.9), $(\forall x \mid R : P \equiv Q) \Rightarrow ((\forall x \mid R : P) \equiv (\forall x \mid R : Q))$. Our proof replaces the whole expression using theorem (3.62).

9.6 Since $\wedge$ is idempotent, Range split for idempotent $\star$ (8.18) specializes for $\wedge$ to

$$(\forall x \mid R \vee Q : P) \equiv (\forall x \mid R : P) \wedge (\forall x \mid Q : P) \quad .$$

However, it is possible to prove this expression without relying on axiom (8.18). Develop such a proof. You may find it useful to trade R and Q into the body.

9.7 Prove Range weakening/strengthening (9.10), $(\forall x \mid Q \vee R : P) \Rightarrow (\forall x \mid Q : P)$. Range splitting may be helpful.

9.8 Prove Body weakening/strengthening (9.11), $(\forall x \mid R : P \wedge Q) \Rightarrow (\forall x \mid R : P)$. Distributivity of $\forall$ over $\wedge$ may be helpful.

9.9 Prove Monotonicity of $\forall$ (9.12), $(\forall x \mid R : Q \Rightarrow P) \Rightarrow ((\forall x \mid R : Q) \Rightarrow (\forall x \mid R : P))$.

9.10 Suppose that instead of One-point rule (8.14) for $\forall$, $(\forall x \mid x = E : P) \equiv P[x := E]$ (provided x does not occur free in E), we choose the axiom $(\forall x \mid x = e : false) \equiv false$. Prove that the one-point rule for $\forall$ still holds (using this new axiom and theorems numbered less than (9.13)).

9.11 Prove Instantiation (9.13), $(\forall x \mid: P) \Rightarrow P[x := E]$. The key is to replace the dummy using Dummy renaming (8.21) so that the dummy occurs neither in P nor in E.

Exercises on existential quantification

9.12 Prove Generalized De Morgan (9.18a), $\neg(\exists x \mid R : \neg P) \equiv (\forall x \mid R : P)$.

9.13 Prove Generalized De Morgan (9.18b), $\neg(\exists x \mid R : P) \equiv (\forall x \mid R : \neg P)$.

9.14 Prove Generalized De Morgan (9.18c), $(\exists x \mid R : \neg P) \equiv \neg(\forall x \mid R : P)$.

9.15 Prove Trading (9.19), $(\exists x \mid R : P) \equiv (\exists x \mid: R \wedge P)$.

9.16 Prove Trading (9.20), $(\exists x \mid Q \wedge R : P) \equiv (\exists x \mid Q : R \wedge P)$.

9.17 Prove Distributivity of $\wedge$ over $\exists$, (9.21), $P \wedge (\exists x \mid R : Q) \equiv (\exists x \mid R : P \wedge Q)$ (provided x does not occur free in P).

9.18 Prove Distributivity of $\exists$ over $\vee$, $(\exists x \mid R : P) \vee (\exists x \mid R : Q) \equiv (\exists x \mid R : P \vee Q)$.

9.19 Prove (9.22), $(\exists x \mid R : P) \equiv P \wedge (\exists x \mid: R)$ (provided $\neg occurs('x', 'P')$).

9.20 Prove Distributivity of $\lor$ over $\exists$ (9.23), $(\exists x \mid : R) \Rightarrow ((\exists x \mid R : P \lor Q) \equiv P \lor (\exists x \mid R : Q))$ (provided x does not occur free in P).

9.21 Prove (9.24), $(\exists x \mid R : false) \equiv false$.

9.22 Since $\lor$ is idempotent, Range split for idempotent $\star$ (8.18) specializes for $\lor$ to

$$(\exists x \mid R \lor Q : P) \equiv (\exists x \mid R : P) \lor (\exists x \mid Q : P) \ .$$

However, it is possible to prove this expression without relying on axiom (8.18). Develop such a proof.

9.23 Prove Range weakening/strengthening (9.25), $(\exists x \mid R : P) \Rightarrow (\exists x \mid Q \lor R : P)$.

9.24 Prove Body weakening/strengthening (9.26), $(\exists x \mid R : P) \Rightarrow (\exists x \mid R : P \lor Q)$.

9.25 Prove Monotonicity of $\exists$ (9.27), $(\forall x \mid R : Q \Rightarrow P) \Rightarrow ((\exists x \mid R : Q) \Rightarrow (\exists x \mid R : P))$.

9.26 Prove $\exists$-introduction (9.28), $P[x := E] \Rightarrow (\exists x \mid : P)$.

9.27 Prove $(\exists x \mid R : P) \Rightarrow Q \equiv (\forall x \mid R : P \Rightarrow Q)$ (provided x does not occur free in Q).

9.28 Prove $(\exists x \mid : R) \Rightarrow ((\forall x \mid R : P) \Rightarrow Q \equiv (\exists x \mid R : P \Rightarrow Q))$ (provided x does not occur free in P).

Exercises on translation to and from predicate logic

9.29 Translate the following English statements into predicate logic.

(a) The natural number 1 is the only natural number that is smaller than positive integer p and divides p.

(b) Some integer is larger than 23.

(c) Adding two odd integers yields an even number. (Use only addition and multiplication; do not use division, mod, or predicates $even.x$ and $odd.x$.)

(d) A positive integer is not negative.

(e) Every positive integer is smaller than the absolute value of some negative integer. (Use $abs.i$ for the absolute value of i.)

(f) Cubes of integers are never even. (Use only addition and multiplication; do not use exponentiation, division, mod, or predicates $even.x$ and $odd.x$.)

(g) Real number i is the largest real solution of the equation $f.i = i + 1$.

(h) For no integer i is $f.i$ both greater than and less than i.

(i) Value $f.j$ is always $j + i$ greater than $f.i$.

(j) Function $f.i$ is non-decreasing as i increases.

(k) No integer is larger than all others.

(l) Every integer is larger than one and smaller than another.

(m) Value $g(f.i)$ is the smallest positive integer i such that $f.i = i$.

9.30 Translate the following predicate-logic formulas into English. In doing so, don't simply make a literal translation; try instead to extract the meaning of each formula and express that as you would in English.

(a) $(\exists k:\mathbb{R} \mid (\forall i:\mathbb{Z} \mid: f.i = k))$

(b) $(\exists z:\mathbb{R} \mid: (\forall i:\mathbb{Z} \mid: f.j = f(j + i\cdot z)))$

(c) $(\forall x:\mathbb{R} \mid x \neq m : f.x > f.m)$

(d) $(\exists x, y:\mathbb{R} \mid: f.x < 0 \wedge 0 < f.y \Rightarrow (\exists z:\mathbb{R} \mid: f.z = 0))$

(e) $(\forall x:\mathbb{Z} \mid: (\exists z:\mathbb{R} \mid: f.x = z))$

(f) $(\forall z:\mathbb{R} \mid: (\exists z:\mathbb{Z} \mid: f.x = z))$

(g) $(\forall z:\mathbb{Z} \mid even.z : (\forall w:\mathbb{Z} \mid odd.w : z.\neq w))$

(h) $(\forall z:\mathbb{Z} \mid even.z : (\exists w:\mathbb{Z} \mid odd.w : z = w + 1))$

9.31 Define suitable predicates and functions and then formalize the sentences that follow.

(a) Messages sent from one process to another are received in the order sent.

(b) Broadcasts made by a process are received by all processes in the order sent.

(c) All messages are received in the same order by all.

9.32 Define suitable predicates and functions and then formalize the sentences that follow.

(a) A student receives a grade for every course in which they registers.

(b) Registration for a course requires passing all its prerequisites.

(c) No student who has received an F in a required course graduates with honors.

9.33 Translate into predicate Logic.

Assuming that each task t requires $work.t$ seconds, the start time $start.t$ for a task t is the earliest time such that all prerequisite tasks in set $prereq.y$ have completed.

9.34 Formalize the following English sentences in predicate logic.

(a) Everybody loves somebody.

(b) Somebody loves somebody.

(c) Everybody loves everybody.

(d) Nobody loves everybody.

(e) Somebody loves nobody.

9.35 Formalize the following English sentences in predicate logic.

(a) You can fool some of the people some of the time.

(b) You can fool all the people some of the time.

(c) You can't fool all the people all the time.

(d) You can't fool a person all the time.

9.36 Show that the following argument is sound by translating it into the predicate calculus and proving that the translation is a theorem: All men are mortal; Socrates is a man; therefore, Socrates is mortal.

Chapter 10

Predicates and Programming

W e turn to some applications of predicate logic in computing: the formal specification of imperative programs (i.e. ones that use assignment statements) and the proof and development of sequences of assignments. Skill with predicate logic can be used to reformulate English specifications, with all their vagueness and ambiguities, as formal specifications. Also, parts of assignments can be *calculated* instead of guessed. Finally, we discuss the conditional statement and conditional expression.

10.1 Specification of programs

Recall from Sec. 1.6 that a *state* is a set of identifier-value pairs. Further, the Hoare triple $\{Q\}\ S\ \{R\}$, where S is a program statement, Q is the precondition, and R is the postcondition, has the interpretation

> Execution of S begun in any state in which Q is *true* is guaranteed to terminate, and R is *true* in the final state.

As a specification notation, $\{Q\}\ S\ \{R\}$ is inadequate. The notation does not indicate which variables may be changed by S. For example, $\{true\}\ S\ \{x = y\}$ says that S should truthify[1] $x = y$, but it does not say which of x and y to change. Also, the notation forces us to name a program S (say), even though there may be no other reason to do so.

A specification of a program should give:

- a precondition Q (say): a boolean expression that describes the initial states for which execution of the program is being defined,

- a list x (say) of variables that may be assigned to, and

- a postcondition R (say): a boolean expression that characterizes the final states, after execution of the program.

We formally denote such a specification by $\{Q\}\ x := ?\ \{R\}$.

---

[1] If you can falsify, you should be able to truthify. We coined this word because alternatives like "establish the truth of" are long-winded and awkward.

A specification can be non-deterministic, which means that, for some initial state, the final state is not unambiguously determined. For example, $\{true\}\ b:=?\ \{b^2 = 25\}$ specifies a program that in any initial state stores a value in b so that $b^2 = 25$. A program that satisfies this specification can assign either -5 or 5 to b.

To formalize an English description of a program we have to define a precondition and a postcondition. In so doing, we are often forced to introduce restrictions that are implicit in the English specification, and we may have to invent variables into which the results of a computation are to be stored. For example, consider the following English specification:

Find an integer approximation to the square root of integer n.

Because of our knowledge of mathematics, we know that n cannot be negative (the output is to be an integer, not a complex number). So a necessary part of the precondition is $0 \leq n$. Next, the integer approximation has to be stored in some variable; we choose d. Finally, we must precisely define what is acceptable as an approximation; we choose the largest integer d such that $d^2 \leq n$. We have derived the formal specification

$$\{0 \leq n\}\ d:=?\ \{d^2 \leq n < (d+1)^2\} \quad .$$

We develop another formal specification. Suppose we want a program that finds the index of a value x in an array $b[0..n-1]$. An informal specification might be "Find x in b". A more precise definition must give conditions on b and n. Can the array segment be empty ($n = 0$)? How should the index of x in b be indicated? Can we assume that x is actually in the array segment? If not, how should its absence be indicated? Here are four possible formal specifications. Each answers these questions in different ways.

$$\{x \in b[0..n-1]\}\ \ i:=?\ \ \{0 \leq i < n\ \wedge\ x = b[i]\}$$
$$\{0 \leq n\}\ \ i:=?\ \ \{(0 \leq i < n\ \wedge\ x = b[i])\ \vee$$
$$(i = n\ \wedge\ x \notin b[0..n-1])\}$$
$$\{0 \leq n\}\ \ i:=?\ \ \{(0 \leq i < n\ \wedge\ x \notin b[1..i-1]);\ \wedge$$
$$(x = b[i]\ \vee\ i = n)\}$$
$$\{0 \leq n\}\ \ i,c:=?\ \ \{(c \equiv x \in b[0..n-1])\ \wedge\ (c \Rightarrow x = b[i])\}$$

The first specification presumes that x is in b and requires the index of x in b to be stored in i. The second does not presume x to be in b, but it requires that i be set to n if x is not in b. The third is similar to the second, but in addition it requires that i should be set to the index of the *first* occurrence of x in b. The fourth uses an extra boolean variable c

to indicate whether x is in b and requires i to be set only if x is in b. Throughout, we have used $x \in b[0..n-1]$ as an abbreviation for

$$(\exists j \mid 0 \leq j < n : x = b[j]) \quad .$$

There may be many ways to formalize an English specification. It takes experience, thought, and care to be able to do it well. Developing a clear and rigorous (if not formal) specification is an important part of the programming task. The more complicated the problem being tackled, the more important are good specifications.

Some specifications use variables that are not actually implemented in the program, usually to refer to the initial or final values of program variables. We call these *rigid* variables. We will use "typewriter font", e.g. X, for rigid variables. For example, here is a specification for an algorithm to add 6 to x.

$$\{x = \mathtt{X}\} \quad x := ? \quad \{x = \mathtt{X} + 6\}$$

This specification means that for all values X, if $x = \mathtt{X}$, then execution of the algorithm should assign to x to truthify $x = \mathtt{X} + 6$. Here, X denotes the initial value of x, but it can just as well be regarded as denoting a final value, as in the following equivalent specification.

$$\{x = \mathtt{X} - 6\} \quad x := ? \quad \{x = \mathtt{X}\}$$

In the following specification, rigid variable C is an array.

$$\{c = \mathtt{C} \wedge 0 \leq n\} \quad c := ? \quad \{(\forall i \mid 0 \leq i < n : c.i = -\mathtt{C}[i])\}$$

Each element of c is to be negated. Note that we allow assignments to an array and not just to its elements. The specification allows (but does not require) all elements of c to be assigned. For example, if $c[i] = 0$, $c[i]$ need not be assigned.

10.2 Reasoning about the assignment statement

DEALING WITH PARTIAL FUNCTIONS

In Sec. 1.6, we defined the (multiple) assignment $x := E$ by the axiom

$$\{R[x := E]\} \quad x := E \quad \{R\} \quad .$$

This definition was given under the assumption that E was total, i.e. that E had a value in all states. Many expressions are not total; for example $10/x$ is defined only if $x \neq 0$, and array reference $c[i]$ is defined only if i is within the array bounds.

For each expression E, we define the predicate

  $dom.`E$'

to be satisfied in exactly those states in which E is defined ($dom.`E$' stands for *domain of* $E$). We do not show how to construct $dom.`E$', but rely on the reader's knowledge of expressions. For example,

$$dom.`\sqrt{x/y}` \equiv y \neq 0 \land x/y \geq 0 \ .$$

We can use different definitions of dom for different purposes. For example, when first writing a program, we assume that type *integer* contains all the integers and use, for example, $dom.`x + y` \equiv true$. Later, when implementing the program on a particular computer whose range [2] of integers is $-2^{16} + 1..2^{16} - 1$, we may want to prove that no overflow occurs, so we restrict the set of values accordingly and use the definition $dom.`x + y` \equiv -2^{16} < x + y < 2^{16}$.

We can now give the more general definition of assignment:

(10.1) $\{dom.`E` \land R[x := E]\} \ \ x := E \ \{R\} \ \ .$

That done, we often omit $dom.`E$' from preconditions when discussing the assignment statement, in order to simplify discussions.

PROOFS OF $\{Q\} \ x := E \ \{R\}$

We claim (without proof) that $R[x := E]$ is the *weakest* precondition [3] such that executing $x := E$ terminates with $R \ \ true$. That is, another precondition Q (say) satisfies $\{Q\} \ x := E \ \{R\}$ iff $Q \Rightarrow R[x := E]$ holds. Therefore, we have the following proof method.

(10.2) **Assignment introduction:** To show that $x := E$ is an implementation of $\{Q\} \ x :=? \ \{R\}$, prove $Q \Rightarrow R[x := E]$.

Here are two examples of Assignment introduction. Consider the specification $\{x > 0\} \ x :=? \ \{x > 1\}$. We want to prove that it is implemented

---

[2] We use $i..j$ for integers i, j to denote the set of integers $i, i + 1, \ldots, j$.

[3] This idea of the weakest precondition of a statement with respect to a postcondition was used by Edsger W. Dijkstra (see Historical note 10.1) in developing a formal definition of a programming language and a methodology for the formal development of sequential programs. In some institutions, this methodology has radically changed how programming is taught. It is the only methodology we know of that has been used to develop new and important algorithms and to better present old algorithms on a non-trivial scale. Sec. 12.6 goes into more detail on this view of programming.

HISTORICAL NOTE 10.1. EDSGER W. DIJKSTRA (1930-)

The citation for Edsger W. Dijkstra's 1972 ACM Turing Award reads, "The working vocabulary of programmers is studded with words originated or force-fully promulgated by E.W. Dijkstra —display, deadly embrace, semaphore, go-to-less programming, structured programming. But his influence on pro-gramming is more pervasive than any glossary can possibly indicate. The pre-cious gift that this Turing Award acknowledges is Dijkstra's *style*: his eloquent insistence and practical demonstration that programs should be composed cor-rectly, and not just debugged into correctness; and his illuminating perception of problems at the foundations of program design. ... We have come to value good programs in much the same way as we value good literature. And at the center of this movement, creating and reflecting patterns no less beautiful than useful, stands E.W. Dijkstra."

This award was made more than two decades ago, before Dijkstra's seminal work on weakest preconditions and the formal development of programs [9], his development of a propositional calculus on which our equational logic **E** is based [10], and all his work on method in mathematics. Dijkstra's influence can be attributed to a penetrating mind; a rare intellectual honesty, which does not allow him to compromise his principles; and a way with words that few computer scientists can match. These factors make him appear caustic, at times, as he says what he believes but not what we want to hear. (A colleague once said, "Dijkstra's right, but you don't say those things.").

Dijkstra loves to write —usually with a fountain pen, and never on a com-puter. He writes letters regularly (not just business letters). His technical pa-pers, trip reports, and essays (e.g. *"Real mathematicians don't prove"* and *On the cruelty of really teaching computing science*), form the "EWD" series. New EWD's —there are almost 1200 EWD's by now— are distributed several times a year through an informal distribution tree.

A native of the Netherlands, Dijkstra has been in the CS Department at the University of Texas at Austin since 1984. One of his pleasures is camping in Texas parks with his wife in their Touring Machine, a Volkswagen camper. While camping, he will walk, bike, and, of course, write. You see, the Touring Machine is equipped with all the equipment he needs: a piece of paper and his Mont Blanc fountain pen.

by the assignment $x := x + 1$. So we prove

$$x > 0 \;\Rightarrow\; (x > 1)[x := x + 1] \quad,$$

by assuming the antecedent and proving the consequent.

$$
\begin{array}{ll}
& (x > 1)[x := x + 1] \\
= & \langle\text{Definition of textual substitution}\rangle \\
& x + 1 > 1 \\
= & \langle\text{Arithmetic}\rangle \\
& x > 0 \quad \text{—Assumption } x > 0
\end{array}
$$

Our second example occurs in an algorithm for summing the elements of array $b[0..n-1]$. Consider the predicate

$$P : 0 \leq i \leq n \ \land \ x = (\Sigma k \mid 0 \leq k < i : b[k]) \quad ,$$

which stipulates that x is the sum of the first i elements of $b[0..n-1]$. We want to show that

(10.3) $\{P \land \mathrm{I} = i \neq n\} \ x, i :=? \ \{P \land i = \mathrm{I} + 1\}$

is implemented by $x, i := x + b[i], i+1$. First, however, we discuss a context in which specification (10.3) might arise, so that the reader can gain some appreciation for this kind of problem. It specifies the body of a loop that accumulates the sum of the elements of $b[0..n-1]$. The requirement that i be increased by 1 ensures that each iteration of the loop makes progress towards termination. The requirement that P be maintained (i.e. that if it is *true* before the iteration it is *true* after) ensures that, upon termination of the loop, x will contain the sum of the first n values of b. Specification (10.3) is illustrative of many specifications of loop bodies.

We proceed by proving that $x, i := x + b[i], i + 1$ truthifies the second conjunct of P, $x = (\Sigma k \mid 0 \leq k < i : b[k])$:

$$P \land \mathrm{I} = i \neq n \ \Rightarrow$$
$$(x = (\Sigma k \mid 0 \leq k < i : b[k]))[x, i := x + b[i], i + 1] \quad .$$

The proof assumes the antecedent and proves the consequent.

$$(x = (\Sigma k \mid 0 \leq k < i : b[k]))[x, i := x + b[i], i + 1]$$
$$= \quad \langle \text{Textual substitution} \rangle$$
$$x + b[i] = (\Sigma k \mid 0 \leq k < i + 1 : b[k])$$
$$= \quad \langle \text{Split off term (8.23)} \rangle$$
$$x + b[i] = (\Sigma k \mid 0 \leq k < i : b[k]) + b[i]$$
$$= \quad \langle \text{Assumption } P \rangle$$
$$x + b[i] = x + b[i] \quad \text{—Identity of } \equiv (3.3)$$

Note how theorem (8.23) was used. In this instance, we want to make use of the conjunct $x = (\Sigma k \mid 0 \leq k < i : b[k])$ of P, and this is best done by splitting off a term of the quantification.

We now prove that $0 \leq i \leq n \ \land \ i = \mathrm{I} + 1$ is truthified, i.e. that the following holds:

$$P \land \mathrm{I} = i \neq n \ \Rightarrow \ (0 \leq i \leq n \ \land \ i = \mathrm{I} + 1)[x, i := x + b[i], i + 1].$$

Again, we assume the antecedent and manipulate the consequent.

$$(0 \leq i \leq n \ \wedge \ i = \mathtt{I} + 1)[x, i := x + b[i], i + 1]$$
$=$ ⟨Textual substitution⟩
$$0 \leq i + 1 \leq n \ \wedge \ i + 1 = \mathtt{I} + 1$$
$=$ ⟨Assumption $i = \mathtt{I}$; Identity of $\wedge$ (3.39)⟩
$$0 \leq i + 1 \leq n$$
$=$ ⟨Assumption $i \neq n$; Arithmetic⟩
$$0 \leq i \leq n \quad \text{—Conjunct of assumption } P$$

REASONING ABOUT SEQUENCES OF ASSIGNMENTS

Suppose we want to find the weakest precondition such that execution of a sequence $x := E; \ y := F$ of assignments will terminate with R *true* :

$$\{?\} \ \ x := E; \ y := F \ \ \{R\} \quad .$$

We know how to find the weakest precondition such that $y := F$ terminates with R *true* : it is $R[y := F]$. We can then find the weakest precondition such that execution of $x := E$ truthifies $R[y := F]$: it is $R[y := F][x := E]$. We illustrate this below. The left column shows the two assignments and the postcondition; the middle column shows the calculated middle assertion, and the rightmost column also shows the calculated precondition.

		$\{R[y := F][x := E]\}$
$x := E$	$x := E$	$x := E$
	$\{R[y := F]\}$	$\{R[y := F]\}$
$y := F$	$y := F$	$y := F$
$\{R\}$	$\{R\}$	$\{R\}$

This method can be generalized to find the weakest precondition for a sequence of assignments in order to truthify R :

$$\{R[x_n := E_n] \cdots [x_2 := E_2][x_1 := E_1]\}$$
$$x_1 := E_1; \ x_2 := E_2; \ \cdots; \ x_n := E_n$$
$$\{R\} \quad .$$

For example, let us find the weakest precondition such that execution of $t := x; \ x := y; \ y := t$ truthifies $x = \mathtt{X} \wedge y = \mathtt{Y}$. We have:

$$(x = \mathtt{X} \wedge y = \mathtt{Y})[y := t][x := y][t := x]$$
$=$ ⟨Textual substitution⟩
$$(x = \mathtt{X} \wedge t = \mathtt{Y})[x := y][t := x]$$
$=$ ⟨Textual substitution⟩
$$(y = \mathtt{X} \wedge t = \mathtt{Y})[t := x]$$
$=$ ⟨Textual substitution⟩
$$y = \mathtt{X} \wedge x = \mathtt{Y}$$

We have discovered that the sequence of assignments swaps x and y .

10.3   Calculating parts of assignments

Consider maintaining

$$P1: \; x = (\Sigma k \mid 0 \le k < i : b[k])$$

using an assignment $i, x := i+1, e$, where we assume that e is unknown. We now show how e can be calculated, instead of guessed. We want to solve for e in

$$\{P1\} \; i, x := i+1, e \; \{P1\} \quad .$$

This Hoare triple is valid exactly when $P1 \Rightarrow P1[i, x := i+1, e]$, so our task is to "solve" this boolean expression for e. We assume the antecedent $P1$ and manipulate the consequent:

$$
\begin{aligned}
& P1[i, x := i+1, e] \\
=\quad & \langle \text{Definition of } P1 \,;\, \text{Textual substitution} \rangle \\
& e = (\Sigma k \mid 0 \le k < i+1 : b[k]) \\
=\quad & \langle \text{Split off term (8.23)} \rangle \\
& e = (\Sigma k \mid 0 \le k < i : b[k]) + b[i] \\
=\quad & \langle \text{Assumption } P1 \rangle \\
& e = x + b[i]
\end{aligned}
$$

Hence, we can use the expression $x + b[i]$ for e.

Here is another example. Consider solving for e in

$$\{P2 : x = (\Sigma k \mid i \le k < n : b[k])\} \; i, x := i-1, e \; \{P2\} \quad .$$

To do this, we have to solve for e in $P2 \Rightarrow P2[i, x := i-1, e]$.

$$
\begin{aligned}
& P2[i, x := i-1, e] \\
=\quad & \langle \text{Definition of } P2 \,;\, \text{Textual substitution} \rangle \\
& e = (\Sigma k \mid i-1 \le k < n : b[k]) \\
=\quad & \langle \text{Split off term (8.23)} \rangle \\
& e = b[i-1] + (\Sigma k \mid i \le k < n : b[k]) \\
=\quad & \langle \text{Assumption } P2 \rangle \\
& e = b[i-1] + x
\end{aligned}
$$

Hence, we can use the expression $b[i-1] + x$ for e.

There is another way to view the task of finding e in

$$P \Rightarrow P[i, x := f, e] \quad .$$

where f is some expression in i. Note that the consequent has the same structure or shape as the antecedent, except that where the antecedent has variables i and x the consequent has expressions. Therefore, if we can manipulate antecedent P until it has the necessary shape but contains expressions where i and x were, we will have identified e.

Here is an example. Consider the function F defined by

$$F.0 = 0$$
$$F.1 = 1$$
$$F(2 \cdot n) = F.n \qquad \text{(for even } n \text{ greater than } 0)$$
$$F(2 \cdot n + 1) = F.n + F(n + 1) \qquad \text{(for odd } n \text{ greater than } 0).$$

Consider the predicate

$$P : C = a \cdot F.n + b \cdot F(n + 1)$$

and suppose we want to find d and e that satisfy

$$\{P \wedge n > 0 \wedge even.n\} \quad n, a, b := n \div 2, d, e \quad \{P\} \ .$$

This requires solving for d and e in $P \wedge n > 0 \wedge even.n \Rightarrow P[n, a, b := n \div 2, d, e])$, which, by Shunting (3.65), we rewrite as

$$n > 0 \wedge even.n \ \Rightarrow \ (P \ \Rightarrow \ P[n, a, b := n \div 2, d, e]) \ .$$

We assume the antecedent $n > 0 \wedge even.n$. To simplify the calculations somewhat, let $k = n \div 2$. We manipulate P with the goal of arriving at an expression with the same shape but with $n \div 2$ instead of n:

$$
\begin{aligned}
& P \\
= \quad & \langle \text{Definition of } P \rangle \\
& C = a \cdot F.n + b \cdot F(n + 1) \\
= \quad & \langle \text{Assumption } even.n \text{, so } n = 2 \cdot k \rangle \\
& C = a \cdot F(2 \cdot k) + b \cdot F(2 \cdot k + 1) \\
= \quad & \langle \text{Definition of } F \text{, twice} \rangle \\
& C = a \cdot F.k + b \cdot (F.k + F(k + 1)) \\
= \quad & \langle \text{Arithmetic} \rangle \\
& C = (a + b) \cdot F.k + b \cdot F(k + 1) \\
= \quad & \langle \text{Definition of } k \text{; Textual substitution} \rangle \\
& P[n, a, b := n \div 2, a + b, b]
\end{aligned}
$$

Hence, we have solved for d and e, and the desired assignment is $n, a, b := n \div 2, a + b, b$, which we can write as $n, a := n \div 2, a + b$.

10.4   Conditional statements and expressions

The conditional statement, call it IF, has the following form in many imperative programming languages:

(10.4) IF : **if** B **then** $S1$ **else** $S2$,

where B is a boolean expression and $S1$ and $S2$ are statements. It is executed as follows: If B is *true*, execute $S1$; otherwise execute $S2$.

Suppose we want execution of the conditional statement begun in a state that satisfies predicate Q to truthify predicate R, i.e. $\{Q\}$ IF $\{R\}$. What must hold in order to guarantee that $\{Q\}$ IF $\{R\}$ is valid? If B is *true*, then $S1$ is executed, so execution of $S1$ must truthify R; on the other hand, if B is *false*, then $S2$ is executed, so execution of $S2$ must truthify R. We can annotate IF to illustrate this and also to indicate that B can be assumed before $S1$ and $\neg B$ can be assumed before $S2$.

> $\{Q\}$
> **if** B **then** $\{Q \wedge B\}$ $S1$ $\{R\}$
>   **else** $\{Q \wedge \neg B\}$ $S2$ $\{R\}$
> $\{R\}$

Thus, we have the following.

(10.5) **Proof method for IF.** To prove $\{Q\}$ IF $\{R\}$, it suffices to prove $\{Q \wedge B\}$ $S1$ $\{R\}$ and $\{Q \wedge \neg B\}$ $S2$ $\{R\}$.

Example. To prove

> $\{true\}$
> **if** $x \leq y$ **then** skip **else** $x, y := y, x$
> $\{x \leq y\}$

we prove the following, both of which are straightforward. [4]

> $\{true \wedge x \leq y\}$ **skip** $\{x \leq y\}$
> $\{true \wedge \neg x \leq y\}$ $x, y := y, x$ $\{x \leq y\}$ □

---

[4] Statement **skip** does nothing, but very fast. In some languages, the effect of **skip** is achieved by writing no statement at all, as in

$$\textbf{if } x \leq y \textbf{ then } \textbf{ else } x, y := y, x \quad .$$

Statement **skip** satisfies $\{R\}$ **skip** $\{R\}$, for all predicates R.

THE ALTERNATIVE STATEMENT

The statement **if** B **then** $S1$ **else** $S2$ can be written in the notation of *guarded commands* as the *alternative statement* shown below.

> **if** $B \to S1$
> [] $\neg B \to S2$
> **fi**

A phrase of the form $B \to S$ is called a *guarded command*. B is the guard at the gate $\to$, making sure that command S is executed only when appropriate. In the guarded command notation, an alternative statement can be written with more than two possible choices. For example, here is an alternative statement, called IFG, with three guarded commands.

(10.6) IFG : **if** $B1 \to S1$
> [] $B2 \to S2$
> [] $B3 \to S3$
> **fi**

Execution of the alternative statement proceeds as follows. If none of the guards is *true*, execution aborts [5]. If at least one guard is *true*, then one *true* guard is chosen and the corresponding command is executed.

There are two key points with the alternative statement.

- Execution aborts if no guard is *true*.

- If more than one guard is *true*, only one of them is chosen (arbitrarily) and its corresponding command is executed.

If more than one guard can be *true*, the alternative statement is said to be *nondeterministic*. Nondeterminism helps in writing some algorithms more cleanly and in allowing symmetry. For example, in the program below, it doesn't matter which of x and y is stored in z when $x = y$. With a nondeterministic alternative statement, the programmer need not make a choice.

> **if** $x \leq y \to z := y$
> [] $y \leq x \to z := x$
> **fi**
> $\{z$ is the maximum of x and $y\}$

---

[5] To abort means to terminate prematurely. When a program aborts, what happens is undefined, although a good implementation will give an error message and then stop execution.

As another example, here is a program that sorts variables w, x, y, z by repeatedly swapping their values. It would be messier to write this program without nondeterminism.

> **while** $\neg(w \le x \le y \le z)$ **do**
>   **if** $w > x \rightarrow w, x := x, w$
> $[\![$ $x > y \rightarrow x, y := y, x$
> $[\![$ $y > z \rightarrow y, z := z, y$
>   **fi**
> $\{w \le x \le y \le z\}$

To prove $\{Q\}$ *IFG* $\{R\}$ (see (10.6)), we have to show that (i) when execution starts, at least one guard is *true* and (ii) each guarded command truthifies R.

(10.7) **Proof method for IFG.** To prove $\{Q\}$ *IFG* $\{R\}$, it suffices to prove

(a) $Q \Rightarrow B1 \vee B2 \vee B3$,
(b) $\{Q \wedge B1\}$ $S1$ $\{R\}$,
(c) $\{Q \wedge B2\}$ $S2$ $\{R\}$, and
(d) $\{Q \wedge B3\}$ $S3$ $\{R\}$.

This method of proof extends to alternative commands with more than three or fewer than three guarded commands in the obvious way.

Conditional expressions

Analogous to the conditional statement **if** B **then** $S1$ **else** $S2$, we have the *conditional expression*

(10.8) **if** B **then** $E1$ **else** $E2$

where B is a boolean expression and $E1$ and $E2$ are expressions of the same type (both yield integers, or booleans, etc.). Evaluation of this expression yields the value of $E1$ if B is *true* and the value of $E2$ otherwise.

Examples of conditional expressions. Consider a state with $x = 5$, $y = 4$, $b = true$, and $c = false$. In this state, we have

(a) (**if** $x = y$ **then** x **else** $x + 2$) $= 7$

(b) (**if** $x \ne y$ **then** x **else** $x + 2$) $= 5$

(c) (**if** $b \vee c$ **then** $x \cdot y$ **else** $x + 2$) $= 20$

(d) (**if** $b \wedge c$ **then** $x \cdot y$ **else** $x + 2$) $= 7$

(e) (**if** b **then** $c \Rightarrow b$ **else** $b \Rightarrow c$) $= true$

(f) (**if** c **then** $c \vee b$ **else** $b \wedge c$) $= false$ $\square$

There are two rules for manipulating the **if-then-else** expression:

(10.9) **Axiom, Conditional:** $B \Rightarrow ((\text{if } B \text{ then } E1 \text{ else } E2) = E1)$

(10.10) **Axiom, Conditional:** $\neg B \Rightarrow ((\text{if } B \text{ then } E1 \text{ else } E2)) = E2)$

Exercises for Chapter 10

10.1 Consider an array segment $b[0..n-1]$, where $0 \leq n$. Let j and k be two integer variables satisfying $0 \leq j \leq k < n$. By $b[j..k]$ we mean the subarray of b consisting of $b[j]$, $b[j+1]$, $\ldots$, $b[k]$. The segment $b[j..k]$ is empty if $j = k+1$.

Translate the following sentences into boolean expressions. For example, the first one can be written as $(\forall i \mid j \leq i \leq k : b[i] = 0)$. Some of the statements may be ambiguous, in which case you should write down all the reasonable possibilities. Simplify the expressions where possible. You may use abbreviations —e.g. use $x \in b[0..n-1]$ for $(\exists i \mid 0 \leq i < n : x = b[i])$.

(a) All elements of $b[j..k]$ are zero.

(b) No values of $b[j..k]$ are zero.

(c) Some values of $b[j..k]$ are zero. (What does "some" mean?)

(d) All zeros of $b[0..n-1]$ are in $b[j..k]$.

(e) Some zeros of $b[0..n-1]$ are in $b[j..k]$.

(f) Those values in $b[0..n-1]$ that are not in $b[j..k]$ are in $b[j..k]$.

(g) It is not the case that all zeros of $b[0..n-1]$ are in $b[j..k]$.

(h) If $b[0..n-1]$ contains a zero, then so does $b[j..k]$.

(i) If $b[j..k]$ contains two zeros then $j = 1$.

(j) Either $b[1..j]$ or $b[j..k]$ contains a zero (or both).

(k) The values of $b[j..k]$ are in ascending order.

(l) If x is in $b[j..k]$ then $x + 1$ is in $b[k+1..n-1]$.

(m) Segment $b[j..k]$ contains at least two zeros.

(n) Every element of $b[0..j]$ is less than x and every value of $b[j+1..k-1]$ exceeds x.

10.2 Define a predicate $perm(b, c, n)$ that means: array segment $b[0..n-1]$ is a permutation of array segment $c[0..n-1]$. (One array segment is a permutation of another if its values can be interchanged (swapped) so the two segments are equal. For example, $(3, 5, 2, 5)$ is a permutation of $(2, 3, 5, 5)$.

10.3 Define a predicate $ascending(b, n)$ that means that array segment $b[0..n-1]$ is sorted (in ascending order).

10.4 Define the term *median* of an array of distinct numbers —a value such that half are lower and half are greater than the value. (You have to make the definition clearer for the case that the size of the set is even.)

10.5 Define the reverse of an array, e.g. the reverse of $(3, 2, 5, 5)$ is $(5, 5, 2, 3)$.

Exercises on program specifications

10.6 Formalize the following English specifications. Be sure to introduce necessary restrictions. Also, if there are ambiguities or vague parts of the English specification, resolve them in some reasonable way (there may not be a single answer). You may use $x \uparrow y$ for the maximum of x and y. Note that $\uparrow$ is symmetric and associative, so $\uparrow$ can be used as a quantifier (see Chap. 8). However, $\uparrow$ over the integers has no identity, so axioms that require an identity cannot be used with it.

(a) Calculate the sum of the elements of $b[j..k-1]$.

(b) Find the maximum value of $b[j..k-1]$.

(c) Find the index of a maximum value of $b[j..k-1]$.

(d) Store in array $c[0..n-1]$ a sorted (in ascending order) permutation of $b[0..n-1]$. Use the predicate $perm(b,c,n)$ to denote that $b[0..n-1]$ is a permutation of $c[0..n-1]$ (see Exercise 10.2). You can also use predicate $ascending(b,n)$ of Exercise 10.3.

(e) Calculate the greatest power of 2 that is not greater than n.

(f) Count how many zeros $b[0..n-1]$ has.

(g) Suppose we have an array integer $b[0..n-1]$. Each of its subsegments $b[i..j-1]$ has a sum. Find the largest such sum. Hints: For this specification, it helps to give the sum of a segment $b[i..j-1]$ a name, say $S_{i,j}$, so that the formula for the sum does not appear everywhere.

(h) Integer array $s[0..n]$ contains the grade of each student on a homework, where a negative number means that no grade was handed in. All the grades handed in turned out to be different. Find the average grade.

(i) Integer array $s[0..n]$ contains the grade of each student on a homework, where a negative number means that no grade was handed in. Find the median (i.e. the number such that half the grades are lower and half higher).

(j) Consider boolean array $bit[0..n-1]$ as a sequence of bits. Think of it as the binary representation of a decimal number, with $bit[0]$ being the least significant bit. Calculate the decimal number.

(k) Array b contains the list of students at Cornell and c the list of people who have part-time jobs in Ithaca. Both lists are alphabetically ordered. Find the first person who is on both lists.

(l) Array b contains the list of students at Cornell and c the list of people who have part-time jobs in Ithaca. Both lists are alphabetically ordered. Make up an alphabetical list of all people who are on both lists.

(m) Array b is sorted. Find the index of the rightmost element (i.e. the element with the highest index) that equals x (also take care of the case that x is not in b in a suitable fashion).

(n) Set x to *true* if integer array b contains a negative value.

10.7 Formalize the following specifications, some of which will require the use of rigid variables to indicate how program variables are to be changed. Be sure to introduce necessary restrictions on the input. Also, if there are ambiguities or vague parts of the English specification, resolve them in any reasonable way that comes to mind (there may not be a single answer).

(a) Double each element of integer array b.

(b) Sort integer array b.

(c) Find the minimum and maximum values of array b.

(d) Reverse b —e.g. change $(3, 2, 5, 5, 1)$ into $(1, 5, 5, 2, 3)$.

(e) Swap arrays b and c.

(f) Integer array b contains the grades of students on a homework (a negative number means the grade was not handed in). Change John's grade (it is $b[j]$) to 80, but if it is a late grade, also subtract 10 percent.

(g) Delete duplicates from array b.

(h) Array b contains red, white, and blue numbers. Put all the red ones first, then the blues, then the whites.

(i) Permute array b so that the elements smaller than x come first and give the index of the last element that is smaller than x.

Exercises on the assignment statement

10.8 Calculate and simplify the weakest precondition for the following (where x and y are integer variables):

$$\{?\} \quad x := x + y;\ y := x - y;\ x := x - y\ \ \{x = X \wedge y = Y\}\quad.$$

10.9 Calculate and simplify the weakest precondition for the following (where x and y are boolean variables):

$$\{?\} \quad x := x \neq y;\ y := x \neq y;\ x := x \neq y\ \ \{(x \equiv X) \wedge (y \equiv Y)\}\quad.$$

10.10 Suppose the number of apples that Mary and John have (represented by m and j, respectively) are related by the formula (C is some constant)

$$P : C = m + 2 \cdot j\quad.$$

Find a solution for e in $\{P \wedge even.m\}\ m, j := m \div 2, e\ \{P\}$.

10.11 The Fibonacci numbers $F.i$ are given by $F.0 = 0$, $F.1 = 1$, and $F.n = F(n-1) + F(n-2)$ for $n \geq 2$. For example, the first few Fibonacci numbers are $0, 1, 1, 2, 3, 5, 8$. The following predicate defines the variables n, a, and b:

$$P : n > 0 \wedge a = F.n \wedge b = F(n-1)$$

Find a solution for e and f in $\{P\}\ n, a, b := n + 1, e, f\ \{P\}$.

Exercises on the conditional statement

10.12 Use method (10.5) to prove that the following annotated program is correct.

$$\{x > 5\}$$
$$\textbf{if } x \leq y \textbf{ then skip else } x, y := y, x$$
$$\{x \leq y\}$$

10.13 Use method (10.5) to prove that the following annotated program is correct.

$$\{x = \text{X}\}$$
$$\textbf{if } x < 0 \textbf{ then } x := -x \textbf{ else skip}$$
$$\{x = abs.\text{X}\}$$

10.14 Use method (10.5) to prove that the following annotated program is correct.

$$\{y > 0 \land z \cdot x^y = X\}$$
if $odd.y$ **then** $z, y := z \cdot x, y - 1$ **else** $x, y := x \cdot x, y/2$
$$\{y \geq 0 \land z \cdot x^y = X\}$$

10.15 Some programming languages have the conditional **if** B **then** S. It is executed as follows: if B is $true$, then execute S; otherwise, do nothing. How does one prove $\{Q\}$ **if** B **then** S $\{R\}$? Hint: Rewrite this statement in terms of the other statement **if** B **then** $S1$ **else** $S2$.

Chapter 11

A Theory of Sets

W e define set theory as an extension of predicate calculus. A *set* is
simply a collection of distinct (different) elements. Examples of sets
are the set of integers, the set of brown cows, and the set of computer science departments. A cornerstone of mathematics, the set is also an essential
ingredient of computer science and finds application in areas such as artifi-
cial intelligence, databases, and programming languages. The study of sets
leads to questions about the existence of many kinds of infinities. Thus,
while appearing simple, set theory is a rich intellectual playground.

11.1   Set comprehension and membership

We start our discussion with *set enumeration* and *set comprehension*, two
methods for describing sets. We define set comprehension in terms of testing
membership in sets. This membership test is the basis for the definition of
equality of sets, as well as for everything else we do with sets.

SET ENUMERATION AND SET COMPREHENSION

One way to describe a set is to list its elements. In the usual syntax, called
set enumeration, the list is delimited by braces { and } and its elements
are separated by commas. For example, $\{5, 2, 3\}$ denotes the set consisting
of the elements 2, 3, and 5. And, if b and c are variables, evaluation of
the expression $\{b, c\}$ in a state yields a set whose elements are the values
of b and c in that state.

Set enumeration has its drawbacks. Consider, for example, describing the
set of even integers between 0 and 9999 in this fashion! A more effective
means of specifying a set is *set comprehension*, which describes a set not
by listing its elements but by stating properties enjoyed (exclusively) by its
elements. For example, the set comprehension

$$\{x: \mathbb{Z} \mid 0 \le x < 5 : 2 \cdot x\}$$

denotes the set of values $2 \cdot x$ for all integers x that satisfy $0 \le x < 5$.
The integers that satisfy $0 \le x < 5$ are $0, 1, 2, 3$ and 4; hence, the set
comprehension above denotes the set $\{0, 2, 4, 6, 8\}$.

We now give the general form of set comprehension. Let R be a predicate, E an expression, x a list of dummies, and t a type. Evaluation of

(11.1) $\{x{:}t \mid R : E\}$

in a state yields the set of values that result from evaluating $E[x := v]$ in the state for each value v in t such that $R[x := v]$ holds in that state. In contexts where the type of the dummy is obvious, the type may be omitted. If E has type $t1$, then the set comprehension has type $set(t1)$.

The notation for set comprehension is similar to that for quantification in (8.6). As in (8.6), boolean expression R is the range and expression E is the body. The notions of scope, free variable, and bound variable apply to set comprehension, without change. Finally, the dummies may have different types, just as in a quantification.

We can define a set enumeration $\{e_0, \ldots, e_{n-1}\}$ to be an abbreviation of a set comprehension:

(11.2) $\{e_0, \ldots, e_{n-1}\} = \{x \mid x = e_0 \vee \cdots \vee x = e_{n-1} : x\}$.

In the following examples of set comprehension, the dummies range over the integers.

$\{i \mid 0 < i < 4 : i\}$	The set $\{1, 2, 3\}$
$\{i \mid 0 < i < 50 \wedge even.i : i\}$	Even positive integers less than 50
$\{i \mid 0 < 2{\cdot}i < 50 : 2{\cdot}i\}$	Even positive integers less than 50
$\{x, y \mid 1 \le x \le 2 \le y \le 3 : x^y\}$	The set $\{1^2, 1^3, 2^2, 2^3\}$
$\{x \mid 0 \le x < 3 : x{\cdot}y\}$	The set $\{0{\cdot}y, 1{\cdot}y, 2{\cdot}y\}$
$\{x \mid 0 \le x < 0 : x{\cdot}y\}$	The empty set $\{\,\}$

The second and third examples denote the same set. The fourth example shows two dummies in one set comprehension. The fifth illustrates the use of a free variable in a set comprehension; the value of the expression depends on the value of y in the state in which the expression is evaluated.

The universe

A theory of sets concerns sets constructed from some collection of elements. There is a theory of sets of integers, a theory of sets of characters, a theory of sets of sets of integers, and so forth. This collection of elements is called the *domain of discourse* or the *universe of values*; it is denoted by $\mathbf{U}$. The universe can be thought of as the type of every set variable in the theory. For example, if the universe is $set(\mathbb{Z})$, then $v{:}set(\mathbb{Z})$.

When several set theories are being used at the same time, there is a different universe for each. The name $\mathbf{U}$ is then overloaded, and we have to distinguish which universe is intended in each case. This overloading is similar to using the constant 1 as a denotation of an integer, a real, the identity matrix, and even (in some texts, alas) the boolean *true* .

SET MEMBERSHIP AND EQUALITY

For an expression e and a set-valued expression [1] S ,

$$e \in S$$

is an expression whose value is the value of the statement " e is a member of S ", or " e is in S ". The expression $\neg(e \in S)$ may be abbreviated by $e \notin S$. For example, $2 \in \{1, 2, 4\}$ is *true* and $3 \notin \{1, 2, 4\}$ is *true* . Symbol $\in$ is treated as a conjunctional operator and has the same precedence as the sign $=$ for equality —see the precedence table on the inside front cover.

Set comprehension is formalized by defining membership in the set it denotes. For expression $F{:}t$ and set $\{x \mid R : E{:}t\}$ (for some type t), we define:

(11.3)   **Axiom, Set membership:** Provided $\neg occurs(\text{`}x\text{'}, \text{`}F\text{'})$,

$$F \in \{x \mid R : E\} \;\equiv\; (\exists x \mid R : F = E) \ .$$

Two sets are equal if they contain the same elements. Thus, for sets S and T we have the following axiom. [2]

(11.4)   **Axiom, Extensionality:** $S = T \;\equiv\; (\forall x \mid : x \in S \;\equiv\; x \in T)$.

Several consequences follow from the definition of set comprehension, set membership and the abbreviation $\{e_0, \ldots, e_{n-1}\}$ for $\{x \mid x = e_0 \lor \ldots \lor x = e_{n-1} : x\}$:

- $\{x \mid false : E\}$ and $\{\ \}$ denote the *empty set*, i.e. the set with no elements. Exercise 11.4 asks you to prove formally that $e \in \{x \mid false : E\} \equiv false$ for all e and E . The empty set is also denoted by $\emptyset$. Note that the set $\{\{\ \}\}$ contains one element: the set $\{\ \}$.

- The expressions $\{x \mid x = e : x\}$ (where x does not occur free in e) and $\{e\}$ yield a *singleton set*, which has one element, the value of e . Note that e yields a value, while $\{e\}$ yields a set containing that value. The expression $e \in \{e\}$ is always *true*; $e = \{e\}$ is not even an expression since the LHS and RHS have different types (t and $set(t)$ for some type t).

---

[1] See Table 11.1 on page 200 for type restrictions on set-theory expressions.

[2] An *extensional* definition of set equality depends only on the contents of the sets. An *intentional* definition would concern how the sets are defined or constructed. For example, was the element 0 added to the set before or after the element 2 ?

- Since $\lor$ is symmetric, the order of elements in a set enumeration is irrelevant. For example, $\{1,3\} = \{x \mid x = 1 \lor x = 3 : x\} = \{x \mid x = 3 \lor x = 1 : x\} = \{3,1\}$.

- Since $\lor$ is idempotent, repetition of elements in a set enumeration has no significance. For example, $\{1,3\} = \{1,1,3\}$. Duplicates may arise when expressions are used to designate elements. For example, evaluation of $\{b,c\}$ in a state in which $b \neq c$ yields a set with two elements, while its evaluation in a state in which $b = c$ yields a set with one element.

- Sets may be elements of other sets. As an example from sports, major league baseball in the U.S. consists of a set of two leagues; each league is a set of two subdivisions, the East and the West; each subdivision is a set of teams; and each team is a set of players.

- Sets may be heterogeneous, i.e. they may contain different kinds of elements. For example, the set $\{\{1, \text{"B"}\}, \{2,3\}, \text{"A"}\}$ has as its elements two sets and the character A. The universe for such a set consists of the integers together with the characters.

Although Leibniz (1.5) can be used to show equality of sets, sometimes axiom Extensionality (11.4) works better. If we can show that an arbitrary element is in S exactly when it is in T , then (11.4) allows us to conclude that $S = T$. We now use the second method to prove the (obvious) theorem

$$(11.5) \quad S = \{x \mid x \in S : x\} \ .$$

According to axiom Extensionality (11.4), it suffices to prove that $v \in S \equiv v \in \{x \mid x \in S : x\}$, for arbitrary v . We have,

$$
\begin{aligned}
& v \in \{x \mid x \in S : x\} \\
= \quad & \langle \text{Definition of membership (11.3)} \rangle \\
& (\exists x \mid x \in S : v = x) \\
= \quad & \langle \text{Trading (9.19), twice} \rangle \\
& (\exists x \mid x = v : x \in S) \\
= \quad & \langle \text{One-point rule (8.14)} \rangle \\
& v \in S
\end{aligned}
$$

THE TRADITIONAL FORM OF SET COMPREHENSION

The traditional mathematical notation for set comprehension is

$$\{x \mid R\}$$

(x is a single variable), which we view as an abbreviation of $\{x \mid R : x\}$. For example,

$\{i \mid 0 < i < 4\}$ is the set $\{1, 2, 3\}$, and
$\{i \mid 0 < i < 50 \land even.i\}$ is all even positive integers less than 50.

The notation $\{x \mid R\}$ is often extended to allow expressions in place of dummy x. Thus, the set of even integers in the range $0..99$ could be written as $\{2 \cdot x \mid 0 \leq x < 100\}$. *We do not use this extension*, because it is ambiguous: it is impossible to tell which variable(s) is the dummy. For example, the value of the expression

$$\{x + y \mid x = y + 1\}$$

depends critically on what the dummies are:

> If x is the dummy, then the set is $\{2 \cdot y + 1\}$;
> If y is the dummy, then the set is $\{2 \cdot x - 1\}$;
> If x and y are the dummies, then it is the set of odd integers.

Actually, the traditional form of set comprehension is sufficient to describe any set that can be described using our more general form (11.1), as the following theorem shows:

(11.6)   Provided $\neg occurs(\text{'}y\text{'}, \text{'}R\text{'})$ and $\neg occurs(\text{'}y\text{'}, \text{'}E\text{'})$,
      $\{x \mid R : E\} \;=\; \{y \mid (\exists x \mid R : y = E)\}$.

We introduced the new notation for set comprehension because it is unambiguous, more suitable for expressing some sets, and more amenable to formal manipulation. Also, we can carry over the definitions of scope, free variables, and bound variables from our notation for quantification. However, we do use the conventional notation when it is more appropriate.

SETS VERSUS PREDICATES

Theorem (11.7) formalizes the connection between sets and predicates: a predicate is a representation for the set of argument-values for which it is *true*.

(11.7) $x \in \{x \mid R\} \;\equiv\; R$

Note that x is used with two different meanings in the LHS of (11.7). The leftmost occurrence of x is free, as are free occurrences of x in the RHS. All occurrences of x in $\{x \mid R\}$ are bound. Since (11.7) is valid, by instantiating free variable x with any expression y we have $y \in \{x \mid R\} \equiv R[x := y]$ for any expression y.

Theorem (11.7) can be stated as the following principle.

(11.8)   **Principle of comprehension.** To each predicate R there corresponds a set comprehension $\{x:t \mid R\}$, which contains the objects in t that satisfy R; R is called a *characteristic predicate* of the set.

Theorem (11.7) tells us that we can define a set not only by using set comprehension but also by giving its characteristic predicate. For example, the following definitions of the set $S = \{3,5\}$ are equivalent:

$$S \;=\; \{x \mid x = 3 \vee x = 5\}$$
$$x \in S \;\equiv\; x = 3 \vee x = 5 \quad \text{(for all } x\text{)}.$$

Henceforth, we use both forms interchangeably, without explicit mention, relying on the more suitable one in each context.

Using (11.7), we can easily prove the following theorems:

(11.9) $\{x \mid Q\} = \{x \mid R\} \;\equiv\; (\forall x \mid : Q \equiv R)$

(11.10) **Theorem.** $\{x \mid Q\} = \{x \mid R\}$ is valid   iff $Q \equiv R$ is valid.

Theorem (11.10) gives us a new method of proving equality of sets: show that their characteristic predicates are equivalent. We now have three general methods for proving set equality:

(11.11) **Methods for proving set equality** $S = T$:
  (a) Use Leibniz directly.
  (b) Use axiom Extensionality (11.4) and prove $v \in S \;\equiv\; v \in T$ for an arbitrary value v.
  (c) Prove $Q \equiv R$ and conclude $\{x \mid Q\} = \{x \mid R\}$.

TABLE 11.1. TYPES OF SET EXPRESSIONS IN THEORY $set(t)$

Expression	Example (with types)	Type of result
Empty set, universe, variable	$\emptyset$ or $\mathbf{U}$ or S	$set(t)$
Set enumeration	$\{e_1:t, \ldots, e_n:t\}$	$set(t)$
Set comprehension	$\{x \mid R:\mathbb{B} : E:t\}$	$set(t)$
	$\{x:t \mid R:\mathbb{B}\}$	$set(t)$
Set membership	$x:t \in S:set(t)$	$\mathbb{B}$
Set equality	$S:set(t) = T:set(t)$	$\mathbb{B}$
Set size	$\# S:set(t)$	$\mathbb{N}$
$\subset, \supset, \subseteq, \supseteq$	$S:set(t) \subseteq T:set(t)$	$\mathbb{B}$
Complement	$\sim S:set(t)$	$set(t)$
$\cup, \cap, -$	$S:set(t) \cup T:set(t)$	$set(t)$
Power set	$(\mathcal{P}\ S):set(t)$	$set(set(t))$

11.2   Operations on sets

We define some useful operations on sets. Table 11.1 contains information concerning the types of set expressions and the table on the inside front cover defines the precedences of operations.

Throughout this chapter, variables S, T, U, V have type $set(t)$ for some type t. With this convention, we do not have to state their types each time they are used. This convention also allows us to write most axioms and theorems without quantification. For example, the definition of cardinality (11.12) given below could be written as

$$(\forall S : set(t) \mid : \#S = (\Sigma x \mid x \in S : 1))\ .$$

However, Metatheorem (9.16) on page 162 allows us to eliminate the quantification and write simply $\#S = (\Sigma x \mid x \in S : 1)$. But in this expression, it is necessary to remember that S has a particular type and that it cannot be replaced by expressions of other types.

CARDINALITY OF FINITE SETS

The *cardinality* or *size* of a finite set S, denoted by $\#S$, is the number of elements in S. It can be defined as follows: [3]

(11.12)   **Axiom, Size:** $\#S = (\Sigma x \mid x \in S : 1)$.

SUBSET AND SUPERSET

Set S is a *subset* of set T if every element of S is an element of T. This is depicted in the *Venn diagram* in this paragraph. With the convention that a circle surrounds the elements of a set, the circle for set S is drawn inside the circle for set T to indicate that every element of S is also an element of T, i.e. S is a subset of T.

S is a *proper subset* of T if it is a subset of T and $S \neq T$ holds. Predicates $S \subseteq T$ and $S \subset T$ denote subset and proper subset:

(11.13)   **Axiom, Subset:** $S \subseteq T \;\equiv\; (\forall x \mid x \in S : x \in T)$
(11.14)   **Axiom, Proper subset:** $S \subset T \;\equiv\; S \subseteq T \wedge S \neq T$

Set T is a *superset* of (*proper superset* of) S if S is a subset of (proper subset of) T. Operators $\supseteq$ and $\supset$ denote superset and proper superset.

---

[3] The notation $|S|$ is sometimes used for $\#S$.

(11.15) **Axiom, Superset:** $T \supseteq S \ \equiv\ S \subseteq T$

(11.16) **Axiom, Proper superset:** $T \supset S \ \equiv\ S \subset T$

Operators $\subset$, $\subseteq$, $\supset$, and $\supseteq$ are conjunctional and have the same precedence as $=$. As with all conjunctional operators, a superimposed slash denotes negation. For example, $S \not\subseteq T$ means $\neg(S \subseteq T)$.

COMPLEMENT

The *complement* of S, written $\sim S$, [4] is the set of elements that are not in S (but are in the universe). In the Venn diagram in this paragraph, we have shown set S and universe $\mathbf{U}$. The non-filled area represents $\sim S$.

(11.17) **Axiom, Complement:** $v \in \sim S \ \equiv\ v \in \mathbf{U} \wedge v \notin S$

For example, for $\mathbf{U} = \{0,1,2,3,4,5\}$, we have

$$\sim\{3,5\} \ = \ \{0,1,2,4\} \quad,$$
$$\sim\mathbf{U} = \emptyset \ , \qquad \sim\emptyset = \mathbf{U} \ .$$

We can easily prove

(11.18) $v \in \sim S \ \equiv\ v \notin S$ \quad (for v in $\mathbf{U}$).

(11.19) $\sim\sim S = S$

SET UNION, INTERSECTION, AND DIFFERENCE

The three operations *union, intersection,* and *difference* are used to construct a set from two other sets. The union of sets S and T, written

---

[4] S^C and $\overline{S}$ are also used to denote set complement.

FIGURE 11.1. Venn Diagrams for Union, Intersection, and Difference

$$S \cup T \qquad\qquad S \cap T \qquad\qquad S - T$$

$S \cup T$, is the set of all elements that are in S or T (or both). The intersection of S and T, written $S \cap T$, is the set of all elements that are in both S and T. The difference of S and T is the set of elements that are in S and not in T. Operators $\cup$, $\cap$, and $-$ have the same precedence. These operators are depicted using Venn diagrams in Fig. 11.1; S and T are the circles, and the shaded portion is the operator applied to S and T.

Formally, these three operations are defined as follows.

(11.20) **Axiom, Union:** $v \in S \cup T \equiv v \in S \vee v \in T$

(11.21) **Axiom, Intersection:** $v \in S \cap T \equiv v \in S \wedge v \in T$

(11.22) **Axiom, Difference:** $v \in S - T \equiv v \in S \wedge v \notin T$

Examples of $\cup$, $\cap$, and $-$

$$\{3,5,6\} \cup \{3,2,1\} = \{3,5,6,2,1\} \ ,$$
$$\{3,5,6\} \cap \{3,2,1\} = \{3\} \ ,$$
$$\{3,5,6\} - \{3,2,1\} = \{5,6\} \ . \qquad \square$$

From definition (11.17) of set complement, we see that the complement of S is the difference of the universe and S: $\sim S = \mathbf{U} - S$.

Sets S and T are *disjoint* if they have no elements in common, i.e. if $S \cap T = \emptyset$.

POWER SET

The *power set* of a set S, denoted by $\mathcal{P}S$, is the set of subsets of S:[5]

(11.23) **Axiom, Power set:** $v \in \mathcal{P}S \equiv v \subseteq S$.

For example, $\mathcal{P}\{3,5\} = \{\emptyset, \{3\}, \{5\}, \{3,5\}\}$.

11.3   Theorems concerning set operations

RELATING SET AND BOOLEAN EXPRESSIONS

The definitions of the set operators reveal a connection between the set operators and the propositional operators. For example, in the definition of $\cup$ (repeated below), as the phrase "$v \in$" of the LHS is distributed

---

[5] 2^S is also used to denote the power set of S.

inward to the operands S and T of the RHS, $\cup$ becomes $\vee$:

$$v \in S \cup T \ \equiv \ v \in S \ \vee \ v \in T \ .$$

This connection suggests that properties of propositional operators may be reflected in similar properties of set operators. The two pairs of valid relations below reinforce this conjecture. The first pair indicates how law of Absorption (3.43a) becomes an absorption law for sets. The second pair indicates how Zero of $\wedge$ (3.40) becomes a zero law for $\cap$.

$$
\begin{aligned}
S \wedge (S \vee T) &\equiv S \qquad &(S,T:\mathbb{B}) \\
S \cap (S \cup T) &= S \qquad &(S,T:set(t))
\end{aligned}
$$

$$
\begin{aligned}
S \wedge false &\equiv false \qquad &(S:\mathbb{B}) \\
S \cap \emptyset &= \emptyset \qquad &(S:set(t)).
\end{aligned}
$$

To arrive at the formal description of the connection between set expressions and boolean expressions, we need the following definition.

(11.24) **Definition.** Let E_s be a set expression constructed from set variables, $\emptyset$, **U** (a universe for all set variables in question), $\sim$, $\cup$, and $\cap$. Then E_p is the expression constructed from E_s by replacing

$\emptyset$ with *false*,	**U** with *true*,
$\cup$ with $\vee$,	$\cap$ with $\wedge$,
$\sim$ with $\neg$.	

The construction is reversible: E_s can be constructed from E_p.

Then we have the following Metatheorem (11.25).

(11.25) **Metatheorem.** For any set expressions E_s and F_s:
(a) $E_s = F_s$ is valid iff $E_p \equiv F_p$ is valid,
(b) $E_s \subseteq F_s$ is valid iff $E_p \Rightarrow F_p$ is valid,
(c) $E_s = \mathbf{U}$ is valid iff E_p is valid.

The proof of this metatheorem, which relies on mathematical induction, is relatively lengthy and would detract from our main task here, which is to survey theorems concerning set operators. Therefore, the proof is presented in Exercises 12.47–12.52 of Chap. 12.

Use of the metatheorem reduces tremendously the work needed to prove validity of various set expressions. Note that the metatheorem does not mention expressions that contain $\equiv$ and $\Rightarrow$. However, any such expression is equivalent to one that contains only $\neg$, $\wedge$, and $\vee$, since $\equiv$ and $\Rightarrow$ can be replaced using Mutual implication (3.80) and Implication (3.59), $p \Rightarrow q \equiv \neg p \vee q$. Therefore, any boolean expression is equivalent to some expression E_p for which E_s can be constructed. Going the other way, set difference can be eliminated from a set expression using the identity $S - T = S \cap \sim T$.

PROPERTIES OF SET OPERATORS

We give some theorems concerning set operators. All the theorems concerning union and intersection given below are proved directly using Metatheorem (11.25). Therefore, they are given the same names as their propositional counterparts. There is no need to memorize these theorems, for you can construct them using Metatheorem (11.25) whenever necessary.

Basic properties of $\cup$

(11.26) **Symmetry of $\cup$:** $S \cup T = T \cup S$

(11.27) **Associativity of $\cup$:** $(S \cup T) \cup U = S \cup (T \cup U)$

(11.28) **Idempotency of $\cup$:** $S \cup S = S$

(11.29) **Zero of $\cup$:** $S \cup \mathbf{U} = \mathbf{U}$

(11.30) **Identity of $\cup$:** $S \cup \emptyset = S$

(11.31) **Weakening:** $S \subseteq S \cup T$

(11.32) **Excluded middle:** $S \cup \sim S = \mathbf{U}$

Basic properties of $\cap$

(11.33) **Symmetry of $\cap$:** $S \cap T = T \cap S$

(11.34) **Associativity of $\cap$:** $(S \cap T) \cap U = S \cap (T \cap U)$

(11.35) **Idempotency of $\cap$:** $S \cap S = S$

(11.36) **Zero of $\cap$:** $S \cap \emptyset = \emptyset$

(11.37) **Identity of $\cap$:** $S \cap \mathbf{U} = S$

(11.38) **Strengthening:** $S \cap T \subseteq S$

(11.39) **Contradiction:** $S \cap \sim S = \emptyset$

Basic properties of combinations of $\cup$ and $\cap$

(11.40) **Distributivity of $\cup$ over $\cap$:**
$$S \cup (T \cap U) = (S \cup T) \cap (S \cup U)$$

(11.41) **Distributivity $\cap$ over $\cup$:**
$$S \cap (T \cup U) = (S \cap T) \cup (S \cap U)$$

Basic properties of combinations of $\cup$ and $\cap$ (cont.)

(11.42) **De Morgan:** (a) $\sim(S \cup T) = \sim S \cap \sim T$

(b) $\sim(S \cap T) = \sim S \cup \sim T$

Some other theorems concerning $\cup$ and $\cap$ are given below. They can be proved directly using predicate logic and set theory.

Additional properties of $\cup$ and $\cap$

(11.43) $S \subseteq T \wedge U \subseteq V \Rightarrow (S \cup U) \subseteq (T \cup V)$

(11.44) $S \subseteq T \wedge U \subseteq V \Rightarrow (S \cap U) \subseteq (T \cap V)$

(11.45) $S \subseteq T \equiv S \cup T = T$

(11.46) $S \subseteq T \equiv S \cap T = S$

(11.47) $S \cup T = \mathbf{U} \equiv (\forall x \mid x \in \mathbf{U} : x \notin S \Rightarrow x \in T)$

(11.48) $S \cap T = \emptyset \equiv (\forall x \mid : x \in S \Rightarrow x \notin T)$

The following theorems concerning set difference can be proved using predicate calculus and set theory. However, Metatheorem (11.25) can be used to advantage if set difference is replaced using Difference (11.22).

Properties of set difference

(11.49) $S - T = S \cap \sim T$

(11.50) $S - T \subseteq S$

(11.51) $S - \emptyset = S$

(11.52) $S \cap (T - S) = \emptyset$

(11.53) $S \cup (T - S) = S \cup T$

(11.54) $S - (T \cup U) = (S - T) \cap (S - U)$

(11.55) $S - (T \cap U) = (S - T) \cup (S - U)$

We turn to theorems concerning subset and superset. Some of these cannot be proved easily using Metatheorem (11.25), so we use predicate logic and set theory directly. Note the relation between $\subseteq$ and $\Rightarrow$. In any given state, one set is a subset of another iff the characteristic predicate for the one implies the characteristic predicate for the other:

Implication versus subset

(11.56) $(\forall x \mid: P \Rightarrow Q) \;\equiv\; \{x \mid P\} \subseteq \{x \mid Q\}$.

Here is a proof of (11.56).

$$\{x \mid P\} \subseteq \{x \mid Q\}$$
$$= \quad \langle \text{Subset } \subseteq (11.13), \text{ where } \neg occurs(`v`,`P,Q`) \rangle$$
$$(\forall v \mid v \in \{x \mid P\} : v \in \{x \mid Q\})$$
$$= \quad \langle (11.7), \text{ twice} \rangle$$
$$(\forall v \mid P[x := v] : Q[x := v])$$
$$= \quad \langle \text{Trading } (9.2); \text{ Dummy renaming } (8.21) \rangle$$
$$(\forall x \mid: P[x := v][v := x] \Rightarrow Q[x := v][v := x])$$
$$= \quad \langle \text{Property of textual substitution, } \neg occurs(`v`,`P,Q`) \rangle$$
$$(\forall x \mid: P \Rightarrow Q)$$

We list below some properties of $\subseteq$ and $\subset$. Properties (11.61) and (11.62) can be viewed as alternative definitions for $\subset$ that make clearer the fact that S satisfying $S \subset T$ is strictly smaller than T.

Properties of subset

(11.57) **Antisymmetry**: $S \subseteq T \wedge T \subseteq S \;\equiv\; S = T$

(11.58) **Reflexivity**: $S \subseteq S$

(11.59) **Transitivity**: $S \subseteq T \wedge T \subseteq U \;\Rightarrow\; S \subseteq U$

(11.60) $\emptyset \subseteq S$

(11.61) $S \subset T \;\equiv\; S \subseteq T \;\wedge\; \neg(T \subseteq S)$

(11.62) $S \subset T \;\equiv\; S \subseteq T \;\wedge\; (\exists x \mid x \in T : x \notin S)$

(11.63) $S \subseteq T \;\equiv\; S \subset T \;\vee\; S = T$

(11.64) $S \not\subset S$

(11.65) $S \subset T \;\Rightarrow\; S \subseteq T$

(11.66) $S \subset T \;\Rightarrow\; T \not\subseteq S$

(11.67) $S \subseteq T \;\Rightarrow\; T \not\subset S$

(11.68) $S \subseteq T \wedge \neg(U \subseteq T) \;\Rightarrow\; \neg(U \subseteq S)$

(11.69) $(\exists x \mid x \in S : x \notin T) \;\Rightarrow\; S \neq T$

(11.70) **Transitivity**: (a) $S \subseteq T \wedge T \subset U \;\Rightarrow\; S \subset U$

$\qquad\qquad\qquad\quad$ (b) $S \subset T \wedge T \subseteq U \;\Rightarrow\; S \subset U$

$\qquad\qquad\qquad\quad$ (c) $S \subset T \wedge T \subset U \;\Rightarrow\; S \subset U$

Transitivity theorems (11.70) can be proved using (11.68) and (11.69). The proof of (11.68) shows a nice use of instantiation (9.13):

$$
\begin{aligned}
& S \subseteq T \land \lnot(U \subseteq T) \\
=\quad & \langle \text{Subset } \subseteq \ (11.13); \text{ Generalized De Morgan } (9.18c)\rangle \\
& S \subseteq T \land (\exists x \mid x \in U : x \notin T) \\
=\quad & \langle \text{Distributivity of } \land \text{ over } \exists,\ (9.21)\rangle \\
& (\exists x \mid x \in U : x \notin T \land S \subseteq T) \\
=\quad & \langle \text{Subset } \subseteq \ (11.13); \text{ Trading } (9.2)\rangle \\
& (\exists x \mid x \in U : x \notin T \land (\forall y \mid : y \in S \Rightarrow y \in T)) \\
\Rightarrow\quad & \langle \text{Monotonicity of } \exists \ (9.27), \text{ using Instantiation } (9.13)\rangle \\
& (\exists x \mid x \in U : x \notin T \land (x \in S \Rightarrow x \in T)) \\
=\quad & \langle \text{Contrapositive } (3.61)\rangle \\
& (\exists x \mid x \in U : x \notin T \land (x \notin T \Rightarrow x \notin S)) \\
=\quad & \langle (3.66),\ p \land (p \Rightarrow q) \equiv p \land q \rangle \\
& (\exists x \mid x \in U : x \notin T \land x \notin S) \\
\Rightarrow\quad & \langle \text{Monotonicity } (9.27), \text{ using } (3.76),\ p \land q \Rightarrow p \rangle \\
& (\exists x \mid x \in U : x \notin S) \\
\Rightarrow\quad & \langle \text{Generalized De Morgan } (9.18c); \text{ Subset } \subseteq \ (11.13)\rangle \\
& \lnot(U \subseteq S)
\end{aligned}
$$

We state three properties of the power set operation. The first property says that the power set of the empty set is a singleton set whose sole element is the empty set. The second property says that S is a member of its power set. The proof of (11.73) requires mathematical induction and must therefore await Chap. 12 —see Exercise 12.15.

Theorems concerning power set $\mathcal{P}$

(11.71) $\mathcal{P}\emptyset = \{\emptyset\}$

(11.72) $S \in \mathcal{P}S$

(11.73) $\#(\mathcal{P}S) = 2^{\#S}$ (for finite set S)

11.4   Union and intersection of families of sets

Union and intersection are symmetric, associative, and idempotent and have identities. Therefore, each is a binary operator $\star$ for which the notation $(\star x \mid R : E)$ is defined, as discussed in Sec. 8.2. Thus, we can use the expressions

(11.74) $(\cup x \mid R : E)$

(11.75) $(\cap x \mid R : E)$

to denote the union and intersection, respectively, of the sets $E[x := v]$ for values v that satisfy $R[x := v]$.

For example,

$$(\cup i \mid 0 \le i < n : \{5^i, 6^i, 7^i\})$$

denotes the set of values 5^i, 6^i, 7^i for i satisfying $0 \le i < n$.

One reason for choosing a particular definition or notation is that it extends easily to other contexts. In this regard, our notation for quantification in Sec. 8.3 was well chosen, for it has allowed immediate use of quantification for the union and intersection of families of sets.

Note that (11.74) and (11.75) satisfy the general axioms of quantification (8.13)–(8.21): Empty range, One-point, Distributivity, Range split, Interchange of dummies, Nesting, and Dummy renaming. In addition, because of the definitions of $\cup$ and $\cap$ in terms of $\vee$ and $\wedge$, other properties of $(\cup x \mid R : E)$ and $(\cap x \mid R : E)$ can be derived from the properties of $(\exists x \mid R : E)$ and $(\forall x \mid R : E)$. We leave this task to the reader.

A set S of sets is called a *partition* of another set T if every element of T is in exactly one of the elements of S. We can state this in another way. Set S *partitions* T if (i) the sets in S are pairwise disjoint and (ii) the union of the sets in S is T, i.e. if

(11.76) **Partition:** $(\forall u, v \mid u \in S \wedge v \in S \wedge u \ne v : u \cap v = \emptyset) \wedge$
$\qquad\qquad (\cup u \mid u \in S : u) = T$.

11.5 The axiom of choice

Given a bag of candy, you can reach in and pick out a piece. In the same way, we would expect to be able to choose some arbitrary element from a nonempty set. We postulate the ability to do so in the following axiom.

(11.77) **Axiom of Choice:** For t a type, there exists a function $f: set(t) \to t$ such that for any nonempty set S, $f.S \in S$.

Thus, f chooses an element from S; it is our formalization of the hand that picks out a piece from a bag of candy.

A more general version of the Axiom of Choice was first formulated by Ernst Zermelo at the beginning of the 20th century. We have stated a simple version only to convey the general idea. Note that the axiom merely states the existence of a choice function; it does not say how to obtain one. The Axiom of Choice seems so obvious that it is often used without mention. However, whether it could be proved from the rest of set theory or had to be postulated turned out to be a central problem of modern mathematics, and it has still not been solved completely. (The problem is with infinite sets, not finite sets.)

When we use the Axiom of Choice, we will point it out. For example, it is used in Chap. 12 to justify a characterization of mathematical induction (see page 231).

11.6   Ill-defined sets and paradoxes

In Sec. 3.3, we proved $\neg P \not\equiv P$, for all predicates P. Therefore, if we could prove $\neg \hat{P} \equiv \hat{P}$ for a particular predicate $\hat{P}$, we could also prove

$$(\neg \hat{P} \not\equiv \hat{P}) \equiv (\neg \hat{P} \equiv \hat{P}) \ ,$$

which is *false*. An inconsistency would have been introduced. Our use of (syntactic) types to restrict expressions ensures that there are no inconsistencies in set theory. However, in an untyped theory of sets, some set comprehensions lead to inconsistencies and have to be prohibited.

For example, suppose our theory of sets is untyped. Consider the set S of all sets that do not contain themselves as elements, which we define by

(11.78) $x \in S \equiv x \notin x$ (for all sets x).

Direct substitution of set S for x in (11.78) yields

$$S \in S \equiv S \notin S$$

which is *false*. An inconsistency arose by introducing the set comprehension $S = \{x \mid x \notin x\}$. We conclude that S is not well defined and refuse to allow it or consider it to denote a set.

This paradoxical set was discovered in 1901 by Bertrand Russell (and independently by Ernst Zermelo), some 25 years after the first publication of a theory of sets by Georg Cantor —see Historical notes 20.1 on page 464 and 11.1 on page 212). Other paradoxes had been known before, but had not been associated with the foundations of mathematics. The oldest paradox, according to Bertrand Russell, was proposed by a Cretan named Epimenides. Epimenides said (in Greek, we think), "All Cretans are liars, and all statements made by Cretans are lies", and then asked whether this statement was true or false. A similar popular paradox concerns the barber who cuts the hair of all people in his small village except those who cut their own: does he cut his own hair? A third paradox asks whether the following statement is true or false:

This statement is false.

The common characteristic of all such paradoxes is self-reference or self-reflectiveness. Something is defined in terms of itself. To eliminate this kind of paradox, it suffices to ban the use of such self-reference. However, if we are not careful in stating the restriction, we may end up banning useful definitions.

Rather than patch the theory of sets by banning contradictions, perhaps we could develop a new theory that, by construction, would not allow the contradictions in the first place. Such a theory (of types) was proposed by Russell in 1901 [33]. In essence, Russell restricted what he would call a set by defining a hierarchy of all possible sets. At the lowest level are the elements that are not sets —the "individuals", like the integers. At the second level are the sets whose elements are in the lowest level. And so on. A set on any level can have as its members only elements from lower levels. Therefore $x \in x$ is *false* : a set x cannot have itself as a member because its members come from lower levels.

Our syntactic type restrictions on expressions serve the same purpose. According to Table 11.1, $e \in \{x \mid R : E\}$ is an expression only if e and E have the same type t (say). And, as long as t cannot contain elements of type $set(t)$, self-reference is prohibited.

11.7 Bags

The elements of a set are distinct. In some situations, we need to deal with collections of elements in which duplicates have significance. For example, when dealing with the collection of names of people in New York City, duplicate names abound, and if we are interested in population counts, we had better not maintain the names as a set. In this situation, the set is not the proper mathematical abstraction to use.

A collection of elements in which an element may occur any (finite) number of times is called a *bag*. [6] Bag comprehension and enumeration have the same forms as set comprehension and enumeration, except that we use delimiters $\{\!|$ and $|\!\}$ instead of $\{$ and $\}$. For example, the bag consisting of the elements 3 , 3 , and 6 is written as $\{\!|3, 3, 6|\!\}$.

The following examples illustrate the difference between bags and sets.

$$\{\!|x{:}\mathbb{N} \mid -2 \leq x \leq 2 : x^2|\!\} = \{\!|4, 1, 0, 1, 4|\!\}$$
$$\{x{:}\mathbb{N} \mid -2 \leq x \leq 2 : x^2\} = \{4, 1, 0\}$$

We define the bag as we did the set, by defining operations on bag-

---

[6] A bag is also called a *multiset*.

HISTORICAL NOTE 11.1. BERTRAND A.W. RUSSELL (1872–1970)

Russell maintained that the whole of pure mathematics could be rigorously deduced from a small number of axioms. In the famous three-volume work *Principia Mathematica* (1910–1913), he and Whitehead went a long way towards showing how this could be done. But Russell was much more than a mathematician; he made fundamental contributions in philosophy and also wrote extensively about education, society, and politics. He wrote a number of popular books as well, for example *The A.B.C. of Atoms* (1923) and *On Education* (1926). He and his (second of four) wives ran a school for young children for five years.

Russell became a Fellow of the Royal Society in 1908, received the Order of Merit in 1949, and won the Nobel prize for literature in 1950.

Russell belongs to a famous English family of dukes and earls, who trace their ancestry back to the fifteenth century. His grandfather, John Russell, was the first earl in the family (Bertrand was the third) and served twice as prime minister of England. In spite of this background and his enormous contributions, Russell was a controversial figure. At the outbreak of World War I, he was fined for writing a leaflet criticizing the sentencing of conscientious objectors. During the War, he was offered a post at Harvard but was refused a passport. And, in 1918, he spent six months in prison for writing a pacifist article (in prison, he wrote *An Introduction to Mathematical Philosophy*). Beginning in 1938, he spent several years in the U.S., teaching at various universities. In 1940, his appointment to teach philosophy at the College of the City of New York was canceled because of his views on morality.

---

comprehension. Three operations are considered primitive: a test of membership of a value v in a finite bag B, $v \in B$; the number of elements in B (if finite), $\#B$; and the number of occurrences of v in a finite bag B, $v \# B$. They are defined as follows.

(11.79) **Axiom, Membership:** $v \in \{\!\!\{x \mid R : E\}\!\!\} \;\equiv\; (\exists x \mid R : v = E)$

(11.80) **Axiom, Size:** $\#\{\!\!\{x \mid R : E\}\!\!\} \;=\; (\Sigma x \mid R : 1)$

(11.81) **Axiom, Number of occurrences:**

$$v\#\{\!\!\{x \mid R : E\}\!\!\} \;=\; (\Sigma x \mid R \wedge v = E : 1)$$

Based on these primitive operations, we define equality of bags, subbag, and proper subbag. Infix operator $\downarrow$, used below, is the minimum of its two operands.

(11.82) **Axiom, Bag equality:** $B = C \;\equiv\; (\forall v \mid : v \# B = v \# C)$

(11.83) **Axiom, Subbag:** $B \subseteq C \;=\; (\forall v \mid : v \# B \leq v \# C)$

(11.84) **Axiom, Proper subbag:** $B \subset C \;=\; B \subseteq C \wedge B \neq C.$

Finally, we define the union, intersection, and difference of bags.

(11.85) **Axiom, Union:** $B \cup C = \{\!| v, i \mid 0 \le i < v \# B + v \# C : v |\!\}$

(11.86) **Axiom, Intersection:**

$$B \cap C = \{\!| v, i \mid 0 \le i < v \# B \downarrow v \# C : v |\!\}$$

(11.87) **Axiom, Difference:**

$$B - C = \{\!| v, i \mid 0 \le i < v \# B - v \# C : v |\!\}$$

We refrain from listing the many properties of the bag operations, under the assumption that the reader can derive them when necessary.

Exercises for Chapter 11

11.1 Define the following sets using set comprehension or one of its abbreviations.

(a) The set of nonnegative integers that are less than 4.
(b) The set of positive integers that are divisible by 3 and less than 7.
(c) The set of letters in the first author's last name.
(d) The names of the two parts of the Congress of the U.S.A.
(e) The set of odd integers.
(f) The set of prime numbers between 10 and 30. You may use *prime.i* as a boolean function that yields the value of "i is a prime".
(g) The set of squares between 0 and 50.
(h) All powers of 2.

11.2 Give English-sentence descriptions of the following sets. The type of all dummies is $\mathbb{Z}$.

(a) $\{x \mid 0 < x \land even.x\}$.
(b) $\{x \mid 0 \le x : 2 * x\}$.
(c) $\{x \mid 0 < x \land (\exists y \mid x = 3 \cdot y)\}$.
(d) $\{x \mid 0 < x : 3 \cdot x\}$.
(e) $\{z \mid (\exists x, y \mid 0 \le x \land 2 \le y \le 3 : z = x^y)\}$.
(f) $\{x, y \mid 0 \le x \land 2 \le y \le 3 : x^y\}$.

11.3 Prove $\{b, b\} \equiv \{b\}$, using the set-enumeration definition of the abbreviation of set comprehension on page 196.

11.4 Prove formally, using Set membership (11.3), that $e \in \{x \mid false : E\} \equiv false$ holds.

11.5 Prove $\{\} = \{x \mid false\}$.

11.6 Using the facts that $\{b\}$ and $\{b, c\}$ are abbreviations for $\{x \mid x = b : x\}$ and $\{x \mid x = b \lor x = c : x\}$, prove $v \in \{b\} \equiv v = b$ and $v \in \{b, c\} \equiv v = b \lor v = c$.

11.7 Use the results of Exercise 11.6 to prove the following four theorems. You can prove them in the order given, or you can prove the last one first and use it to prove the first three.

(a) $\{b\} = \{c\} \equiv b = c$,
(b) $\{b\} = \{c,d\} \equiv b = c = d$,
(c) $\{b,c\} = \{b,e\} \equiv c = e$,
(d) $\{b,c\} = \{d,e\} \equiv (b = d \wedge c = e) \vee (b = e \wedge c = d)$.

11.8 Prove (11.6), $\{x \mid R : E\} = \{y \mid (\exists x \mid R : y = E)\}$ (under the condition $\neg occurs('y', 'R, E')$).

11.9 Prove (11.7), $x \in \{x \mid R\} \equiv R$.

11.10 Prove (11.9), $\{x \mid Q\} = \{x \mid R\} \equiv (\forall x \mid : Q \equiv R)$.

11.11 Prove (11.10), that $\{x \mid Q\} = \{x \mid R\}$ is valid iff $Q \equiv R$ is valid.

Exercises on the set operators

11.12 Prove the following theorems concerning union and intersection.
(a) Theorem (11.43), $S \subseteq T \wedge U \subseteq V \Rightarrow (S \cup U) \subseteq (T \cup V)$. To prove this theorem, you may use the theorem of Exercise 4.4.
(b) Theorem (11.44), $S \subseteq T \wedge U \subseteq V \Rightarrow (S \cap U) \subseteq (T \cap V)$. To prove this theorem, you may use the theorem of Exercise 4.5.
(c) Theorem (11.45), $S \subseteq T \equiv S \cup T = T$.
(d) Theorem (11.46), $S \subseteq T \equiv S \cap T = S$.
(e) Theorem (11.47), $S \cup T = \mathbf{U} \equiv (\forall x \mid x \in \mathbf{U} : x \notin S \Rightarrow x \in T)$.
(f) Theorem (11.48), $S \cap T = \emptyset \equiv (\forall x \mid x \in \mathbf{U} : x \in S \Rightarrow x \notin T)$.

11.13 Prove the following theorems concerning set difference. Where possible, make use of Metatheorem (11.25).
(a) Theorem (11.49), $S - T = S \cap \sim T$.
(b) Theorem (11.50), $S - T \subseteq S$.
(c) Theorem (11.51), $S - \emptyset = S$.
(d) Theorem (11.52), $S \cap (T - S) = \emptyset$.
(e) Theorem (11.53), $S \cup (T - S) = S \cup T$.
(f) Theorem (11.54), $S - (T \cup U) = (S - T) \cap (S - U)$.
(g) Theorem (11.55), $S - (T \cap U) = (S - T) \cup (S - U)$.

11.14 Prove $\#\{x \mid P\} = (\Sigma x \mid P : 1)$.

11.15 Prove (11.69), $(\exists x \mid x \in S : x \notin T) \Rightarrow S \neq T$.

11.16 Prove the following theorems concerning $\subseteq$ and $\subset$.
(a) Antisymmetry of subset, (11.57), $S \subseteq T \wedge T \subseteq S \equiv S = T$.
(b) Reflexivity of subset (11.58), $S \subseteq S$.
(c) Transitivity of subset (11.59), $S \subseteq T \wedge T \subseteq U \Rightarrow S \subseteq U$.
(d) (11.60), $\emptyset \subseteq S$.
(e) (11.61), $S \subset T \equiv S \subseteq T \wedge \neg(T \subseteq S)$.
(f) (11.62), $S \subset T \equiv S \subseteq T \wedge (\exists x \mid x \in T : x \notin S)$.
(g) (11.63), $S \subseteq T \equiv S \subset T \vee S = T$.
(h) (11.64), $S \not\subseteq S$.

(i) (11.65), $S \subset T \Rightarrow S \subseteq T$.

(j) (11.66), $S \subset T \Rightarrow T \nsubseteq S$.

(k) (11.67), $S \subseteq T \Rightarrow T \not\subset S$.

(l) (11.70a), $S \subseteq T \wedge T \subset U \Rightarrow S \subset U$.

(m) (11.70b), $S \subset T \wedge T \subseteq U \Rightarrow S \subset U$.

(n) (11.70c), $S \subset T \wedge T \subset U \Rightarrow S \subset U$.

11.17 Prove (11.71), $\mathcal{P}\emptyset = \{\emptyset\}$.

11.18 Prove (11.72), $S \in \mathcal{P}S$.

11.19 Prove $S = {\sim}T \equiv S \cup T = \mathbf{U} \wedge S \cap T = \emptyset$.

Chapter 12

Mathematical Induction

$$\mathbf{T}\!\!\!\!$$ he set $\mathbb{N}$ of natural numbers $\{0, 1, 2, \ldots\}$ is infinite. Proving properties of such an infinite set often requires a technique that is of fundamental importance in mathematics and computer science: *mathematical induction*. We explore this technique in this chapter. We also investigate induction over sets other than $\mathbb{N}$. We show how properties of an inductively defined function can be proved using induction, and we show how a loop can be analyzed using induction.

12.1 Induction over the natural numbers

Consider the following boolean expression, which we view as a boolean function $P(n{:}\mathbb{N})$ of its free variable n.

(12.1) $P.n : \ (\Sigma i \mid 1 \le i \le n : 2 \cdot i - 1) \ = \ n^2$

For example, for n equal to 2 and 3, respectively, it states $1 + 3 = 2^2$ and $1 + 3 + 5 = 3^2$.

We can prove $(\forall n{:}\mathbb{N} \mid 0 \le n : P.n)$ as follows. First prove $P.0$. Then prove that for all $n \ge 0$, if $P.0, \ldots, P(n-1)$ hold, then so does $P.n$:

(12.2) $(\forall n{:}\mathbb{N} \mid 0 < n : P.0 \land P.1 \land \cdots \land P(n-1) \ \Rightarrow \ P.n)$.

Having proved $P.0$ and (12.2), we claim that $P.n$ holds for all natural numbers n. This is because in principle —given enough time and space— we can now prove $P.N$ for any given N by proving, in turn, $P.1$, $P.2$, $\ldots$, and finally $P.N$:

- From $P.0$ and $P.0 \Rightarrow P.1$ (which is (12.2) instantiated with $n := 1$), by Modus ponens (3.77) we conclude $P.1$.

- From $P.0 \land P.1$ and $P.0 \land P.1 \Rightarrow P.2$ (which is (12.2) instantiated with $n := 2$), by Modus ponens (3.77) we conclude $P.2$.

  $\cdots$

- From $P.0 \land \cdots \land P(N-1)$ and $P.0 \land \cdots \land P(N-1) \ \Rightarrow \ P.N$ (which is (12.2) instantiated with $n := N$) by Modus ponens (3.77) we conclude $P.N$.

Of course, we do not really have to prove $P.N$ in this fashion; it suffices to know that in principle we can do so. The proofs of $P.0$ and (12.2) are all we need to conclude that $P.n$ holds for all natural numbers.

The technique just described for proving $P.n$ for all natural numbers n is called *mathematical induction over the natural numbers*. It can be used to prove many properties, and not just the particular one defined in (12.1). The technique is formalized as a single axiom in the predicate calculus as follows, where $P:\mathbb{N} \to \mathbb{B}$.

(12.3)   **Axiom, Mathematical Induction over** $\mathbb{N}$:
$$(\forall n{:}\mathbb{N}\,|: (\forall i \mid 0 \leq i < n : P.i) \Rightarrow P.n) \Rightarrow (\forall n{:}\mathbb{N}\,|: P.n).$$

The consequent of (12.3) trivially implies the antecedent. Hence, by mutual implication (3.80), we can rewrite (12.3) as follows.

(12.4)   **Mathematical Induction over** $\mathbb{N}$:
$$(\forall n{:}\mathbb{N}\,|: (\forall i \mid 0 \leq i < n : P.i) \Rightarrow P.n) \equiv (\forall n{:}\mathbb{N}\,|: P.n).$$

Whether we use (12.3) or (12.4) depends on our purpose. For proving universal quantifications by induction, the first is usually the best. For proving properties of induction, the second may be easier to use, because $\equiv$ is symmetric and $\Rightarrow$ is not.

The case $P.0$ is included in (12.3), as we show by manipulating the antecedent of (12.3).

$$
\begin{aligned}
&(\forall n \mid 0 \leq n : (\forall i \mid 0 \leq i < n : P.i) \Rightarrow P.n) \\
=\quad &\langle \text{Split off term (8.23)} \rangle \\
&((\forall i \mid 0 \leq i < 0 : P.i) \Rightarrow P.0) \,\wedge \\
&(\forall n \mid 1 \leq n : (\forall i \mid 0 \leq i < n : P.i) \Rightarrow P.n) \\
=\quad &\langle \text{Empty range (8.13)} \rangle \\
&(true \Rightarrow P.0) \wedge (\forall n \mid 1 \leq n : (\forall i \mid 0 \leq i < n : P.i) \Rightarrow P.n) \\
=\quad &\langle \text{Left identity of } \Rightarrow \text{ (3.73); Change of dummy (8.22)} \rangle \\
&P.0 \wedge (\forall n \mid 0 \leq n : (\forall i \mid 0 \leq i < n+1 : P.i) \Rightarrow P(n+1))
\end{aligned}
$$

Thus, we can rewrite (12.3) in the following form, which is the form we generally use when proving properties by induction.

(12.5)   **Mathematical Induction over** $\mathbb{N}$:
$$P.0 \wedge (\forall n{:}\mathbb{N}\,|: (\forall i \mid 0 \leq i \leq n : P.i) \Rightarrow P(n+1)) \Rightarrow$$
$$(\forall n{:}\mathbb{N}\,|: P.n).$$

Conjunct $P.0$ of (12.5) is called the *base case* of the mathematical induction. The second conjunct of the antecedent,

(12.6) $(\forall n{:}\mathbb{N}\,|: (\forall i \mid 0 \leq i \leq n : P.i) \Rightarrow P(n+1))$, is called the *inductive case*, and $(\forall i \mid 0 \le i \le n : P.i)$ is called the *inductive hypothesis*.

When proving $(\forall n : \mathbb{N} \mid: P.n)$ by induction, we often prove the base case and inductive case separately and then assert, in English, that $P.n$ holds for all natural numbers n. The proof of the inductive case is typically done by proving $(\forall i \mid 0 \le i \le n : P.i) \Rightarrow P(n+1)$ for arbitrary $n \ge 0$. Further, $(\forall i \mid 0 \le i \le n : P.i) \Rightarrow P(n+1)$ is usually proved by assuming $(\forall i \mid 0 \le i \le n : P.i)$ and then proving $P(n+1)$. The standard phraseology for such proofs is: "prove $P(n+1)$ using inductive hypothesis $P.0 \wedge \ldots \wedge P.n$".

(12.7) **Example of a proof by induction.** We prove (12.1) for all natural numbers.

Base case $P.0$.

$$(\Sigma i \mid 1 \le i \le 0 : 2 \cdot i - 1)$$
$$= \quad \langle \text{Identity of } + \text{ (8.13) —since the range is empty} \rangle$$
$$0$$
$$= \quad \langle \text{Arithmetic} \rangle$$
$$0^2$$

Inductive case. For arbitrary $n \ge 0$, we prove $P(n+1)$ using inductive hypothesis $P.0 \wedge \ldots \wedge P.n$. To prove $P(n+1)$, we transform its LHS

$$(\Sigma i \mid 1 \le i \le n+1 : 2 \cdot i - 1)$$

into its RHS $(n+1)^2$ (see (12.1)):

$$(\Sigma i \mid 1 \le i \le n+1 : 2 \cdot i - 1)$$
$$= \quad \langle \text{Split off term (8.23)} \rangle$$
$$(\Sigma i \mid 1 \le i \le n : 2 \cdot i - 1) + 2 \cdot (n+1) - 1$$
$$= \quad \langle \text{Inductive hypothesis } P.n \rangle$$
$$n^2 + 2 \cdot (n+1) - 1$$
$$= \quad \langle \text{Arithmetic} \rangle$$
$$(n+1)^2 \qquad \qquad \qquad \qquad \square$$

In the proof above, only $P.n$ of the inductive hypothesis was used. When only $P.n$ is used, the proof is called a proof by *weak* induction. When other conjuncts of the inductive hypothesis are used, it is called a proof by *strong* induction. But don't worry about the difference between weak and strong induction; for the inductive case, just prove $P.0 \wedge \cdots \wedge P.n \Rightarrow P(n+1)$ in whatever way you can.

Study carefully the proof of $P.n \Rightarrow P(n+1)$ given above, for it employs a technique that is used often: The LHS of $P(n+1)$ is manipulated to "expose $P.n$", that is, to make it possible to make use of inductive hypothesis $P.n$. Here, splitting off a term exposes $P.n$.

INDUCTION STARTING AT OTHER INTEGERS

Formula (12.5) describes induction over the natural numbers $0, 1, \ldots$. Actually, induction can be performed over any subset $n_0, n_0 + 1, n_0 + 2, \ldots$ of the integers. The only difference in such a proof is the starting point and thus the base case; it is either $P.0$ or $P.n_0$. The statement of induction over $n_0, n_0 + 1, n_0 + 2, \ldots$ is given as follows.

(12.8)   **Mathematical Induction over** $\{n_0, n_0 + 1, \ldots\}$:

$P.n_0 \wedge (\forall n \mid n_0 \leq n : (\forall i \mid n_0 \leq i \leq n : P.i) \Rightarrow P(n+1)) \Rightarrow$
$(\forall n \mid n_0 \leq n : P.n)$

(12.9)   **Example of a proof by induction.** Prove $2 \cdot n + 1 < 2^n$, for $n \geq 3$.

Here, we prove $(\forall n \mid 3 \leq n : P.n)$, where $P.n$ is $2 \cdot n + 1 < 2^n$.

Base case $P.3$. $P.3$ is $2 \cdot 3 + 1 < 2^3$, which is valid.

Inductive case. For arbitrary $n \geq 3$ we prove $P(n+1)$ using inductive hypothesis $P.n$.

$$
\begin{aligned}
& 2^{n+1} \\
= \quad & \langle \text{Arithmetic} \rangle \\
& 2 \cdot 2^n \\
> \quad & \langle \text{Inductive hypothesis } P.n \rangle \\
& 2 \cdot (2 \cdot n + 1) \\
= \quad & \langle \text{Arithmetic} \rangle \\
& 2 \cdot (n+1) + 1 + 2 \cdot n - 1 \\
> \quad & \langle \text{Arithmetic} - 2 \cdot n - 1 > 0, \text{ because } 3 \leq n \rangle \\
& 2 \cdot (n+1) + 1 \qquad\qquad\qquad\qquad\qquad\qquad\qquad \square
\end{aligned}
$$

Here, to prove the inductive case, we transformed the RHS of $P(n+1)$ into the LHS. Instead, we could have transformed the LHS into the RHS, or $P(n+1)$ into *true* .

The following example shows how a proof by induction can be done informally, with $P.n$ written in English.

Example of a proof by induction. Consider a currency consisting of 2-cent and 5-cent coins. Show that any amount above 3 cents can be represented using these coins.

We write $P.n$ in English as

$P.n$: Some bag of 2-cent and 5-cent coins has the sum n .

Our task is to prove $(\forall n \mid 4 \leq n : P.n)$.

Base case $P.4$. A bag with two 2-cent coins has the sum 4.

Inductive case. We prove $P(n+1)$ using inductive hypothesis $P.n$. $P.n$ means that there is a bag of 2-cent and 5-cent coins whose sum is n. We have two cases: either the bag contains a 5-cent coin or it does not:

Case (a) The bag contains a 5-cent coin. Replacing the 5-cent coin by three 2-cent coins yields a bag of coins whose sum is one greater, so $P(n+1)$ holds.

Case (b) The bag contains only 2-cent coins. It has at least two 2-cent coins, since $4 \leq n$. Replacing two 2-cent coins by a 5-cent coin yields a bag whose sum is one greater, so $P(n+1)$ holds.

In this proof, the arguments are in English. We now show how to formalize the proof. The difficulty is in defining $P.n$. We need names for the numbers of 2-cent and 5-cent coins in the bag. Existential quantification can be used to create these names; we write $P.n$ as

$$P.n: (\exists h, k \mid 0 \leq h \land 0 \leq k : 2 \cdot h + 5 \cdot k = n) \quad .$$

Exercise 12.3 asks you to finish this formal proof. $\qquad\qquad\square$

HINTS ON PROVING BY INDUCTION

The first step in proving a universal quantification by induction is to put the formula in the form

$$(\forall n \mid n_0 \leq n : P.n) \quad .$$

This means identifying n_0 and $P.n$. $P.n$ may be a mathematical statement or it may be in English; it does not matter. What does matter is that $P.n$ be of type $\mathbb{B}$ —i.e. a true-false statement— and that it be precise. Without identifying $P.n$ correctly, the proof cannot be completed.

Typically, the inductive case

$$(\forall n \mid n_0 \leq n : (\forall i \mid n_0 \leq i \leq n : P.i) \Rightarrow P(n+1))$$

is done by proving $(\forall i \mid n_0 \leq i \leq n : P.i) \Rightarrow P(n+1)$ for arbitrary $n \geq n_0$. And this step is typically (though it need not be) proved by assuming $P.n_0, \ldots, P.n$ and proving $P(n+1)$. This kind of proof often requires manipulating $P(n+1)$ in some fashion.

The goal in manipulating $P(n+1)$ is to make it possible to use the conjuncts $P.0, \ldots, P.n$ of the inductive hypothesis. We call this *exposing* the inductive hypothesis. For instance, in Example (12.9), we rewrote 2^{n+1} as $2 \cdot 2^n$ so that $P.n : 2 \cdot n + 1 < 2^n$ could be used. And, in Example (12.7), we exposed $P.n$ by splitting off a term. This technique of exposing some of the $P.i$ is the key in many proofs by induction.

(12.10) **Heuristic:** In proving $(\forall i \mid n_0 \leq i \leq n : P.i) \Rightarrow P(n+1)$ by assuming $(\forall i \mid n_0 \leq i \leq n : P.i)$, manipulate or restate $P(n+1)$ in order to expose at least one of $P.0, \ldots, P.n$.

12.2 Inductive definitions

Thus far, we have defined functions and operations directly in terms of the propositional and predicate calculi. For example, we might define exponentiation b^n for $b:\mathbb{Z}$ and $n:\mathbb{N}$ as

$$b^n = (\Pi i \mid 1 \leq i \leq n : b) \ .$$

An alternative style of definition uses *recursion*. For example, we can define b^n with two axioms:

(12.11) $b^0 = 1$,

$\qquad b^{n+1} = b \cdot b^n \qquad$ (for $n \geq 0$).

Such a definition consists of two parts. First, there is (at least) one *base case*, which defines the function directly for one (or more) argument. Second, there is (at least) one *inductive* case, which defines the function for all other arguments recursively —the definition contains function applications of the same function. Such a definition bears a resemblance to mathematical induction, in which there is a base case and an inductive case. For this reason, such recursive definitions are called *inductive definitions.*

Using the above definition with $b = 2$, the base case indicates that $2^0 = 1$ and the inductive case yields, in order,

$\qquad 2^1 = 2 \cdot 2^0 = 2$,

$\qquad 2^2 = 2 \cdot 2^1 = 4$,

$\qquad 2^3 = 2 \cdot 2^2 = 8$, etc.

Note that Definition (12.11) is equivalent to

(12.12) $b^0 = 1$,

$\qquad b^n = b \cdot b^{n-1} \qquad$ (for $n \geq 1$) .

Definition (12.11) defines b^{n+1} in terms of b^n, while (12.12) defines b^n in terms of b^{n-1}. The range of n in the inductive or recursive case differs in the two definitions.

Because of their resemblance to induction, inductive definitions lend themselves to proofs by induction. We now give examples.

Example of proof by induction. Prove by mathematical induction that for all natural numbers m and n, $b^{m+n} = b^m \cdot b^n$.

The theorem to be proved is the first line of the following calculation.

$$(\forall m, n: \mathbb{N} \mid : b^{m+n} = b^m \cdot b^n)$$
$$= \quad \langle \text{Nesting (8.20)} \rangle$$
$$(\forall n: \mathbb{N} \mid : (\forall m: \mathbb{N} \mid : b^{m+n} = b^m \cdot b^n))$$

We prove the last formula by proving $(\forall n: \mathbb{N} \mid : P.n)$, where $P.n$ is

$$P.n : (\forall m: \mathbb{N} \mid : b^{m+n} = b^m \cdot b^n)$$

Base case $P.0$. For arbitrary m, we have:

$$b^{m+0} = b^m \cdot b^0$$
$$= \quad \langle \text{Identity of addition; Definition (12.11) of } b^0 \rangle$$
$$b^m = b^m \cdot 1$$
$$= \quad \langle \text{Identity of multiplication} \rangle$$
$$true$$

Inductive case. For arbitrary m, we prove $b^{m+(n+1)} = b^m \cdot b^{n+1}$ using inductive hypothesis $(\forall m \mid : b^{m+n} = b^m \cdot b^n)$.

$$b^{m+(n+1)}$$
$$= \quad \langle \text{Arithmetic} \rangle$$
$$b^{(m+1)+n}$$
$$= \quad \langle \text{Inductive hypothesis } P.n \text{, with } m := m+1. \rangle$$
$$b^{m+1} \cdot b^n$$
$$= \quad \langle \text{Definition (12.11)} \rangle$$
$$b \cdot b^m \cdot b^n$$
$$= \quad \langle \text{Associativity and symmetry of } \cdot \rangle$$
$$b^m \cdot (b \cdot b^n)$$
$$= \quad \langle \text{Definition (12.11)} \rangle$$
$$b^m \cdot b^{n+1}$$

$\square$

Example of a proof by induction. Consider function *factorial* n, written $n!$, which is inductively defined by

$$(12.13) \quad 0! = 1,$$
$$n! = n \cdot (n-1)! \quad (\text{for } n > 0).$$

We prove that $n! = 1 \cdot 2 \cdot \ldots \cdot n = (\Pi i \mid 1 \leq i \leq n : i)$. For $P.n$, we have

$$P.n : \quad n! = (\Pi i \mid 1 \leq i \leq n : i) \quad .$$

Base case $P.0$. We have

$$0! = (\Pi i \mid 1 \leq i \leq 0 : i) = (\Pi i \mid false : i) = 1 \quad ,$$

HISTORICAL NOTE 12.1. FIBONACCI NUMBERS

We use Fibonacci numbers in discussing mathematical induction, but they are quite interesting and useful in their own right. They were introduced by Leonardo Fibonacci (*Filius Bonaccii*, i.e. son of Bonaccio) of Pisa in 1202, in connection with the following problem. Beginning with one fertile pair of rabbits, how many pairs will be produced in a year if each fertile pair produces a new pair each month and a pair becomes fertile after one month?

Over 400 years later, Kepler, unaware of Fibonacci's work, discovered the same sequence of numbers in studying the arrangement of leaves and flowers in plant life. Fibonacci numbers have been observed in other places in nature, as well. For example, consider the bee. The male bee develops from an unfertilized egg and therefore has no father and one mother, while the female has a father and a mother. So, the male bee has 0 fathers and 1 mother, 1 grandfather and 1 grandmother, 1 great grandfather and 2 great grandmothers, and 2 great^2 grandfathers and 3 great^2 grandmothers. In general, he has F_{n+1} great^n grandfathers and F_{n+2} great^n grandmothers.

In the latter part of the 19th century, E. Lucas gave the sequence the name "Fibonacci numbers" and proved many properties of them. In 1844, G. Lamé used the Fibonacci numbers in studying Euclid's algorithm for finding the greatest common divisor of two positive integers. Since then, Fibonacci numbers have been used in various places in algorithms and computer science. For example, they have been suggested for use in algorithms for sorting using magnetic tapes and in a scheme for allocating memory in a computer.

---

the last equality following from Empty range (8.13).

Inductive case. We prove $P(n+1)$ assuming inductive hypothesis $P.n$:

$$(n+1)!$$
$$= \quad \langle \text{Definition (12.13)} \rangle$$
$$(n+1) \cdot n!$$
$$= \quad \langle \text{Inductive hypothesis } P.n \rangle$$
$$(n+1) \cdot (\Pi i \mid 1 \leq i \leq n : i)$$
$$= \quad \langle \text{Split off term (8.23) (in reverse)} \rangle$$
$$(\Pi i \mid 1 \leq i \leq n+1 : i) \qquad\qquad \Box$$

Our next examples concern the *Fibonacci numbers*, which are defined as follows (see also Historical note 12.1) for $n : \mathbb{N}$:

(12.14) $F_0 = 0, \ F_1 = 1,$
$\qquad F_n = F_{n-1} + F_{n-2} \qquad$ (for $n > 1$).

The first few Fibonacci numbers are $0, 1, 1, 2, 3, 5, 8, 13$. Observe that except for the first two, each is the sum of the previous pair.

The Fibonacci numbers are intimately connected with the number $\phi =$

HISTORICAL NOTE 12.2. THE GOLDEN RATIO

The golden ratio $(1+\sqrt{5})/2 = 1.6180339887\ldots$ has been known for a long time. Euclid called it the *extreme and mean value*, and Renaissance writers called it the *divine proportion*. It is also called the *golden mean*: the perfect moderate course or position that avoids extremes; the *happy medium*. The name ϕ for the golden ratio is said to have come from the name of the Greek artist Phideas, who used it in his sculptures.

<u>$B \ A$</u>   The *golden section* is the division of a length such that the smaller length A and larger length B satisfy $A/B = B/(A+B)$, as illustrated at the beginning of this paragraph. The ratio B/A is then the golden ratio. The main measurements of many buildings in antiquity and the middle ages have the golden section built into them —the Parthenon on the Acropolis in Athens, for example. Today, the golden section is still in use in the fine arts and in industrial design.

In a *golden rectangle*, illustrated to the left, the vertical and horizontal side lengths A and B satisfy $B/A = \phi$. Drawing another vertical line at distance A from the left side splits the rectangle into two: one is a square and the other is again a golden rectangle.

The architect Corbusier developed a scale for the human body based on the golden section: A is from the head to the navel and B from the navel to the foot. Further, the length from the naval down splits in golden-section form at the knee; from the naval up, at the throat. Do you have these golden proportions? Measure yourself.

$(1+\sqrt{5})/2$, and its twin $\hat{\phi} = (1-\sqrt{5})/2$. ϕ is called the *golden ratio* (see Historical note 12.2). As proved in Exercise 12.21, ϕ and $\hat{\phi}$ satisfy

(12.15) $\phi^2 = \phi + 1$ and $\hat{\phi}^2 = \hat{\phi} + 1$.

Example of a proof by induction. Prove the remarkable fact that for $n \geq 1$,

$$\phi^{n-2} \leq F_n \leq \phi^{n-1} \ .$$

We prove only $(\forall n \mid n \geq 1 : P.n)$, where $P.n$ is given in (12.16), and leave the other part to the reader. The proof by induction proceeds as follows.

(12.16) $P.n : F_n \leq \phi^{n-1}$

Base case $P.1$. Since both sides of $P.1$ reduce to 1, $P.1$ holds.

Base case $P.2$. $P.2 \equiv 1 \leq (1+\sqrt{5})/2$, which is *true*. Hence, $P.2$ holds.

Inductive case. For arbitrary $n \geq 2$, We assume inductive hypothesis $P.i$ for $1 \leq i \leq n$ and prove $P(n+1)$:

$$F_{n+1}$$
$$= \quad \langle \text{Definition (12.14) of Fibonacci numbers} \rangle$$
$$F_n + F_{n-1}$$
$$\leq \quad \langle \text{Inductive hypothesis, } P.n \text{ and } P(n-1) \rangle$$
$$\phi^{n-1} + \phi^{n-2}$$
$$= \quad \langle \text{Arithmetic —factor out } \phi^{n-2} \rangle$$
$$\phi^{n-2} \cdot (\phi + 1)$$
$$= \quad \langle (12.15), \ \phi^2 = \phi + 1 \,; \text{Arithmetic} \rangle$$
$$\phi^n$$

This is the first proof that relies on more than one of $P.0, \ldots, P.n$. The recursive definition of F_n forces the use of $P(n-1)$ and $P.n$. $\qquad \square$

Example of a proof by induction. Prove

(12.17) $\quad F_{n+m} = F_m \cdot F_{n+1} + F_{m-1} \cdot F_n \qquad$ (for $n \geq 0$ and $m \geq 1$).

We prove $(\forall n : \mathbb{N} \,|: P.n)$ by induction, where $P.n$ is

$$P.n : \ (\forall m \mid 1 \leq m : F_{n+m} = F_m \cdot F_{n+1} + F_{m-1} \cdot F_n) \ .$$

Base case $P.0$. Since $F.0 = 0$ and $F.1 = 1$, the body of $P.0$ reduces to $F_m = F_m \cdot 1 + F_{m-1} \cdot 0$, which further reduces to $F_m = F_m$. Hence, the base case holds.

Inductive case. We assume inductive hypothesis $P.n$ and prove $P(n+1)$. Since the RHS of the body of $P(n+1)$ is more complicated than its LHS, we transform its RHS into its LHS. For arbitrary $m \geq 1$, we have,

$$F_m \cdot F_{n+1+1} + F_{m-1} \cdot F_{n+1}$$
$$= \quad \langle \text{Definition (12.14) of Fibonacci numbers} \rangle$$
$$F_m \cdot (F_{n+1} + F_n) + F_{m-1} \cdot F_{n+1}$$
$$= \quad \langle \text{Arithmetic} \rangle$$
$$(F_m + F_{m-1}) \cdot F_{n+1} + F_m \cdot F_n$$
$$= \quad \langle \text{Definition (12.14) of Fibonacci numbers} \rangle$$
$$F_{m+1} \cdot F_{n+1} + F_m \cdot F_n$$
$$= \quad \langle \text{Inductive hypothesis } P.n \,, \text{ with } m := m + 1 \rangle$$
$$F_{n+m+1} \qquad\qquad\qquad\qquad\qquad\qquad\qquad\qquad\qquad\qquad \square$$

The exercises give more interesting facts about the Fibonacci numbers.

12.3   Peano arithmetic

An inductive definition of the positive integers was first suggested by Peano in 1889 (see Historical note 12.3). This definition and the theory that ensues from it has been called *Peano arithmetic*.[1] Here, we define the natural numbers in a similar manner.

(12.18)   **Definition.** The set of natural numbers $\mathbb{N}$, expressed in terms of 0 and a function S (for *successor*), $S : \mathbb{N} \to \mathbb{N}$, is defined as follows.

    (a)   0 is a member of $\mathbb{N}$: $0 \in \mathbb{N}$.

    (b)   If n is in $\mathbb{N}$, then so is $S.n$: $n \in \mathbb{N} \Rightarrow S.n \in \mathbb{N}$.

    (c)   The element 0 is not the successor of any natural number: $(\forall n : \mathbb{N} \mid : S.n \neq 0)$.

    (d) S is one-to-one, i.e. $(\forall n, m : \mathbb{N} \mid : S.n = S.m \Rightarrow n = m)$.

    (e)   If a subset N of $\mathbb{N}$ (i) contains 0 and (ii) contains the successors of all its elements, then $N = \mathbb{N}$:
$$N \subseteq \mathbb{N} \wedge 0 \in N \wedge (\forall n \mid n \in N : S.n \in N) \Rightarrow N = \mathbb{N} \ .$$

Each part of the definition is necessary to define $\mathbb{N}$ unambiguously. Parts (a) and (b) would seem to define the set of natural numbers —we could use the notation $n + 1$ for $S.n$. However, by themselves, these two parts are satisfied by many other sets, for example the integers and the real numbers. Part (c) rules out cases like the following: $\mathbb{N}$ is $\{0, 1, 2\}$ and $S.0 = 1$, $S.1 = 2$, $S.2 = 0$. Part (d) rules out cases like $\mathbb{N}$ is $\{0, 1, 2\}$ and $S.0 = 1$, $S.1 = 2$, $S.2 = 1$.

Part (e) is actually a form of weak induction, but expressed in terms of sets instead of predicates. To see this, define predicate $P.n$ by $P.n \equiv n \in N$ and let N be a subset of $\mathbb{N}$. Then we manipulate part (e) as shown below. Compare the last formula of the manipulation with axiom (12.3) for induction.

---

[1] And one who dabbles in this arithmetic might be called a Peano player.

$$N \subseteq \mathbb{N} \wedge 0 \in N \wedge (\forall n \mid n \in N : S.n \in N) \Rightarrow N = \mathbb{N}$$
$$= \quad \langle \text{Assumption } N \subseteq \mathbb{N} \, ; \text{ replace } N \text{ using definition of } P \rangle$$
$$P.0 \wedge (\forall n \mid P.n : P(S.n)) \Rightarrow (\forall n \mid n \in \mathbb{N} : P.n)$$
$$= \quad \langle \text{Denote } S.n \text{ by } n+1 \rangle$$
$$P.0 \wedge (\forall n \mid P.n : P(n+1)) \Rightarrow (\forall n \mid n \in \mathbb{N} : P.n)$$

We can introduce conventional notation for the integers by using n for $n + 0$, $n + 1$ for $S.n$, and $n + m + 1$ for $S(n + m)$. We can now prove various properties of addition —for example, that addition is symmetric and associative. We do this in Chap. 15.

12.4   Induction and well-founded sets

Thus far, we have been exploring mathematical induction over natural numbers using relation $<$. We now generalize the notion of mathematical induction to deal with sets other than $\mathbb{N}$ and other relations. For example, we can use mathematical induction to prove properties of the negative integers $\mathbb{Z}^-$ with relation $>$; to prove properties of Pascal programs with the relation "program p' is a subprogram of program p"; and to prove properties of binary trees with the relation "tree t' is a subtree of tree t".

Let $\prec$ be a boolean function of two arguments of type U (say), i.e. a function of type $U \times U \to \mathbb{B}$. We want to determine the cases in which $\langle U, \prec \rangle$ *admits induction* —that is, in which mathematical induction over $\langle U, \prec \rangle$ is sound. Not every pair $\langle U, \prec \rangle$ admits induction, and we characterize those that do.

We write the principle of mathematical induction over $\langle U, \prec \rangle$ as follows (omitted ranges are *true*; also $x{:}U$ and $y{:}U$).

(12.19) **Mathematical induction over** $\langle U, \prec \rangle$:

$$(\forall x \mid : P.x) \equiv (\forall x \mid : (\forall y \mid y \prec x : P.y) \Rightarrow P.x)$$

In the case $\langle U, \prec \rangle = \langle \mathbb{N}, < \rangle$, (12.19) reduces to the induction over $\mathbb{N}$, (12.4). To see this, rewrite (12.19), substituting $\mathbb{N}$ for U and $<$ for $\prec$:

$$(\forall x{:}\mathbb{N} \mid : P.x) \equiv (\forall x{:}\mathbb{N} \mid : (\forall y{:}\mathbb{N} \mid y < x : P.y) \Rightarrow P.x).$$

This expression is like (12.4) (see page 218), except for renaming of dummies and the interchange of the LHS and RHS.

We want to show that mathematical induction has two characterizations. These require the notion of a *minimal element* of a nonempty subset S of U:

(12.20) **Definition.** Element y is a *minimal element of* S if $y \in S$ and $(\forall x \mid x \prec y : x \notin S)$.

Examples of minimal elements

(a) For $\langle \mathbb{N}, < \rangle$, the minimal element of any nonempty subset of $\mathbb{N}$ is its smallest element, in the usual sense.

(b) For $\langle \mathbb{N}, \leq \rangle$, no nonempty subset of $\mathbb{N}$ has a minimal element, because $i \leq i$ holds for all natural numbers i.

(c) Consider $\langle \mathbb{N}, pdiv \rangle$, where $i \, pdiv \, j$ means "i is a divisor of j and $i < j$" ("$pdiv$" stands for $proper \, divisor$). Then the subset $S = \{5, 15, 3, 20\}$ has two minimal elements, 5 and 3, since they have no proper divisors in S.

(d) Consider $\langle P, pdiv \rangle$, where P is the set of prime numbers and $pdiv$ is as in the previous example. All elements of $\langle P, pdiv \rangle$ are minimal, since their only proper divisor is 1 and 1 is not a prime and consequently is not in the set.   □

We use this notion of minimal element to define well foundedness.

(12.21) **Definition.** $\langle U, \prec \rangle$ is $well \, founded$ if every nonempty subset of U has a minimal element, i.e. if for all subsets S of U,

$$ S \neq \emptyset \;\equiv\; (\exists x \,|: x \in S \land (\forall y \mid y \prec x : y \notin S)) \quad . $$

Examples of well founded $\langle U, \prec \rangle$

(a) $\langle \mathbb{N}, < \rangle$ is well founded: the minimal element of any nonempty set of the natural numbers is its smallest element, in the usual sense.

(b) $\langle \mathbb{Z}, < \rangle$ is not well founded. To see this, take $S = \mathbb{Z}$; $\mathbb{Z}$ has no smallest integer.

(c) Let U be the set of all boolean expressions, and let $x \prec y$ mean "x is a proper subexpression of y", i.e. x is a subexpression of y but x and y are (syntactically) different. Note that a constant or variable contains no proper subexpression. Since any boolean expression contains at least one constant or variable, $\langle U, \prec \rangle$ is well founded. □

We now prove a remarkable fact: well foundedness of $\langle U, \prec \rangle$ and mathematical induction over $\langle U, \prec \rangle$ are equivalent. That is, we can perform induction over $\langle U, \prec \rangle$ iff $\langle U, \prec \rangle$ is well founded. The proof is simple. It rests on the fact that for any subset S of U we can construct the expression $P.z \equiv z \notin S$, and for any boolean expression $P.z$ we can construct the set $S = \{z \mid \neg P.z\}$.

(12.22) **Theorem.** $\langle U, \prec \rangle$ is well founded iff it admits induction.

$Proof.$ For any subset S of U and corresponding expression $P.z \equiv z \notin S$, we change the formula of (12.21) into the formula of (12.19):

$$S \neq \emptyset \;\equiv\; (\exists x \mid : x \in S \land (\forall y \mid y \prec x : y \notin S))$$

$=\quad \langle (3.11),\; X \equiv Y \equiv \neg X \equiv \neg Y \;;\; \text{Double negation (3.12)} \rangle$

$$S = \emptyset \;\equiv\; \neg(\exists x \mid : x \in S \land (\forall y \mid y \prec x : y \notin S))$$

$=\quad \langle \text{De Morgan (9.18b)}; \text{De Morgan (3.47a)} \rangle$

$$S = \emptyset \;\equiv\; (\forall x \mid : x \notin S \lor \neg(\forall y \mid y \prec x : y \notin S))$$

$=\quad \langle\, P.z \;\equiv\; z \notin S \;\text{—replace occurrences of}\; S \,\rangle$

$$(\forall x \mid : P.x) \;\equiv\; (\forall x \mid : P.x \lor \neg(\forall y \mid y \prec x : P.y))$$

$=\quad \langle \text{Law of implication (3.59)} \rangle$

$$(\forall x \mid : P.x) \;\equiv\; (\forall x \mid : (\forall y \mid y \prec x : P.y) \Rightarrow P.x) \qquad \square$$

There is a third characterization of well foundedness, in terms of the *decreasing finite chain* property. Consider a chain of relations $<$ using only natural numbers x_i :

$$x_n < \cdots < x_3 < x_2 < x_1 < x_0 \qquad \text{(for some } n \text{)}.$$

Any such chain is finite, since for any natural number i there are a finite number of natural numbers smaller than i. In contrast, the set of all negative integers has the infinite chain $\ldots < -3 < -2 < -1$.

Consider again $\langle U, \prec \rangle$, and define predicate $DCF.x$:

(12.23) $DCF.x$: "every decreasing chain beginning with x is finite".

The following property of $DCF.x$ is based on our understanding of finiteness. Suppose for every y satisfying $y \prec x$ that every decreasing chain starting with y is finite, i.e. $DCF.y$ holds. Then $DCF.x$ holds as well, since a chain beginning with x is one longer than some chain beginning with y. We formalize this property as follows.

(12.24) **Axiom, Finite chain property:**

$$(\forall x \mid : (\forall y \mid y \prec x : DCF.y) \;\Rightarrow\; DCF.x)$$

A relation $\prec$ over a set U is called *noetherian* if every decreasing chain beginning with any x in U is finite (the name *noetherian* honors Emmy Noether; see Historical note 12.4):

(12.25) **Definition.** $\langle U, \prec \rangle$ is *noetherian* iff $(\forall x : U \mid : DCF.x)$.

We now characterize well foundedness in terms of finite decreasing chains.

(12.26) **Theorem.** $\langle U, \prec \rangle$ is well founded iff $\langle U, \prec \rangle$ is noetherian.

Proof. The proof is by mutual implication.

LHS $\Rightarrow$ RHS. Assume that $\langle U, \prec \rangle$ is well founded. Then, by (12.22), $\langle U, \prec \rangle$ admits induction. So we have

$$(\forall x \mid : (\forall y \mid y \prec x : DCF.y) \;\Rightarrow\; DCF.x) \qquad \text{—(12.24)}$$

$=\quad \langle \text{Induction over } \langle U, \prec \rangle \text{ (12.19), with } P := DCF \rangle$

$$(\forall x \mid : DCF.x) \quad \text{—this is the RHS}$$

HISTORICAL NOTE 12.4. EMMY NOETHER (1882–1935)

E.T. Bell says that Noether was one of the most creative abstract algebraists in the world [3]. She had an impressive knowledge of areas that David Hilbert and Felix Klein needed in their work, and, during World War I, she arrived in Göttingen to study and work with them.

Unfortunately, Noether's sex was a handicap for her advancement. When she was ready to take her habilitation exam (a second doctorate, required in Germany before one could teach), most of the non-mathematical members of the Faculty were opposed. If she passed, she could later become a professor and a member of the University Senate, and *that* couldn't be allowed! Hilbert had no patience with this, saying that her sex should not matter, for the Senate was not a bathhouse. She finally did habilitate in 1919 and received a special professorship in 1922. But the professorship carried no salary; in fact, during most of her 15 or so years in Göttingen she received no official salary.

In 1934, Noether, a Jew, was forced by the Nazis to give up her work in Göttingen and leave the country. She obtained a position at Bryn Mawr College in Pennsylvania and died there a year later, following an operation. Einstein wrote a letter to the New York Times, calling her "the most significant creative mathematical genius [of her sex] thus far produced ... ." (See [32, p. 208].)

RHS $\Rightarrow$ **LHS.** We prove this part informally, since DCF is defined informally. Assume that $\langle U, \prec \rangle$ is noetherian, so that every decreasing chain is finite. With this assumption, we show that every nonempty subset S of U has a minimal element. Let subset S be nonempty. Choose an arbitrary element x_0 of S (by Axiom of Choice (11.77)). Construct a descending chain beginning with x_0, choosing at each step i some element x_{i+1} satisfying $x_{i+1} \prec x_i$. Since every decreasing chain is finite, the construction of the descending chain stops with an x_n for which there is no element y in S that satisfies $y \prec x_n$. Element x_n is a minimal element of S.   □

In the following sections, we give other examples of induction over well founded sets. First, however, we state some theorems concerning induction (the proofs are left as exercises).

(12.27) **Theorem.** If $\langle U, \prec \rangle$ admits induction, then $\prec$ is irreflexive, that is, $x \not\prec x$ holds for every x in U.

(12.28) **Theorem.** If $\langle U, \prec \rangle$ admits induction, then for all x, y in U, $x \prec y \Rightarrow y \not\prec x$.

12.5   Induction for inductive definitions

We now know that $\langle U, \prec \rangle$ admits induction iff $\langle U, \prec \rangle$ is well founded, which also means that every decreasing chain is finite. Any inductive definition gives rise to a set U and a relation $\prec$ that admits induction.

Suppose an inductive definition is given by three (say) cases (as in the first example below). Then in proving some property P by induction, it has to be shown that each case satisfies P. If the proof of a case does not require an inductive hypothesis, then that case is a base case; otherwise, it is an inductive case. In the first example given below, the first case is a base case and the second two are inductive cases.

Inductively defined expressions

Let U be the set of finite expressions defined inductively as follows.

(a) A digit $0, 1, 2, 3, 4, 5, 6, 7, 8, 9$ is an expression.
(b) If E_0 and E_1 are expressions, then so is $E_0 + E_1$.
(c) If E is an expression, then so is (E).

For two expressions E_0 and E_1, define 'E_0' $\prec$ 'E_1' to mean that E_0 is a proper subexpression of E_1, i.e. a subexpression of E_1 that is not E_1 itself. For example, the proper subexpressions of $(3+5)$ are $3+5$, 3, and 5.

Above, the term "finite" means that we are considering only expressions that can be written with a finite number of symbols. For example, the sequence $(((\cdots 1 \cdots)))$ that involves an infinite number of parentheses is not an expression. Restricting consideration to finite expressions allows us to claim that $\langle U, \prec \rangle$ has the finite decreasing chain property, which means that $\langle U, \prec \rangle$ admits induction.

In summary, the inductive definition of the set U of expressions E gives rise to a pair $\langle U, \prec \rangle$ that admits induction.

(12.29) **Theorem.** Each expression of U contains the same number of left and right parentheses.

Proof. Let $L.`E`$ and $R.`E`$ denote the number of left and right parentheses in E, respectively. Our task is to prove

$$P.`X`: \quad L.`X` = R.`X`$$

for arbitrary expressions X. We prove this by induction by proving $P.`X`$ under the assumption that $P.`Y`$ holds for all proper subexpressions Y of X. We proceed by case analysis, investigating each kind of expression:

Case (a) For any digit d, $L.`d` = 0 = R.`d`$, so $P.`d`$ holds.

Case (b) The proper subexpressions of an expression $E_0 + E_1$ are E_0 and E_1. We prove $P.`E_0` \land P.`E_1` \Rightarrow P.`E_0 + E_1`$.

$$P.`E_0 + E_1`$$
$$= \quad \langle \text{Definition of } P \rangle$$
$$L.`E_0 + E_1` = R.`E_0 + E_1`$$
$$= \quad \langle \text{Definition of } L \text{ and } R \rangle$$
$$L.`E_0` + L.`E_1` = R.`E_0` + R.`E_1`$$
$$\Leftarrow \quad \langle \text{Arithmetic} \rangle$$
$$L.`E_0` = R.`E_0` \land L.`E_1` = R.`E_1`$$
$$= \quad \langle \text{Definition of } P \rangle$$
$$P.`E_0` \land P.`E_1`$$

Case (c) Consider an expression (E). We have $L.`(E)` = 1 + L.`E`$ and $R.`(E)` = 1 + R.`E`$. From this, we can conclude $P.`E` \Rightarrow P.`(E)`$. □

Such detail is not really needed on such a trivial problem. We went to such great lengths for three reasons. First, so you could see in detail how induction is applied on a pair other than $\langle \mathbb{N}, < \rangle$. Second, so you could see how the structure of the definition of expressions was reflected in the structure of the proof: each kind of expression was a separate case. Third, so you could see that, because the base case was submerged in the inductive definition, the base case did not have to be mentioned separately. We simply had to prove $(\forall y \mid y \prec x : P.y) \Rightarrow P.x$ for arbitrary x.

INDUCTIVELY DEFINED BINARY TREES

Our second example of proof by induction over a pair $\langle U, \prec \rangle$ concerns binary trees. We define the (finite) set of binary trees inductively, as follows.

(12.30) **Definition.**

$\emptyset$ is a *binary tree*, called the *empty tree*.

(d, l, r) is a *binary tree*, for $d : \mathbb{Z}$ and l, r binary trees.

We consider only *finite* binary trees, which means that we consider only trees that can be written using a finite number of symbols. A tree $(d, \emptyset, \emptyset)$ is often abbreviated as (d). Let t be a nonempty binary tree, i.e. $t = (d, l, r)$. The value d is called the *root* of t, l is called the *left subtree* of t, and r is called the *right subtree* of t. The three components of t are referenced using $t.d$, $t.l$, and $t.r$.

In computer science, nonempty trees are drawn as shown in Fig. 12.1. The values d in the tree are called *nodes* of the tree. The middle tree in Fig. 12.1 has nodes 3, 4, and 2. Nodes with two empty subtrees are called *leaves*, and the others are called *internal* nodes. The leaves of the rightmost tree of Fig. 12.1 are 6 and 7, and its interior nodes are 3, 4, and 2.

For a tree (d, l, r), d is called the *parent* of the roots of subtrees l and r, the roots of l and r are called d's *left and right children*, and the roots of l and r are called each other's *siblings*. In the rightmost tree in Fig. 12.1, 3's child is 4 and 4's children are 6 and 2. Root 3 has no parent. An empty subtree $\emptyset$ is usually interpreted as being the absence of a subtree. In other words, 3 has no right child and 6 has no children.

We define $\#t$, the number of nodes in tree t, inductively as follows.

(12.31) $\#\emptyset = 0$

$\qquad \#(d, l, r) = 1 + \#l + \#r$.

The root of a tree is on level 0, its children are on level 1, its grandchildren are on level 2, and so on. The height of a tree t is defined inductively as follows (where $b \uparrow c$ is the maximum of b and c).

(12.32) $height.\emptyset = 0$

$\qquad height(d, l, r) = 1 + (height.l \uparrow height.r)$.

For example, the height of the empty tree is 0 and the heights of the three trees of Fig. 12.1 are 1, 2, and 4, respectively.

A binary tree is *complete* if every node has either 0 or 2 children. The empty tree and the first two trees of Fig. 12.1 are complete, but the rightmost tree of Fig. 12.1 is not.

Now consider U to be the set of finite binary trees and $\prec$ to be the proper-subtree relationship. $\langle U, \prec \rangle$ has the finite decreasing chain property, since a finite binary tree has only a finite number of proper subtrees. Therefore $\langle U, \prec \rangle$ admits induction.

(12.33) **Theorem.** The maximum number of nodes in a tree with height n is $2^n - 1$.

Proof. We prove the theorem by mathematical induction over $\langle U, \prec \rangle$. We consider the two kinds of trees as given by definition (12.30).

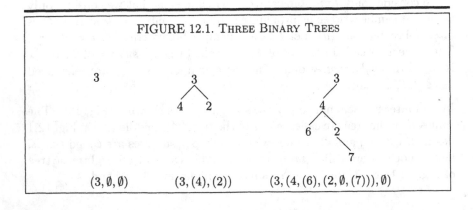

FIGURE 12.1. THREE BINARY TREES

$(3, \emptyset, \emptyset)$ $\qquad\qquad$ $(3, (4), (2))$ $\qquad\qquad$ $(3, (4, (6), (2, \emptyset, (7))), \emptyset)$

Case (a) The empty tree has 0 nodes and height 0, and $2^0 - 1 = 0$.

Case (b) Consider a tree $t = (d, l, r)$ with height n (say), so $n > 0$. First, since t has the maximum number of nodes for its height, both subtrees have the same height, $n - 1$ (if not, one could add nodes to the subtree with smaller height without changing *height.t*). We calculate $\#t$:

$$\#(d, l, r)$$
$$= \quad \langle \text{Definition } (12.31) \rangle$$
$$1 + \#l + \#r$$
$$= \quad \langle \text{Inductive hypothesis, twice} \rangle$$
$$1 + 2^{n-1} - 1 + 2^{n-1} - 1$$
$$= \quad \langle \text{Arithmetic} \rangle$$
$$2^n - 1 \qquad\qquad\qquad\qquad \square$$

The proofs of the following theorems are left as exercises.

(12.34) **Theorem.** The minimum number of nodes of a tree of height n is n.

(12.35) **Theorem.** The maximum number of leaves in a tree of height n is 2^{n-1}; the maximum number of internal nodes is $2^{n-1} - 1$.

(12.36) **Theorem.** The minimum number of leaves in a tree of height n is 1; if $n > 0$, the minimum number of internal nodes is $n - 1$.

(12.37) **Theorem.** Every nonempty complete tree has an odd number of nodes.

LEXICOGRAPHIC ORDERING OF PAIRS OF NATURAL NUMBERS

Let $\mathbb{N} \times \mathbb{N}$ denote the set of pairs $\langle i, j \rangle$ of natural numbers. For example, $\mathbb{N} \times \mathbb{N}$ contains $\langle 0, 0 \rangle$, $\langle 0, 3 \rangle$, and $\langle 999, 1 \rangle$. We define binary relations $<$ and $>$ over $\mathbb{N} \times \mathbb{N}$ (thus overloading $<$ and $>$), called the *lexicographic ordering of pairs of natural numbers*, as follows.

(12.38) $\langle b, c \rangle < \langle b', c' \rangle \quad \equiv \quad b < b' \lor (b = b' \land c < c')$

$\langle b', c' \rangle > \langle b, c \rangle \quad \equiv \quad \langle b, c \rangle < \langle b', c' \rangle$

Examples of lexicographic ordering

(a) $\langle \ 1, 0 \rangle > \langle 0, \ 99 \rangle > \cdots > \langle 0, \ 2 \rangle > \langle 0, 1 \rangle > \langle 0, 0 \rangle$
(b) $\langle \ 2, 0 \rangle > \langle 1, 999 \rangle > \cdots > \langle 1, \ 2 \rangle > \langle 1, 1 \rangle > \langle 1, 0 \rangle$
(c) $\langle \ 3, 0 \rangle > \langle 2, \ 57 \rangle > \cdots > \langle 2, \ 2 \rangle > \langle 2, 1 \rangle > \langle 2, 0 \rangle$
(d) $\langle 22, 9 \rangle > \langle 14, 85 \rangle > \cdots > \langle 6, 11 \rangle > \langle 4, 9 \rangle > \langle 0, 0 \rangle$ $\qquad \square$

This ordering is similar to the dictionary ordering of words —IN, IT, TO, for example. The only difference between the two orderings is that, with the dictionary ordering, the number of different symbols is finite, 26 , while with $\mathbb{N}$ the number of symbols is infinite. This means that an infinite number of pairs follow $\langle 1, 0 \rangle$ in the ordering $>$: $\langle 1, 0 \rangle > \langle 0, b \rangle$ for all natural numbers b. For any b, the longest decreasing chain that begins with $\langle 0, b \rangle$ has length $b + 1$, but there is no upper bound on the length of decreasing chains beginning with $\langle 1, 0 \rangle$. Nevertheless, we can prove the following theorem by induction on b.

(12.39) **Theorem.** For natural numbers b, c, every decreasing chain beginning with $\langle b, c \rangle$ is finite.

Thus, by Theorem (12.26), $\langle \mathbb{N} \times \mathbb{N}, < \rangle$ is well founded and admits induction.

Note the difference between $\langle \mathbb{N} \times \mathbb{N}, < \rangle$ and $\langle \mathbb{N}, < \rangle$. In $\langle \mathbb{N}, < \rangle$, the length of a finite chain beginning with b is bounded above by $b + 1$; in $\langle \mathbb{N} \times \mathbb{N}, < \rangle$, there is no upper bound on the length of a chain beginning with $\langle 1, 0 \rangle$, although all such chains are finite in length.

Another ordering that has this unboundedness property is $\langle \mathbb{Z}, \prec \rangle$, with $b \prec c$ defined as follows: (i) $\prec$ does not hold between negative integers; (ii) a negative integer is "bigger" than any nonnegative integer; and (iii) the conventional ordering holds between natural numbers:

(12.40) $b \prec c \;\equiv\; 0 \le b < c \lor c < 0 \le b$

Here are examples:

$$0 \prec 2 \prec 15 \prec -3$$

It is readily seen that every decreasing chain is finite, so that $\langle \mathbb{Z}, \prec \rangle$ admits induction. We put $\langle \mathbb{Z}, \prec \rangle$ to use in proving termination of a loop in the next Sec. 12.6.

12.6   The correctness of loops

We introduce a theorem concerning the while loop **while** B **do** S. The proof of the theorem will show how correctness of a loop is inextricably intertwined with induction. This section builds on Chap. 10.

We prefer to write a while loop using the syntax

(12.41) **do** $B \to S$ **od**

where boolean expression B is called the *guard* and statement S is called the *repetend*[2].

---

[2] Repetend: the thing repeated.

Loop (12.41) is executed as follows: If B is *false*, then execution of the loop terminates; otherwise S is executed and then the process is repeated.

Each execution of repetend S is called an *iteration*. Thus, if B is initially *false*, then 0 iterations occur.

We will be analyzing the following loop (with initialization). Its execution requires exactly n iterations. As shown by the annotation, given $0 \le n$ initially, execution stores the value $n \cdot x$ in p. We have labeled the assertions P, Q, and R for later reference.

(12.42) $\{Q: \ 0 \le n\}$

  $i, p := 0, 0;$

  $\{P: \ 0 \le i \le n \ \wedge \ p = i \cdot x\}$

  do $i \ne n \to i, p := i + 1, p + x$ **od**

  $\{R: \ p = n \cdot x\}$

We now state and prove the fundamental invariance theorem for loops. This theorem refers to an assertion P that holds before and after each iteration (provided it holds before the first). Such a predicate is called a *loop invariant*. In algorithm (12.42), loop invariant P is $0 \le i \le n \ \wedge \ p = i \cdot x$.

(12.43)   **Fundamental invariance theorem.** Suppose

- $\{P \wedge B\}\ S\ \{P\}$ holds —i.e. execution of S begun in a state in which P and B are *true* terminates with P *true* — and

- $\{P\}$ **do** $B \to S$ **od** $\{true\}$ —i.e. execution of the loop begun in a state in which P is *true* terminates. [3]

Then $\{P\}$ **do** $B \to S$ **od** $\{P \wedge \neg B\}$ holds.

Proof. By the second hypothesis, the loop terminates, say in $n \ge 0$ iterations. It remains to show that $P \wedge \neg B$ holds upon termination. B is *false* upon termination because the loop can terminate only when B becomes *false*. We prove that P is *true* upon termination of the n iterations by proving (by induction) that it is *true* after i iterations, $0 \le i \le n$.

P is *true* before execution of the loop, so P is *true* after 0 iterations. Hence the base case holds. For the inductive case, assume P is *true* after i ($i < n$) iterations. Iteration $i + 1$ is executed with P and B *true* and consists of executing S. By the first hypothesis of the theorem, P holds after iteration $i + 1$. Hence the inductive case holds. □

---

[3] The formalization of the argument for termination is given on page 240.

Example. We use Theorem (12.43) to prove the following Hoare triple, where we have labeled the guard B and the invariant P.

(12.44) $\{\text{invariant } P : 0 \le i \le n \ \land \ p = i \cdot x\}$
   do $B : i \ne n \rightarrow i, p := i + 1, p + x$ **od**
   $\{P \ \land \ i = n\}$

We prove the first hypothesis of the theorem, $\{P \land B\}$ $i, p := i + 1, p + x$ $\{P\}$. To do this, we calculate the precondition $P[i, p := i + 1, p + x]$ and show that it is implied by $P \land B$.

$$
\begin{aligned}
&\quad P[i, p := i + 1, p + x] \\
&= \quad \langle \text{Definition of } P \text{; textual substitution} \rangle \\
&\quad 0 \le i + 1 \le n \ \land \ p + x = (i + 1) \cdot x \\
&= \quad \langle \text{Arithmetic} \rangle \\
&\quad -1 \le i < n \ \land \ p = i \cdot x \\
&\Leftarrow \quad \langle \text{Arithmetic} \rangle \\
&\quad i \ne n \ \land \ 0 \le i \le n \ \land \ p = i \cdot x \\
&= \quad \langle \text{Definition of } B \text{ and } P \rangle \\
&\quad B \land P
\end{aligned}
$$

Next, we prove the second hypothesis of the theorem. Since initially $i \le n$ and each iteration increases i by 1, after a finite number of iterations $i = n$ and the loop guard is *false*.

Hence, by Theorem (12.43), we conclude that (12.44) holds.   □

Theorem (12.43) concerns a loop with a precondition and postcondition, in isolation. Usually, we need to show something about a loop in a given context, since the loop may have initialization and a postcondition R that differs from $P \land \neg B$:

$\{Q\}$ *initialization*; **do** $B \rightarrow S$ **od** $\{R\}$

Hence, there is more to prove concerning the loop than simply the two points given in Theorem (12.43). With a loop annotated in this fashion, we need to prove four points:

(12.45)   **Checklist for proving loop correct**
   (a) P is *true* before execution of the loop.
   (b) P is a loop invariant: $\{P \land B\}$ S $\{P\}$.
   (c)   Execution of the loop terminates.
   (d) R holds upon termination: $P \land \neg B \Rightarrow R$.

Example of the use of Checklist (12.45). We prove that the annotation in program (12.42) is correct; we repeat the annotated program here.

$\{\ 0 \leq n\}$
$i, p := 0, 0;$
$\{\text{invariant } P:\ 0 \leq i \leq n \ \wedge\ p = i \cdot x\}$
do $i \neq n \rightarrow i, p := i + 1, p + x$ **od**
$\{R:\ p = n \cdot x\}$

Proving point (a) requires proving $0 \leq n \ \Rightarrow\ P[i, p := 0, 0]$; proving point (d) requires proving $\neg B \wedge P \ \Rightarrow\ R$. We leave these two proofs to the reader. Since we already proved the other two points, we conclude that the program is correct.   □

Loop invariants are crucial to understanding loops —so crucial that all but the most trivial loops should be documented with the invariants used to prove their annotations correct. In fact, (a first approximation to) the invariant should be developed *before* the loop is written and should act as a guide to the development of the loop. For example, since the fourth point for proving correctness of the loop is $P \wedge \neg B \Rightarrow R$, given P and R one can derive the loop guard by solving this expression for B.

Finding a suitable loop invariant is the most difficult part of writing most loops. However, a few simple ways of finding an approximation to the invariant work in many instances. In the example used above, invariant P is derived from result assertion R by replacing n by a fresh variable i and imposing suitable bounds on i. In the next example and all the exercises, we indicate how the invariant is obtained.

Example of a proof using Checklist (12.45). We prove correct an algorithm for division, which finds the quotient q and remainder r when b is divided by c (where $c > 0$). The annotated algorithm is given below. Invariant P is obtained by deleting conjunct $r < c$ from R.

(12.46) $\{Q:\ b \geq 0\ \wedge\ c > 0\}$
$q, r := 0, b;$
$\{\text{invariant } P:\ b = q \cdot c + r\ \wedge\ 0 \leq r\}$
do $r \geq c \rightarrow q, r := q + 1, r - c$ **od**
$\{R:\ b = q \cdot c + r\ \wedge\ 0 \leq r < c\}$

We prove the correctness of this annotated program. We prove point (a) of Checklist (12.45) by proving that $Q \ \Rightarrow\ P[q, r := 0, b]$.

$\qquad P[q, r := 0, b]$
$=\qquad \langle \text{Definition of } P\,; \text{textual substitution} \rangle$

$$b = 0 \cdot c + b \ \land \ 0 \leq b$$
$$\Leftarrow \quad \langle \text{Arithmetic; definition of } Q \rangle$$
$$Q$$

We next prove point (b), $\{P \land B\} \ S \ \{P\}$, by proving that $P \land B \Rightarrow P[q, r := q + 1, r - c]$.

$$P[q, r := q + 1, r - c]$$
$$= \quad \langle \text{Definition of } P \text{ and textual substitution} \rangle$$
$$b = (q + 1) \cdot c + (r - c) \ \land \ 0 \leq r - c$$
$$= \quad \langle \text{Arithmetic} \rangle$$
$$b = q \cdot c + r \ \land \ r \geq c$$
$$\Leftarrow \quad \langle \text{Definition of } P \text{ and } B \rangle$$
$$P \land B$$

For point (c), note that each iteration decreases r by c ($c > 0$), so that after a finite number of iterations $r < c$ is achieved.

Point (d), $P \land \neg B \Rightarrow R$, is trivial. $\square$

Proving termination of loops

Consider the following loop. [4]

(12.47) $\{0 \leq i = \mathtt{I}\}$
  $\{\text{invariant } P : 0 \leq i\}$
  do $0 \neq i \rightarrow$ **if** $true \rightarrow i := i - 1$
        $[\!]\ i \neq 1 \rightarrow i := i - 2$
      fi
  od
  $\{R : i = 0\}$

It is readily seen that invariant P is initially $true$, that the repetend maintains P, and that $P \land \neg(0 \neq i) \Rightarrow R$.

We can argue that the loop terminates as follows. (i) Integer expression i is decreased by at least 1 at each iteration, and (ii) as long as there is another iteration to be performed, $i > 0$ holds. Since $0 \leq i = \mathtt{I}$ holds initially, the loop terminates after at most $\mathtt{I}$ iterations.

More generally, we can prove the following theorem. In the program scheme within the theorem, we have added a comment to indicate the

---

[4] See page 189 for a definition of the alternative statement **if** $\cdots$ **fi**.

bound function that is used in proving termination of the loop.

(12.48) **Theorem.** To prove that

$$\{\text{invariant} : P\}$$
$$\{\text{bound function} : T\}$$
$$\textbf{do } B \rightarrow S \textbf{ od}$$

terminates, it suffices to find a *bound function* T, i.e. an integer expression T that is an upper bound on the number of iterations still to be performed. Thus, bound function T satisfies:

(a) T decreases at each iteration: that is, for v a fresh variable, $\{P \wedge B\}\ v := T; S\ \{T < v\}$.

(b) As long as there is another iteration to perform, $T > 0$: $P \wedge B \Rightarrow T > 0$.

Proof. We prove the theorem by induction on the initial value of T.

Base case $T \leq 0$. Since P is initially *true*, from $P \wedge B \Rightarrow T > 0$ (which equivales $P \wedge T \leq 0 \Rightarrow \neg B$), we conclude $\neg B$, so the loop terminates after 0 iterations.

Inductive case $T > 0$. We assume as inductive hypothesis that the theorem holds for all initial values of $T \leq k$ for some arbitrary integer $k \geq 0$; we prove the theorem for $T = k + 1$. If B is initially *false*, then the loop terminates immediately and the theorem holds. If B is initially *true*, then execution of one iteration decreases T so that $T \leq k$ (while maintaining P); by the inductive hypothesis, further execution of the loop terminates in at most k iterations. □

Example use of Theorem (12.48). Consider program (12.42) on page 237. We write the part of it that is germane to this discussion, annotated with the bound function:

$$\{\text{invariant } P : 0 \leq i \leq n\ \wedge\ p = i \cdot x\}$$
$$\{\text{bound function } T : n - i\}$$
$$\textbf{do } i \neq n \rightarrow i, p := i + 1, p + x \textbf{ od}$$

Each iteration increases i by 1 and thus decreases $n - i$. Second, we prove $P \wedge B \Rightarrow T > 0$, by transforming $P \wedge B$ to $T > 0$.

$$0 \leq i \leq n\ \wedge\ p = i \cdot x\ \wedge\ i \neq n$$
$$\Rightarrow \quad \langle \text{Weakening} \rangle$$
$$i < n$$
$$= \quad \langle \text{Arithmetic} \rangle$$
$$0 < n - i$$
$$= \quad \langle \text{Definition of } T \rangle$$
$$0 < T$$

Hence, by Theorem (12.48), the loop terminates.   □

Example use of Theorem (12.48). We prove that the loop of program (12.46) on page 239 terminates. Here is the pertinent part of that program:

$$\{\text{invariant } P:\ b = q \cdot c + r\ \land\ 0 \leq r\}$$
$$\{\text{bound function } T:\ r\}$$
$$\textbf{do } r \geq c \rightarrow q, r := q + 1, r - c \textbf{ od}$$

T is decreased by each iteration, since $c > 0$.[5] Second $P \land B$, along with $c > 0$, implies $r \geq c > 0$, so point (b) of Theorem (12.48) also holds. Therefore, by Theorem (12.48), the loop terminates in at most r iterations.

Note that r is not the exact number of iterations still to perform, but only an upper bound on the number of iterations. For T we could have taken the smaller expression $r - c$ as well.   □

Finally, we revisit program (12.47). Use expression i for T. The invariant $0 \leq i$ is *not* needed to prove $P \land \neg B \Rightarrow R$. P is used only in proving point (b) of (12.48): $P \land B \Rightarrow T > 0$. For example, if we changed invariant P to *true*, the *only* part that would not be provable would be this point (b). This might seem strange, but note that if we changed the invariant to *true* and also replaced the repetend by $i := i - 2$, termination could no longer be guaranteed, although every other part concerning correctness would be provable.

TERMINATION PROOFS USING OTHER WELL-FOUNDED SETS

Thus far, we have proved termination of loops using a bound function. And our proof that the bound function was sufficient to show termination was based on mathematical induction over the natural numbers. We now present a loop for which this method of proof does not work.

Let $choose(x)$ store an arbitrary natural number in variable x. Statement $choose(x)$ is *nondeterministic*: its execution need not always store the same value in x. One execution of $choose(x)$ may store 0 in x, another may store 99, and another 16180339887. Now consider the following loop, where $i{:}\mathbb{Z}$ (thus, initially i contains an integer).

---

[5] In principle, $c > 0$ should be a conjunct of the invariant. Note, however, that c is not changed by the algorithm. Cluttering up the invariant with the many facts about variables that remain unchanged would be counterproductive, and we use the mathematician's license to leave the obvious unstated.

(12.49) {Q : $true$}
          {invariant P : $true$}
          do $i \neq 0 \rightarrow$ if $i < 0$ then $choose(i)$ else $i := i - 1$ od
          {R : $i = 0$}

It is easy to see that this loop terminates. Its first iteration ensures $i \geq 0$, and thereafter each iteration decreases i by 1 until $i = 0$. However, our previous method of proof of termination cannot be used to prove termination, because there is no a priori upper bound on the number of iterations. If initially $i < 0$, then the number of iterations is determined by the value chosen for i during the first iteration, and the value chosen for i is unbounded.

We outline briefly how one can prove termination of a loop do $B \rightarrow S$ od with invariant P, using a pair $\langle U, \prec \rangle$ that admits induction. Since the pair admits induction, every decreasing chain is finite. Consider an expression $T{:}U$. Suppose that each iteration of the loop changes T to a smaller value:

$$\{P \wedge B\}\ v := T;\ S\ \{v \prec T\}$$

Suppose further that $P \wedge B \Rightarrow (\exists u{:}U \mid : u \prec T)$. Since every decreasing chain is finite, in a finite number of iterations, T will become a minimal element of U, in which case B is $false$ and the loop terminates.

In the case of program (12.47), to prove termination, we can choose $\langle \mathbb{Z}, \prec \rangle$, where $\prec$ is defined by (12.40) on page 236.

Exercises for Chapter 12

12.1 State an induction principle for proving properties of the negative integers.

12.2 What is wrong with the following proof that all people in any group have red hair? The proof is by induction on the number of people. For the base case, consider a group of 0 people. Since the group is empty, each person in it has red hair. For the inductive case, for arbitrary $n \geq 0$ we prove that $n + 1$ people have red hair using the fact that n people have red hair. So consider a group of $n + 1$ people. Remove one of them. By the inductive hypothesis, all those left in the group have red hair; take one (with red hair) of them out and place the first one removed back in; the group still consists of n people, and they all have red hair. Hence, the original group of $n + 1$ people had red hair.

12.3 Prove by induction that the following boolean expression holds for all n, $4 \leq n$:

$$P.n :\ (\exists h, k \mid 0 \leq h \wedge 0 \leq k : 2 {\cdot} h + 5 {\cdot} k = n)$$

12.4 Prove the following arithmetic identities by induction on n.

(a) For $n \geq 0$, $(\Sigma i \mid 1 \leq i \leq n : i) = n {\cdot} (n + 1)/2$.
(b) For $n \geq 0$, $(\Sigma i \mid 0 \leq i < n : 2 {\cdot} i + 1) = n^2$.

(c) For $n \geq 0$, $(\Sigma i \mid 0 \leq i < n : 2^i) = 2^n - 1$.

(d) For $n \geq 0$, $(\Sigma i \mid 0 \leq i < n : 3^i) = (3^n - 1)/2$.

(e) For $n \geq 0$, $(\Sigma i \mid 0 \leq i \leq n : i^2) = n \cdot (n+1) \cdot (2 \cdot n + 1)/6$.

(f) For $n \geq 0$, $(\Sigma i \mid 1 \leq i \leq n : i \cdot 2^i) = (n-1) \cdot 2^{n+1} + 2$.

(g) For $n \geq 0$, $(\Sigma i \mid 1 \leq i \leq n : 3i^2 - 3i + 1) = n^3$.

12.5 Prove $(\Sigma i \mid 0 \leq i \leq n : r^i) = (r^{n+1} - 1)/(r - 1)$ for r a real number, $r \neq 1$, and n a natural number. Use induction.

12.6 Prove $(\Sigma i \mid 0 \leq i \leq n : i \cdot r^i) = (n \cdot r^{n+2} - (n+1) \cdot r^{n+1} + r)/(r-1)^2$ for r a real number, $r \neq 1$, and n a natural number. Use induction.

12.7 A convex polygon is a polygon in which the line joining any two points on its perimeter is in the polygon. Prove by induction that, for $n \geq 3$, the sum of the angles of a convex polygon with n sides is $(n-2) \cdot 180°$. Use the fact that the sum of the angles in a triangle is $180°$.

12.8 Prove by induction on n that $2 \cdot n + 1 < 2^n$ for $n \geq 3$.

12.9 Prove by induction on n that $n^2 \leq 2^n$ for $n \geq 4$.

12.10 Prove by induction that $2^{2 \cdot n} - 1$ is divisible by 3, for $n \geq 0$.

12.11 Prove by induction that $4^n - 1$ is divisible by 3, for $n \geq 0$.

12.12 Prove by induction that $10^i - 1$ is divisible by 9, for $i \geq 0$. Use this to show that 9 divides a decimal integer $r_{n-1} \cdots r_1 r_0$ (where all the r_i satisfy $0 \leq r_i < 10$) if and only if 9 divides the sum of the r_i.

12.13 Prove by induction that for $x \neq y$, $x^n - y^n$ is divisible by $x - y$, for $n \geq 0$. Hint: subtract and add $x \cdot y^n$ to $x^{n+1} - y^{n+1}$.

12.14 Prove by induction that any amount greater than 14 can be obtained using 3-cent and 8-cent coins.

12.15 Prove (11.73), $\#(\mathcal{P}S) = 2^{\#S}$.

Exercises on equivalence of weak and strong induction

Induction (12.3) describes what is known as *strong* induction. *Weak* induction allows only $P.n$ as the inductive hypothesis:

Weak induction over $\mathbb{N}$:

$$P.0 \land (\forall n : \mathbb{N} \mid : P.n \Rightarrow P(n+1)) \Rightarrow (\forall n : \mathbb{N} \mid : P.n)$$

Exercises 12.16–12.19 are devoted to proving that weak induction and strong induction are equivalent. We give some abbreviations that will be used in the exercises. First, let *WS* and *SS* stand for the *Weak induction Step* and *Strong induction Step*:

$$WS : (\forall n \mid 0 \leq n : P.n \Rightarrow P(n+1)) \quad ,$$

$$SS : (\forall n \mid 0 \leq n : (\forall i \mid 0 \leq i \leq n : P.i) \Rightarrow P(n+1)) \quad .$$

Next, let A mean that all $P.n$ are *true*:

$$A : (\forall n \mid 0 \leq n : P.n) \quad .$$

Finally, let WI and SI denote Weak Induction and Strong Induction:

$$WI : P.0 \wedge WS \Rightarrow A \ ,$$
$$SI : P.0 \wedge SS \Rightarrow A \ .$$

The equivalence of weak and strong induction is then written as

$$(12.50) \quad (\forall P|: WI) \equiv (\forall P|: SI) \ ,$$

where $P : \mathbb{N} \to \mathbb{B}$. Formula (12.50) says that weak induction holds *for all predicates* P iff strong induction holds *for all predicates* P. This quantification is necessary for a correct statement of equivalence of the two kinds of induction. Formula (12.50) is different from $(\forall P|: WI \equiv SI)$ which is not valid.

Thus far, we have not encountered quantification over predicates. The predicate calculus with quantification over conventional variables, as in Chap. 9, is called the *first-order* predicate calculus. A calculus with quantification over predicates, as in (12.50), is a *second-order* predicate calculus. In the second-order predicate calculus, all the theorems of the first-order calculus are valid, and there are additional axioms and theorems to deal with the new kind of quantification.

12.16 We begin the proof of (12.50). The first step, the object of this first exercise, is to prove $WS \Rightarrow SS$.

12.17 Prove that $SI \Rightarrow WI$ (see the previous exercise).

12.18 Prove $(\forall P|: WI) \Rightarrow (\forall P|: SI)$ by setting aside its antecedent and proving its consequent. This means proving that strong induction holds for all P under the assumption that weak induction holds for all P. Here are some hints. Introduce a predicate $Q.n$:

$$Q.n : (\forall i \mid 0 \leq i \leq n : P.i) \ .$$

Since $(\forall n \mid 0 \leq n : P.n) \equiv (\forall n \mid 0 \leq n : Q.n)$, we can rewrite strong induction (for arbitrary P) as

$$P.0 \wedge ((\forall n \mid 0 \leq n : Q.n) \Rightarrow P(n+1)) \Rightarrow (\forall n \mid 0 \leq n : Q.n) \ .$$

Now prove this formulation of mathematical induction by assuming the two conjuncts of its antecedent and proving its consequent by (weak) induction.

12.19 Prove (12.50) of Exercise 12.16, using the results of the previous two exercises.

Exercises on Fibonacci numbers

12.20 Prove by strong induction that $F_n < 2^n$ for $n \leq 0$.

12.21 Prove properties (12.15).

12.22 Prove that $\phi^{n-2} \leq F_n$ for $n \geq 1$. (Note that $\phi^{1-2} = \phi^{-1} = 1/\phi$.)

12.23 Prove that, for all $n \geq 0$, $F_n = (\phi^n - \hat{\phi}^n)/\sqrt{5}$.

12.24 Using the results of the previous exercise, prove the *Binet formula* $F_n = (\phi^n - \hat{\phi}^n)/(\phi - \hat{\phi})$ for $n \geq 0$.

12.25 Prove that the following identities hold. Hint: Substitute appropriately in (12.17).

$$F_{2 \cdot n} = F_n \cdot F_{n+1} + F_{n-1} \cdot F_n \quad \text{for } n \geq 1,$$
$$F_{2 \cdot n+1} = F_{n+1}^2 + F_n^2 \quad \text{for } n \geq 0$$

12.26 Prove that $F_n^2 = F_{n-1} \cdot F_{n+1} - (-1)^n$ for $n \geq 1$.

12.27 Prove that $(\Sigma i \mid 0 \leq i \leq n : F_i) = F_{n+2} - 1$ for $n \geq 0$.

12.28 Prove that $(\Sigma i \mid 0 \leq i \leq n : F_i^2) = F_n \cdot F_{n+1}$ for $n \geq 0$.

12.29 Prove that, for $n \geq 0$, $F_{3 \cdot n}$ is even, $F_{3 \cdot n+1}$ is odd, and $F_{3 \cdot n+2}$ is odd.

Other exercises on proofs by induction

12.30 The *greatest common divisor* of two natural numbers p and q, written $p \text{ gcd } q$, is the largest natural number that divides both. For example, $10 \text{ gcd } 0 = 10$, $12 \text{ gcd } 10 = 2$, and $1 \text{ gcd } 8 = 1$. Prove by induction that $F_n \text{ gcd } F_{n+1} = 1$ for all $n \geq 0$.

12.31 Prove that the two definitions (12.11) and (12.12) of exponentiation are equivalent, i.e. b^n has the same value in both definitions.

12.32 Juris Jones maintains that he is exactly one-third Latvian. Prove that he is lying. Hint: Relate this problem to the following set S and show that $1/3$ is not in S.

> 0 is in S;
>
> 1 is in S;
>
> If x and y are in S, then so is $(x + y)/2$.

12.33 Define the value $n!$ for $n \geq 0$ by

$$0! = 1 ,$$
$$(n+1)! = (n+1) \cdot n! \quad \text{for } n \geq 0.$$

Prove by induction that, for $n \geq 0$, $n! = (\Pi i \mid 1 \leq i \leq n : i)$.

12.34 Prove by induction that $n! > 2^{n-1}$ for $n \geq 3$. See Exercise 12.33 for a recursive definition of $n!$.

12.35 Prove by induction that $(\Sigma i \mid 0 \leq i \leq n : i \cdot i!) = (n+1)! - 1$ for $n \geq 0$. See Exercise 12.33 for a recursive definition of $n!$.

12.36 Define the values m_n for $n \geq 0$ recursively by

$$m_0 = 0 ,$$
$$m_{n+1} = 2 \cdot m_n + 1 \quad \text{for } n \geq 0.$$

Prove by induction that $m_n = 2^n - 1$ for $n \geq 0$.

12.37 *The ring of lights.* Suppose we have a ring of 2^N lights, for some $N \geq 0$, each of which can be on or off. The lights repeatedly change their state, in synchrony, according to the following rule: If the follower of a light (in clockwise order) is off, the light switches (from off to on or from on to off); if the follower is on, the light does not switch but remains the same. Show that after a certain number of steps all the lights will be on —the number of steps to achieve this depends on N but not on the initial state of the ring of lights.

Hint: Number the lights $0, 1, \ldots 2^N - 1$, so that the follower of light i is light $i + 1$ (modulo 2^N). Let boolean $L(i, j)$ denote the state of light i at (i.e. just before) step j, according to $L(i, j) \equiv$ (light i is on at step j). Prove by induction on n that for arbitrary i and j,

$$L(i.j + 2^n) \equiv L(i, j) \equiv L(i + 2^n, j) \ .$$

12.38 Consider the following game, played with a non-empty bag S of positive real numbers. Operation avg removes two elements of S (at random) and inserts two copies of the average of the two removed elements. The game terminates when all numbers in S are equal. Does the game always terminate?

12.39 Define inductively the function $dom.e$ of page 182 for integer expressions using integers, integer variables, array references $b[i]$, binary addition, subtraction, multiplication, integer division, and unary subtraction.

12.40 Prove theorem (12.27).

12.41 Prove theorem (12.28).

Exercises on loops

12.42 Each algorithm below is annotated with a precondition, loop invariant, and postcondition. Prove the algorithm correct using Checklist (12.45).

(a) This algorithm stores in c the Fibonacci number F_n, for $n \geq 0$. In addition to the definition of Fibonacci numbers given in (12.14), we define $F_{-1} = 1$, so that F_1 satisfies the recursive definition ($F_1 = F_0 + F_{-1}$). Invariant P arises by replacing n in R by a fresh variable, placing suitable bounds on k, and then adding the extra conjunct $b = F_{k-1}$ —for reasons that cannot be made entirely clear at this time.

> $\{Q : n \geq 0\}$
> $k, b, c := 0, 1, 0;$
> $\{\text{invariant } P : 0 \leq k \leq n \ \wedge \ b = F_{k-1} \ \wedge \ c = F_k\}$
> **do** $k \neq n \rightarrow k, b, c := k + 1, c, b + c$ **od**
> $\{R : c = F_n\}$

(b) This algorithm stores in x the sum of the n elements of array $b[0..n - 1]$, for $n \geq 0$. Invariant P is developed by replacing n in R by a fresh variable k and placing suitable bounds on k.

> $\{Q : n \geq 0\}$
> $x, k := 0, 0;$
> $\{\text{invariant } P : 0 \leq k \leq n \ \wedge \ x = (\Sigma i \mid 0 \leq i < k : b[i])\}$
> **do** $k \neq n \rightarrow x, k := x + b[k], k + 1$ **od**

$$\{R: \ x = (\Sigma i \mid 0 \leq i < n : b[i])\}$$

(c) This algorithm stores in x the sum of the n elements of array $b[0..n-1]$, for $n \geq 0$. Invariant P is developed by replacing 0 in R by a fresh variable k and placing suitable bounds on k.

$$\{Q: \ n \geq 0\}$$
$$x, k := 0, n;$$
$$\{\text{invariant } P: \ 0 \leq k \leq n \ \wedge \ x = (\Sigma i \mid k \leq i < n : b[i])\}$$
$$\textbf{do } k \neq 0 \rightarrow x, k := x + b[k-1], k-1 \ \textbf{od}$$
$$\{R: \ x = (\Sigma i \mid 0 \leq i < n : b[i])\}$$

(d) This algorithm finds the greatest common divisor $X \ \textbf{gcd} \ Y$ of two natural numbers X and Y —i.e. the largest natural number that divides both X and Y. ($\textbf{gcd}$ is discussed in Sec. 15.4 on page 316). The algorithm uses $x \ \textbf{mod} \ y$, which is the remainder of x divided by y. You can use the following properties of $x \ \textbf{gcd} \ y$: (0) $x \ \textbf{gcd} \ y = y \ \textbf{gcd} \ x$, (1) $x \ \textbf{gcd} \ 0 = x$, and (2) $x \ \textbf{gcd} \ y = y \ \textbf{gcd} \ (x \ \textbf{mod} \ y)$. Property (2) holds because, if $x = q \cdot y + r$, if an integer divides both x and y then it also divides r, and if an integer divides both y and r then it also divides x.

$$\{Q: \ 0 \leq X \wedge 0 \leq Y\}$$
$$x, y := X, Y;$$
$$\{\text{invariant } P: \ x \ \textbf{gcd} \ y = X \ \textbf{gcd} \ Y \ \wedge \ 0 \leq x \ \wedge \ 0 \leq y\}$$
$$\textbf{do } y \neq 0 \rightarrow x, y := y, x \ \textbf{mod} \ y \ \textbf{od}$$
$$\{R: \ X \ \textbf{gcd} \ Y = x\}$$

Exercises on the proof of Metatheorem duality

Metatheorem duality (2.3a) on page 32 states that a propositional formula P is valid iff $\neg P_D$ is valid, where P_D is the dual of P, and that $P \equiv Q$ is valid iff $P_D \equiv Q_D$ is valid. The following exercises prove these claims.

We begin by defining expression $\overline{P}$ corresponding to a boolean expression P.

$\overline{P}$ is constructed from P by replacing in P each variable q (say) by $\neg q$ and interchanging symbols as given in the following table. Note that only operator $\neg$ remains unchanged.

true	and	*false*
$\wedge$	and	$\vee$
$\equiv$	and	$\not\equiv$
$\Rightarrow$	and	$\not\Leftarrow$
$\Leftarrow$	and	$\not\Rightarrow$

Examples.

P	$\overline{P}$
$p \lor q$	$\lnot p \land \lnot q$
$p \Rightarrow q$	$\lnot p \not\Leftarrow \lnot q$ (or $\lnot(\lnot p \Leftarrow \lnot q)$)
$p \equiv q$	$\lnot p \not\equiv \lnot q$
$\lnot p \land \lnot q \equiv r$	$\lnot\lnot p \lor \lnot\lnot q \not\equiv \lnot r$

12.43 Define expression $\overline{P}$, illustrated above, inductively.

12.44 Using your inductive definition of $\overline{P}$ from Exercise 12.43, prove that for any boolean expression P,

$$\lnot P \equiv \overline{P} \ .$$

12.45 The dual of P is similar to $\overline{P}$, the only difference being that variables are left unchanged (and not replaced by their negations):

Examples of expressions P, $\overline{P}$, and P_D.

expression P	$\overline{P}$	dual P_D
$p \lor q$	$\lnot p \land \lnot q$	$p \land q$
$p \Rightarrow q$	$\lnot p \not\Leftarrow \lnot q$	$p \not\Leftarrow q$
$p \equiv q$	$\lnot p \not\equiv \lnot q$	$p \not\equiv q$
$\lnot p \land \lnot q \equiv r$	$\lnot\lnot p \lor \lnot\lnot q \not\equiv \lnot r$	$\lnot p \lor \lnot q \not\equiv r$

Prove that if an expression is a theorem, then so is the negation of its dual —i.e. if P is a theorem, then so is $\lnot P_D$.

12.46 Prove that if $P \equiv Q$ is a theorem, then so is $P_D \equiv Q_D$.

Exercises on proving Metatheorem (11.25)

Consider a set expression E_s constructed from set variables, $\{\,\}$, **U** (the universe for all set variables in question), $\sim$, $\cup$, and $\cap$. Let E_p be the proposition constructed from E_s by replacing $\emptyset$, **U**, $\sim$, $\cup$, and $\cap$ with *false*, *true*, $\lnot$, $\lor$, and $\land$, respectively. Note that the transformation is reversible: E_s can be constructed from E_p.

We wish to prove Metatheorem (11.25) on page 204.

Metatheorem. For any set expressions E_s and F_s,

(12.51) $E_s = F_s$ is valid iff $E_p \equiv F_p$ is valid,

(12.52) $E_s \subseteq F_s$ is valid iff $E_p \Rightarrow F_p$ is valid,

(12.53) $E_s = $ **U** is valid iff E_p is valid.

12.47 The first step in the proof is to introduce another translation of expression E_s. Let E_c be a copy of E_s in which each occurrence of a set variable P, $\{\,\}$, or **U** is replaced by the set $\{x \mid P\}$, $\{x \mid false\}$, or $\{x \mid true\}$, respectively. Thus, each set variable and constant of E_s is replaced by a set comprehension that exhibits the characteristic predicate of the set —note that in E_c the identifiers P are interpreted as predicates rather than sets.

For example, for $E_s = P \cup \textbf{U}$, we have $E_c = \{x \mid P\} \cup \{x \mid true\}$.

We introduced E_c because we can prove that, for any predicate E_c,

(12.54) $x \in E_c \equiv E_p$ (for all x),

which is equivalent to

(12.55) $E_c \equiv \{x \mid E_p\}$.

Thus, E_p is the characteristic predicate for the set E_c .

Your task in this exercise is to prove (12.54) by induction on the structure of E_c .

12.48 Using the results of Exercise 12.47, prove that for any expression E_c and F_c ,

(12.56) $E_c = F_c \equiv E_p \equiv F_p$.

12.49 Prove the following theorem. Make use of (12.56) from Exercise 12.48.

(12.57) $E_c = \mathbf{U} \equiv E_p$

12.50 Prove the following theorem by mutual implication.

(12.58) $E_c \subseteq F_c \equiv E_p \Rightarrow F_p$.

12.51 Use the validity of (12.56)–(12.58) to argue that the following hold (trivial)

$E_c = F_c$ is valid iff $E_p \equiv F_p$ is valid.

$E_c = \mathbf{U}$ is valid iff E_p is valid.

$E_c \subseteq F_c$ is valid iff $E_p \Rightarrow F_p$ is valid.

12.52 Finally, Metatheorem (11.25) can be proved. Note that (11.25) is in terms of E_s , while the theorems proved in Exercise 12.51 are in terms of E_c . Find the connection between them that allows (11.25) to be proved (and prove it).

Chapter 13

A Theory of Sequences

$\mathbf{A}$ sequence is a finite list of elements from some set. In this chapter, we develop a theory of sequences by defining them inductively and then defining various operations on them.

There are a variety of reasons for studying sequences. First, the theory of sequences provides an excellent opportunity to practice proofs by induction in a setting other than the natural numbers. Second, the theory of sequences serves as a basis for reasoning about lists in Lisp and arrays in imperative languages, allowing us to make our reasoning about programs written in these languages clearer and more precise. Third, the theory is the basis for the important study of *formal languages*, which, among other things, has led to methods for the automatic generation of parts of compilers.

13.1 The basic theory of sequences

Let A be a nonempty set that does not include an element ϵ. We define inductively the set $seq(A)$ of finite sequences over A. Throughout this chapter, variables a, b, c, d are of type A, while w, x, y, z are of type $seq(A)$.

(13.1) **Axiom, Empty sequence:** $\epsilon \in seq(A)$

(13.2) **Axiom, Prepend:** $c \triangleleft x \in seq(A)$

(13.3) **Axiom, Nonempty sequence:** $c \triangleleft x \neq \epsilon$

(13.4) **Axiom, Equality:** $b \triangleleft x = c \triangleleft y \equiv b = c \wedge x = y$

The first two axioms define the members of set $seq(A)$. Constant ϵ is called the *empty sequence*; it contains no elements. Operator $\triangleleft$ is called the *prepend* operator, because it "prepends" an element to a sequence. [1]

Axioms (13.3) and (13.4) define equality and inequality of sequences.

---

[1] To append an element to a sequence means to add the element at the end of the sequence. There is no word for adding an element to the beginning of a sequence, so we have coined the word "prepend". The Oxford English Dictionary defines the (obsolete) word "prependant" as "hanging down in front".

Operator $\triangleleft$ is taken to be *right associative*, so that

$$b \triangleleft c \triangleleft x = b \triangleleft (c \triangleleft x) \quad .$$

Left associativity would not make sense, because $b \triangleleft c$ is not defined for c an element; $b \triangleleft x$ is defined only for $b\!:\!A$ and $x\!:\!seq(A)$. This explains why the sequence ϵ is placed at the end of $3 \triangleleft 6 \triangleleft 4 \triangleleft \epsilon$; $3 \triangleleft 6 \triangleleft 4$ is not an expression.

In this text, we abbreviate a sequence $3 \triangleleft 6 \triangleleft 4 \triangleleft \epsilon$ by the *tuple* $\langle 3, 6, 4 \rangle$. A tuple is simply a list of expressions, separated by commas and delimited by $\langle$ and $\rangle$. Sometimes, we prefix the term tuple with the length of the tuple in question. For example, we may talk of the 2-tuple $\langle 6, 4 \rangle$.

Note that ϵ is not considered an element of the sequence $3 \triangleleft 6 \triangleleft 4 \triangleleft \epsilon$. Also, $\epsilon = \langle \rangle$. Further, an element c is different from the *singleton* sequence consisting of c, which is written as $c \triangleleft \epsilon$, or $\langle c \rangle$.

INDUCTION OVER SEQUENCES

Define relation *istail* for sequences x and y by

$$\langle y, x \rangle \in istail \equiv (\exists c \mid : x = c \triangleleft y) \quad .$$

Thus, y is a tail of x iff deleting the first element of x results in y.

Since we are considering only finite sequences —i.e. sequences with a finite number of elements— the length of any chain

$$\langle x_1, x_0 \rangle \in istail,$$
$$\langle x_2, x_1 \rangle \in istail,$$
$$\langle x_3, x_2 \rangle \in istail,$$
$$\cdots$$

is finite. Hence, $\langle seq(A), istail \rangle$ satisfies finite chain property (12.24) and is noetherian (12.25), which in turn implies that it is well founded and admits induction. The induction principle, according to (12.19), is

$$(\forall x \mid : P.x) \equiv (\forall x \mid : (\forall y \mid \langle y, x \rangle \in istail : P.y) \Rightarrow P.x)$$

Since $\langle x, c \triangleleft x \rangle \in istail$, we can express this induction without referring to relation *istail*, and as a form of (weak) induction. For any predicate $P.x$,

(13.5)   **Axiom, Induction over sequences:**
$$(\forall x \mid : P.x) \equiv P.\epsilon \wedge (\forall c, x \mid : P.x \Rightarrow P(c \triangleleft x))$$

We present two theorems. Theorem Decomposition (13.6), the more important one, gives us a means of proving properties of an arbitrary sequence x by case analysis: either $x = \epsilon$ or $x = b \triangleleft y$ for some b and y.

```
                   Theorems for sequences
(13.6)  Decomposition:  x = ε ∨ (∃b, y |: x = b ◁ y) .

(13.7)  c ◁ x ≠ x
```

The proof of Decomposition (13.6) is by Induction (13.5). We prove $(\forall x \mid: P.x)$, where $P.x$ is $x = \epsilon \lor (\exists b, y \mid: x = b \lhd y)$.

Base case $P[x := \epsilon]$.

$$P[x := \epsilon]$$
$$= \quad \langle \text{Definition of } P \text{; textual substitution} \rangle$$
$$\epsilon = \epsilon \lor (\exists b, y \mid: \epsilon = b \lhd y)$$
$$= \quad \langle \text{Reflexivity of equality; Zero of } \lor \text{ (3.29)} \rangle$$
$$true$$

Inductive case. For arbitrary element c and sequence x, we assume inductive hypothesis $P.x$ and prove $P(c \lhd x)$.

$$P(c \lhd x)$$
$$= \quad \langle \text{Definition of } P \rangle$$
$$c \lhd x = \epsilon \lor (\exists b, y \mid: c \lhd x = b \lhd y)$$
$$= \quad \langle \text{Nonempty sequence (13.3); Identity of } \lor \text{ (3.30)} \rangle$$
$$(\exists b, y \mid: c \lhd x = b \lhd y)$$
$$\Leftarrow \quad \langle \text{Range strengthening (9.25)} \rangle$$
$$(\exists b, y \mid b = c \land y = x : c \lhd x = b \lhd y)$$
$$= \quad \langle \text{One-point rule (8.14), twice} \rangle$$
$$c \lhd x = c \lhd x \quad \text{—Reflexivity of equality}$$

We now prove (13.7), $c \lhd x \neq x$, by induction. Thus, we prove $(\forall x \mid: P.x)$ where $P.x$ is $(\forall c \mid: c \lhd x \neq x)$.

Base case $P.\epsilon$. $P.\epsilon$ is Nonempty sequence (13.3) with x instantiated with ϵ .

Inductive case. We assume inductive hypothesis $P.x$ and prove $P(d \lhd x)$ for arbitrary d. For arbitrary c, we have,

$$\neg(c \lhd d \lhd x = d \lhd x)$$
$$= \quad \langle \text{Equality (13.4)} \rangle$$
$$\neg(c = d \land d \lhd x = x)$$
$$= \quad \langle \text{Inductive hypothesis } P.x \rangle$$
$$\neg(c = d \land false)$$
$$= \quad \langle \text{Zero of } \land \text{ (3.40); Negation of } false \text{ (3.13)} \rangle$$
$$true$$

13.2   Extending the theory with new operations

Thus far, we have presented a rather bare theory of sequences. Our next step is to make the theory more convenient to use, by defining new operations. The operations we define and analyze are:

- $head.x$ and $tail.x$, the first element of x and the rest of x.

- $c \in x$, a test for membership of c in x.

- $x \triangleright c$, which appends element c to x.

- $x \mathbin{\char94} y$, which catenates two sequences together.

- $x \subseteq y$, a predicate equal to "x is a subsequence of y".

- $isprefix(x, y)$ a predicate equal to "x is a prefix of y".

- $isseg(x, y)$ a predicate equal to "x is a subsegment of y".

HEAD AND TAIL

Functions $head$ and $tail$, defined below by axioms (13.8) and (13.9), provide a convenient way to refer to the elements of a sequence. Note that $head$ and $tail$ are applied only to nonempty sequences.

(13.8)   **Axiom, Head:** $head(c \triangleleft x) = c$

(13.9)   **Axiom, Tail:** $tail(c \triangleleft x) = x$

Examples. Let $x = 3 \triangleleft 6 \triangleleft 4 \triangleleft \epsilon$, i.e. $x = \langle 3, 6, 4 \rangle$. Then

$head.x = 3$	$tail.x = 6 \triangleleft 4 \triangleleft \epsilon$	$(= \langle 6, 4 \rangle)$
$head(tail.x) = 6$	$tail(tail.x) = 4 \triangleleft \epsilon$	$(= \langle 4 \rangle)$
$head(tail(tail.x)) = 4$	$tail(tail(tail.x)) = \epsilon$	$(= \langle \rangle)$

$\square$

Operations $c \triangleleft x$, $head.x$, and $tail.x$ are found in many functional programming languages. For example, they are written in Lisp and Scheme as $(cons\ c\ x)$, $(car\ x)$, and $(cdr\ x)$, respectively.

MEMBERSHIP

The following axioms define the membership relation for sequences: element c is in sequence x iff c is one of the elements of x. The first axiom indicates that no element is in the empty sequence ϵ; the second gives a more positive, recursive, statement about when an element is in a sequence.

(13.10)   **Axiom, Membership:** $b \in \epsilon \equiv false$

(13.11) **Axiom, Membership:** $b \in c \triangleleft x \equiv b = c \lor b \in x$

APPEND

Being able only to prepend an element to a sequence is a bit limiting. We now define binary infix operator *append*, $\triangleright$. Expression $x \triangleright c$ yields a sequence consisting of the elements of x followed by element c.

(13.12) **Axiom, Append:** $\epsilon \triangleright c = c \triangleleft \epsilon$

(13.13) **Axiom, Append:** $(b \triangleleft x) \triangleright c = b \triangleleft (x \triangleright c)$

We show how this definition is used in calculating the result of appending an element to a sequence.

$$
\begin{aligned}
& (a \triangleleft b \triangleleft c \triangleleft \epsilon) \triangleright d \quad \text{—which is } \langle a, b, c \rangle \triangleright d \\
= \quad & \langle \text{Append (13.13), with } b, x, c := a, b \triangleleft c \triangleleft \epsilon, d \rangle \\
& a \triangleleft ((b \triangleleft c \triangleleft \epsilon) \triangleright d) \\
= \quad & \langle \text{Append (13.13), with } b, x, c := b, c \triangleleft \epsilon, d \rangle \\
& a \triangleleft b \triangleleft ((c \triangleleft \epsilon) \triangleright d) \\
= \quad & \langle \text{Append (13.13), with } b, x, c := c, \epsilon, d \rangle \\
& a \triangleleft b \triangleleft c \triangleleft (\epsilon \triangleright d) \\
= \quad & \langle \text{Append (13.12)} \rangle \\
& a \triangleleft b \triangleleft c \triangleleft d \triangleleft \epsilon \quad \text{—which is } \langle a, b, c, d \rangle
\end{aligned}
$$

Here are some theorems concerning $\triangleright$.

Theorems for $\triangleright$

(13.14) **Nonempty sequence:** $x \triangleright c \neq \epsilon$

(13.15) **Equality:** $x \triangleright b = y \triangleright c \equiv x = y \land b = c$

(13.16) **Membership in $\triangleright$:** $b \in (x \triangleright c) \equiv b \in x \lor b = c$

We prove Nonempty sequence (13.14) by induction, where we write (13.14) as $(\forall x \mid : P.x)$ with $P.x : (\forall c \mid : \epsilon \neq x \triangleright c)$.

Base case $P.\epsilon$. For arbitrary c, we have

$$
\begin{aligned}
& \epsilon \triangleright c \\
= \quad & \langle \text{Append (13.12)} \rangle \\
& c \triangleleft \epsilon \\
\neq \quad & \langle \text{Nonempty sequence (13.3)} \rangle \\
& \epsilon
\end{aligned}
$$

Inductive case. We assume inductive hypothesis $P.x$ and prove $P(d \triangleleft x)$ for arbitrary d. For arbitrary c, we have,

$$
\begin{aligned}
&(d \triangleleft x) \triangleright c \\
= \quad &\langle \text{Append } (13.13) \rangle \\
&d \triangleleft (x \triangleright c) \\
\neq \quad &\langle \text{Nonempty sequence } (13.3) \rangle \\
&\epsilon
\end{aligned}
$$

CATENATION

Evaluation of the expression $x \,\widehat{}\, y$ yields a sequence consisting of the elements of sequence x followed by the elements of sequence y. Operation $\widehat{}\,$ is inductively defined as follows.

(13.17) **Axiom, Left identity of** $\widehat{}\,$**:** $\epsilon \,\widehat{}\, x = x$

(13.18) **Axiom, Mutual associativity:** $(b \triangleleft y) \,\widehat{}\, x = b \triangleleft (y \,\widehat{}\, x)$

We have the following theorems. Mutual associativity theorem (13.22) allows us to write expressions like $x \,\widehat{}\, y \triangleright c$ without parentheses. Due to (13.13) and (13.18), we can now write sequences like $a \triangleleft b \triangleleft c \,\widehat{}\, x \,\widehat{}\, y \triangleright c \triangleright d$ and associate in any way we please.

Membership (13.24) is often taken as the definition of membership in a sequence. To make it easier to read, we have used the notation $\langle b \rangle$ instead of $b \triangleleft \epsilon$.

Theorems for $\triangleright$ **and** $\widehat{}\,$

(13.19) **Right identity of** $\widehat{}\,$**:** $x \,\widehat{}\, \epsilon = x$

(13.20) **Associativity of** $\widehat{}\,$**:** $x \,\widehat{}\, (y \,\widehat{}\, z) = (x \,\widehat{}\, y) \,\widehat{}\, z$

(13.21) **Membership:** $b \in x \,\widehat{}\, y \;\equiv\; b \in x \lor b \in y$

(13.22) **Mutual associativity:** $(x \,\widehat{}\, y) \triangleright c \doteq x \,\widehat{}\, (y \triangleright c)$

(13.23) **Empty catenation:** $x \,\widehat{}\, y = \epsilon \;\equiv\; x = \epsilon \land y = \epsilon$

(13.24) **Membership:** $b \in x \;\equiv\; (\exists y, z \mid : x = y \,\widehat{}\, \langle b \rangle \,\widehat{}\, z)$

SUBSEQUENCE

One sequence x is a *subsequence* of another sequence y if eliminating zero or more elements from y yields x. For example, three subsequences of $y = \langle 2, 3, 8, 5, 2 \rangle$ are $\langle \rangle$, $\langle 3, 5 \rangle$, and $\langle 3, 5, 2 \rangle$. The sequence $\langle 8, 3 \rangle$ is not a subsequence of y. We use the predicate $x \subseteq y$ for "x is a subsequence of y". Formally, we define the subsequence relation, as well as the proper-subsequence relation, as follows.

(13.25) **Axiom, Empty subsequence:** $\epsilon \subseteq y$

(13.26) **Axiom, Subsequence:** $\neg(c \triangleleft x \subseteq \epsilon)$

(13.27) **Axiom, Subsequence:** $c \triangleleft x \subseteq c \triangleleft y \;\equiv\; x \subseteq y$

(13.28) **Axiom, Subsequence:** $b \neq c \;\Rightarrow\; (b \triangleleft x \subseteq c \triangleleft y \;\equiv\; b \triangleleft x \subseteq y)$

(13.29) **Axiom, Proper subsequence:** $x \subset y \;\equiv\; x \subseteq y \land x \neq y$

The following theorems can be proved concerning subsequences.

Theorems for subsequence

(13.30) **Reflexivity of $\subseteq$:** $x \subseteq x$

(13.31) $x \subseteq c \triangleleft x$

(13.32) $x \subset c \triangleleft x$

(13.33) $x \subseteq \epsilon \;\equiv\; x = \epsilon$

(13.34) $x \subseteq y \;\Rightarrow\; (\forall c \,|: c \in x \;\Rightarrow\; c \in y)$

(13.35) $c \triangleleft x \subseteq y \;\equiv$
$\qquad y \neq \epsilon \land ((c = head.y \land x \subseteq tail.y) \lor c \triangleleft x \subseteq tail.y)$

PREFIXES AND SEGMENTS

A sequence x is *prefix* of y if y begins with x. For example, $\langle 2, 3 \rangle$ is a prefix of $\langle 2, 3, 8, 1 \rangle$. Similarly, x is a *segment* of y iff x appears somewhere within y as a subsequence of adjacent elements. For example, $\langle 3, 5, 8 \rangle$ is a segment of $\langle 2, 3, 5, 8, 6 \rangle$, but $\langle 3, 8 \rangle$ is not. We define relations $isprefix(x, y)$ and $isseg(x, y)$ as follows.

(13.36) **Axiom, Empty prefix:** $isprefix(\epsilon, y)$

(13.37) **Axiom, Not Prefix:** $isprefix(c \triangleleft x, \epsilon) \equiv false$

(13.38) **Axiom, Prefix:** $isprefix(c \triangleleft x, d \triangleleft y) \equiv c = d \wedge isprefix(x, y)$

(13.39) **Axiom, Segment:** $isseg(x, \epsilon) \equiv x = \epsilon$

(13.40) **Axiom, Segment:**
$$isseg(x, c \triangleleft y) \equiv isprefix(x, c \triangleleft y) \vee isseg(x, y)$$

The definitions of *isprefix* and *isseg* are rather cumbersome to use. However, we can use operation $\hat{\ }$ to provide characterizations of *isprefix* and *isseg* that are easier to use in reasoning about them.

Characterization of *isprefix* and *isseg*

(13.41) $isprefix(x, y) \equiv (\exists z \mid : x \hat{\ } z = y)$

(13.42) $isseg(x, y) \equiv (\exists w, z \mid : w \hat{\ } x \hat{\ } z = y)$

13.3 Extending the theory to use integers

In order to define the length of a sequence (the number of elements in it) and to refer to elements directly (for example, using $x.0$, $x.1$, ... to reference the elements of x), we need integers. A theory of integers is introduced later, in Chap. 15; we will use integers here, assuming knowledge of the few properties of the integers that we will need.

THE LENGTH OF A SEQUENCE

The length of a sequence x, denoted by $\#x$, is the number of elements in x. The length is defined by two axioms.

(13.43) **Axiom, Length:** $\#\epsilon = 0$

(13.44) **Axiom, Length:** $\#(c \triangleleft x) = 1 + \#x$

From these axioms, we can prove the following properties.

Properties of length #

(13.45) **Singleton length:** $\#(c \lhd \epsilon) = 1$

(13.46) **Length of ^ :** $\#(x \,\hat{}\, y) = \#x + \#y$

(13.47) **Length of subsequence:** $x \subseteq y \;\Rightarrow\; \#x \le \#y$

COUNT

Operation $c \,\#\, x$ yields the number of occurrences of element c in sequence x, analogous to the corresponding operation on bags (see page 212). Operation $c \,\#\, x$ is defined as follows.

(13.48) **Axiom, Count:** $c \,\#\, \epsilon = 0$

(13.49) **Axiom, Count:** $c \,\#\, (c \lhd x) = 1 + (c \,\#\, x)$

(13.50) **Axiom, Count:** $b \ne c \;\Rightarrow\; b \,\#\, (c \lhd x) = b \,\#\, x$

We can use # to characterize the membership relation:

Characterization of membership

(13.51) $c \in x \;\equiv\; c \,\#\, x > 0$

REFERRING TO ELEMENTS OF A SEQUENCE

In the sequence $x = \langle 4, 6, 1 \rangle$, we refer to the first element 4 by $x.0$, the second element 6 by $x.1$, and the third element by $x.2$. That is, we use function-application notation to refer to elements. We define this notation as follows.

(13.52) **Axiom, Element reference:** $(c \lhd x).0 = c$

(13.53) **Axiom, Element reference:**

$$(\forall n \mid 0 \le n < \#x : (c \lhd x)(n+1) = x.n)$$

Note that $x.n$ is not defined if $n \ge \#x$. In particular $\epsilon.n$ is undefined for all natural numbers n because $\#\epsilon = 0$. Also, we now have two notations for referring to the first element of a sequence: $x.0$ and $head.x$.

Catenation is related to our notation for referring to elements by the following two theorems.

Referencing elements of a catenation

(13.54) $(\forall n: \mathbb{N} \mid n < \#x : (x \,\hat{}\, y).n = x.n)$

(13.55) $(\forall n: \mathbb{N} \mid \#x \leq n < \#(x \,\hat{}\, y) : (x \,\hat{}\, y).n = y(n - \#x))$

Our notation for referring to an element of a sequence suggests a relation between sequences and functions. In fact, we could have *defined* a sequence $x: seq(A)$ of length n to be a function $x: \mathbb{N} \to A$ where $x.i$ is defined only for i satisfying $0 \leq i < n$. With this definition, element references $x.i$ would have been primitives, while *head.x*, *tail.x*, etc., would have been defined in terms of these primitives.

REFERENCING A SEGMENT OF A SEQUENCE

Given sequence x and two integers i, j satisfying $0 \leq i \leq j + 1 \leq \#x$, the notation $x[i..j]$ refers to the segment of x consisting of $x.i$, $x(i+1)$, ..., $x.j$. For example,

$$\langle 3, 5, 6, 8 \rangle [0..1] = \langle 3, 5 \rangle$$
$$\langle 3, 5, 6, 8 \rangle [1..3] = \langle 5, 6, 8 \rangle$$
$$\langle 3, 5, 6, 8 \rangle [1..1] = \langle 5 \rangle$$
$$\langle 3, 5, 6, 8 \rangle [2..1] = \langle \rangle \quad .$$

Note that $\#(x[i..j]) = j - i + 1$. So, $x[i..i]$ is a singleton and $x[i..i-1]$ is the empty segment beginning at $x.i$. In particular, $\epsilon[0.. - 1] = \epsilon$.

This notation can be defined inductively as follows.

(13.56) **Axiom, Empty reference:** $x[0.. - 1] = \epsilon$

(13.57) **Axiom, Prefix reference:**
$$(\forall j: \mathbb{N} \mid 0 \leq j < \#x : (c \triangleleft x)[0..j] = c \triangleleft x[0..j - 1])$$

(13.58) **Axiom, Segment reference:**
$$(\forall i, j: \mathbb{N} \mid 1 \leq j \leq \#x : (c \triangleleft x)[i..j] = x[i - 1..j - 1])$$

As an example of the use of the notation, we show how $\langle 3, 5, 6, 8, 9 \rangle [1..2]$ can be calculated.

$$(3 \triangleleft 5 \triangleleft 6 \triangleleft 8 \triangleleft 9 \triangleleft \epsilon)[1..2] \quad \text{—which is } \langle 3, 5, 6, 8, 9 \rangle [1..2]$$
$$= \quad \langle \text{Segment reference (13.58)} \rangle$$

$$(5 \vartriangleleft 6 \vartriangleleft 8 \vartriangleleft 9 \vartriangleleft \epsilon)[0..1]$$
$$= \quad \langle \text{Segment reference } (13.57) \rangle$$
$$5 \vartriangleleft (6 \vartriangleleft 8 \vartriangleleft 9 \vartriangleleft \epsilon)[0..0]$$
$$= \quad \langle \text{Segment reference } (13.57) \rangle$$
$$5 \vartriangleleft 6 \vartriangleleft (8 \vartriangleleft 9 \vartriangleleft \epsilon)[0.. - 1]$$
$$= \quad \langle \text{Segment reference } (13.56) \rangle$$
$$5 \vartriangleleft 6 \vartriangleleft \epsilon \quad \text{—which is } \langle 5, 6 \rangle$$

As an abbreviation, we write $x[i..]$ for the segment $x[i..\#x - 1]$ and $x[..i]$ for the segment $x[0..i]$.

Alter

Most programming languages allow assignment to an element $x.i$ of a sequence or array x using the assignment statement $x[i] := c$. When viewing a sequence as a function, it is advantageous to view $x[i] := c$ as an assignment to x itself, and not simply as an assignment to one of its elements. But to do this, we need a notation for the function that is being assigned to x.

The notation $(x; i:c)$ denotes a function or sequence that is the same as sequence x except that its value at index i is c. Thus, the assignment $x[i] := c$ could be written as $x := (x; i:c)$. For example,

$$(\langle 3, 5, 6 \rangle; 0:7) = \langle 7, 5, 6 \rangle$$
$$(\langle 3, 5, 6 \rangle; 1:7) = \langle 3, 7, 6 \rangle$$
$$(\langle 3, 5, 6 \rangle; 2:7) = \langle 3, 5, 7 \rangle \qquad \qquad \square$$

We define function *alter* in non-inductive fashion as follows.

(13.59) **Axiom:** $(\forall i:\mathbb{N} \mid i < \#x : (x; \ i:c) = x[0..i - 1] \ ^\frown \langle c \rangle \ ^\frown x[i + 1..])$

Exercise 13.24 asks for an inductive definition for *alter* and a proof that it is equivalent to (13.59). We also have the following theorem.

Alternative definition of *alter*

(13.60) $(\forall i, j:\mathbb{N}, x, c \mid: (x; i:c)[j] = \textbf{if } i = j \textbf{ then } c \textbf{ else } x.j)$

Discussion

This chapter illustrates how one builds a theory of a set of objects by:

- Inductively defining the set of objects.

- Deriving from the inductive definition a principle of induction that can be used to prove things about the objects.

- Defining convenient functions on the objects and proving properties about them.

Our theory of sequences introduced several notations for referring to elements of a sequence. For example, the second element of x could be referred to by $head(tail.x)$ or by $x.2$. Each notation has some context where it is useful —or else it should not have been created. For example, Lisp aficionados will prefer using $head.x$ (i.e. $(car\ x)$), $tail.x$ $((cdr\ x))$ and $c \lhd x$ $((cons\ c\ x))$. But those using imperative languages like Pascal, as well as the Lispers when dealing with arrays in Lisp, will use the notation $x.0$ to refer to the first element of x. An array, after all, is simply a variable that contains a sequence of fixed length. Our theory of sequences provides the basic rules for reasoning about sequences in many different languages; just the notation may change.

A number of concepts dealing with sequences have been relegated to exercises:

- For sequences containing elements that are all the same, see Exercises 13.26–13.27.

- For the *reverse* of a sequence, see Exercises 13.28–13.31.

- For *permutations* of a sequence, see Exercises 13.32–13.36.

- For palindromes, see Exercise 13.37.

Exercises for Chapter 13

13.1 Prove Equality (13.15), $x \rhd b = y \rhd c \equiv b = c \land x = y$ by induction. Hint. Rewrite this as $(\forall x\,|\colon P.x)$. In both the base case and the inductive case, a case analysis on y will be used: $y = \epsilon$ or $y = e \lhd z$ for some e, z.

13.2 Prove Membership in $\rhd$ (13.16), $b \in (x \rhd c) \equiv b \in x \lor b = c$.

13.3 Prove Right identity of $\hat{\ }$ (13.19), $x \hat{\ } \epsilon = x$.

13.4 Prove Associativity of $\hat{\ }$ (13.20), $x \hat{\ } (y \hat{\ } z) = (x \hat{\ } y) \hat{\ } z$.

13.5 Prove Membership in $\hat{\ }$ (13.21), $b \in (x \hat{\ } y) \equiv b \in x \lor b \in y$.

13.6 Prove Mutual associativity (13.22), $(x \hat{\ } y) \rhd c = x \hat{\ } (y \rhd c)$.

13.7 Prove Empty catenation (13.23), $x \hat{\ } y = \epsilon \equiv x = \epsilon \land y = \epsilon$.

13.8 Prove Membership (13.24), $b \in x \equiv (\exists y, z\,|\colon x = y \hat{\ } \langle b \rangle \hat{\ } z)$.

13.9 Prove $x \subseteq y \equiv x \subset y \lor x = y$.

13.10 Prove Reflexivity of $\subseteq$, (13.30), $x \subseteq x$.

13.11 Prove (13.31), $x \subseteq c \triangleleft x$.

13.12 Prove (13.32) $x \subset c \triangleleft x$.

13.13 Prove (13.33), $x \subseteq \epsilon \equiv x = \epsilon$.

13.14 Prove (13.34), $x \subseteq y \Rightarrow (\forall c \mid: c \in x \Rightarrow c \in y)$. This is perhaps the messiest proof in the text, and we don't like it. Our proof is by induction on y. The inductive case has a case analysis based on $x = \epsilon$ and $x = b \triangleleft x'$ for some b and x'. The proof in the case $x = b \triangleleft x'$ has a three-case analysis.

13.15 Prove (13.35), $c \triangleleft x \subseteq y \equiv y \neq \epsilon \wedge ((c = head.y \wedge x \subseteq tail.y) \vee c \triangleleft x \subseteq tail.y)$.

13.16 Prove (13.41), $isprefix(x,y) \equiv (\exists z \mid: x \,\hat{}\, z = y)$.

13.17 Prove (13.42), $isseg\,(x,y) \equiv (\exists w, z \mid: w \,\hat{}\, x \,\hat{}\, z = y)$.

13.18 Prove Singleton length (13.45), $\#(c \triangleleft \epsilon) = 1$.

13.19 Prove Length of $\hat{}\,$ (13.46), $\#(x \,\hat{}\, y) = \#x + \#y$.

13.20 Prove Length of subsequence (13.47), $x \subseteq y \Rightarrow \#x \leq \#y$.

13.21 Prove (13.51), $c \in x \equiv c \# x > 0$.

13.22 Prove (13.54), $(\forall n: \mathbb{N} \mid n < \#x : (x \,\hat{}\, y).n = x.n)$.

13.23 Prove (13.55), $(\forall n: \mathbb{N} \mid \#x \leq n < \#(x \,\hat{}\, y) : (x \,\hat{}\, y).n = y(n - \#x))$.

13.24 Give an inductive definition for function *alter* and prove that the inductive definition is equivalent to axiom (13.59).

13.25 Prove (13.60), $(\forall i, j: \mathbb{N} \mid: (x; i: c)[j] = \textbf{if } i = j \textbf{ then } c \textbf{ else } x.j)$.

Exercises on relation same

13.26 Define inductively a boolean function $same: seq(A) \rightarrow \mathbb{B}$ with meaning "all the elements of sequence x are the same".

13.27 Prove the following theorem concerning function *same* of Exercise 13.26.
$$same.x \equiv (\forall b, c \mid b \in x \wedge c \in x : b = c)$$

Exercises on the reverse of a sequence

13.28 Give an inductive definition, using $\triangleleft$ and $\triangleright$, of the *reverse rev.x* of a sequence x. Function $rev: seq(A) \rightarrow seq(A)$ yields the elements of x, but in reverse order. For example, $rev.\langle 3, 5, 6, 2 \rangle = rev.\langle 2, 6, 5, 3 \rangle$.

13.29 Prove $rev(x \triangleright b) = b \triangleleft rev.x$.

13.30 Prove $rev(rev.x) = x$.

13.31 Prove $rev(x \mathbin{\char`\^} y) = rev.y \mathbin{\char`\^} rev.x$.

Exercises on permutations of a sequence

13.32 Sequence x is a permutation of sequence y, written $perm(x, y)$, if y can be constructed by rearranging the order of the elements of x. For example, $\langle 2, 5, 1, 4 \rangle$ is a permutation of $\langle 5, 4, 2, 1 \rangle$.

Define $perm(x, y)$ using catenation, by defining $perm(\epsilon, y)$ and $perm(c \vartriangleleft x, y)$ (for all d, x). Define the latter predicate in a manner similar to characterization (13.42) of *isseg* .

13.33 Prove $perm(x, x)$ (for all x).

13.34 Prove that *perm* is symmetric: $perm(x, y) = perm(y, x)$.

13.35 Prove that *perm* is transitive: $perm(x, y) \wedge perm(y, z) \Rightarrow perm(x, z)$. This exercise, together with the two previous ones, shows that *perm* is an equivalence relation (see Definition (14.33) on page 276).

13.36 Prove $perm(x, rev.x)$.

Exercise on palindromes

13.37 Let A be the set of lowercase letters 'a', ... 'z'. The *palindromes* are the elements of $seq(A)$ that read the same forwards and backwards. For example, *noon*, is a palindrome, as is the following (if the blanks and punctuation are removed): *a man, a plan, a canal, panama!*. And, the empty sequence is a palindrome.

Using $pal.x$ to mean that x is a palindrome, we define the palindromes as follows.

$$pal.\epsilon \;\equiv\; true$$
$$pal(c \vartriangleleft \epsilon) \;\equiv\; true$$
$$pal(b \vartriangleleft x \vartriangleright c) \;\equiv\; b = c \wedge pal.x$$

Prove that $pal.x \equiv rev.x = x$ for all sequences x , where $rev.x$ is defined in Exercise 13.28.

Chapter 14

Relations and Functions

W e study *tuples*, *cross products*, *relations*, and *functions*. The n-tuple, or sequence of length n, is the mathematical analogue of the Pascal-like record: it is a list of values (but without names). The cross product is the mathematical analogue of the Pascal record type: it is a set of tuples, corresponding to the set of records that may be associated with a variable of a record type.

In everyday life, we often deal with relationships between objects. There is the relationship between parent and child, between name and address, between position and wage, and so on. In mathematics, such relationships are modeled using relations, which are simply sets of tuples with the same length. And, a function can be viewed as a restricted kind of relation. The theory of relations and functions that we present here is essential to much of mathematics. Further, as can be seen in Sec. 14.5, the theory of relations finds application in the very practical area of computerized databases.

14.1   Tuples and cross products

For expressions b and c, the 2-tuple $\langle b, c \rangle$ is called an *ordered pair*, or simply a *pair*. In some notations, parentheses are used around a pair instead of angle brackets.

Ordered pairs are frequently useful. For example, a pair can be used to denote a point in the plane, with the first component being the horizontal coordinate and the second component being the vertical coordinate of the point. As another example, the set of pairs $\langle name, address \rangle$, where *name* is the name of a student at Cornell and *address* is their address, can represent the correspondence between students and the addresses to which their grades should be sent.

It is possible to define ordered pairs using sets, as follows. For any expressions b and c,

(14.1)   **Ordered pair:** $\langle b, c \rangle = \{\{b\}, \{b, c\}\}$.

Thus, the pair can be formally defined in terms of set theory. We do not explore the use of definition (14.1) here, but relegate it to exercises. Instead, we use the following definition of pair equality.

(14.2)   **Axiom, Pair equality:** $\langle b, c \rangle = \langle b', c' \rangle \;\equiv\; b = b' \wedge c = c'$

CROSS PRODUCTS

The *cross product* or *Cartesian product* (named after the French mathematician René Descartes; see Historical note 14.1) $S \times T$ of two sets S and T is the set of all pairs $\langle b, c \rangle$ such that b is in S and c is in T.

(14.3)   **Axiom, Cross product:** $S \times T = \{b, c \mid b \in S \wedge c \in T : \langle b, c \rangle\}$

For example, $\mathbb{Z} \times \mathbb{Z}$ denotes the set of integral points in the plane, $\mathbb{R} \times \mathbb{R}$ denotes the set of all points in the plane, and $\{2, 5\} \times \{1, 2, 3\}$ is the set $\{\langle 2, 1 \rangle, \langle 2, 2 \rangle, \langle 2, 3 \rangle, \langle 5, 1 \rangle, \langle 5, 2 \rangle, \langle 5, 3 \rangle\}$.

Here are some properties of the cross product.

Theorems for cross product

(14.4)   **Membership:** $\langle x, y \rangle \in S \times T \;\equiv\; x \in S \wedge y \in T$

(14.5) $\langle x, y \rangle \in S \times T \;\equiv\; \langle y, x \rangle \in T \times S$

(14.6) $S = \emptyset \;\Rightarrow\; S \times T = T \times S = \emptyset$

(14.7) $S \times T = T \times S \;\equiv\; S = \emptyset \vee T = \emptyset \vee S = T$

(14.8)   **Distributivity of $\times$ over $\cup$:**
$$S \times (T \cup U) = (S \times T) \cup (S \times U)$$
$$(S \cup T) \times U = (S \times U) \cup (T \times U)$$

(14.9)   **Distributivity of $\times$ over $\cap$:**
$$S \times (T \cap U) = (S \times T) \cap (S \times U)$$
$$(S \cap T) \times U = (S \times U) \cap (T \times U)$$

(14.10) **Distributivity of $\times$ over $-$:**
$$S \times (T - U) = (S \times T) - (S \times U)$$

(14.11) **Monotonicity:** $T \subseteq U \;\Rightarrow\; S \times T \subseteq S \times U$

(14.12) $S \subseteq U \wedge T \subseteq V \;\Rightarrow\; S \times T \subseteq U \times V$

(14.13) $S \times T \subseteq S \times U \wedge S \neq \emptyset \;\Rightarrow\; T \subseteq U$

Theorems for cross product (continued)

(14.14) $(S \cap T) \times (U \cap V) = (S \times U) \cap (T \times V)$

(14.15) For finite S and T, $\#(S \times T) = \#S \cdot \#T$

We can extend the notion of a cross product from two sets to n sets. For example, $\mathbb{Z} \times \mathbb{N} \times \{3, 4, 5\}$ is the set of triples $\langle x, y, z \rangle$ where x is an integer, y is a natural number, and z is 3, 4, or 5. And, $\mathbb{R} \times \mathbb{R} \times \mathbb{R}$ is the set of all points in three-dimensional space. The theorems shown for the cross product of two sets extend to theorems for the cross product of n sets in the expected way, so we don't discuss them further.

The n-tuple and cross product are directly related to the record and record type in a programming language like Pascal. Suppose type T and variable v are declared in Pascal as follows:

$T = $ **record** x:*integer*; r:*real*; y:*integer* **end** ;
var v:T .

Then T denotes the cross product $\mathbb{Z} \times \mathbb{R} \times \mathbb{Z}$ —i.e. T stands for the set of all tuples $\langle x, r, y \rangle$ where x, r, and y have the appropriate types. Also, variable v may be associated with any such tuple. The difference between the record and the tuple is that the record names the components, while the tuple does not. We could call a record a *named tuple*.

In Chap. 8, we discussed the type of a function. A function of two arguments of types $t1$ and $t2$ and a result of type $t3$ was given type $t1 \times t2 \to t3$. We see now the reason for the use of $\times$ in describing this type. As a tuple, the arguments $a1$ and $a2$ (say) of a function application $f(a1, a2)$ form an element of the set $t1 \times t2$.

14.2 Relations

A *relation* on a cross product $B_1 \times \cdots \times B_n$ is simply a subset of $B_1 \times \cdots \times B_n$. Thus, a relation is a set of n-tuples (for some fixed n). [1] A *binary* relation over $B \times C$ is a subset of $B \times C$. The term *binary* is used because a member of a binary relation is a 2-tuple. [2] If B and C are the same, so that the relation is on $B \times B$, we call it simply a (binary) relation on B.

---

[1] The reader is invited to skip ahead to Sec. 14.5, where databases are discussed. There, the idea of a relation as a set of n-tuples is made concrete.

[2] Prefix *bi*, from Latin, means *two*.

HISTORICAL NOTE 14.1. RENÉ DESCARTES (1596–1650)

Descartes, the greatest of French philosophers, is known for his *Discourse on Method*. The first part of *Discourse* outlines the essentials of Descartes's philosophy. Important for him are four points: (1) never accept anything as true that is not clearly known to be so; (2) break problems into small, simple parts, (3) start with the simplest and easiest things to know and build up knowledge from them in small, orderly steps; and (4) make sure enumerations are complete, so that nothing is omitted. Other parts of *Discourse* apply his method in different fields, e.g. optics and analytic geometry. In later works, application of the method leads Descartes to his first certitude, his famous "I think, therefore I am.", as well as to proofs for himself of the existence of God and the separateness of the soul and the body.

Discourse was finished in 1637, but the beginnings of Descartes's great discoveries in analytic geometry came by way of a sort of spiritual conversion, some twenty years earlier. In 1616, while in the army, he had three vivid dreams, which filled him with "enthusiasm" and revealed to him, among other things, "the foundations of a wonderful science". This was the application of algebra to geometry, with the use of coordinates in the plane as a link between the two. Descartes is the first to describe curves by equations, to classify curves, and to use algebra to discover new curves and theorems about curves. He made the study of geometries of higher dimensions possible. Through Descartes, as E.T. Bell puts it, "algebra and analysis [became] our pilots to the uncharted seas of 'space' and its 'geometry'."

At the age of eight, Descartes was precocious but very frail. Consequently, the rector of his school let him stay in bed as late as he wanted, even until noon. Descartes continued the practice almost all his life — suppress your envy, college students! At the age of 53, he was persuaded to come to Sweden to teach Queen Christina. Unfortunately for Descartes, *she* wanted to start lessons at 5AM. Less than five months later, he caught a chill coming home from one of her lessons one bitter January morning and died a few weeks later.

In the following sections, we will be focusing on binary relations. Therefore, from now on, we abbreviate "binary relation" by "relation".

If a relation is not too large, we can define it by listing its pairs. For example, the following relation consists of two pairs, each containing the names of a coauthor of this book and his spouse.

$$\{\langle \text{David, Elaine}\rangle, \langle \text{Fred, Mimi}\rangle\}$$

The "less than" relation $<$ over the natural numbers is also a binary relation. We could try to list its pairs,

$$\{\langle 0,1\rangle, \langle 0,2\rangle, \langle 0,3\rangle, \dots$$
$$\langle 1,2\rangle, \langle 1,3\rangle, \langle 1,4\rangle, \dots$$
$$\dots\} \quad ,$$

but because there are an infinite number of pairs, a better presentation would use set comprehension:

$$\{i, j : \mathbb{N} \mid j - i \text{ is positive} : \langle i, j \rangle\} \quad .$$

(1, 1)
(1, 3)
(2, 5)
(2, 1)
(5, 3)

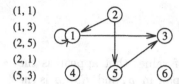

Finally, a binary relation can be described by a *directed graph*. The graph has one *vertex* for each element of the set, and there is a directed edge from vertex b (say) to vertex c iff $\langle b, c \rangle$ is in the binary relation. Thus, the theory of relations and the theory of directed graphs are related. We return to graph theory in Chap. 19.

Examples of (binary) relations

(a) The *empty relation* on $B \times C$ is the empty set, $\emptyset$.

(b) The *identity relation* ι_B on B is $\{x \mid x \in B : \langle x, x \rangle\}$.

(c) Relation *parent* on the set of people is the set of pairs $\langle b, c \rangle$ such that b is a parent of c. Relation *child* on the set of people is the set of pairs $\langle b, c \rangle$ such that b is a child of c. Relation *sister* on the set of people is the set of pairs $\langle b, c \rangle$ such that b is a sister of c.

(d) Relation *pred* (for predecessor) on $\mathbb{Z}$ is the set of pairs $\langle b-1, b \rangle$ for integers b, $pred = \{b : \mathbb{Z} \mid \langle b-1, b \rangle\}$. Relation *succ* (for successor) is defined by $succ = \{b : \mathbb{Z} \mid \langle b+1, b \rangle\}$.

(e) Relation *sqrt* on $\mathbb{R}$ is the set $\{b, c : \mathbb{R} \mid b^2 = c : \langle b, c \rangle\}$.

(f) An algorithm P can be viewed as a relation on states. A pair $\langle b, c \rangle$ is in the relation iff some execution of P begun in state b terminates in state c.   □

Two completely different notations are used for membership in a relation. Conventionally, we view $b < c$ as an expression that evaluates to *true* or *false* depending on whether or not b is less than c. Alternatively, $<$ is a relation, a set of pairs, so it is sensible to write $\langle b, c \rangle \in <$. In general, for any relation ρ:

$$\langle b, c \rangle \in \rho \text{ and } b \rho c \text{ are interchangeable notations.}$$

One notation views ρ as a set of pairs; the other views ρ as a binary boolean function written as an infix operator. By convention, the precedence of a name ρ of a relation that is used as a binary infix operator is the same as the precedence of $=$; furthermore, ρ is considered to be conjunctional. For example,

$$b \rho c \rho d \;\equiv\; b \rho c \land c \rho d \quad .$$

In this chapter, we use small Greek letters for names of arbitrary relations, to distinguish them from names of other entities (see the front inside cover).

The *domain* $Dom.\rho$ and *range* $Ran.\rho$ of a relation ρ on $B \times C$ are defined by

(14.16) $Dom.\rho = \{b{:}B \mid (\exists c \mid : b\,\rho\,c)\}$

(14.17) $Ran.\rho = \{c{:}C \mid (\exists b \mid : b\,\rho\,c)\}$

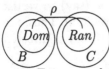

$Dom.\rho$ is just the set of values that appear as the first component of some pair in ρ, and $Ran.\rho$ is the set of values that appear as the second component of some pair in ρ. B and $Dom.\rho$ need not be the same. For example, let B be the set of people and let ρ be the relation *parent* given above. Then $Dom.\rho$ is the set of people who have children, and not the set of all people.

OPERATIONS ON RELATIONS

Suppose ρ and σ are relations on $B \times C$. Since a relation is a set, $\rho \cup \sigma$, $\rho \cap \sigma$, $\rho - \sigma$, and $\sim\rho$ (where the complement is taken relative to universe $B \times C$) are also relations on $B \times C$. We now introduce two other important operations on relations: the inverse of a relation and the product $\circ$ of two relations.

The *inverse* ρ^{-1} of a relation ρ on $B \times C$ is the relation defined by

(14.18) $\langle b, c \rangle \in \rho^{-1} \equiv \langle c, b \rangle \in \rho$ (for all $b{:}B$, $c{:}C$).

For example, the inverse of relation *parent* (see page 269) is relation *child*, the inverse of *pred* is *succ*, and the identity relation is its own inverse.

The following theorem gives useful properties of the inverse; its proof is left to the reader.

FIGURE 14.1. ILLUSTRATION OF PRODUCT RELATION

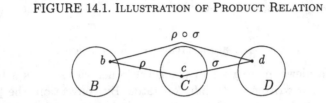

$b\,(\rho \circ \sigma)\,d$ holds iff $b\,\rho\,c\,\sigma\,d$ holds for some c.

(14.19) **Theorem.** Let ρ and σ be relations.

    (a) $Dom(\rho^{-1}) = Ran.\rho$.

    (b) $Ran(\rho^{-1}) = Dom.\rho$.

    (c) If ρ is a relation on $B \times C$, then ρ^{-1} is a relation on $C \times B$.

    (d) $(\rho^{-1})^{-1} = \rho$.

    (e) $\rho \subseteq \sigma \;\equiv\; \rho^{-1} \subseteq \sigma^{-1}$.

Let ρ be a relation on $B \times C$ and σ be a relation on $C \times D$. The *product* of ρ and σ, denoted by $\rho \circ \sigma$, is the relation defined by

$$(14.20) \quad \langle b, d \rangle \in \rho \circ \sigma \;\equiv\; (\exists c \mid c \in C : \langle b, c \rangle \in \rho \wedge \langle c, d \rangle \in \sigma) \;,$$

or, using the alternative notation, by

$$(14.21) \quad b \,(\rho \circ \sigma)\, d \;\equiv\; (\exists c \mid : b \,\rho\, c \,\sigma\, d) \;.$$

The product relation is illustrated in Fig. 14.1.

Examples of product

(a) Let ρ and σ both be relation *parent*. Then $\langle b, d \rangle \in \rho \circ \sigma$ iff there is a person c such that b is a parent of c and c is a parent of d. Thus, $b \,\rho \circ \sigma\, d$ iff b is d's grandparent, so *parent $\circ$ parent* is the relation *grandparent*.

(b) The relation *sister $\circ$ father* denotes the relation *paternal aunt*: b is a paternal aunt of d means b is a sister of a person who is the father of d.

(c) The relation *succ $\circ$ pred* is the identity relation $\imath_{\mathbb{Z}}$.   □

We list below a number of theorems for $\circ$. These theorems hold for all binary relations ρ, σ, and θ.

Theorems for relation product

(14.22) **Associativity of $\circ$:** $\rho \circ (\sigma \circ \theta) = (\rho \circ \sigma) \circ \theta$

(14.23) **Distributivity of $\circ$ over $\cup$:** $\rho \circ (\sigma \cup \theta) = \rho \circ \sigma \cup \rho \circ \theta$
$$(\sigma \cup \theta) \circ \rho = \sigma \circ \rho \cup \theta \circ \rho$$

(14.24) **Distributivity of $\circ$ over $\cap$:** $\rho \circ (\sigma \cap \theta) \subseteq \rho \circ \sigma \cap \rho \circ \theta$
$$(\sigma \cap \theta) \circ \rho \subseteq \sigma \circ \rho \cap \theta \circ \rho$$

Since relation product is associative, we may omit parentheses in a sequence of products.

We prove Associativity of $\circ$ (14.22). By Axiom of Extensionality (11.4), it suffices to prove that any arbitrary element is in the LHS exactly when it is in the RHS, which we now do. For an arbitrary pair $\langle a, d \rangle$ we have:

$$a \; \rho \circ (\sigma \circ \theta) \; d$$
$$= \quad \langle \text{Definition (14.21) of } \rho \circ (\sigma \circ \theta) \rangle$$
$$(\exists b \,|: a \, \rho \, b \;\wedge\; b \, (\sigma \circ \theta) \, d)$$
$$= \quad \langle \text{Definition (14.21) of } \sigma \circ \theta \,; \text{ Nesting (8.20)} \rangle$$
$$(\exists b, c \,|: a \, \rho \, b \;\wedge\; b \, \sigma \, c \;\wedge\; c \, \theta \, d)$$
$$= \quad \langle \text{Nesting (8.20); Definition (14.21) of } \rho \circ \sigma \rangle$$
$$(\exists c \,|: a \, (\rho \circ \sigma) \, c \;\wedge\; c \, \theta \, d)$$
$$= \quad \langle \text{Definition (14.21) of } (\rho \circ \sigma) \circ \theta \rangle$$
$$a \; (\rho \circ \sigma) \circ \theta \; d$$

Relation $\rho \circ \rho$ is often written as ρ^2. In fact, for ρ defined on a set B, for any natural number n we define ρ composed with itself n times, or ρ^n, as follows.

(14.25) $\rho^0 = \imath_B$ (the identity relation on B; see example (b) on
                     page 269)

      $\rho^{n+1} = \rho^n \circ \rho$ (for $n \geq 0$)

For example, we have: $parent^2$ is $parent \circ parent$, (i.e. *grandparent*), $parent^3$ is $parent^2 \circ parent$ (*great-grandparent*), and so forth. We also have $b \, pred^i \, c \equiv b + i = c$.

We have the following two theorems. Their proofs, by mathematical induction, are left as exercises.

Theorems for powers of a relation

(14.26) $\rho^m \circ \rho^n = \rho^{m+n}$ (for $m \geq 0, n \geq 0$)

(14.27) $(\rho^m)^n = \rho^{m \cdot n}$ (for $m \geq 0, n \geq 0$)

When the set over which a relation is constructed is infinite, then all the powers ρ^i may be different. For example, relation $pred^i$ is distinct for each natural number i. However, if the set is finite, then there are only a finite number of possible relations for the ρ^i:

(14.28) **Theorem.** For ρ a relation on finite set B of n elements,

$$(\exists i, j \mid 0 \leq i < j \leq 2^{n^2} : \rho^i = \rho^j) \quad .$$

Proof. Each relation ρ^i is a member of the power set $\mathcal{P}(B \times B)$. $B \times B$ has n^2 elements. By theorem (11.73), $\mathcal{P}(B \times B)$ has 2^{n^2} elements, so there are 2^{n^2} different relations on B. The sequence $\rho^0, \rho^1, \ldots, \rho^k$ for $k = 2^{n^2}$ has $2^{n^2} + 1$ elements, so at least two of them are the same.   □

We leave the proof of the following theorem to the reader. The theorem states that if the sequence $\rho^0, \rho^1, \ldots$ begins repeating, it repeats forever with the same period.

(14.29) **Theorem.** Let ρ be a relation on a finite set B. Suppose $\rho^i = \rho^j$ and $0 \leq i < j$. Then

$$(a) \rho^{i+k} = \rho^{j+k} \qquad \text{(for } k \geq 0 \text{)}$$
$$(b) \rho^i = \rho^{i+p \cdot (j-i)} \qquad \text{(for } p \geq 0 \text{)} \ .$$

CLASSES OF RELATIONS

A few classes of relations that enjoy certain properties are used frequently, and it is best to memorize them. Table 14.1 defines classes of relations ρ over some set B. Each class is defined in two ways: first in terms of the property that the elements of such a relation satisfy and then in terms of operations on sets. The definition in terms of the properties, which is often the first one thought of, mentions set members in some way. The definition in terms of operations on sets is more succinct and is often easier to work with. Exercise 14.25 asks you to prove that these alternative definitions are equivalent.

Examples of classes of relations

(a) Relation $\leq$ on $\mathbb{Z}$ is reflexive, since $b \leq b$ holds for all integers b. It is not irreflexive. Relation $<$ on $\mathbb{Z}$ is not reflexive, since $2 < 2$ is false. It is irreflexive.

TABLE 14.1. CLASSES OF RELATIONS ρ OVER SET B

Name	Property	Alternative
(a) reflexive	$(\forall b \mid : b \rho b)$	$\iota_B \subseteq \rho$
(b) irreflexive	$(\forall b \mid : \neg(b \rho b))$	$\iota_B \cap \rho = \emptyset$
(c) symmetric	$(\forall b, c \mid : b \rho c \equiv c \rho b)$	$\rho^{-1} = \rho$
(d) antisymmetric	$(\forall b, c \mid : b \rho c \wedge c \rho b \Rightarrow b = c)$	$\rho \cap \rho^{-1} \subseteq \iota_B$
(e) asymmetric	$(\forall b, c \mid : b \rho c \Rightarrow \neg(c \rho b))$	$\rho \cap \rho^{-1} = \emptyset$
(f) transitive	$(\forall b, c, d \mid : b \rho c \wedge c \rho d \Rightarrow b \rho d)$	$\rho = (\cup i \mid i > 0 : \rho^i)$

(b)   Consider relation *square* on $\mathbb{Z}$ that is defined by b *square* c iff
$b = c \cdot c$. It is not reflexive because it does not contain the pair $\langle 2, 2 \rangle$.
It is not irreflexive because it does contain the pair $\langle 1, 1 \rangle$. Thus, a
relation that is not reflexive need not be irreflexive.

(c)   Relation $=$ on the integers is symmetric, since $b = c \equiv c = b$.
Relation $<$ is not symmetric.

(d)   Relation $\leq$ is antisymmetric since $b \leq c \wedge c \leq b \Rightarrow b = c$. Relation
$<$ is antisymmetric: since $b < c \wedge c < b$ is always *false*, we have
$b < c \wedge c < b \Rightarrow b = c$ for all b, c. Relation $\neq$ is not antisymmetric.

(e)   Relation $<$ is asymmetric, since $b < c$ implies $\neg(c < b)$. Relation
$\leq$ is not asymmetric.

(f)   Relation $<$ is transitive, since if $b < c$ and $c < d$ then $b < d$. Relation *parent* is not transitive. However, relation *ancestor* is transitive,
where b *ancestor* c holds if b is an ancestor of c.   □

The *closure* of a relation ρ with respect to some property (e.g. reflexivity) is the smallest relation that both has that property and contains
ρ. To construct a closure, add pairs to ρ, but not too many, until it has
the property. For example, the reflexive closure of $<$ over the integers is
the relation constructed by adding to relation $<$ all pairs $\langle b, b \rangle$ for b an
integer. Therefore, $\leq$ is the reflexive closure of $<$.

The construction of a closure does not always make sense. For example,
the irreflexive closure of a relation containing $\langle 1, 1 \rangle$ doesn't exist, since it
is precisely the *presence* of this pair that makes the relation not irreflexive.
Three properties for which constructing closures makes sense are given in
the following definition.

(14.30)   **Definition.** Let ρ be a relation on a set. The *reflexive* (*symmetric*,
transitive) *closure of* ρ is the relation ρ' that satisfies:
(a) ρ' is reflexive (symmetric, transitive);
(b) $\rho \subseteq \rho'$;
(c)   If ρ'' is reflexive (symmetric, transitive) and $\rho \subseteq \rho''$, then
$\rho' \subseteq \rho''$.

We use the following notations: $r(\rho)$ is the reflexive closure of ρ;
$s(\rho)$, the symmetric closure; ρ^+, the transitive closure; and ρ^*,
the reflexive transitive closure.

Examples of closures

(a)   The reflexive closure $r(<)$ of $<$ on the integers is $\leq$.

(b)   The symmetric closure $s(parent)$ of *parent* is *parent* $\cup$ *child*, since
if $\langle b, c \rangle$ is in the symmetric closure, then so is $\langle c, b \rangle$.

(c) The transitive closure *parent*$^+$ of *parent* is *ancestor*, since whenever $\langle b, c \rangle$ and $\langle c, d \rangle$ are in the transitive closure, then so is $\langle b, d \rangle$.

(d) The reflexive transitive closure *parent** of *parent* is the relation *ancestor-or-self*. $\qquad\qquad\qquad\qquad\qquad\qquad\qquad\qquad\qquad\qquad\qquad$ □

The following theorem is almost so trivial that it needs no proof, although we do ask you to prove it in the exercises.

(14.31) **Theorem.** A reflexive relation is its own reflexive closure; a symmetric relation is its own symmetric closure; and a transitive relation is its own transitive closure.

In Definition (14.30), we defined a closure of a relation ρ in terms of three properties enjoyed by ρ and its closure. An alternative definition shows how to construct the closure from the set. Here, we state the constructive formulations as a theorem.

(14.32) **Theorem.** Let ρ be a relation on a set B. Then,

(a) $r(\rho) = \rho \cup \imath_B$

(b) $s(\rho) = \rho \cup \rho^{-1}$

(c) $\rho^+ = (\cup i \mid 0 < i : \rho^i)$

(d) $\rho^* = \rho^+ \cup \imath_B$.

Proof. We prove the more difficult part (14.32c) and leave the others to the reader. To prove (14.32c), we have to show that it satisfies the three parts of Def. (14.30). We first show that ρ^+ is transitive. For arbitrary elements b, c, d of B, we have,

$$\langle b, c \rangle \in (\cup i \mid 1 \leq i : \rho^i) \;\wedge\; \langle c, d \rangle \in (\cup i \mid 1 \leq i : \rho^i)$$
$\qquad =\qquad$ ⟨Definition of $\cup$ (11.20), twice⟩
$$(\exists j \mid ... : \langle b, c \rangle \in \rho^j) \;\wedge\; (\exists k \mid ... : \langle c, d \rangle \in \rho^k)$$
$\qquad =\qquad$ ⟨Distributivity of $\wedge$ over $\exists$ (9.21); Nesting (8.20)⟩
$$(\exists j, k \mid ... : \langle b, c \rangle \in \rho^j \;\wedge\; \langle c, d \rangle \in \rho^k)$$
$\qquad \Rightarrow\qquad$ ⟨Definition of product (14.20)⟩
$$(\exists j, k \mid ... : \langle b, d \rangle \in \rho^j \circ \rho^k)$$
$\qquad =\qquad$ ⟨Theorem (14.26) —and inserting ranges of j and k⟩
$$(\exists j, k \mid 1 \leq j \wedge 1 \leq k : \langle b, d \rangle \in \rho^{j+k})$$
$\qquad =\qquad$ ⟨One-point rule (8.14)⟩
$$(\exists i, j, k \mid i = j + k \wedge 1 \leq j \wedge 1 \leq k : \langle b, d \rangle \in \rho^i)$$
$\qquad \Rightarrow\qquad$ ⟨Arithmetic; predicate calculus to eliminate j, k⟩
$$(\exists i \mid 2 \leq i : \langle b, d \rangle \in \rho^i)$$
$\qquad \Rightarrow\qquad$ ⟨Definition of $\cup$ (11.20); Range weakening⟩
$$\langle b, d \rangle \in (\cup i \mid : \rho^i)$$

Hence, $(\cup i \mid : \rho^i)$ is transitive.

Next, Part (b) of Def. (14.30), $\rho \subseteq (\cup i \mid 0 < i : \rho^i)$, follows easily from $\rho^1 = \rho$ and properties of $\cup$.

Finally, we show that Part (c) of Def. (14.30) holds:

if ρ'' is transitive and $\rho \subseteq \rho''$, then $\rho^+ \subseteq \rho''$.

where $\rho^+ = (\cup i \mid 0 < i : \rho^i)$. Thus, we assume ρ'' is transitive and $\rho \subseteq \rho''$ and prove $\rho^+ \subseteq \rho''$. Any pair $\langle b, c \rangle$ in ρ^+ satisfies $b \, \rho^i \, c$ for some positive integer i. Hence, we prove the following by induction.

$$(\forall i \mid 0 < i : \rho^i \subseteq \rho'') \quad .$$

Base case $i = 1$. The base case follows from the definition $\rho^1 = \rho$.

Inductive case. For $i > 1$, we assume the inductive hypothesis $\rho^i \subseteq \rho''$ and prove $\rho^{i+1} \subseteq \rho''$. For arbitrary b, d we have,

$\quad \langle b, d \rangle \in \rho^{i+1}$
$= \quad \langle$Def. of power (14.25); Def. of product (14.20)$\rangle$
$\quad (\exists c \mid : \langle b, c \rangle \in \rho^i \wedge \langle c, d \rangle \in \rho)$
$\Rightarrow \quad \langle$Induction hypothesis $\rho^i \subseteq \rho''$; Assumption $\rho \subseteq \rho'' \rangle$
$\quad (\exists c \mid : \langle b, c \rangle \in \rho'' \wedge \langle c, d \rangle \in \rho'')$
$\Rightarrow \quad \langle$Assumption ρ'' is transitive$\rangle$
$\quad \langle b, d \rangle \in \rho'' \hfill \square$

EQUIVALENCE RELATIONS

Another important class of relations is the class of *equivalence* relations.

(14.33) **Definition.** A relation is an *equivalence relation* iff it is reflexive, symmetric, and transitive.

For example, equality $=$ is an equivalence relation, while $<$ is not.

An equivalence relation ρ on a set B partitions B into non-empty disjoint subsets. Elements that are equivalent under ρ are placed in the same partition element, and elements that are not equivalent are placed in different partition elements. For example, the relation *sameeye* over the set of people is defined by

$$\langle b, c \rangle \in \textit{sameeye} \equiv b \text{ and } c \text{ have the same eye color.}$$

This relation partitions the people into the subset of people with blue eyes, the subset with brown eyes, etc. Having a correspondence between equivalence relations and partitions is a useful bridge between the theory of relations and the theory of sets. The purpose of this subsection is to prove formally that this correspondence exists.

We begin by defining the subsets determined by an equivalence relation.

(14.34) **Definition.** Let ρ be an equivalence relation on B. Then $[b]_\rho$, the *equivalence class* of b, is the subset of elements of B that are equivalent (under ρ) to b:

$$x \in [b]_\rho \;\equiv\; x \, \rho \, b \ .$$

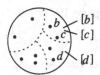 In what follows, we eliminate the subscript ρ and write $[b]_\rho$ as $[b]$ when it is obvious from the context what relation is meant. The diagram to the left illustrates the partition of a set by an equivalence relation ρ. Assuming $b \, \rho \, c$, b and c are in the same partition element and $[b] = [c]$. Assuming $\neg(b \, \rho \, d)$, b and d are in different partition elements and $[b] \cap [d] = \emptyset$.

Examples of equivalence classes

(a) Consider the relation $b \overset{4}{=} c$ on the integers $0..9$:
$$b \overset{4}{=} c \;\equiv\; b - c \text{ is a multiple of } 4 \ .$$
We have,
$$[0] = [4] = [8] = \{0, 4, 8\}$$
$$[1] = [5] = [9] = \{1, 5, 9\}$$
$$[2] = [6] = \{2, 6\}$$
$$[3] = [7] = \{3, 7\}$$

(b) Consider relation ρ defined on the set of people by $b \, \rho \, c$ iff b and c are female and either b and c are the same person or b is c's sister. Relation ρ is reflexive, symmetric, and transitive, so it is an equivalence relation. For a female b, $[b]$ consists of b and b's sisters, while the equivalence class for a male contains only that male.   □

We now prove the following theorem.

(14.35) **Theorem.** Let ρ be an equivalence relation on B and let b, c be members of B. The following three predicates are equivalent:
(a) $b \, \rho \, c$,
(b) $[b] \cap [c] \neq \emptyset$, and
(c) $[b] = [c]$.

That is, $(b \, \rho \, c) \;=\; ([b] \cap [c] \neq \emptyset) \;=\; ([b] = [c])$.

Proof. We can prove, in turn, (a) $\Rightarrow$ (b), (b) $\Rightarrow$ (c), and (c) $\Rightarrow$ (a). Mutual implication and transitivity of $\Rightarrow$ then give the equivalence of all three. We prove (a) $\Rightarrow$ (b) and leave the other two cases to the reader:

$$b \, \rho \, c$$
$$= \quad \langle \text{Identity of } \wedge \text{ (3.39)}; \ \rho \text{ is reflexive} \rangle$$
$$b \, \rho \, b \ \wedge \ b \, \rho \, c$$
$$= \quad \langle \text{Definition (14.34), twice} \rangle$$
$$b \in [b] \ \wedge \ b \in [c]$$
$$= \quad \langle \text{Definition of } \cap \text{ (11.21)} \rangle$$
$$b \in [b] \cap [c]$$
$$\Rightarrow \quad \langle \text{The empty set } \emptyset \text{ does not contain } b \text{ (see page 197)} \rangle$$
$$[b] \cap [c] \neq \emptyset \qquad \qquad \qquad \qquad \qquad \qquad \qquad \square$$

Theorem (14.35) allows us to show, and quite easily, that the sets $[b]_\rho$ for an equivalence relation ρ on B form a partition of B. First, none of the sets is empty, since each element b is in $[b]_\rho$. Second, the union of the sets $[b]_\rho$ is B, since each element b is in the set $[b]_\rho$. Third, we show below that if two sets $[b]_\rho$ and $[c]_\rho$ are not the same, then they are disjoint. The proof relies on the fact that (b) and (c) of Theorem (14.35) are equivalent:

$$[b]_\rho \neq [c]_\rho \ \equiv \ [b] \cap [c] = \emptyset$$
$$= \quad \langle (3.11), \ \neg p \equiv q \equiv p \equiv \neg q \rangle$$
$$\neg([b]_\rho \neq [c]_\rho) \ \equiv \ \neg([b] \cap [c] = \emptyset)$$
$$= \quad \langle \text{Double negation (3.12)} \rangle$$
$$[b]_\rho = [c]_\rho \ \equiv \ [b] \cap [c] \neq \emptyset$$
$$= \quad \langle (14.35b) \ \equiv \ (14.35c) \rangle$$
$$true$$

Thus, an equivalence relation on B induces a partition of B, where each partition element consists of equivalent elements.

We could ask the question in the other direction: does a partition of B define an equivalence relation on B? The next theorem answers this question affirmatively.

(14.36) **Theorem.** Let P be the set of sets of a partition of B. The following relation ρ on B is an equivalence relation:

$$b \, \rho \, c \ = \ (\exists p \mid p \in P : b \in p \ \wedge \ c \in p) \quad .$$

Proof. We must show that ρ is reflexive, symmetric, and transitive. Reflexivity follows from the fact that each element is in one of the sets in P. Symmetry follows from the definition of ρ in terms of $\wedge$, which is symmetric. Thus, to prove $b \, \rho \, c \ \equiv \ c \, \rho \, b$, apply the definition of ρ to the LHS, use Symmetry of $\wedge$ (3.36), and then apply the definition of ρ in the other direction. We prove transitivity as follows (in the proof, the range of dummies p and q is P):

$$b \, \rho \, c \; \wedge \; c \, \rho \, d$$
$$= \quad \langle \text{Definition of } \rho \,, \text{ twice} \rangle$$
$$(\exists p \mid : b \in p \, \wedge \, c \in p) \; \wedge \; (\exists q \mid : c \in q \, \wedge \, d \in q)$$
$$= \quad \langle \text{Distributivity of } \wedge \text{ over } \exists \ (9.21); \text{ Nesting } (8.20) \rangle$$
$$(\exists p, q \mid : b \in p \, \wedge \, c \in p \, \wedge \, c \in q \, \wedge \, d \in q)$$
$$\Rightarrow \quad \langle \, c \text{ is in only one element of the partition,}$$
$$\text{so } c \in p \, \wedge \, c \in q \; \Rightarrow \; p = q \, \rangle$$
$$(\exists p, q \mid : b \in p \, \wedge \, c \in p \, \wedge \, c \in q \, \wedge \, d \in q \wedge p = q)$$
$$\Rightarrow \quad \langle \text{Trading } (9.20), \text{ One-point rule } (8.14) \rangle$$
$$(\exists p \mid : b \in p \, \wedge \, c \in p \, \wedge \, c \in p \, \wedge \, d \in p)$$
$$\Rightarrow \quad \langle \text{Idempotency of } \wedge \ (3.38) \rangle$$
$$(\exists p \mid : b \in p \, \wedge \, c \in p \, \wedge \, d \in p)$$
$$\Rightarrow \quad \langle \text{Monotonicity of } \exists \ (9.27) \rangle$$
$$(\exists p \mid : b \in p \, \wedge \, d \in p)$$
$$= \quad \langle \text{Definition of } \rho \, \rangle$$
$$b \, \rho \, d \qquad \qquad \qquad \qquad \qquad \qquad \qquad \square$$

14.3 Functions

We have used functions throughout this text, but in a rather informal manner. We regarded a function f as a rule for computing a value v (say) from another value w (page 13), so that function application $f(w)$ or $f.w$ denotes value $v: f.w = v$. The fundamental property of function application, stated in terms of inference rule Leibniz (page 13), is:

$$\frac{X = Y}{f.X = f.Y} \ .$$

It is this property that allows us to conclude theorems like $f(b+b) = f(2 \cdot b)$ and $f.b + f.b = 2 \cdot f.b$.

This definition of function is different from that found in many programming languages. It is possible in some programming languages to define a function f that has the side effect of changing a parameter or global variable, so that in evaluating $f.b + f.b$, the value of the first and second function applications $f.b$ are different. But this means that $f.b + f.b = 2 \cdot f.b$ no longer holds! Eschewing such side effects when programming enables the use of basic mathematical laws for reasoning about programs involving function application.

For the rest of this section, we deal only with functions of one argument. This restriction is not serious, because a function $f(p_1, \ldots, p_n)$ can be viewed as a function $\hat{f}.p$ with a single argument that is an n-tuple. Thus, a function application $f(a_1, \ldots, a_n)$ would be written as $\hat{f}.\langle a_1, \ldots, a_n \rangle$.

In addition to thinking of a function as a rule for computing a value, we can regard a function as a binary relation on $B \times C$ that contains all pairs $\langle b, c \rangle$ such that $f.b = c$. However, a relation f can have distinct values c and c' that satisfy $b\,f\,c$ and $b\,f\,c'$, but a function cannot.

(14.37) **Definition.** A binary relation f on $B \times C$ is called a function iff it is *determinate*:

> **Determinate:** $(\forall b, c, c' \mid b\,f\,c \,\wedge\, b\,f\,c' : c = c')$.

Further, we distinguish between two kinds of functions:

(14.38) **Definition.** A function f on $B \times C$ is *total* if

> **Total:** $B = Dom.f$;

otherwise it *partial*. We write $f : B \rightarrow C$ for the type of f if f is total and $f : B \rightsquigarrow C$ if f is partial.

In some texts, the word *function* means either a total or a partial function; in others, *function* means only total function. In this section, we are careful to state exactly which we mean. In the rest of the text, we are not so careful. For example, elsewhere, we use the notation $f : B \rightarrow C$ for all functions, total or partial.

The reason for distinguishing between total and partial functions is that dealing with partial functions can be messy. What, for example, is the value of $f.b = f.b$ if $b \notin Dom.f$, so that $f.b$ is undefined? The choice of value must be such that our rules of manipulation —the propositional and predicate calculi— hold even in the presence of undefined values, and this is not so easy to achieve. However, for a partial function $f : B \rightsquigarrow C$, one can always restrict attention to its total counterpart, $f : Dom.f \rightarrow C$.

Examples of functions as relations

(a)   Binary relation $<$ is not a function, because it is not determinate —both $1 < 2$ and $1 < 3$ hold.

(b)   Identity relation ι_B over B is a total function $\iota_B : B \rightarrow B$; $\iota.b = b$ for all b in B.

(c)   Total function $f : \mathbb{N} \rightarrow \mathbb{N}$ defined by $f(n) = n + 1$ is the relation $\{\langle 0, 1 \rangle, \langle 1, 2 \rangle, \langle 2, 3 \rangle, \ldots\}$.

(d)   Partial function $f : \mathbb{N} \rightsquigarrow \mathbb{Q}$ defined by $f(n) = 1/n$ is the relation $\{\langle 1, 1/1 \rangle, \langle 2, 1/2 \rangle, \langle 3, 1/3 \rangle, \ldots\}$. It is partial because $f.0$ is not defined.

(e)   Function $f : \mathbb{Z}^+ \rightarrow \mathbb{Q}$ defined by $f(b) = 1/b$ is total, since $f.b$ is defined for all elements of $\mathbb{Z}^+$, the positive integers. However, $g : \mathbb{N} \rightsquigarrow \mathbb{Q}$ defined by $g.b = 1/b$ is partial because $g.0$ is not defined.

(f) The partial function f that takes each lower-case character to the next character can be defined by a finite number of pairs: $\{\langle 'a','b'\rangle,$ $\langle 'b','c'\rangle, \ldots, \langle 'y','z'\rangle\}$. It is partial because there is no pair whose first component is 'z'. □

When partial and total functions are viewed as binary relations, functions can inherit operations and properties of binary relations. For example, two functions (partial or total) are equal exactly when, viewed as relations, their sets of pairs are equal.

On page 271, we defined the product of two relations. Therefore, the product $(f \circ g)$ of two total functions has already been defined. We now manipulate $(f \circ g).b = d$ to determine what this means in terms of f and g separately.

$$(f \circ g).b = d$$
$$= \quad \langle\text{Viewing } f \circ g \text{ as a relation}\rangle$$
$$b\,(f \circ g)\,d$$
$$= \quad \langle\text{Definition (14.20) of the product of relations}\rangle$$
$$(\exists c\,|: b\,f\,c \,\wedge\, c\,g\,d)$$
$$= \quad \langle\text{Viewing relation pairs in terms of function application}\rangle$$
$$(\exists c\,|: f.b = c \,\wedge\, g.c = d)$$
$$= \quad \langle\text{Trading (9.19)}\rangle$$
$$(\exists c \mid c = f.b : g.c = d)$$
$$= \quad \langle\text{One-point rule (8.14)}\rangle$$
$$g(f.b) = d$$

Hence, $(f \circ g).b = g(f.b)$. That seems backward! We would rather see $f(g.b) = d$ in the RHS of this equality, so that we don't have to switch the order of f and g when switching between relational notation and functional notation. We therefore introduce a new symbol •, called *composition*:

(14.39) **Definition.** For functions f and g, $f \bullet g \equiv g \circ f$.

Then, with the above calculation, we have proved the following theorem.

(14.40) **Theorem.** Let $g: B \to C$ and $f: C \to D$ be total functions. Then the composition $f \bullet g$ of f and g is the total function defined by

$$(f \bullet g).b = f(g.b) .$$

The theory of binary relations tells us that function composition is associative: $(f \bullet g) \bullet h = f \bullet (g \bullet h)$. Powers of a function $f: B \to B$ are defined as well. Thus, f^0 is the identity function: $f^0.b = b$. And, for $n \geq 0$, $f^{n+1}.b = f(f^n.b)$.

Inverses of total functions

We now investigate the inverse of a total function. Every relation ρ has an inverse ρ^{-1}, which is defined by $\langle c, b \rangle \in \rho^{-1} \equiv \langle b, c \rangle \in \rho$. However, for total function f, relation f^{-1} need not be a function. For example, consider the total function $f:\mathbb{Z} \to \mathbb{N}$ given by $f(b) = b^2$. We have

$$f(-2) = 4 \text{ and } f.2 = 4 \text{, i.e. } \langle -2, 4 \rangle \in f \text{ and } \langle 2, 4 \rangle \in f.$$

Therefore, $\langle 4, 2 \rangle \in f^{-1}$ and $\langle 4, -2 \rangle \in f^{-1}$, so f^{-1} is not determinate and is not a function.

Some terminology helps characterize the total functions with inverses that are functions.

(14.41) **Definition.** Total function $f:B \to C$ is *onto* or *surjective* if $Ran.f = C$. Total function f is *one-to-one* or *injective* if

$$(\forall b, b':B, c:C \mid: b \, f \, c \wedge b' \, f \, c \equiv b = b') \quad .$$

Function f is *bijective* if it is one-to-one and onto.

A function can be made *onto* by changing its type. For example, function $f:\mathbb{N} \to \mathbb{N}$ defined by $f(b) = b+1$ is not onto, since $f.b \neq 0$ for all natural numbers b. However, function $f:\mathbb{N} \to \mathbb{Z}^+$ defined by $f(b) = b+1$ is onto.

(14.42) **Theorem.** Let f be a total function, and let f^{-1} be its relational inverse. Then f^{-1} is a function, i.e. is determinate, iff f is one-to-one. And, f^{-1} is total iff f is onto.

Proof. We first show that f^{-1} is determinate iff f is one-to-one —the left part of Fig. 14.2 illustrates this property.

$$
\begin{aligned}
&(\forall c, b, b' \mid c \, f^{-1} \, b \wedge c \, f^{-1} \, b' : b = b') \; - f^{-1} \text{ is determinate} \\
= \quad &\langle \text{Definition of } f^{-1} \rangle \\
&(\forall c, b, b' \mid b \, f \, c \wedge b' \, f \, c : b = b') \; - f \text{ is one-to-one}
\end{aligned}
$$

Next, we prove that f^{-1} is total iff f is onto —the right part of Fig. 14.2 illustrates this property.

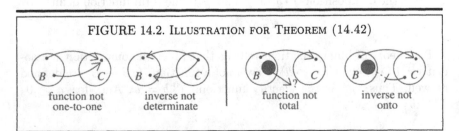

FIGURE 14.2. Illustration for Theorem (14.42)

| function not one-to-one | inverse not determinate | function not total | inverse not onto |

$$Dom(f^{-1}) = C - f^{-1} \text{ is total}$$
$$= \quad \langle \text{Definition of } f^{-1} \rangle$$
$$Ran.f = C - f \text{ is onto} \qquad \qquad \Box$$

From the theory of relations, we also borrow the notion of an identity. If total function $f:B \to C$ has an inverse, then $f^{-1} \cdot f = \imath_B$, the identity function on B, while $f \cdot f^{-1} = \imath_C$.

(14.43) **Definition.** Let $f:B \to C$. A *left inverse* of f is a function $g:C \to B$ such that $g \cdot f = \imath_B$. A *right inverse* of f is a function $g:C \to B$ such that $f \cdot g = \imath_C$. Function g is an *inverse* of f if it is both a left and a right inverse.

In some situations, it helps to distinguish between left and right inverses. Historical note 14.2 shows how these concepts are useful in providing security in electronic message-passing systems, where a third party should not be able to understand an intercepted message between two people.

Examples of left and right inverses

(a) Consider $abs:\mathbb{Z} \to \mathbb{N}$ defined by **if** $b < 0$ **then** $-b$ **else** b. Then, for any natural number b, $abs(\imath_\mathbb{N}.b) = abs.b = b$. Therefore, $abs \cdot \imath_\mathbb{N} = \imath_\mathbb{N}$, so $\imath_\mathbb{N}$ is a right inverse of abs.

(b) Consider $abs:\mathbb{Z} \to \mathbb{N}$ defined by **if** $b < 0$ **then** $-b$ **else** b. Define $neg:\mathbb{N} \to \mathbb{Z}$ by $neg.b = -b$. Then $abs(neg.b) = abs(-b) = b$. Therefore, $abs \cdot neg = \imath_\mathbb{N}$, so neg is a right inverse of abs (see example (a) above).

(c) Look at the first two examples. Both $\imath:\mathbb{N} \to \mathbb{N}$ and $neg:\mathbb{N} \to \mathbb{Z}$ are one-to-one. By theorem (14.45) below, they have left inverses. The two examples above show that $abs:\mathbb{Z} \to \mathbb{N}$ is a left inverse of both functions. $\qquad \Box$

(14.44) **Theorem.** Function $f:B \to C$ is onto iff f has a right inverse.

Proof. Consider relation f^{-1}. Theorem (14.42) says that f^{-1} is total iff f is onto. However, relation f^{-1} may not be determinate. We show how to construct from f^{-1} a function g such that

(a) g is determinate,
(b) g is total iff f is onto, and
(c) $f \cdot g = \imath_C$ (iff f is onto).

This function g, then, is the right inverse of f —iff f is onto.

HISTORICAL NOTE 14.2. Message Encryption and Authentication

Some functions are time-consuming to evaluate, but this fact can be exploited to implement message encryption and authentication in computer networks. In *private key cryptography*, two users agree secretly on a function E and its inverse E^{-1}. Then, to communicate a message m, the sender sends an encrypted text $m' = E.m$ and the receiver of m' computes $E^{-1}(m') = E^{-1}(E.m) = m$. Provided E^{-1} is difficult to compute from $E.m$, an intruder cannot easily infer m from m'.

In *public-key cryptography*, each user U to whom messages can be sent selects functions E_U and D_U having the following properties.

(i) $D_U(E_U.m) = m$, so that D_U is a left inverse of E_U and E_U is a right inverse of D_U.

(ii) It is prohibitively expensive to compute D_U given E_U.

A message m is sent to U in the encrypted form $m' = E_U.m$. U decrypts m' by calculating $D_U.m' = D_U(E_U.m) = m$. E_U can be made publicly available without compromising messages encrypted using E_U because, according to (ii), knowing E_U doesn't help an intruder compute $D_U.m'$. The name *public key cryptography* is apt because encryption scheme E_U is made public.

In some situations, U wants to be certain who sent a message. For example, a request for an electronic funds transfer should be honored only if it comes from the account owner. A *digital signature* can be implemented using a public key cryptosystem if E_U and D_U also satisfy:

(iii) $E_U(D_U.m) = m$, so that E_U is a left inverse of D_U and D_U is a right inverse of E_U.

The signer can use $D_U.t$ as a signature, for some text t. For example, to construct a signed message m, user U might send $m'' = D_U(t \char94 m)$, where t is U's name. The receiver has access to E_U and can compute $E_U.m'' = E_U(D_U(t \char94 m))$. Provided U has not revealed D_U to anyone, by (ii), no one else knows D_U. Thus, no one else can construct a message m'' such that $E_U.m''$ produces U's name.

---

f^{-1} For each c in C, there may be several values $b_0, b_1, \ldots$ such that $f.b_i = c$, as illustrated to the left. Therefore, $c \; f^{-1} \; b_i$ holds for all i. For each such c, arbitrarily

g choose one element, say b_0, and define $g.c = b_0$. Hence g is determinate and satisfies (a). Next, (b) is satisfied —g is total iff f is onto— because $Dom.g = Dom.f^{-1}$ and f^{-1} is total iff f is onto. Finally, we show that (c) is satisfied by computing $(f \bullet g).c$, assuming f is onto.

$$(f \bullet g).c$$
$$= \quad \langle \text{Definition of } \bullet \rangle$$

$$f(g.c)$$
$$= \quad \langle \text{Since } f \text{ is total, at least one pair } \langle c, b \rangle \text{ is in } f^{-1}.$$
$$\text{Therefore, there is one pair } \langle c, b \rangle \text{ in } g. \rangle$$
$$f.b$$
$$= \quad \langle \, \langle c, b \rangle \in g \Rightarrow \langle c, b \rangle \in f^{-1}. \text{ Therefore, } \langle b, c \rangle \in f. \rangle$$
$$c \qquad\qquad\qquad\qquad\qquad\qquad\qquad\qquad\qquad\qquad\qquad \square$$

(14.45) **Theorem.** Let $f : B \to C$ be total. Then f is one-to-one iff f has a left inverse.

We leave the proof of this theorem and the following one to the reader.

(14.46) **Theorem.** Let $f : B \to C$ be total. The following statements are equivalent.

(a) f is one-to-one and onto.

(b) There is a function $g : C \to B$ that is both a left and a right inverse of f.

(c) f has a left inverse and f has a right inverse.

14.4 Order relations

An *order relation* compares (some) members of a set. A typical order relation is relation $<$ on the integers. However, an order relation need not allow comparison of every pair of members of a set. For example, with relation *parent*, some pairs are comparable but not others. For example, neither Schneider *parent* Gries nor Gries *parent* Schneider holds.

(14.47) **Definition.** A binary relation ρ on a set B is called a *partial order on B* if it is reflexive, antisymmetric, and transitive. In this case, the pair $\langle B, \rho \rangle$ is called a *partially ordered set* or *poset*.

We use the symbol $\preceq$ for an arbitrary partial order, sometimes writing $c \succeq b$ instead of $b \preceq c$.

Examples of partial orders

(a) $\langle \mathbb{N}, \leq \rangle$ is a poset and $\leq$ is a partial order on $\mathbb{N}$.

(b) Let B be a set. Then $\langle \mathcal{P}B, \subseteq \rangle$ is a poset and $\subseteq$ is a partial order on $\mathcal{P}B$, since $\subseteq$ is reflexive, antisymmetric, and transitive (Theorems (11.57)–(11.59) on page 207).

(c) Consider the set C of courses offered at Cornell University. Define the relation $\preceq$ by $c1 \preceq c2$ if courses $c1 = c2$ or if $c1$ is a prerequisite for $c2$. Then $\langle C, \preceq \rangle$ is a poset and $\preceq$ is a partial order on C.

(d) Let P be the set of loops in a Pascal program. Define $\preceq$ on P by $l1 \preceq l2$ if loops $l1$ and $l2$ are the same or if $l1$ is nested within $l2$. Then $\langle P, \preceq \rangle$ is a poset and $\preceq$ is a partial order on P.

(e) In constructing a house, certain jobs have to be done before other jobs. Let J be the set of jobs to be done, and let $\preceq$ on J be defined by $j1 \preceq j2$ if $j1$ and $j2$ are the same or if $j1$ has to be completed before $j2$ can be started. Then $\langle J, \preceq \rangle$ is a poset. The scheduling of jobs in such situations, including redefining and ordering jobs in order to reduce time to completion, is sometimes referred to as PERT (Program Evaluation and Review Technique).   □

If set B of poset $\langle B, \preceq \rangle$ is finite (and small enough), then relation $\preceq$ can be depicted in a *Hasse diagram*, as illustrated in Fig. 14.3. The Hasse diagram on the left in Fig. 14.3 describes the poset $\langle 1..9, | \rangle$, where $b \mid c \equiv$ "b divides c". For example, $2 \mid 4$ holds, but not $2 \mid 5$. In general, in the Hasse diagram for poset $\langle B, \preceq \rangle$, if $b \preceq c$ holds, then b appears below c. Further, a line is drawn from b up to c iff

$$b \preceq c \quad \text{and} \quad \text{no element } d \text{ (other than } b, c) \text{ satisfies } b \preceq d \preceq c.$$

An element b is connected to an element c by a series of lines in the Hasse diagram iff $b \preceq c$. The Hasse diagram is a minimal description of the poset, in that as few lines as possible are drawn. Thus, the Hasse diagram for a partial order ρ actually presents the smallest relation ρ' such that $\rho = (\rho')^*$. Relation ρ' is called the *transitive reduction* of partial order ρ.

Deleting all pairs $\langle b, b \rangle$ from relation $\leq$ on the integers results in relation $<$. Similarly, deleting all pairs $\langle b, b \rangle$ from a subset relation $\subseteq$ gives the relation $\subset$. Such an operation can be applied to any partial order $\preceq$ to yield a relation $\prec$. We give a name to the class of relations that result from this operation.

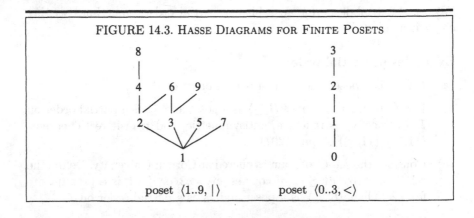

FIGURE 14.3. Hasse Diagrams for Finite Posets

poset $\langle 1..9, | \rangle$ poset $\langle 0..3, < \rangle$

(14.48) **Definition.** Relation $\prec$ is a *quasi order* or *strict partial order* if
$\prec$ is transitive and irreflexive.

Being irreflexive means that $b \prec c$ and $c \prec b$ do not both hold (for all
b, c). Therefore, the antecedent of the definition of antisymmetry, $b \prec c \land$
$c \prec b \Rightarrow b = c$ (for all b, c), is *false*, so all quasi orders are antisymmetric.
The operation of adding pairs $\langle b, b \rangle$ to a quasi order or deleting such pairs
from a partial order does not harm the transitivity property of the relation.
Hence, we see that adding in all pairs $\langle b, b \rangle$ to a quasi order makes it into
a partial order, and deleting all pairs $\langle b, b \rangle$ from a partial order makes it
into a quasi order. Thus, we have the following theorem.

(14.49) **Theorem.** If ρ is a partial order over a set B, then $\rho - \iota_B$ is a
quasi order. If ρ is a quasi order over a set B, then $\rho \cup \iota_B$ is a
partial order.

Given $\prec$, then, its reflexive closure $\preceq$ is computed by adding all pairs $\langle b, b \rangle$
to $\prec$. Given $\preceq$, its *reflexive reduction* $\prec$ is computed by eliminating all
pairs $\langle b, b \rangle$ from $\preceq$. The same Hasse diagram can be used to represent
both a partial order and its corresponding quasi order; we just have to
know which is intended by the diagram.

TOTAL ORDERS AND TOPOLOGICAL SORT

Thus far, we have dealt with partial orders, so all elements need not be
comparable. We now investigate the class of *total orders*. Again, we define
the class of total orders in two ways: in terms of membership and in terms
of operations on sets.

(14.50) **Definition.** A partial order $\preceq$ over B is called a *total* or *linear*
order if

$$(\forall b, c \mid : b \preceq c \lor b \succeq c) \quad ,$$

i.e. iff $\preceq \cup \preceq^{-1} = B \times B$. In this case, the pair $\langle B, \preceq \rangle$ is called
a *linearly ordered set* or a *chain*.

Examples of total orders and chains

(a) $\leq$ over the natural numbers is a total order, and $\langle \mathbb{N}, \leq \rangle$ is a chain.

(b) $\leq$ over the reals is a total order, and $\langle \mathbb{R}, \leq \rangle$ is a chain.

(c) Let set S contain more than one element. Then $\subseteq$ over $\mathcal{P}S$ is not a
total order. For example, if b and c are distinct elements in S, then
neither $\{b\} \subseteq \{c\}$ nor $\{c\} \subseteq \{b\}$ holds.

(d) Let C be the set of courses at Cornell. Let $b \preceq c$ mean that $b = c$
or b is a prerequisite for c. Relation $\preceq$ is a partial order but not a
total order.   □

A linear order $\preceq$ over B can be given simply by listing the elements of B in the order imposed by $\preceq$: b precedes c in the sequence iff $b \prec c$ holds. For example, the linear order $\leq$ on the integers $1..9$ can be presented as $\langle 1, 2, 3, 4, 5, 6, 7, 8, 9 \rangle$.

It is possible to extend any partial order $\preceq$ to a total order. That is, we can construct a total order $\preceq'$ such that $\preceq \; \subseteq \; \preceq'$. For example, consider the partially ordered set $\langle 1..9, | \rangle$ of Fig. 14.3. The two linear orders $\langle 1, 2, 3, 4, 5, 6, 7, 8, 9 \rangle$ and $\langle 1, 2, 3, 4, 5, 6, 7, 9, 8 \rangle$ both contain relation $|$. Since 8 and 9 are incomparable, their relative placement does not matter in the extension of partial order $|$ to a total order. This example illustrates that there may be several ways to extend a partial order to a total order.

We now present an algorithm, called *topological sort*, for extending a partial order $\preceq$ over a finite set B to a total order. This algorithm has many applications, including the ubiquitous spreadsheet, where quasi order $\prec$ is given by the way in which values in the spreadsheet depend on each other. For example, an entry that is to contain $c := d + 3$ has to be computed before an entry that is to contain $b := c + 2$, since the formula for b depends on the formula for c. Therefore, "$c := d+3$" $\prec$ "$b := c+2$". Other applications of topological sort are found in code optimization within compilers.

We use topological sort to illustrate a method of presenting (or developing) algorithms. We start with the specification of the algorithm. We then write a simple, but inefficient, algorithm, whose correctness is easy to see. Finally, we replace some variables of the algorithm by fresh ones and obtain a more efficient algorithm. This *data refinement* or *coordinate transformation* requires replacing the statements and expressions that use the old variables by ones that have the same effect on the new variables. The replacement is done independently of the algorithm itself; correctness of the resulting algorithm is ensured if each replacement of a local statement has certain properties.

Topological sort is the subject of some legal maneuvers concerning software patents. See Historical note 14.3.

The algorithm begins with a variable B containing set B and terminates with sequence variable s being the linear order of B. So, the precondition Q and postcondition R of the algorithm are:

Q : $B = $ B
R : s is a linear order of B that contains $\prec$.

We assume that B $= 0..K$ for some natural number K. Thus, the elements of set B have been labeled in some fashion, and the labels of the elements are manipulated rather than the elements themselves.

HISTORICAL NOTE 14.3. SOFTWARE PATENTS

The U.S. Constitution gives Congress the power "to promote the Progress of Science and useful Arts, by securing for limited Times to Authors and Inventors the exclusive Right to their respective Writings and Discoveries." So, Congress passed laws to allow the copyrighting of written material (and, later, music, art, records, films, etc.) and the patenting of "any new and useful process, machine, manufacture, or composition of matter, or any new and useful improvement thereof." Patents and copyrights protect the rights of inventors and writers and create incentive for advances in technology.

Software does not fit the framework of these laws very well. Is an algorithm just an idea or concept, like the unpatentable mathematical theorem? Or is it a real invention, like the carrot peeler or frisbee? Needless to say, the development of computers and software has led to a morass of legal and economic problems in regard to patents.

Patents are being granted for software. In fact, the patent office has 145 examiners who deal with patent applications related to computer applications and systems, including software-related patents. Their workload is so high that they would like to grow in 1993 to 200. But the League for Programming Freedom argues forcefully for eliminating software patents [41], and the issue of software patents is being hotly debated (see [34] for references).

In 1968, Knuth published in an undergraduate text [26] a topological sort in a general setting that required only n steps for a set of n pairs. Two years later, in 1970, two people filed for a patent for a version of topological sort in a business application. Their version was slower than Knuth's, requiring up to n^2 steps. The patent office told them they could not patent topological sort, because lots of people knew about it. The filers appealed. The judge for the appeal said that the patent office cannot simply *say* that people know about it; evidence of *prior art* has to be given. Such evidence had not been given, so, in 1983 the filers got their patent. Believe it or not, the whole process took thirteen years. Lotus was then sued for infringement of the patent (even though Lotus probably used Knuth's faster algorithm —who would use the slow one?). As of Spring 1993, the case is in the courts. Some prior art has been found that seems to be directly related to the patent —topological sort in a business application— but no one knows what will happen. Millions of dollars ride on the outcome.

The basic idea of the algorithm is this. Start with $s = \epsilon$. Then, repeatedly choose a minimal element (with respect to $\prec$) of subset B of B, delete it from B, and append it to s. A minimal element of B has nothing smaller than it in B, so all elements that precede it in the partial order $\prec$ are already in s. Hence, s becomes a linear extension of $\preceq$.

We can write this algorithm as follows.

$\{Q: \ B = \mathbf{B}\}$

$s := \epsilon;$

$\{$invariant $P: \ (B$ and $\{b \mid b \in s\}$ partition $\mathbf{B}) \ \wedge$
$\qquad\qquad\quad (\forall b, c{:}\mathbf{B} \mid b \prec c \wedge c \in s : b$ precedes c in $s)\}$

do $B \neq \emptyset \rightarrow$
$\qquad$ Choose a minimal element b (say) of B;
$\qquad B, s := B - \{b\}, s \ ^\frown \langle b \rangle$
od

$\{R: \ (\forall b, c{:}\mathbf{B} \mid b \prec c : b$ precedes c in $s)\}$

The correctness of the algorithm should be checked using the points of Checklist (12.45) for proving a loop correct: (a) Does the initialization truthify invariant P? (b) Does the repetend maintain the truth of P? (c) Does the loop terminate? (d) Does $P \wedge B = \emptyset \Rightarrow R$ hold? Each of these questions can be answered affirmatively, so the algorithm implements the specification.

A naive implementation for choosing the minimum element of B would require checking every pair b, c. The algorithm would then end up requiring a number of steps that is at least quadratic in the number of such pairs. We can do better by developing data structures that allow this choice to be done more efficiently.

We need an efficient way to find minimal elements of B. So, let sequence variable m contain the minimal elements of B.

When a minimal element b of B is deleted from B, other elements may become minimal elements of B. We want to determine these new minimal elements quickly and add them to sequence m. This requires knowing the elements c that satisfy $b \prec c$, as well as the number of predecessors (according to $\prec$) that such a c has. For this purpose, we introduce two arrays N and S, so we have three new variables:

$\qquad$ **var** $m : seq(0..K);$
$\qquad$ **var** $N :$ **array** $0..K$ **of** $integer;$
$\qquad$ **var** $S :$ **array** $0..K$ **of** $set(0..K);$

$N[c]$ is the number of elements of B that precede c in relation $\prec$, and $S[b]$ is the set of elements of B that succeed b.

We describe the relation between variable B, which is being replaced, and the three new variables in the following *coupling invariant*.

$\qquad$ (sequence m contains the minimal elements of B) $\wedge$
$\qquad (\forall c \mid c \in B : N[c] = (\Sigma b \mid b \in B \wedge b \prec c : 1)) \ \wedge$
$\qquad (\forall b \mid b \in B : S[b] = \{c \mid c \in B \wedge b \prec c\})$

The new variables are initialized as follows.

for $c \in B$ **do** $N[c] := (\Sigma b \mid b \in B \wedge b \prec c : 1)$;
for $b \in B$ **do** $S[b] := \{c \mid c \in B \wedge b \prec c\}$;
$m := \epsilon$; **for** $c \in 0..K$ **do if** $N[c] = 0$ **then** $m := m \,\hat{}\, \langle c \rangle$

We now show how the various statements and expressions of the algorithm can be rewritten to make them efficient and to maintain the definitions of the three new variables.

(a)   Expression $B \neq \emptyset$ can be replaced by $m \neq \epsilon$, since every finite nonempty set over quasi order $\prec$ has a minimal element.

(b)   The statement "Choose a minimal element b of B" can be replaced by $b := m.0$.

(c)   The replacement for $B := B - \{b\}$, where $b = m.0$, may have to change all three new variables. Here is its replacement:

$$m := m[1..]; \qquad -m.0 \text{ is being removed from } B$$
$$\textbf{for } c \in S[b] \textbf{ do } N[c] := N[c] - 1;$$
$$\textbf{if } N[c] = 0 \textbf{ then } m := m \,\hat{}\, \langle c \rangle$$
$$\textbf{od}$$

The algorithm that results from making these replacements in the original algorithm is shown below. This algorithm takes time proportional to the number of pairs $b \prec c$ because, in total, the statement $N[c] := N[c] - 1$ is executed exactly once for each such pair.

for $c \in B$ **do** $N[c] := (\Sigma b \mid b \in B \wedge b \prec c : 1)$;
for $b \in B$ **do** $S[b] := \{c \mid c \in B \wedge b \prec c\}$;
$m := \epsilon$; **for** $c \in 0..K$ **do if** $N[c] = 0$ **then** $m := m \,\hat{}\, \langle c \rangle$;
$s := \epsilon$;
do $m \neq \epsilon \rightarrow$
    $b := m.0$;
    $s := s \,\hat{}\, \langle b \rangle$;
    $m := m[1..]$; $-m.0$ is being removed from B
    for $c \in S[b]$ **do** $N[c] := N[c] - 1$;
        if $N[c] = 0$ **then** $m := m \,\hat{}\, \langle c \rangle$
        od
od

More on posets

This subsection defines some special elements of a poset (e.g. maximal element and least upper bound of a subset of a poset). These elements play important roles in further analysis of posets and their application, but much of this is beyond the scope of this text. Thus, we restrict ourselves to giving some definitions and theorems and making a few remarks.

Throughout, $\langle U, \preceq \rangle$ denotes an arbitrary poset and $\prec$ is the quasi order corresponding to partial order $\preceq$.

(14.51) **Definition.** Let S be a nonempty subset of poset $\langle U, \preceq \rangle$.

   (a)   Element b of S is a *minimal element of S* if no element of S is smaller than b, i.e. if $b \in S \wedge (\forall c \mid c \prec b : c \notin S)$.

   (b)   Element b of S is the *least element of S* if $b \in S \wedge$ $(\forall c \mid c \in S : b \preceq c)$.

   (c)   Element b is a *lower bound of S* if $(\forall c \mid c \in S : b \preceq c)$. (Element b need not be in S.)

   (d)   Element b is the *greatest lower bound of S*, written $glb.S$, if b is a lower bound and if every lower bound c satisfies $c \preceq b$.

We already defined "minimal element" on page 228, where we proved that $\langle U, \prec \rangle$ admits induction iff $\langle U, \prec \rangle$ is well founded. A set may have more than one minimal element, as the examples below show. However, a set has at most one least element. Minimal elements and least elements of a set belong to the set. Lower bounds need not belong to the set.

Examples of minimal and least elements

(a)   Set $\mathbb{N}$ of poset $\langle \mathbb{N}, \leq \rangle$ has minimal element 0 and least element 0.

(b)   Set $\mathbb{R}$ of poset $\langle \mathbb{R}, \leq \rangle$ has no minimal or least element. But subset $\{x \mid 0 \leq x\}$ has 0 as its minimal and least element.

(c)   Consider $\langle \mathbb{N}, \mid \rangle$, where $i \mid j$ means "i divides j". Subset $\{3, 5, 7, 15, 20\}$ has three minimal elements, 3, 5, and 7, but it has no least element. Subset $\{2, 4, 6, 8\}$ has minimal and least element 2.

(d)   Consider poset $\langle \mathcal{P}\{b, c\}, \subseteq \rangle$, with Hasse diagram in Fig. 14.4. The elements of $\mathcal{P}\{b, c\}$ are $\emptyset$, $\{b\}$, $\{c\}$, and $\{b, c\}$. Its minimal and least element is $\emptyset$. The minimal and least element of subset $\{\{b\}\}$ is $\{b\}$. Subset $\{\{b\}, \{c\}\}$; has two minimal elements, $\{b\}$ and $\{c\}$, and no least element.   □

Examples of lower bounds and greatest lower bounds

(a) Consider poset $\langle \mathbb{R}, \leq \rangle$. Subset $S = \{x \mid 0 < x < 1\}$ has 0 and all nonnegative numbers for lower bounds. Its greatest lower bound is 0. But S has no least element. On the other hand, subset $T = \{x \mid 0 \leq x < 1\}$ has the same lower bounds and greatest lower bounds, but it has a least element: 0.

(b) Consider poset $\langle \mathcal{P}\{b, c\}, \subseteq \rangle$, with Hasse diagram in Fig. 14.4. Subset $\{\{b\}\}$ has $\emptyset$ and $\{b\}$ for lower bounds, and its greatest lower bound is $\{b\}$. Subset $\{\{b, c\}\}$ has every element of $\mathcal{P}\{b, c\}$ as lower bound, and itself as its greatest lower bound. $\qquad\qquad\qquad\qquad\square$

We give a simple condition for minimal elements to exist.

(14.52) **Theorem.** Every finite nonempty subset S of poset $\langle U, \preceq \rangle$ has a minimal element.

Proof. Choose any element x_0 of S and construct a decreasing chain of elements of S: $x_n \ldots \prec x_2 \prec x_1 \prec x_0$ (for some n) until no longer possible. Antisymmetry of $\preceq$ implies that all elements of the chain are distinct. Since S is finite, this chain is finite. Element x_n is a minimal element of S. $\qquad\qquad\qquad\qquad\square$

The following theorem follows directly from the definitions. The proof is left to the reader.

(14.53) **Theorem.** Let B be a nonempty subset of poset $\langle U, \preceq \rangle$.

(a) A least element of B is also a minimal element of B (but not necessarily vice versa).

(b) A least element of B is also a greatest lower bound of B (but not necessarily vice versa).

(c) A lower bound of B that belongs to B is also a least element of B.

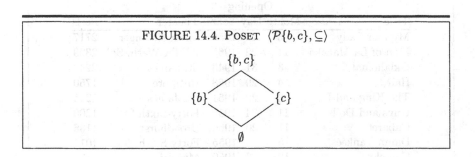

FIGURE 14.4. POSET $\langle \mathcal{P}\{b, c\}, \subseteq \rangle$

We now define maximal elements, greatest elements, and upper bounds of a poset.

(14.54) **Definition.** Let S be a nonempty subset of poset $\langle U, \preceq \rangle$.

    (a) Element b of S is a *maximal element of S* if no element of S is larger than b, i.e. if $b \in S \land (\forall c \mid b \prec c : c \notin S)$.

    (b) Element b of S is the *greatest element of S* if $b \in S \land (\forall c \mid c \in S : c \preceq b)$.

    (c) Element b is an *upper bound of S* if $(\forall c \mid c \in S : c \preceq b)$. (An upper bound of S need not be in S.)

    (d) Element b is the *least upper bound of S*, written *lub.S*, if b is an upper bound and if every upper bound c satisfies $b \preceq c$.

There is a symmetry between Definitions (14.51) and (14.54). In fact, one can easily see the following about a subset S of U. An element is a maximal element of S with respect to relation $\preceq$ iff it is a minimal element of S with respect to $\succeq$. Similar statements can be made concerning greatest elements, upper bounds, and least upper bounds. Thus, any results concerning minimal elements, least elements, and lower bounds have their counterparts concerning maximal elements, greatest elements, and upper bounds.

14.5   Relational Databases

An n-ary relation over the cross product $B_1 \times \cdots \times B_n$ is simply a subset of the n-tuples of $B_1 \times \cdots \times B_n$. Such an n-ary relation can be presented

---

TABLE 14.2. POPULAR AMERICAN BROADWAY MUSICALS (*PABM*)

Title	Opening Month	Day	Year	Theater	Perfs
My Fair Lady	3	15	1956	Mark Hellinger	2717
Man of La Mancha	11	22	1965	ANTA Wash. Sq.	2329
Oklahoma!	3	31	1943	St. James	2248
Hair	4	29	1968	Biltmore	1750
The King and I	3	29	1951	St. James	1246
Guys and Dolls	11	24	1950	Forty-Sixth St.	1200
Cabaret	11	20	1966	Broadhurst	1166
Damn Yankees	5	5	1955	Forty-Sixth St.	1019
Camelot	12	3	1960	Majestic	878
West Side Story	9	26	1957	Winter Garden	732

as an n-column table. Each row of the table corresponds to an n-tuple of the relation, and each column corresponds to one of the dimensions, or components, of the cross product on which the relation the based. For example, the relation defined by Table 14.2 is a subset of the following cross product *PABM* ,

$$PABM = Title \times Month \times Day \times Year \times Theater \times Perfs$$

where

Title is the set of titles for Broadway shows;
Month is the set 1..12 corresponding to the months of the year;
Day is the set 1..31 corresponding to the days of the months;
Year is the set $\mathbb{Z}^+$ of positive integers;
Theater is the set of theaters in and around Broadway, NYC;
Perfs is the set $\mathbb{Z}^+$ of positive integers.

Table 14.3 contains another table or relation, *MC* .

A *database* is a collection of information, or data, about some area of interest. When the database is accessed in a way that appears to users as if it consists of a set of relations, it is called a *relational database*. The designer of such a database decides on a set of cross products (or tables) that together comprise the database. Each cross product is defined by a *relational scheme*, which is conventionally denoted by

$$rel\text{-}name(attribute_1, attribute_2, \ldots, attribute_n)$$

where *rel-name* is the name associated with the cross product and each attribute is a name for a set of values. Thus, our database of Broadway

TABLE 14.3. MUSICAL CREATORS (*MC*)			
Title	Book	Lyrics	Music
My Fair Lady	Lerner	Lerner	Loewe
Man of La Mancha	Wasserman	Darion	Leigh
Oklahoma!	Hammerstein	Hammerstein	Rodgers
Hair	Ragni & Rado	Ragni & Rado	MacDermot
The King and I	Hammerstein	Hammerstein	Rodgers
Guys and Dolls	Swerling & Burrows	Loesser	Loesser
Cabaret	Masteroff	Ebb	Kander
Damn Yankees	Abbott & Wallop	Adler & Ross	Adler & Ross
Camelot	Lerner	Lerner	Loewe
West Side Story	Laurents	Sondheim	Bernstein

musicals is defined by the following two relational schemes.

$$PABM(Title, Month, Day, Year, Theater, Perfs)$$
$$MC(Title, Book, Lyrics, Music)$$

The relational schemes shown above are not the only possible relational-database design for this application. We might have a database with a single relational scheme that combines the information in $PABM$ and MC:

$$ALL(Title, Month, Day, Year, Theater, Perfs, Book, Lyrics,$$
$$Music) .$$

Or, we might use a larger collection of simpler relations:

(14.55) $Where(Title, Theater)$
   $When(Title, Month, Day, Year)$
   $Author(Title, Book)$
   $Run(Title, Perfs)$
   $Lyricist(Title, Lyrics)$
   $Composer(Title, Music)$.

All three of these collections of schemes contain the same information. Further, as we see below, all can be used to answer the same questions, or *queries*, as they are called in the database world. In choosing a database design, the designer takes into consideration factors such as the speed at which various queries can be answered, the amount of space needed for the scheme, and ease of modification of the database. Database $PABM\text{-}MC$ of this chapter is small enough so that questions of space and speed are not relevant. However, some databases contain millions of records, and for such databases economy of space, quick access, and ease of modification are important. Consider, for example, the IRS's database of people in the U.S. and their income-tax returns, or the database of financial accounts and transactions that a bank must maintain.

OPERATIONS FOR CONSTRUCTING QUERIES

Users of a relational database formulate queries, or questions, about the database. Processing a query produces a set of tuples that answer the query. For example, consider a query that requests all shows that opened at the Mark Hellinger on 3/15/56. Applying this query to $PABM$ of Table 14.2 would generate a subset consisting of a single 6-tuple:

(14.56) $\langle$My Fair Lady, 3, 15, 1956, Mark Hellinger, 2717$\rangle$.

A query to find shows that opened at the Forty-Sixth St. would produce a relation with two elements:

$\langle$Guys and Dolls, 11, 24, 1950, Forty-Sixth St., 1200$\rangle$

$\langle$Damn Yankees, 5, 5, 1955, Forty-Sixth St., 1019$\rangle$.

Three basic operators used in constructing queries are: *selection* (denoted by σ); *projection* (denoted by π), and natural join (denoted by $\bowtie$). We now discuss these three operators.

Operation $\sigma(R, F)$ selects the set of tuples of relation R that satisfy predicate F. Here, predicate F may contain the names of the fields of relation R. Thus,

$$\sigma(R, F) = \{t \mid t \in R \wedge F\} .$$

For example, $\sigma(PABM, \textit{Theater} = \text{Mark Hellinger})$ is the set consisting of the single tuple (14.56).

The operations allowed in F depends on the particular database system being used. Some systems are quite primitive, but, in principle, any operation that could be applied to a field name could be applied to an entry in a field. For example, $\sigma(PABM, \textit{Perfs} > 2,000)$ would select the three tuples of $PABM$ with more than 2,000 performances.

Projection operator π allows irrelevant information to be discarded in answering a query. Suppose, for example, that we are interested in the titles (only) of shows that opened at the Forty-Sixth St. and not in the dates of their opening. By itself, query

$$\sigma(PABM, \textit{Theater} = \text{Forty-Sixth St.})$$

produces 6-tuples containing dates and numbers of performances, as well as titles. Operator π allows us to identify the desired fields, causing the unnamed fields to be suppressed. For $A_1, \ldots, A_m$ a subset of the names of the fields of relation R,

$$\pi(R, A_1, \ldots, A_m) = \{t \mid t \in R : \langle t.A_1, t.A_2, \ldots, t.A_m \rangle\}$$

Thus, $\pi(R, A_1, \ldots, A_m)$ is like R, but it has fewer dimensions (columns), because the attributes of R that do not appear as an argument to π are not in the projection. (A projection of R may also have fewer rows than R, by virtue of having fewer dimensions —in deleting a dimension, previously distinct tuples may become identical.) For example, the following query lists the titles (only) of shows that opened at the Forty-Sixth St. Note that evaluation of $\sigma(\ldots)$ constructs the desired set of tuples; then, evaluation of $\pi(\ldots)$ discards the unwanted fields. Here, the first argument of π is not one of the original relations but a relation that was constructed using σ.

$$\pi(\sigma(PABM, \textit{Theater} = \text{Forty-Sixth St.}), \textit{Titles})$$

The database consisting of *PABM* and *MC* contains enough information to determine who wrote the lyrics for the show having 2717 performances (Lerner). However, the query that will produce the answer cannot be solely in terms of σ and π and relations *PABM* and *MC*, because the information needed to determine the answer is split across the two relations. We need a way to join the relations together. One operation to accomplish this is called the *natural join*, denoted by the infix symbol $\bowtie$. Relation *PABM* $\bowtie$ *MC* has all the attributes that *PABM* and *MC* have, but if an attribute appears in both, then it appears only once in the result; further, only those tuples that agree on this common attribute are included. An example will make this clear. Suppose relation *Where* of (14.55) contains the three tuples

⟨My Fair Lady, Mark Hellinger⟩
⟨Oklahoma!, St. James⟩
⟨Hair, Biltmore⟩   .

Suppose relation *Author* of (14.55) contains the three tuples

⟨My Fair Lady, Lerner⟩
⟨Oklahoma!, Hammerstein⟩
⟨The King and I, Hammerstein⟩   .

Then the natural join *Where* $\bowtie$ *Author* includes

⟨My Fair Lady, Mark Hellinger, Lerner⟩
⟨Oklahoma!, St. James, Hammerstein⟩   .

Using natural join, a query to find out who wrote the lyrics for the show that had 2717 performances is

$$\pi(\sigma(PABM \bowtie MC, Perfs = 2717), Lyrics) \quad .$$

With natural join, we can now revisit the various database schemes for representing our small database. Observe that

$$ALL = (PABM \bowtie MC) =$$
$$(When \bowtie Where \bowtie Run \bowtie Author \bowtie Lyricist \bowtie Composer) \quad .$$

Thus, these three relational schemes are equivalent. Not all such relational schemes are equivalent. For example, the relational scheme

(14.57)   *Where1*(*Composer, Theater*)
When1(*Composer, Month, Day, Year*)
Author1(*Composer, Book*)
Run1(*Title, Perfs*)

$Lyricist1\,(Composer, Lyrics)$

$What1\,(Composer, Title)$

is not equivalent to ALL because

$$When1 \bowtie Where1 \bowtie Run1 \bowtie Author1 \bowtie Lyricist1 \bowtie What1$$

contains, among others, the tuple

⟨Camelot, 12, 3, 1960, Mark Hellinger, 2717, Lerner, Lerner, Lowe⟩

which is not in ALL. The problem is that (14.57) decomposes ALL in such a way that the common attribute in each relation ($Composer$) is not a *key*: it does not uniquely identify a tuple in ALL.

DISCUSSION

Databases are an important application of the theory of relations. There is a rich theory concerning ways to decompose relational schemes in ways that information will not be lost. There are also ways to rearrange computations automatically in queries in order to reduce time and space requirements. For example, in many cases, deleting extraneous fields before constructing a cross product will produce the same result as deleting the fields after constructing the cross product, but the former is far more efficient. There is also research on ways to make the implementation of a relational database and its operations efficient.

Exercises for Chapter 14

14.1 Using Definition (14.1) of an ordered pair, what are the pairs ⟨1, 1⟩ and ⟨1, 2⟩ ?

14.2 Using Definition (14.1), prove that $\{b\} = (\cap y \mid y \in \langle b, c\rangle : y)$.

14.3 Using Definition (14.1), prove the following concerning the value of the second component of a pair $\langle b, c\rangle$: If $b \neq c$, then $\{c\} = (\cup y \mid y \in \langle b, c\rangle : y) - (\cap y \mid y \in \langle b, c\rangle : y)$.

14.4 Using Definition (14.1), prove the following concerning the value of the second component of a pair $\langle b, c\rangle$: If $b = c$, then $\{\,\} = (\cup y \mid y \in \langle b, c\rangle : y) - (\cap y \mid y \in \langle b, c\rangle : y)$.

Exercises on cross products

14.5 Prove Membership (14.4), $\langle x, y\rangle \in S \times T \equiv x \in S \wedge y \in T$.

14.6 Prove theorem (14.5), $\langle x, y\rangle \in S \times T \equiv \langle y, x\rangle \in T \times S$.

14.7 Prove theorem (14.6), $S = \emptyset \;\Rightarrow\; S \times T = T \times S = \emptyset$.

14.8 Prove theorem (14.7), $S \times T = T \times S \;\equiv\; S = \emptyset \vee T = \emptyset \vee S = T$.

14.9 Prove Distributivity of $\times$ over $\cup$ (14.8), $(S \cup T) \times U = (S \times U) \cup (T \times U)$ and $S \times (T \cup U) = (S \times T) \cup (S \times U)$.

14.10 Prove Distributivity of $\times$ over $\cap$ (14.9), $(S \cap T) \times U = (S \times U) \cap (T \times U)$ and $S \times (T \cap U) = (S \times T) \cap (S \times U)$.

14.11 Prove Distributivity of $\times$ over $-$ (14.10), $S \times (T - U) = (S \times T) - (S \times U)$.

14.12 Prove Monotonicity (14.11), $T \subseteq U \;\Rightarrow\; S \times T \subseteq S \times U$.

14.13 Prove theorem (14.12), $S \subseteq U \wedge T \subseteq V \;\Rightarrow\; S \times T \subseteq U \times V$.

14.14 Prove theorem (14.13), $S \times T \subseteq S \times U \wedge S \neq \emptyset \;\Rightarrow\; T \subseteq U$.

14.15 Prove theorem (14.14), $(S \cap T) \times (U \cap V) = (S \times U) \cap (T \times V)$.

14.16 Prove theorem (14.15), For finite S and T, $\#(S \times T) = \#S \cdot \#T$.

Exercises on relations

14.17 Prove Theorem (14.19).

14.18 Let ρ and σ be relations on set $B = \{b, c, d, e\}$:
$$\rho = \{\langle b,b\rangle, \langle b,c\rangle, \langle c,d\rangle\}$$
$$\sigma = \{\langle b,c\rangle, \langle c,d\rangle, \langle d,b\rangle\}$$
Compute $\rho \circ \sigma$, $\sigma \circ \rho$, ρ^2, and ρ^3.

14.19 Prove Distributivity of $\circ$ over $\cup$ (14.23), $\rho \circ (\sigma \cup \theta) = \rho \circ \sigma \cup \rho \circ \theta$ and $(\sigma \cup \theta) \circ \rho = \sigma \circ \rho \cup \theta \circ \rho$.

14.20 Prove Distributivity of $\circ$ over $\cap$ (14.24), $\rho \circ (\sigma \cap \theta) \subseteq \rho \circ \sigma \cap \rho \circ \theta$ and $(\sigma \cap \theta) \circ \rho \subseteq \sigma \circ \rho \cap \theta \circ \rho$.

14.21 Prove theorem (14.26), $\rho^m \circ \rho^n = \rho^{m+n}$, by induction.

14.22 Prove theorem (14.27), $(\rho^m)^n = \rho^{m \cdot n}$, by induction.

14.23 Prove Theorem (14.29a).

14.24 Prove Theorem (14.29b).

14.25 Each of the six classes given in Table 14.1 is defined in two different ways. Prove that the two ways are equivalent.

14.26 The following argument purports to prove that every symmetric and transitive relation is an equivalence relation. What is wrong with it?

Let R be symmetric and transitive. To show that R is an equivalence relation, we have to show that R is also reflexive. Because R is symmetric, if $\langle x,y\rangle \in R$, then $\langle y,x\rangle \in R$. Because R is transitive, if $\langle x,y\rangle \in R$ and $\langle y,x\rangle \in R$, then $\langle x,x\rangle \in R$. Therefore, R is reflexive.

14.27 Which of the properties Table 14.1(a)–Table 14.1(f) holds for each of the following relations?

(a) $b \rho c$ iff b and c are both positive or both negative, for integers b, c.

(b) $b \rho c$ iff $b - c$ is a multiple of 5, for integers b, c.

(c) $\emptyset$ for a non-empty set B.

(d) ι_B, the identity relation on a nonempty set B.

(e) $\iota_B \times \iota_B$, where ι_B is the identity relation on a set B.

(f) $=$ over the integers $\mathbb{Z}$.

(g) $<$ over the integers $\mathbb{Z}$.

(h) $\leq$ over the integers $\mathbb{Z}$.

(i) $b \rho c$ iff b is the father of c.

(j) $b \rho c$ iff b is the father of c or vice versa.

(k) $b \rho c$ iff b is c or the father of c.

14.28 Find a smallest nonempty set and a relation on it that is neither reflexive nor irreflexive.

14.29 Find a smallest nonempty set and a relation on it that is neither symmetric nor antisymmetric.

14.30 Prove Theorem (14.31): A reflexive relation is its own reflexive closure; a symmetric relation is its own symmetric closure; and a transitive relation is its own transitive closure.

14.31 Prove Theorem (14.32), parts (a), (b), and (d).

14.32 Consider binary relations over a set B. A property of a relation on B is *preserved* under some set operation if applying the operation to the relation results in a relation with the same property. For example, the union of two symmetric relations is symmetric, so $\cup$ preserves relational symmetry. Fill in the entries of the following table with Y if the operation in the column preserves the property for the row and with N otherwise. For each N, give a counterexample.

	$\rho \cup \sigma$	$\rho \cap \sigma$	$\rho - \sigma$	$(B \times B) - \rho$
Reflexivity				
Irreflexivity				
Symmetry				
Antisymmetry				
Transitivity				

14.33 Prove the part (b) $\Rightarrow$ (c) of Theorem (14.35).

14.34 Prove the part (c) $\Rightarrow$ (a) of Theorem (14.35).

Exercises on functions

14.35 Prove that the composition of two total functions is a total function.

14.36 Prove that the composition of a partial function and a total function (in either order) or of two partial functions is a partial function.

14.37 Prove formally that function composition is associative (using the notion that a function is a binary relation).

14.38 Prove Theorem (14.45). In this proof construct a left inverse g by *adding* pairs to f^{-1} to make it total, instead of deleting pairs, as in the proof of Theorem (14.44).

14.39 Prove Theorem (14.46).

Exercises on posets

14.40 Prove Theorem (14.53).

Chapter 15

A Theory of Integers

W e have used laws of integer arithmetic for manipulating integer expressions in several places in this text. We now study a theory of integers. Many of the properties of the integers will be familiar to you, but new ones will also emerge.

We can proceed in two different ways. We can start with the inductive definition of the integers à la Peano (see page 227) and begin proving theorems about the integers from this definition. Alternatively, we can postulate various axioms that the integers should satisfy.

We choose the second alternative. To start, we define an *integral domain*: a set of elements on which binary operators + and · have certain properties. We then introduce notions of positive and negative and a relational operator < to obtain an *ordered* integral domain. Finally, we add the *well-ordering principle* for the positive elements, yielding the integers as we know them.

Once we have the integers, we study operations $min(x, y)$, $max(x, y)$, and $abs.x$ (the absolute value of x). We also study division, greatest common divisors, and prime numbers. Finally, we look at various representations of the integers.

15.1 Integral domains

Let D be a set (type) of elements, two of which are 0 and 1, and let + and · be binary operators on D. Assume D is *closed* with respect to + and · , i.e. for any a and b in D, $a + b$ and $a \cdot b$ are also in D. D (together with + and ·) is called an *integral domain* if the following axioms hold. [1] Throughout, variables a, b, c, d are of type D.

(15.1) **Axiom, Associativity:** $(a + b) + c = a + (b + c)$

$$(a \cdot b) \cdot c = a \cdot (b \cdot c)$$

---

[1] Because of Symmetry (15.2), we could have fewer axioms. Only one Additive identity axiom, Multiplicative identity axiom, Distributivity axiom, Additive inverse axiom, and Cancellation axiom are needed. Also, in the Cancellation axiom, we could have used $\Rightarrow$ instead of $\equiv$.

(15.2)   **Axiom, Symmetry:** $a + b = b + a$
$$a \cdot b = b \cdot a$$

(15.3)   **Axiom, Additive identity:** $0 + a = a$
$$a + 0 = a$$

(15.4)   **Axiom, Multiplicative identity:** $1 \cdot a = a$
$$a \cdot 1 = a$$

(15.5)   **Axiom, Distributivity:** $a \cdot (b + c) = a \cdot b + a \cdot c$
$$(b + c) \cdot a = b \cdot a + c \cdot a$$

(15.6)   **Axiom, Additive inverse:** $(\exists x \colon D \mid : x + a = 0)$
$$(\exists x \colon D \mid : a + x = 0)$$

(15.7)   **Axiom, Cancellation:** $c \neq 0 \;\Rightarrow\; (c \cdot a = c \cdot b \;\equiv\; a = b)$
$$c \neq 0 \;\Rightarrow\; (a \cdot c = b \cdot c \;\equiv\; a = b)$$

The set $\mathbb{Z}$ of integers satisfies these axioms and is, therefore, an integral domain. Other integral domains exist as well, e.g. the rational numbers $\mathbb{Q}$ and the real numbers $\mathbb{R}$. A less familiar integral domain is the set of irrational numbers of the form $a + b \cdot \sqrt{5}$, where a and b are integers.

We have the following theorems.

Theorems for integral domains

(15.8)   **Cancellation:** $a + b = a + c \;\equiv\; b = c$

(15.9)   **Zero:** $a \cdot 0 = 0$

(15.10)   **Unique identity:** $a + z = a \;\equiv\; z = 0$
$$a \neq 0 \;\Rightarrow\; (a \cdot z = a \;\equiv\; z = 1)$$

(15.11) $a \cdot b = 0 \;\equiv\; a = 0 \vee b = 0$

We prove Cancellation (15.8) by mutual implication.

LHS $\Rightarrow$ RHS. We assume the LHS and prove the RHS. This proof rests on Axiom (15.6), which says there is a witness x satisfying $x + a = 0$:

$$b$$
$$= \quad \langle \text{Additive identity (15.3)} \rangle$$
$$0 + b$$

$$
\begin{aligned}
= \quad & \langle \text{Additive inverse (15.6), with witness } x \rangle \\
& x + a + b \\
= \quad & \langle \text{Assumption } a + b = a + c \rangle \\
& x + a + c \\
= \quad & \langle \text{Additive inverse (15.6)} \rangle \\
& 0 + c \\
= \quad & \langle \text{Additive identity (15.3)} \rangle \\
& c
\end{aligned}
$$

RHS $\Rightarrow$ **LHS.** We assume the antecedent $b = c$ and prove the consequent $a + b = a + c$:

$$
\begin{aligned}
& a + b \\
= \quad & \langle \text{Assumption } b = c \rangle \\
& a + c
\end{aligned}
$$

As another example, we prove (15.9), $a \cdot 0 = 0$. In the proof, Cancellation (15.8) is used to add something to both sides of the equation so that Distributivity (15.5) can be used.

$$
\begin{aligned}
& a \cdot 0 = 0 \\
= \quad & \langle \text{Cancellation (15.8), with } a, b, c := a \cdot a, a \cdot 0, 0 \rangle \\
& a \cdot a + a \cdot 0 = a \cdot a + 0 \\
= \quad & \langle \text{Distributivity (15.5)} \rangle \\
& a \cdot (a + 0) = a \cdot a + 0 \\
= \quad & \langle \text{Additive identity of } + \text{ (15.3), twice} \rangle \\
& a \cdot a = a \cdot a \; \text{—which is Reflexivity (1.2), with } x := a \cdot a
\end{aligned}
$$

SUBTRACTION

Additive inverse (15.6) indicates that for any element a there exists an element x satisfying $x + a = 0$. This element x is unique:

Unique additive inverse

(15.12) $x + a = 0 \land y + a = 0 \;\Rightarrow\; x = y$

We prove this theorem as follows. For arbitrary a, x, y in D we have,

$$
\begin{aligned}
& x + a = 0 \;\land\; y + a = 0 \\
\Rightarrow \quad & \langle \text{Transitivity of equality (1.4)} \rangle \\
& x + a = y + a \\
= \quad & \langle \text{Cancellation (15.8)} \rangle \\
& x = y
\end{aligned}
$$

Since the x that satisfies $x + a = 0$ is unique, we can define it as a function of a: $-a$. We can also define subtraction now.

(15.13) **Axiom, Unary minus:** $a + (-a) = 0$

(15.14) **Axiom, Subtraction:** $a - b = a + (-b)$

The familiar laws of subtraction now follow.

Some theorems for subtraction

(15.15) $x + a = 0 \equiv x = -a$

(15.16) $-a = -b \equiv a = b$

(15.17) $-(-a) = a$

(15.18) $-0 = 0$

(15.19) $-(a + b) = (-a) + (-b)$

(15.20) $-a = (-1) \cdot a$

(15.21) $(-a) \cdot b = a \cdot (-b)$

(15.22) $a \cdot (-b) = -(a \cdot b)$

(15.23) $(-a) \cdot (-b) = a \cdot b$

(15.24) $a - 0 = a$

(15.25) $(a - b) + (c - d) = (a + c) - (b + d)$

(15.26) $(a - b) - (c - d) = (a + d) - (b + c)$

(15.27) $(a - b) \cdot (c - d) = (a \cdot c + b \cdot d) - (a \cdot d + b \cdot c)$

(15.28) $a - b = c - d \equiv a + d = b + c$

(15.29) $(a - b) \cdot c = a \cdot c - b \cdot c$

A simple corollary of Theorem (15.23) is $(-1) \cdot (-1) = 1$. We prove theorem (15.23). For arbitrary a, b in D, we have

$$
\begin{aligned}
&(-a) \cdot (-b) \\
= \quad &\langle (15.21),\ (-a) \cdot b = a \cdot (-b) \rangle \\
&a \cdot (-(-b)) \\
= \quad &\langle (15.17),\ -(-a) = a \rangle \\
&a \cdot b
\end{aligned}
$$

ORDERED DOMAINS

We usually list the integers in the order

$$\ldots, -3, -2, -1, 0, 1, 2, 3, \ldots$$

and write $b < c$ if integer b occurs before integer c in this list. We now restrict attention to integral domains that have such an order. To define an order, we first define a predicate $pos.b$ for b in domain D, with interpretation "b appears after 0 in the order", or "b is positive". Note that this interpretation is not the real definition of $pos.b$, but only the interpretation we want $pos.b$ to have.

Predicate $pos.b$ is defined by four axioms. The first says that the sum of two positive elements is positive. The second says that the product of two positive elements is positive. The third says that 0 is not positive. The fourth says that for any non-zero element b, exactly one of b and $-b$ is positive.

(15.30) **Axiom, Addition:** $pos.a \land pos.b \Rightarrow pos(a+b)$

(15.31) **Axiom, Multiplication:** $pos.a \land pos.b \Rightarrow pos(a \cdot b)$

(15.32) **Axiom:** $\neg pos.0$

(15.33) **Axiom:** $b \neq 0 \Rightarrow (pos.b \equiv \neg pos(-b))$

An integral domain D with predicate pos that satisfies axioms (15.30)–(15.33) is called an *ordered domain*, and the ordering is a linear order or total order (see Definition (14.50) on page 287). The integers are an ordered domain, as are the rational numbers and the real numbers (and many others). In all ordered domains, we have the following two theorems, the first of which says that the square of a non-zero element is positive.

Theorems for pos

(15.34) $b \neq 0 \Rightarrow pos(b \cdot b)$

(15.35) $pos.a \Rightarrow (pos.b \equiv pos(a \cdot b))$

We prove (15.34). For arbitrary nonzero b in D, we prove $pos(b \cdot b)$ by case analysis: either $pos.b$ or $\neg pos.b$ holds (see (15.33)).

Case $pos.b$. By axiom (15.31) with $a, b := b, b$, $pos(b \cdot b)$ holds.

Case $\neg pos.b \land b \neq 0$. We have the following.

$$
\begin{array}{ll}
& pos(b \cdot b) \\
= & \langle (15.23), \text{ with } a, b := b, b \rangle \\
& pos((-b) \cdot (-b)) \\
\Leftarrow & \langle \text{Multiplication (15.31)} \rangle \\
& pos(-b) \land pos(-b) \\
= & \langle \text{Idempotency of } \land \ (3.38) \rangle
\end{array}
$$

$$pos(-b)$$
$$= \quad \langle \text{Double negation (3.12) —note that } b \neq 0 \,; (15.33) \rangle$$
$$\neg pos.b \quad \text{—the case under consideration}$$

A corollary of this theorem is that $1\ (=1 \cdot 1)$ is positive, so -1 is negative.

We are finally ready to define the conventional inequality relations, which are predicates over pairs of elements of D.

(15.36) **Axiom, Less:** $a < b \equiv pos(b-a)$

(15.37) **Axiom, Greater:** $a > b \equiv pos(a-b)$

(15.38) **Axiom, At most:** $a \leq b \equiv a < b \lor a = b$

(15.39) **Axiom, At least:** $a \geq b \equiv a > b \lor a = b$

Now we can prove that the positive elements are greater than 0 (i.e. $pos.b \equiv b > 0$) and the negative elements are less than 0. A host of other theorems follow, a few of which are given below. Theorem (15.44), the law of Trichotomy, says that exactly one of $a < b$, $a = b$, and $a > b$ is *true*. According to the discussion on page 46, the first conjunct of (15.44) is *true* iff one or three of its equivalents are *true*, and the second conjunct is *true* iff fewer than three of them are *true*.

Some theorems for arithmetic relations

(15.40) **Positive elements:** $pos.b \equiv 0 < b$

(15.41) **Transitivity:** (a) $a < b \land b < c \Rightarrow a < c$

(b) $a \leq b \land b < c \Rightarrow a < c$

(c) $a < b \land b \leq c \Rightarrow a < c$

(d) $a \leq b \land b \leq c \Rightarrow a \leq c$

(15.42) **Monotonicity:** $a < b \equiv a + d < b + d$

(15.43) **Monotonicity:** $0 < d \Rightarrow (a < b \equiv a \cdot d < b \cdot d)$

(15.44) **Trichotomy:** $(a < b \ \equiv \ a = b \ \equiv \ a > b) \land$
$\neg(a < b \ \land \ a = b \ \land \ a > b)$

(15.45) **Antisymmetry:** $a \leq b \land b \leq a \equiv a = b$

(15.46) **Reflexivity:** $a \leq a$

(15.47) $a = b \equiv (\forall z : D \mid : z \leq a \ \equiv \ z \leq b)$

We prove the first of the Transitivity theorems (15.41a). The proof uses $(b-a) + (c-b) = c - a$, which is proved in an exercise.

$$a < b \wedge b < c$$
$$= \quad \langle \text{Axiom Less (15.36)} \rangle$$
$$pos.(b-a) \wedge pos(c-b)$$
$$\Rightarrow \quad \langle \text{Addition (15.30)} \rangle$$
$$pos((b-a)+(c-b))$$
$$= \quad \langle \text{Arithmetic —see Exercise 15.22} \rangle$$
$$pos(c-a)$$
$$= \quad \langle \text{Axiom Less (15.36)} \rangle$$
$$a < c$$

WELL-ORDERED DOMAINS

We began this chapter with integral domains. We then postulated additional properties (in terms of predicate *pos*) that hold only for some integral domains, the ordered domains. We now give one more property, which is enjoyed (essentially) only by the ordered domain of integers.

A subset D' of an ordered domain is called *well ordered* if each nonempty subset S of D' contains a minimal element (according to relation $<$):

(15.48) $\quad S \neq \emptyset \equiv (\exists b \mid b \in S : (\forall c \mid c < b : c \notin S)) \quad$ for all $S \subseteq D'$.

According to Definition (12.21) on page 229, a pair $\langle D', < \rangle$ is *well founded* if it satisfies (15.48). But if $<$ is also a total order on D', then D' is called well ordered. So, a well order is simply a well-founded set that is totally ordered.

We state the following axiom concerning the natural numbers.

(15.49) **Axiom, Well ordering:** The set $\mathbb{N}$ of natural numbers is well ordered (under the ordering $<$ defined in (15.36)).

Thus, any subset of the natural numbers contains a minimal element. For example, the minimal element of the set of odd natural numbers is 1. Note that Well ordering (15.49) does not hold for the set of all integers, since the subset consisting of the negative numbers has no minimal element. In Chap. 20, we show that any infinite set contains (in a sense described in that chapter) the natural numbers. Thus, the natural numbers are the smallest infinite set and the integers are the smallest infinite ordered domain.

In Sec. 12.1, we introduced mathematical induction over the integers. We also justified induction, by giving an argument why, having proved $(\forall i \mid 0 \leq i < n : P.i) \Rightarrow P.n$ for all $n{:}\mathbb{N}$, we could in theory prove $P.N$ for any natural number N. Here, we have taken a different tack and postulated the well-ordering property for the natural numbers. And this property is enough to guarantee that $\langle \mathbb{N}, < \rangle$ admits induction.

The following theorem shows that the real numbers and the rationals are not well ordered.

(15.50) **Theorem.** In a well-ordered domain, there is no element between 0 and 1.

Proof. The proof is by contradiction. Assume element c from the well-ordered domain satisfies $0 < c < 1$. By Well ordering (15.49), the set of elements between 0 and 1 has a minimal element m (say). We have,

$$0 < m < 1$$
$$= \quad \langle \text{Remove abbreviation; Idempotency of } \wedge \ (3.38) \rangle$$
$$0 < m \ \wedge \ m < 1 \ \wedge \ m < 1$$
$$\Rightarrow \quad \langle (15.43), \text{ twice} \rangle$$
$$0 \cdot m < m \cdot m \ \wedge \ m \cdot m < 1 \cdot m \ \wedge \ m < 1$$
$$= \quad \langle \text{Zero of } \cdot \ (15.9); \text{ Multiplicative identity } (15.4) \rangle$$
$$0 < m \cdot m \ \wedge \ m \cdot m < m \ \wedge \ m < 1$$
$$= \quad \langle \text{Introduce abbreviation} \rangle$$
$$0 < m \cdot m < m < 1$$

The last line contradicts the fact that m is the smallest element between 0 and 1, so the assumption that there exists a c satisfying $0 < c < 1$ is *false*. $\qquad \square$

QUANTIFICATION FOR + AND ·

Arithmetic operators $+$ and $\cdot$ are symmetric and associative. The identity of $+$ is 0 and the identity of $\cdot$ is 1. Hence, $+$ and $\cdot$ are candidate operators for $\star$ in Sec. 8.2. The expressions $(+x \mid R : P)$ and $(\cdot x \mid R : P)$ are conventionally written as

$$(\Sigma x \mid R : P) \quad \text{and} \quad (\Pi x \mid R : P) \ .$$

The first is read as "the sum of P for x in the range R"; the second as "the product of P for x in the range R".

Axioms (8.13)–(8.21) hold for $(\Sigma x \mid R : P)$ and $(\Pi x \mid R : P)$ and will not be repeated here. In addition, we have the following distributive law.

(15.51) **Axiom, Distributivity:** For finite R and $\neg occurs('x', 'Q')$,
$$Q \cdot (\Sigma x \mid R : P) = (\Sigma x \mid R : Q \cdot P)$$

15.2   Exploring minimum and maximum

We define the *minimum* and *maximum* of two numbers in an ordered domain and explore the properties of these operators. Our treatment differs from the usual one, since we are concerned with simplifying the manipulation of expressions containing these operators. Our treatment is based on work found in [14].

The minimum of x and y is the smaller of the two; the maximum is the larger. For example, 3 is the minimum and 5 is the maximum of 3 and 5. Using $x \downarrow y$ and $x \uparrow y$ to denote the minimum and maximum of x and y, we can define $\downarrow$ and $\uparrow$ in terms of relation $\leq$ as follows.

(15.52) $x \downarrow y \;=\;$ **if** $x \leq y$ **then** x **else** y

   $x \uparrow y \;=\;$ **if** $x \leq y$ **then** y **else** x

The above definitions are by cases. Therefore, manipulation of expressions containing $\downarrow$ or $\uparrow$ are likely to require case analysis, because the definitions force us to handle the two cases $x \leq y$ and $x > y$ separately. We formulate a definition of $\downarrow$ and $\uparrow$ that avoids this problem —at the expense of using quantification.

Operators $\downarrow$ and $\uparrow$ are defined to satisfy the following properties.

(15.53) **Axiom, Definition of $\downarrow$ and $\uparrow$:**

   $(\forall z \mid: z \leq x \downarrow y \;\equiv\; z \leq x \;\wedge\; z \leq y)$

   $(\forall z \mid: z \geq x \uparrow y \;\equiv\; z \geq x \;\wedge\; z \geq y)$

Definitions (15.52) constructively define $\downarrow$ and $\uparrow$: they show how to compute them. Definitions (15.53) do not show how to compute $\downarrow$ and $\uparrow$; but they provide a way to manipulate expressions containing $\downarrow$ and $\uparrow$.

Having defined $\downarrow$ and $\uparrow$, we investigate their properties. We list below some theorems that follow from Definition (15.53). These theorems justify our belief that $\downarrow$ and $\uparrow$ are indeed definitions of minimum and maximum.

Theorems for minimum and maximum

(15.54) **Symmetry:** $x \downarrow y = y \downarrow x$

   $x \uparrow y = y \uparrow x$

(15.55) **Associativity:** $(x \downarrow y) \downarrow z = x \downarrow (y \downarrow z)$

   $(x \uparrow y) \uparrow z = x \uparrow (y \uparrow z)$

Theorems for minimum and maximum (continued)

(15.56) **Idempotency:** $x \downarrow x = x$
$$x \uparrow x = x$$

(15.57) $x \downarrow y \leq x \;\wedge\; x \downarrow y \leq y$
$$x \uparrow y \geq x \;\wedge\; x \uparrow y \geq y$$

(15.58) $x \leq y \;\equiv\; x \downarrow y = x$
$$x \geq y \;\equiv\; x \uparrow y = x$$

(15.59) $x \downarrow y = x \;\vee\; x \downarrow y = y$
$$x \uparrow y = x \;\vee\; x \uparrow y = y$$

The next theorems describe the interaction between $\downarrow$ (and $\uparrow$), addition, and multiplication.

Distributivity of + and · over $\downarrow$ and $\uparrow$

(15.60) **Distributivity:** $c + (x \downarrow y) = (c+x) \downarrow (c+y)$
$$c + (x \uparrow y) = (c+x) \uparrow (c+y)$$

(15.61) **Distributivity:** $c \geq 0 \;\Rightarrow\; c \cdot (x \downarrow y) = (c \cdot x) \downarrow (c \cdot y)$
$$c \geq 0 \;\Rightarrow\; c \cdot (x \uparrow y) = (c \cdot x) \uparrow (c \cdot y)$$

(15.62) **Distributivity:** $c \leq 0 \;\Rightarrow\; c \cdot (x \uparrow y) = (c \cdot x) \downarrow (c \cdot y)$
$$c \leq 0 \;\Rightarrow\; c \cdot (x \downarrow y) = (c \cdot x) \uparrow (c \cdot y)$$

Proving theorems about $\downarrow$

Because of the similarity in the definitions of $\downarrow$ and $\uparrow$, the theorems (and their proofs) for $\uparrow$ are similar to those for $\downarrow$. Therefore, we deal only with theorems concerning $\downarrow$.

In order to prove Symmetry (15.54), $x \downarrow y = y \downarrow x$, we need to be able to manipulate equations of the form $z = x \downarrow y$. Theorem (15.47) and its obvious counterpart, both repeated here, provide help.

$$a = b \;\equiv\; (\forall z \,|: z \leq a \;\equiv\; z \leq b)$$
$$a = b \;\equiv\; (\forall z \,|: z \geq a \;\equiv\; z \geq b)$$

By (15.47) with $a, b := x \downarrow y, y \downarrow x$, we have

$$x \downarrow y = y \downarrow x \;\equiv\; (\forall z \mid : z \leq x \downarrow y \;\equiv\; z \leq y \downarrow x) \quad.$$

Hence, symmetry can be proved by proving that the RHS of this equivalence is a theorem. To prove the RHS, we prove that the body of the quantification holds for arbitrary z:

$$
\begin{aligned}
& z \leq x \downarrow y \\
=\;& \quad \langle \text{Definition } (15.53) \rangle \\
& z \leq x \;\wedge\; z \leq y \\
=\;& \quad \langle \text{Symmetry of } \wedge \; (3.36) \rangle \\
& z \leq y \;\wedge\; z \leq x \\
=\;& \quad \langle \text{Definition } (15.53), \text{ with } x, y := y, x \rangle \\
& z \leq y \downarrow x
\end{aligned}
$$

Hence, we have proved that $\downarrow$ is symmetric.

Next, we prove the first theorem of (15.57), $x \downarrow y \leq x \;\wedge\; x \downarrow y \leq y$. The point to note about (15.57) is that it is the RHS of the body of (15.53), with the substitution $z := x \downarrow y$. We proceed as follows.

$$
\begin{aligned}
& x \downarrow y \leq x \;\wedge\; x \downarrow y \leq y \\
=\;& \quad \langle (15.53), \text{ with } z := x \downarrow y \rangle \\
& x \downarrow y \leq x \downarrow y \quad \text{—Reflexivity (15.46)}
\end{aligned}
$$

QUANTIFICATION FOR $\downarrow$ AND $\uparrow$

Operators $\downarrow$ and $\uparrow$ over the integers are symmetric and associative, so they are examples of operator $\star$ of Sec. 8.2. That is, we can write quantifications

(15.63) $(\downarrow i \mid R : E)$ and $(\uparrow i \mid R : E)$

to express the minimum and maximum of the values found by evaluating E with i ranging over values that satisfy R. For example, $(\downarrow i \mid 0 \leq i \leq 10 : b[i])$ is the minimum of the array elements $b[0], \ldots, b[10]$.

Formulas (15.63) satisfy general laws of quantification (8.14)–(8.21). Note also that $\downarrow$ and $\uparrow$ are idempotent, so Range split for idempotent $\star$ (8.18) holds for them. However, $\downarrow$ and $\uparrow$ do not have identities in all ordered domains. Therefore, they do not satisfy Empty-range (8.13), and when using range-split axioms, no range should be *false*.[2]

---

[2] In some cases, $\downarrow$ and $\uparrow$ have identities. For example, if the set under consideration is the natural numbers, then the identity of $\uparrow$ is 0, but $\downarrow$ has no identity.

In addition, we have the following distributive laws. Theorems (15.70) say that the minimum of a set is no larger than any of its elements and that the maximum is no smaller than its elements.

Distributive properties of $\downarrow$ and $\uparrow$

(15.64) **Distributivity of $+$ over $\downarrow$:** Provided $\neg occurs(`x`, `E`)$,
$$(\exists x \,|\!: R) \;\Rightarrow\; E + (\downarrow x \mid R : P) = (\downarrow x \mid R : E + P)$$

(15.65) **Distributivity of $+$ over $\uparrow$:** Provided $\neg occurs(`x`, `E`)$,
$$(\exists x \,|\!: R) \;\Rightarrow\; E + (\uparrow x \mid R : P) = (\uparrow x \mid R : E + P)$$

(15.66) **Distributivity of $\cdot$ over $\downarrow$:** Provided $\neg occurs(`x`, `E`)$,
$$(\exists x \,|\!: R) \land E \geq 0 \;\Rightarrow\; E \cdot (\downarrow x \mid R : P) = (\downarrow x \mid R : E \cdot P)$$

(15.67) **Distributivity of $\cdot$ over $\uparrow$:** Provided $\neg occurs(`x`, `E`)$,
$$(\exists x \,|\!: R) \land E \geq 0 \;\Rightarrow\; E \cdot (\uparrow x \mid R : P) = (\uparrow x \mid R : E \cdot P)$$

(15.68) **Distributivity of $\downarrow$ over $\uparrow$:** Provided $\neg occurs(`x`, `E`)$,
$$E \downarrow (\uparrow x \mid R : P) = (\uparrow x \mid R : E \downarrow P)$$

(15.69) **Distributivity of $\uparrow$ over $\downarrow$:** Provided $\neg occurs(`x`, `E`)$,
$$E \uparrow (\downarrow x \mid R : P) = (\downarrow x \mid R : E \uparrow P)$$

(15.70) Provided $\neg occurs(`x`, `E`)$,
$$R[x := E] \;\Rightarrow\; E = E \uparrow (\downarrow x \mid R : x)$$
$$R[x := E] \;\Rightarrow\; E = E \downarrow (\uparrow x \mid R : x)$$

15.3 Exploring absolutes

Consider any ordered domain. For x in that domain, we can define the absolute value of x, written $abs.x$, by

(15.71) $abs.x \;=\; x \uparrow - x$.

For example, $abs.5 = 5$ and $abs(-5) = 5$. Note how this definition avoids case analysis, which is used in the usual definition of abs:

$$abs.x \;=\; \textbf{if } x < 0 \textbf{ then } -x \textbf{ else } x \quad .$$

Because abs is defined in terms of $\uparrow$, all properties of abs can be derived from the properties of $\uparrow$. We list below a few properties of abs.

Theorems for absolute value

(15.72) $abs.x = abs(-x)$

(15.73) **Triangle inequality:** $abs(x + y) \leq abs.x + abs.y$

(15.74) $abs(abs.x) = abs.x$

(15.75) $abs(x \cdot y) = abs.x \cdot abs.y$

(15.76) $-(abs.x + abs.y) \leq x + y \leq abs.x + abs.y$

15.4 Divisibility, common divisors, and primes

We now restrict our attention to the integers. Throughout this section, variables a, b, c, d are of type $\mathbb{Z}$, unless otherwise stated.

The equation $5 \cdot x = 10$ has the integral solution $x = 2$, but the equation $5 \cdot x = 11$ has no integral solution —no integer x satisfies it. If an equation $c \cdot x = b$ with integer coefficients b and c has an integral solution, we say that b is *divisible* by c. We introduce relation $c \mid b$ with meaning " c divides b", or " b is divisible by c". Operator $\mid$ has the same precedence as $=$ and is viewed as a conjunctional operator. Formally, $\mid$ is defined for integer operands as follows.

(15.77) $c \mid b \;\equiv\; (\exists d \mid : c \cdot d = b)$

Some properties of relation $\mid$ are captured in the following theorems.

Theorems concerning divisibility

(15.78) $c \mid c$

(15.79) $c \mid 0$

(15.80) $1 \mid b$

(15.81) $c \mid 1 \;\Rightarrow\; c = 1 \lor c = -1$

(15.82) $d \mid c \land c \mid b \;\Rightarrow\; d \mid b$

(15.83) $b \mid c \land c \mid b \;\equiv\; b = c \lor b = -c$

(15.84) $b \mid c \;\Rightarrow\; b \mid c \cdot d$

(15.85) $b \mid c \;\Rightarrow\; b \cdot d \mid c \cdot d$

(15.86) $1 < b \land b \mid c \;\Rightarrow\; \neg(b \mid (c + 1))$

Given natural numbers b and c, $c \neq 0$, we conventionally think of dividing b by c to yield a quotient q and remainder r. We express this property without using division $/$, which we have not formally defined, in the following theorem.

(15.87) **Theorem.** Given integers b, c with $c \neq 0$, there exist (unique) integers q and r such that

$$b = q \cdot c + r \qquad \text{where } 0 \leq r < c.$$

Proof. We prove the existence of q and r by giving an algorithm to compute them. In fact, we already gave such an algorithm for the case $b \geq 0$ and $c > 0$ on page 239. We repeat the algorithm here; the other cases are left to the reader.

(15.88) $\{Q: b \geq 0 \ \wedge \ c > 0\}$
        $q, r := 0, b;$
        $\{$invariant $P: b = q \cdot c + r \ \wedge \ 0 \leq r\}$
        $\{$bound function : $r\}$
        do $r \geq c \rightarrow q, r := q + 1, r - c$ **od**
        $\{R: b = q \cdot c + r \ \wedge 0 \leq r < c\}$ □

(15.89) **Corollary.** For given b, c, the values q and r of Theorem (15.87) are unique.

We define operators $\div$ and **mod** for operands b and c, $c \neq 0$, by

(15.90) $b \div c = q, \quad b \bmod c = r, \quad$ where $b = q \cdot c + r$ and $0 \leq r < c$.

(Operators $\div$ and **mod** have the same precedence as   .) This means that

(15.91) $b = c \cdot (b \div c) + b \bmod c \qquad$ (for $c \neq 0$).

THE GREATEST COMMON DIVISOR

The *greatest common divisor* b **gcd** c of integers b and c that are not both zero is the greatest integer that divides both.

(15.92) $b \textbf{ gcd } c = (\uparrow d \mid d \mid b \wedge d \mid c : d) \qquad$ (for b, c not both 0)
        $0 \textbf{ gcd } 0 = 0$

The first line of (15.92) does not define 0 **gcd** 0; since all integers divide 0, 0 has no maximum divisor. We define 0 **gcd** 0 to be 0, so that **gcd** is a total function over $\mathbb{Z} \times \mathbb{Z}$. Infix operator **gcd** has the same precedence as multiplication —see the precedence table on the inside front cover.

Here are examples of gcd: $1 \textbf{ gcd } 5 = 1$, $0 \textbf{ gcd } 5 = 5$, $24 \textbf{ gcd } 30 = 6$, and $-24 \textbf{ gcd } 30 = -24 \textbf{ gcd } -30 = 24 \textbf{ gcd } -30 = 6$.

The greatest common divisor is used in reducing a fraction to lowest terms: to reduce p/q to lowest terms, divide p and q by their **gcd**. Thus,

$$(15.93) \quad \frac{24}{30} = \frac{24/6}{30/6} = \frac{4}{5} \quad \text{(recall } 30 \textbf{ gcd } 24 = 6 \text{).}$$

The *least common multiple* $b \textbf{ lcm } c$ of b and c is the smallest positive integer that is a multiple of both b and c:

$$(15.94) \quad b \textbf{ lcm } c = (\downarrow k:\mathbb{Z}^+ \mid b \mid k \wedge c \mid k : k) \quad \text{(for } b \neq 0 \text{ and } c \neq 0 \text{)}$$
$$b \textbf{ lcm } c = 0 \quad \text{(for } b = 0 \text{ or } c = 0 \text{)}$$

For example, $1 \textbf{ lcm } 6 = -1 \textbf{ lcm } 6 = 6$, $3 \textbf{ lcm } 9 = 9$, and $12 \textbf{ lcm } 18 = 36$. The least common multiple $b \textbf{ lcm } c$ is used when adding fractions with denominators b and c. For example,

$$(15.95) \quad \frac{5}{12} + \frac{5}{18} = \frac{15}{36} + \frac{10}{36} = \frac{25}{36} \quad \text{(recall } 12 \textbf{ lcm } 18 = 36 \text{).}$$

There is an obvious similarity in the definitions of **gcd** and **lcm**. We prove later, when we have the tools, that $b \cdot c = (b \textbf{ gcd } c) \cdot (b \textbf{ lcm } c)$.

For the moment, however, let us turn our attention to the greatest common divisor. We have the following properties of **gcd**.

Properties of gcd

(15.96) **Symmetry** : $b \textbf{ gcd } c = c \textbf{ gcd } b$

(15.97) **Associativity** : $(b \textbf{ gcd } c) \textbf{ gcd } d = b \textbf{ gcd } (c \textbf{ gcd } d)$

(15.98) $b \textbf{ gcd } b = abs.b$

(15.99) **Zero** : $1 \textbf{ gcd } b = 1$

(15.100) $0 \textbf{ gcd } b = abs.b$

(15.101) $b \textbf{ gcd } c = (abs.b) \textbf{ gcd } (abs.c)$

(15.102) $b \textbf{ gcd } c = b \textbf{ gcd } (b+c) = b \textbf{ gcd } (b-c)$

Properties of gcd (continued)

(15.103) $b = a \cdot c + d \;\Rightarrow\; b \gcd c = c \gcd d$

(15.104) **Distributivity** : $0 \le d \;\Rightarrow\; d \cdot (b \gcd c) = (d \cdot b) \gcd (d \cdot c)$

Property (15.101) indicates that we can reduce the problem of finding (or analyzing in some way) the **gcd** of two integers to the problem of finding the **gcd** of two natural numbers. So from now on, we restrict ourselves to the case that b and c are natural numbers.

Property (15.103) is particularly important, because it will be used several times later on. So be sure you understand it. It rests on the fact that, if the antecedent $b = a \cdot c + d$ holds, then any integer that divides b and c also divides c and d, and vice versa.

(15.105) **Definition.** Natural numbers b and c are *relatively prime*, denoted [3] by $b \perp c$, if their *gcd* is 1: $b \perp c \equiv b \gcd c = 1$.

For example, $4 \perp 33$ holds, since the only positive divisor of 4 and 33 is 1, but $4 \perp 34$ does not, since 2 divides both 4 and 34.

We now present an algorithm for finding the greatest common divisor of two positive integers $b > 0$ and $c > 0$. This algorithm is called *Euclid's algorithm*, in honor of Euclid, who presented it over 2,000 years ago (see Historical note 15.1).

$\{Q : 0 < b \;\wedge\; 0 < c\}$
$x, y := b, c;$
$\{\text{invariant } P :\; x \gcd y = b \gcd c \;\wedge\; 0 < x \;\wedge\; 0 < y\}$
$\{\text{bound function} :\; x \uparrow y\}$
do $x \ne y \rightarrow$ **if** $x > y \rightarrow x := x - y$
$\qquad\qquad\quad [\!]\; y > x \rightarrow y := y - x$
$\qquad\qquad$ **fi**
od
$\{R : x = y = b \gcd c\}$

We prove the correctness of this algorithm as follows.

- Initialization $x, y := b, c$ truthifies invariant P.

- Upon termination, $x = y$; together with the invariant, this yields $x \gcd x = b \gcd c$, and property (15.98) of **gcd** gives result R.

---

[3] The notation $b \perp c$ is not standard in mathematics. Graham *et al.* [16] call for its introduction, saying that its use makes many formulas clearer.

HISTORICAL NOTE 15.1. EUCLID (ABOUT 300 B.C.)

Little is known about the life of Euclid. He did teach in his own school in Alexandria at the time of Ptolemy I. Euclid is best known for writing *Elements*, a text consisting of 13 books that taught geometry and the theory of numbers. *Elements* incorporates many discoveries made by other people, and Euclid is viewed mainly as a compilator and expositor —but a great one. In fact, *Elements* is one of the most successful scientific books, ever: for over 2,000 years, geometry was learned only from Euclid's *Elements*.

Elements develops geometry through a series of definitions, explanations, axioms, and theorems and their proofs. It is the first book to follow this deductive method, recognized as the basic method in mathematics ever since. Since the development of algebra, which was unknown at the time of Euclid, the formulation and theory of geometry has changed radically, and Euclid's postulates are no longer widely used. Well, 2,000 years is enough for any book to be a best seller. We will be happy if this text lasts 20!

- Each iteration decreases the bound function, and the bound function is bounded below by 0. Hence, the algorithm terminates.

- Each iteration maintains loop invariant P. To show this, we should show that in each case $x > y$ and $x < y$, execution of the repetend maintains P. We show only the case $x > y$, because the other is similar. In the case $x > y$, we have to prove

$$\{P \wedge x > y\} \ x := x - y \ \{P\} \quad .$$

According to Assignment introduction (10.2) on page 182, we can prove this by proving

$$P \wedge x > y \Rightarrow P[x := x - y] \quad .$$

We assume the antecedent and prove the consequent.

$$
\begin{aligned}
&P[x := x - y] \\
=\quad &\langle \text{Definition of } P \,; \text{Textual substitution} \rangle \\
&(x - y) \ \textbf{gcd} \ y = b \ \textbf{gcd} \ c \ \wedge \ 0 < x - y \ \wedge \ 0 < y \\
=\quad &\langle \text{Assumption } x > y \,; \text{Conjunct } 0 < y \text{ of } P \rangle \\
&(x - y) \ \textbf{gcd} \ y = b \ \textbf{gcd} \ c \ \wedge \ true \ \wedge \ true \\
=\quad &\langle \text{Identity of } \wedge \ (3.39) \rangle \\
&(x - y) \ \textbf{gcd} \ y = b \ \textbf{gcd} \ c \\
=\quad &\langle \ x = 1 \cdot y + (x - y) \,, \text{ so, by } (15.103), \\
&\qquad x \ \textbf{gcd} \ y = y \ \textbf{gcd} \ (x - y) \ \rangle \\
&x \ \textbf{gcd} \ y = b \ \textbf{gcd} \ c \quad \text{—First conjunct in assumption } P
\end{aligned}
$$

At each iteration, Euclid's algorithm subtracts the smaller of the two values from the larger. This algorithm is slower than need be, and we now develop a faster one. Consider equation (15.91):

$$b = c \cdot (b \div c) + b \bmod c \quad \text{(for } c \neq 0\text{)}.$$

Theorem (15.103) instantiated with $a, d := b \div c, b \bmod c$ is

(15.106) $b = c \cdot (b \div c) + b \bmod c \Rightarrow b \gcd c = c \gcd (b \bmod c)$.

Since (15.91) is the antecedent of (15.106), we conclude that the consequent $b \gcd c = c \gcd (b \bmod c)$ is valid as well. Therefore, we can write the following, which could be viewed as an inductive definition of $\gcd$ —note that $c > b \bmod c$, so that the recursion is suitably defined.

(15.107) $b \gcd 0 = b$

   $b \gcd c = c \gcd (b \bmod c) \quad$ for $c > 0$

This inductive definition could be viewed as a recursive algorithm, or we can write the following iterative version.

$\{Q : 0 \leq b \ \wedge \ 0 \leq c\}$
$x, y := b, c;$
$\{\text{invariant } P : \ x \gcd y = b \gcd c \ \wedge \ 0 \leq x \ \wedge \ 0 \leq y\}$
$\{\text{bound function} : \ y\}$
$\textbf{do } 0 \neq y \rightarrow x, y := y, x \bmod y \textbf{ od}$
$\{R : x = b \gcd c\}$

We prove the correctness of this algorithm as follows.

- Initialization $x, y := b, c$ truthifies invariant P.

- Upon termination, $y = 0$; together with the invariant, this yields $x \gcd 0 = b \gcd c$, and Identity of $\gcd$ (15.100) gives result R.

- Each iteration decreases bound function y (since $x \bmod y < y$), and the bound function is bounded below by 0. Hence, the algorithm terminates.

- Each iteration maintains P —this can be proved using the property $b \gcd c = c \gcd (b \bmod c)$.

Surprisingly, this algorithm takes the most time when b and c are consecutive Fibonacci numbers! (Another interesting connection between gcd and π is discussed in Historical note 15.2.) The number of iterations of this loop has been shown to be bounded above by [4] $\lceil 4.8 \log_{10}(b \uparrow c) - .32 \rceil$.

---

[4] The *ceiling* of real number x, written $\lceil x \rceil$, is the smallest integer i that is at least x. For example, $\lceil 2.9 \rceil = 3$, $\lceil 3 \rceil = 3$, and $\lceil 3.1 \rceil = 4$.

Similarly, the *floor* of x, written $\lfloor x \rfloor$, is the largest integer that is at most x. Thus, $\lfloor 2.9 \rfloor = 2$, $\lfloor 3 \rfloor = 3$, and $\lfloor 3.1 \rfloor = 3$.

HISTORICAL NOTE 15.2. PERCENTAGE OF RELATIVELY PRIME PAIRS

If you choose two positive integers at random, the chance that they will be relatively prime is $6/\pi^2$ —a startling relation between π, the ratio of the circumference of the circle to its diameter, and primes. This fact was proved by E. Cesàro in 1881. The proof requires stating precisely what "random" means and also requires some mathematics that is beyond the scope of this text. However, we can give the idea here (see [26] for a full proof).

Let p be the probability that $b \gcd c = 1$. For any positive integer d, consider the probability that $b \gcd c = d$. This happens when b is a multiple of d, c is a multiple of d, and $(b/d) \gcd (c/d) = 1$.

The probability that d divides b is $1/d$. Therefore, the probability that $b \gcd c = d$ is $(1/d)\cdot(1/d)\cdot p$, i.e. p/d^2. Summing these probabilities over all possible values of d yields

$$1 = (\Sigma d \mid 1 \leq d : p/d^2) = p\cdot(1 + \frac{1}{4} + \frac{1}{9} + \frac{1}{16} + \cdots) .$$

The summation is known to have the value $\pi^2/6$, so $p = 6/\pi^2$.

Hence, this algorithm is logarithmic in the size of b and c. The analysis of the running time falls outside the scope of this text —see pp. 316–33 of [26].

The following theorem will be useful later; it says that witnesses x and y exist that satisfy the equation $x\cdot b + y\cdot c = b \gcd c$.

(15.108) $(\exists x, y \mid : x\cdot b + y\cdot c = b \gcd c)$ (for all $b, c : \mathbb{N}$)

Proof. The proof of (15.108) is by induction on c. We prove $(\forall c \mid : P.c)$ where $P.c$ is $(\forall b \mid : (\exists x, y \mid : x\cdot b + y\cdot c = b \gcd c))$. In each case, we exhibit the necessary x and y.

Base case $c = 0$. Choose $x = 1$ and $y = 0$.

Inductive case $c > 0$. We assume inductive hypotheses $P.i$ for $0 \leq i < c$ and prove $P.c$. That is, for arbitrary b, we prove

(15.109) $(\exists x, y \mid : x\cdot b + y\cdot c = b \gcd c)$.

Since $0 \leq b \bmod c < c$ is valid (see (15.90) and (15.91)), the inductive hypothesis indicates that there exist witnesses $\hat{x}, \hat{y}$ that satisfy

(15.110) $\hat{x}\cdot c + \hat{y}\cdot(b \bmod c) = c \gcd (b \bmod c)$.

We play with the RHS of the body of (15.109) until we get it into a shape that allows us to determine what to choose for x and y.

$b \gcd c$
$= \quad \langle(15.107)\rangle$
$c \gcd (b \bmod c)$

$$
\begin{aligned}
= \quad &\langle(15.110)\rangle \\
&\hat{x} \cdot c + \hat{y} \cdot (b \bmod c) \\
= \quad &\langle(15.91) \text{ yields } b \bmod c = b - (b \div c) \cdot c \rangle \\
&\hat{x} \cdot c + \hat{y} \cdot (b - (b \div c) \cdot c) \\
= \quad &\langle\text{Arithmetic}\rangle \\
&\hat{x} \cdot c + \hat{y} \cdot b - \hat{y} \cdot (b \div c) \cdot c \\
= \quad &\langle\text{Arithmetic}\rangle \\
&\hat{y} \cdot b + (\hat{x} - \hat{y} \cdot (b \div c)) \cdot c
\end{aligned}
$$

Comparing the last expression with the LHS of (15.109), we see that $x = \hat{y}$ and $y = \hat{x} - \hat{y} \cdot (b \div c)$ are the witnesses that substantiate (15.109). $\square$

It is clear that any divisor of $b \gcd c$ divides b and c. Expression (15.108) tells us also that any divisor of b and c is also a divisor of $b \gcd c$, and we have

$$(15.111) \quad k \mid b \wedge k \mid c \;\equiv\; k \mid (b \gcd c) \quad .$$

PRIME NUMBERS

A non-zero integer $p > 1$ is *prime* if the only positive integers that divide p are 1 and p; otherwise, p is *composite*.

Throughout this section, the identifier p, sometimes subscripted (e.g. p_2), denotes a prime number. The first eleven prime numbers are

$$2, 3, 5, 7, 11, 13, 17, 19, 23, 29, 31 \quad .$$

An important consequence can be drawn from the existence of x and y satisfying expression (15.108) above.

(15.112) **Theorem.** For p a prime, $p \mid b \cdot c \Rightarrow p \mid b \vee p \mid c$.

Proof. The theorem can be rewritten as $p \mid b \cdot c \wedge \neg(p \mid b) \Rightarrow p \mid c$. We assume the antecedent and prove the consequent. Since p is a prime, from the assumption $\neg(p \mid b)$ we conclude that the only common divisors of p and b are ± 1, so $b \gcd p = 1$. Therefore, by (15.108), there exist witnesses x and y that satisfy

$$
\begin{aligned}
&x \cdot b + y \cdot p = 1 \\
= \quad &\langle\text{Multiply both sides by } c; \text{ Multiplicative identity (15.4)}\rangle \\
&x \cdot b \cdot c + y \cdot p \cdot c = c \\
\Rightarrow \quad &\langle\text{Assumption } p \mid b \cdot c; \; p \mid y \cdot p \cdot c \text{ by Def of } \mid \rangle \\
&p \mid c
\end{aligned}
$$

$\square$

The argument used to prove Theorem (15.112) can also be used to prove the following generalization concerning relatively prime numbers.

HISTORICAL NOTE 15.3. THE FASCINATING PRIME NUMBERS

The basic theory of primes and composites was known to Euclid, who proved (15.112), as well as the existence of an infinite number of primes. Eratosthenes, some 50 years later, gave his algorithm for computing all primes, the *Sieve of Eratosthenes*. It goes like this. Write down the sequence of odd positive integers greater than 1. Then cross out every third integer, every fifth integer, and so on. At each step, choose the first number that has not yet been crossed out as the next prime, and cross out all multiples of it.

Just before 1900, it was proved that for each n, the number of primes less than n is approximately $n/(ln\, n)$ for n large, so there are approximately 72,382 primes less than 1,000,000 (we now know there are exactly 78,498). It is also known that the gaps between successive primes can be arbitrarily large, but relatively little is known about the behavior of these gaps.

In the 17th century, Father Marin Mersenne studied integers of the form $2^p - 1$ (for p a prime). They are now called *Mersenne numbers*; some are prime and some are not. This study was continued by many others, without computers —imagine trying to find the factors of a number like $2^{127} - 1$, which has approximately 40 decimal digits, without a computer! Computers made primality testing easier. In 1952, a computer found that the Mersenne numbers were primes for $p = 521$, 607, 1,279, 2,203, and 2,281. However, that was just the start. In 1984, the largest known Mersenne prime was $2^{216091} - 1$. At 75 characters per line and 60 lines on a page, its 65,050-digit decimal representation would take over 7 double-sided pages. Larger primes have been found since then.

To the layman, a lot of math (like primality testing and factoring large numbers) may seem a frivolous waste of time. However, this research often pays off unexpectedly years later. Factoring and primality testing have become important because of the need to make electronic communications secure (see Historical note 14.2). In 1978, a cryptosystem was developed based on the fact that it is easy to multiply two large numbers together but very difficult to factor the result into primes. Even for computers, if large enough integers are chosen, the task is intractable. So, what used to be an esoteric playground for mathematicians has become applicable research.

(15.113) **Theorem.** $b \perp c \ \wedge \ c\,|\,(b \cdot d) \ \Rightarrow \ c\,|\,d$.

Prime numbers are important because they are the basic building blocks for the positive integers. This fact is embodied in the

(15.114) **Fundamental Theorem of Arithmetic.** Every positive integer n can be written in a unique way as a product of primes:

$$n = p_0 \cdot \ldots \cdot p_{m-1} \qquad \text{where } p_0 \leq \cdots \leq p_{m-1}.$$

Proof. The proof is by induction on n.

Base case: 1 is the product of zero primes: $1 = (\Pi i \mid false : p_i)$.

Inductive case. We assume, as the inductive hypothesis, that the theorem holds for all positive integers less than n, where $n > 1$, and we prove the theorem for n. Two cases arise: n is prime and n is composite.

Case n is prime. Then n is a product of itself.

Case n is composite. Then $n = b \cdot c$ for some positive integers b and c, which are both less than n. By the inductive hypothesis, b and c can be written as products of primes: say $b = p_0 \cdot \ldots \cdot p_{h-1}$ and $c = q_0 \cdot \ldots \cdot q_{k-1}$. Therefore, $n = b \cdot c = p_0 \cdot \ldots \cdot p_{h-1} \cdot q_0 \cdot \ldots \cdot q_{k-1}$. Due to symmetry and associativity of multiplication, the primes in the RHS can be ordered to the required factorization of n.

The proof that the factorization of composite n is unique is given in Lemma (15.115).   □

(15.115) **Lemma.** The factorization of primes is unique (up to reordering of the factors).

Proof. The proof is by induction; we prove $(\forall n \mid 1 < n : P.n)$, where inductive hypothesis $P.n$ is "the prime factorization of n is unique".

Base case. The product $(\Pi i \mid false : p_i) = 1$ is unique.

Inductive case. We assume the theorem holds for positive integers less than n, where $n > 1$, and prove it *true* for n. Suppose

$$n = p_0 \cdot \ldots \cdot p_{m-1} = q_0 \cdot \ldots \cdot q_{h-1}$$
$$\text{where } p_0 \leq \cdots \leq p_{m-1} \text{ and } q_0 \leq \cdots \leq q_{h-1}$$

and the p_i and q_i are primes. We prove below that $p_0 = q_0$. Then, by Cancellation (15.7), $p_1 \cdot \ldots \cdot p_{m-1} = q_1 \cdot \ldots \cdot q_{h-1}$ and, by the inductive hypothesis, the representation of this integer is unique, so $m = h$ and $p_1 = q_1, \ldots, p_{h-1} = q_{h-1}$.

We now prove $p_0 = q_0$. We have,

$$
\begin{aligned}
& p_0 \cdot \ldots \cdot p_{m-1} = q_0 \cdot \ldots \cdot q_{h-1} \quad \text{—the assumption} \\
\Rightarrow \quad & \langle \text{Def. (15.77), with } c, b, d := p_0, q_0 \ldots q_{h-1}, p_1 \ldots p_{m-1} \rangle \\
& p_0 \mid q_0 \cdot \ldots \cdot q_{h-1} \\
= \quad & \langle (15.112) - p_0 \text{ is prime} \rangle \\
& p_0 \mid q_k \quad (\text{for some } k) \\
= \quad & \langle \text{Assumption that } p_0 \text{ and } q_k \text{ are primes} \rangle \\
& p_0 = q_k \quad (\text{for some } k) \\
= \quad & \langle q_0 \leq q_k \rangle \\
& p_0 \geq q_0
\end{aligned}
$$

By a similar argument, $p_0 \leq q_0$, so, by antisymmetry of $\leq$, $p_0 = q_0$. $\Box$

We now prove that there are an infinite number of primes, using the same idea that Euclid used in his proof long long ago.

(15.116) **Theorem.** There are an infinite number of primes.

Proof. For any natural number k, we give an algorithm to construct the first $k + 1$ primes, in order. The first prime is 2. Now assume that the smallest k primes $p_0, p_1, \ldots, p_{k-1}$, in order, have been constructed. We show how to construct prime p_k. Consider the integer

(15.117) $M = p_0 \cdot p_1 \cdot \ldots \cdot p_{k-1} + 1$.

By (15.86) and the fact that each p_i divides $M - 1$, none of the p_i divides M. Hence, there is a prime bigger than p_{k-1} that divides M (it could be M itself). Choose for p_k the smallest prime in $(p_{k-1} + 1)..M$. $\Box$

CONGRUENCES

In the U.S., we use a 12-hour clock, so that after 12 (noon or midnight) comes 1 again. Thus, in describing hours, we throw away multiples of 12. If we began counting hours on the first day of the year, we would equate the hours 2, 14 (which is 2PM), 26 (2AM the next day), 38 (2PM the next day), etc. We call two integers *congruent modulo* 12 if they differ by an integral multiple of 12. Many Europeans use a 24-hour clock: they count hours modulo 24.

(15.118) **Definition.** Integers b and c are *congruent modulo* m, written $b \overset{m}{=} c$, iff $m \mid (c - b)$. Relation $\overset{m}{=}$ is called *congruence*,[5] and m is called the *modulus* of the congruence. We read $b \overset{m}{=} c$ as "b is congruent mod m to c" or "b and c are congruent mod m".

A first property to note is that $\overset{m}{=}$ is an equivalence relation —it is reflexive, symmetric, and transitive (the proof is left to the reader.) In addition, operator $\overset{m}{=}$ satisfies a number of properties that are similar to those of $=$.

---

[5] The standard notation for $b \overset{m}{=} c$ is $b \equiv c \pmod{m}$. We do not use this standard because $\equiv$ already plays a role in our propositional calculus.

Congruence theorems

(15.119) **Alternative definition of $\stackrel{m}{=}$:**
$$b \stackrel{m}{=} c \;\equiv\; b \bmod m = c \bmod m$$

(15.120) **Addition:** $b \stackrel{m}{=} c \;\equiv\; b+d \stackrel{m}{=} c+d$

(15.121) **Negation:** $b \stackrel{m}{=} c \;\equiv\; -b \stackrel{m}{=} -c$

(15.122) **Multiplication:** $b \stackrel{m}{=} c \;\Rightarrow\; b\cdot d \stackrel{m}{=} c\cdot d$

(15.123) **Powers:** $b \stackrel{m}{=} c \;\Rightarrow\; b^n \stackrel{m}{=} c^n$ (for $n \geq 0$)

(15.124) **Cancellation:** $d \perp m \;\Rightarrow\; (b\cdot d \stackrel{m}{=} c\cdot d \;\equiv\; b \stackrel{m}{=} c)$

(15.125) **Cancellation:** $b\cdot d \stackrel{m\cdot d}{=} c\cdot d \;\equiv\; b \stackrel{m}{=} c$ (for $d \geq 0$)

(15.126) $d \perp m \;\Rightarrow\; (\exists x \,|: d\cdot x \stackrel{m}{=} b)$

(15.127) $d \perp m \;\Rightarrow\; (d\cdot x \stackrel{m}{=} b \wedge d\cdot y \stackrel{m}{=} b \;\Rightarrow\; x \stackrel{m}{=} y)$

Theorem (15.119) provides an alternative definition of congruence. The rest of the theorems show that $\stackrel{m}{=}$ enjoys many, but not all, of the properties of $=$. For example, Cancellation (15.7) does not hold in full generality —compare (15.7) with (15.122). The implication in (15.122) does not go in the other direction, as the following counterexample shows: $4\cdot 2 \stackrel{6}{=} 1\cdot 2$ does not imply $4 \stackrel{6}{=} 1$. The cancellation of the 2 does not work because 2 is a factor of modulus 6. Theorem (15.124) provides a weaker cancellation law, while theorem (15.125) indicates we can cancel if we cancel in the modulus as well.

Theorem (15.126) gives conditions under which there is a solution x to the equation $d\cdot x \stackrel{m}{=} b$, while (15.127) says that all solutions to it are congruent mod m.

Many other theorems hold concerning congruences, and congruences have many applications; we are only providing a brief overview of the concept. We leave the proofs of all but one of these theorems to the reader. Here is a proof of theorem (15.119), which is based on the fact that dividing b by m leaves a unique remainder.

We prove (15.119), by mutual implication.

LHS $\Rightarrow$ RHS. The LHS is equivalent to the fact that there is a witness d that satisfies $d\cdot m = c - b$. We assume the LHS and prove the antecedent.

$$c \bmod m$$
$$= \quad \langle (15.91) \rangle$$
$$c - m \cdot (c \div m)$$
$$= \quad \langle \text{Assumption } d \cdot m = c - b \rangle$$
$$d \cdot m + b - m \cdot ((d \cdot m + b) \div m)$$
$$= \quad \langle \text{Arithmetic} \rangle$$
$$b - m \cdot (b \div m)$$
$$= \quad \langle (15.91) \rangle$$
$$b \bmod m$$

RHS $\Rightarrow$ LHS.

$$b \bmod m = c \bmod m$$
$$= \quad \langle (15.91), \text{ twice} \rangle$$
$$b - (b \div m) \cdot m = c - (c \div m) \cdot m$$
$$= \quad \langle \text{Arithmetic} \rangle$$
$$b - c = (b \div m - c \div m) \cdot m$$
$$\Rightarrow \quad \langle \text{Definition of } | \quad (15.77) \rangle$$
$$m \mid (b - c)$$
$$= \quad \langle \text{Definition } (15.118) \rangle$$
$$b \overset{m}{=} c$$

15.5 Common representations of natural numbers

There are many ways to represent the natural numbers $0, 1, 2, \ldots$. Three age-old representations are depicted in Table 15.1.

TABLE 15.1. Primitive Representations of the Natural Numbers

integer	tally	encoded tally	roman number
0			
1	|	|	I
2	||	||	II
3	|||	|||	III
4	||||	||||	IV
5	|||||	ﬀﬀ	V
6	||||||	ﬀﬀ |	VI
7	|||||||	ﬀﬀ ||	VII
8	||||||||	ﬀﬀ |||	VIII
9	|||||||||	ﬀﬀ ||||	IX
10	||||||||||	ﬀﬀ ﬀﬀ	X
11	|||||||||||	ﬀﬀ ﬀﬀ |	XI

Column 2 represents the integer n by n strokes $|$. Column 3 uses an improvement that allows one to see more easily how many strokes there are. Column 4 contains the roman numerals. Note that there is no representation of 0 in the roman-numeral system. The tally systems, on the other hand, have a representation for 0: the absence of strokes.

Actually, column 1 itself uses a representation of the integers: the *decimal* representation. This representation is so ubiquitous that we tend to think of this column *as being* the integers. Nevertheless, it is just one among many representations. In the decimal representation $d_{k-1} \ldots d_1 d_0$ of n, the d_i are called *digits*. The d_i satisfy the following properties.

$$0 \le d_i < 10 \qquad \text{for } 0 \le i < k$$
$$n = (\Sigma \, i \mid 0 \le i < k : d_i \cdot 10^i)$$

Digit d_0 is the *least-significant digit* and d_{k-1} is the *most-significant digit*. Note that the natural number 0 can be represented by any sequence of 0's, including the empty sequence (i.e. with $k = 0$).

The decimal system uses ten different symbols: $0, 1, 2, 3, 4, 5, 6, 7, 8, 9$. Here, 10 is called the base of the number system, so the decimal system is the base-10 system. For any integer b, $2 \le b$, we can use the base b system to represent the natural numbers. The first 19 natural numbers in several different bases are given in Table 15.2.

The binary system (base 2) is used heavily in computers because it is easy to use an electronic signal to represent a binary unit, or *bit*, 0 or 1 — see Sec. 5.2. The octal system (base 8) [6] and hexadecimal system (base 16) are also used because integers have shorter representations in them and the translation between them and binary is trivial. For example, to translate from octal to binary, just replace each octal unit by its binary equivalent —e.g. $73_8 = 111\,011_2$. Note how we indicate the base using a subscript.

For an integer b, $2 \le b$, the base b representation of a natural number n is a sequence of "b-units" $d_{k-1} \ldots d_1 d_0$ where the b-units d_i satisfy

(15.128) $0 \le d_i < b \qquad \text{for } 0 \le i < k$
$$n = (\Sigma \, i \mid 0 \le i < k : d_i \cdot b^i) \quad .$$

Given the base b representation of n, it is easy to compute n; simply calculate the sum given in (15.128).

We now present an algorithm that, given $n \ge 0$ and a base b, produces the base b representation of n. Thus, the algorithm stores values in integer variable k and array $d[0..k-1]$ to truthify (15.128).

---

[6] Why send a Christmas card on Halloween? Because DEC 25 = OCT 31.

$$k, x := 0, n;$$
$$\textbf{do } x > 0 \rightarrow x, d[k] := x \div b, x \textbf{ mod } b;$$
$$k := k + 1$$
od

The invariant of the loop of the algorithm is

$$P : 0 \leq k \ \wedge \ 0 \leq x \ \wedge$$
$$(\forall i \mid 0 \leq i < k : 0 \leq d[i] < b) \ \wedge$$
$$n = x \cdot b^k + (\Sigma\, i \mid 0 \leq i < k : d[i] \cdot b^i)$$

It is easy to see that P is truthified by the initialization, that upon termination the result holds, and that the loop terminates (each iteration decreases x and x is bounded below by 0). It is also easy to see that the first three conjuncts are maintained by the repetend.

We now prove that the last conjunct $P4$ (say) of P is maintained by the repetend. The key to this proof is theorem (15.91), which we rewrite here with the variables we will be needing:

$$(15.129) \quad x = (x \div b) \cdot b + x \textbf{ mod } b \qquad (\text{for } b \neq 0).$$

TABLE 15.2. Natural Numbers in Different Bases

binary (base 2)	ternary (base 3)	octal (base 8)	decimal (base 10)	hexadecimal (base 16)
0	0	0	0	0
1	1	1	1	1
10	2	2	2	2
11	10	3	3	3
100	11	4	4	4
101	12	5	5	5
110	20	6	6	6
111	21	7	7	7
1000	22	10	8	8
1001	100	11	9	9
1010	101	12	10	A
1011	102	13	11	B
1100	110	14	12	C
1101	111	15	13	D
1110	112	16	14	E
1111	120	17	15	F
10000	121	20	16	10
10001	122	21	17	11
10010	200	22	18	12

To see that $P4$ is maintained by the repetend, annotate the repetend with assertions as follows.

$$\{P\}$$
$$\{A1: \ n = ((x \div b) \cdot b + x \bmod b) \cdot b^k +$$
$$(\Sigma \, i \mid 0 \leq i < k : d[i] \cdot b^i)\}$$
$$x, d[k] := x \div b, x \bmod b;$$
$$\{A2: \ n = (x \cdot b + d[k]) \cdot b^k + (\Sigma \, i \mid 0 \leq i < k : d[i] \cdot b^i)\}$$
$$\{P4[k := k + 1]\}$$
$$k := k + 1$$
$$\{P4\}$$

Implication $P \Rightarrow A1$ follows from (15.129). The Hoare triple $\{A1\}$ $x, d[k] := x \div b, x \bmod b \ \{A2\}$ is valid because $A1$ is $A2$ with x and $d[k]$ replaced by $x \div b$ and $x \bmod b$. $A2 \Rightarrow P4[k := k + 1]$ is shown below. And $\{P4[k := k + 1]\} \ k := k + 1 \ \{P4\}$ follows by definition of the assignment. Here, now, is the proof of $A2 \Rightarrow P4[k := k + 1]$.

$$P4[k := k + 1]$$
$$= \quad \langle \text{Definition of } P4 \text{ and textual substitution} \rangle$$
$$n = x \cdot b^{k+1} + (\Sigma \, i \mid 0 \leq i < k + 1 : d[i] \cdot b^i)$$
$$= \quad \langle \text{Split off term (8.23)} \rangle$$
$$n = x \cdot b^{k+1} + d[k] \cdot b^k + (\Sigma \, i \mid 0 \leq i < k : d[i] \cdot b^i)$$
$$= \quad \langle \text{Factor out } b^k \rangle$$
$$n = (x \cdot b + d[k]) \cdot b^k + (\Sigma \, i \mid 0 \leq i < k : d[i] \cdot b^i)$$

A REPRESENTATION OF THE POSITIVE INTEGERS

The Fundamental Theorem of Arithmetic can be restated to give another representation of the positive integers: Any positive integer n can be written uniquely in the form

(15.130) $\quad n = (\Pi p \mid p \text{ a prime} : p^{n_p})$ $\quad$ (each $n_p \geq 0$).

For example, $126 = 2 \cdot 3^2 \cdot 7$, so for 126 we have

$$126 = 2^1 \cdot 3^2 \cdot 5^0 \cdot 7^1 \cdot 11^0 \cdot 13^0 \cdot 17^0 \cdot \ldots \ .$$

The RHS of (15.130) is a product of infinitely many primes, but for any given n, all but a finite number of exponents are 0, so the corresponding factors are 1. Therefore, we can view it as a finite product instead of an infinite product. Suppose we list the primes by size, $p_0, p_1, p_2, \ldots$, with p_0 being the smallest. Then, for any positive integer n, we can think of the sequence of exponents in the RHS of (15.130) as a *representation* $\overline{n}$ of n. For example, we have $\overline{126} = \langle 1, 2, 0, 1, 0, 0, \ldots \rangle$. This gives us a different number system for positive integers. It is easy to see that multiplying two positive integers is done in this number system by adding their representations —where addition for these representations is done component-wise. For example,

$$\overline{126 \cdot 2} = \langle 2, 2, 0, 1, 0, 0, \ldots \rangle = \langle 1, 2, 0, 1, 0, 0, \ldots \rangle + \langle 1, 0, 0, 0, \ldots \rangle.$$

Let $\bar{b}$ denote the representation of b, and let $\bar{b}_p$ denote the exponent of p in the unique factorization of b. Then the first four theorems below follow directly from this representation.

Theorems on representation of positive integers

(15.131) $(\overline{b \cdot c})_p \equiv \bar{b}_p + \bar{c}_p$ (for all primes p)

(15.132) $b \mid c \equiv (\forall p \mid : \bar{b}_p \leq \bar{c}_p)$

(15.133) $(\overline{b \operatorname{\mathbf{gcd}} c})_p \equiv \bar{b}_p \downarrow \bar{c}_p$ (for all primes p)

(15.134) $(\overline{b \operatorname{\mathbf{lcm}} c})_p \equiv \bar{b}_p \uparrow \bar{c}_p$ (for all primes p)

(15.135) $b \cdot c = (b \operatorname{\mathbf{gcd}} c) \cdot (b \operatorname{\mathbf{lcm}} c)$ (for natural numbers b, c)

Theorem (15.135) provides the relationship between **gcd** and **lcm**. We prove it as follows. For $b = 0$ or $c = 0$, (15.135) follows from $b \operatorname{\mathbf{lcm}} 0 = 0$. To prove (15.135) in the case $b > 0$ and $c > 0$, we have to show that the corresponding components of the representations of $b \cdot c$ and $(b \operatorname{\mathbf{gcd}} c) \cdot (b \operatorname{\mathbf{lcm}} c)$ are equal. For any prime p, we have

$$(\overline{b \cdot c})_p$$
$$= \quad \langle (15.131) \rangle$$
$$\bar{b}_p + \bar{c}_p$$
$$= \quad \langle \text{One of } \bar{b}_p, \bar{c}_p \text{ is the min; the other the max} \rangle$$
$$(\bar{b}_p \uparrow \bar{c}_p) + (\bar{b}_p \downarrow \bar{c}_p)$$
$$= \quad \langle (15.133); (15.134) \rangle$$
$$(\overline{b \operatorname{\mathbf{gcd}} c})_p + (\overline{b \operatorname{\mathbf{lcm}} c})_p$$
$$= \quad \langle (15.131) \rangle$$
$$((b \operatorname{\mathbf{gcd}} c) \cdot (b \operatorname{\mathbf{lcm}} c))_p$$

Exercises for Chapter 15

15.1 Prove that the set of numbers $a + b \cdot \sqrt{5}$ for a, b in $\mathbb{Z}$ form an integral domain (except for Cancellation (15.7), which holds but is harder to prove). Assume that the reals and integers are integral domains.

15.2 Prove that if (i) 0 is a left identity of $+$ and (ii) $+$ is symmetric, then 0 is a right identity of $+$.

15.3 Prove that if (i) 1 is a left identity of $\cdot$ and (ii) $\cdot$ is symmetric, then 1 is a right identity of $\cdot$.

15.4 Prove that Right distributivity (15.5), $(b+c)\cdot a = b\cdot a + c\cdot a$, follows from associativity and symmetry of $+$ and $\cdot$ and Left distributivity (15.5), $a\cdot(b+c) = a\cdot b + a\cdot c$.

15.5 Prove Unique identity theorems (15.10).

15.6 Prove theorem (15.11), $a\cdot b = 0 \equiv a = 0 \lor b = 0$. Hint: Use mutual implication. Because of the disjunction in the RHS, both of the proofs may require a case analysis.

15.7 Prove the following theorems (a, b, c, d are arbitrary elements of D).
   (a) $(a+b)\cdot(c+d) = ac + bc + ad + bd$
   (b) $a\cdot(b+c)\cdot d = a\cdot b\cdot d + a\cdot c\cdot d$

15.8 Prove theorem (15.15), $x + a = 0 \equiv x = -a$.

15.9 Prove theorem (15.16), $-a = -b \equiv a = b$.

15.10 Prove theorem (15.17), $-(-a) = a$.

15.11 Prove theorem (15.18), $-0 = 0$.

15.12 Prove theorem (15.19), $-(a+b) = (-a) + (-b)$.

15.13 Prove theorem (15.20), $-a = (-1)\cdot a$.

15.14 Prove theorem (15.21), $(-a)\cdot b = a\cdot(-b)$.

15.15 Prove theorem (15.22), $a\cdot(-b) = -(a\cdot b)$.

15.16 Prove theorem (15.24), $a - 0 = a$.

15.17 Prove theorem (15.25), $(a-b) + (c-d) = (a+c) - (b+d)$.

15.18 Prove theorem (15.26), $(a-b) - (c-d) = (a+d) - (b+c)$.

15.19 Prove theorem (15.27), $(a-b)\cdot(c-d) = (a\cdot c + b\cdot d) - (a\cdot d + b\cdot c)$.

15.20 Prove theorem (15.28), $a - b = c - d \equiv a + d = b + c$.

15.21 Prove theorem (15.29), $(a-b)\cdot c = a\cdot c - b\cdot c$.

15.22 Prove $(b-a) + (c-b) = c - a$.

15.23 Use theorems (15.29) and earlier to prove $(-1)\cdot(-1) = 1$.

15.24 Let D consist only of 0 and 1, let multiplication $\cdot$ be defined as usual on this set, and let addition be defined as usual except that $1 + 1 = 0$. Prove that D is an integral domain.

15.25 Let D contain only 0, let $1 = 0$, and let $0 + 0 = 0\cdot 0 = 0$. Is this an integral domain? If not, why not?

Exercises on ordered domains

In these exercises, you may use the hint "Arithmetic" for relations involving addition and subtraction of elements of an integral domain.

15.26 Prove (15.35), $pos.a \Rightarrow (pos.b \equiv pos(a \cdot b))$.

15.27 Prove Positive elements (15.40), $pos.b \equiv 0 < b$.

15.28 Prove Transitivity (15.41b).

15.29 Prove Transitivity (15.41c).

15.30 Prove Transitivity (15.41d).

15.31 Prove Monotonicity (15.42), $a < b \equiv a + d < b + d$.

15.32 Prove Monotonicity (15.43), $0 < d \Rightarrow (a < b \equiv a \cdot d < b \cdot d)$.

15.33 Prove Trichotomy (15.44), $(a < b \equiv a = b \equiv a > b) \land \neg(a < b \land a = b \land a > b)$.

15.34 Prove Antisymmetry (15.45), $a \leq b \land b \leq a \equiv a = b$.

15.35 Prove Reflexivity (15.46), $a \leq a$.

15.36 Prove theorem (15.47), $a = b \equiv (\forall z : D \mid : z \leq a \equiv z \leq b)$. Use mutual implication. The proof of LHS $\Rightarrow$ RHS can be done by starting with the antecedent, using reflexivity of $\land$, and then instantiating twice and simplifying.

15.37 Prove the following additional properties of the arithmetic relations on an ordered domain (for arbitrary b, c, d in the ordered domain).
  (a) $b - c < b - d \equiv c > d$
  (b) $b < 0 \Rightarrow (b \cdot c > b \cdot d \equiv c < d)$
  (c) $0 < d \land b \cdot d < c \cdot d \Rightarrow b < c$
  (d) $d + d + d = 0 \Rightarrow d = 0$
  (e) $b < c \Rightarrow b \cdot b \cdot b < c \cdot c \cdot c$

15.38 Show that $a \cdot a - a \cdot b + b \cdot b \geq 0$ for D an ordered domain.

15.39 Prove theorem (8.24), $b \leq c \leq d \Rightarrow (b \leq i < d \equiv b \leq i < c \lor c \leq i < d)$, on page 152.

Exercises on minimum and maximum

In these exercises, you may use the hint "Arithmetic" for relations involving addition, subtraction, and multiplication of elements of an integral domain.

15.40 Prove Associativity of $\downarrow$ (15.55), $(x \downarrow y) \downarrow z = x \downarrow (y \downarrow z)$.

15.41 Prove Idempotency of $\downarrow$ (15.56), $x \downarrow x = x$.

15.42 Prove (15.58), $x \leq y \equiv x \downarrow y = x$. A possible first step is to use the theorem $b = c \equiv b \leq c \land c \leq b$.

15.43 Prove (15.59), $x \downarrow y = x \ \lor \ x \downarrow y = y$. This is most easily done using (15.58)

15.44 Prove Distributivity of $+$ over $\downarrow$, (15.60), $c+(x \downarrow y) = (c+x) \downarrow (c+y)$, using (15.47).

15.45 Prove Distributivity of $\cdot$ over $\downarrow$, (15.61), $c \geq 0 \ \Rightarrow \ c \cdot (x \downarrow y) = (c \cdot x) \downarrow (c \cdot y)$.

15.46 Prove Distributivity of $\cdot$ over $\uparrow$, (15.62), $c \leq 0 \ \Rightarrow \ c \cdot (x \uparrow y) = (c \cdot x) \downarrow (c \cdot y)$.

15.47 Prove (15.70): Provided $\neg occurs(`x`, `E`)$,
$$R[x := E] \ \Rightarrow \ E = E \uparrow (\downarrow x \mid R : x) \ .$$

15.48 Write down the general laws (8.14)–(8.21) and (8.18), but particularized for $\star$ being the operator $\downarrow$.

15.49 *The California problem.* Consider a nonempty set of couples (each comprising a male and a female). The oldest male is the same age as the oldest female. If two of the original couples swap partners temporarily, the younger members of the two new pairs are the same age. Prove that the partners of each couple are the same age. Hint: The key to solving this problem without case analysis is to formalize the situation properly.

Exercises on absolutes

In these exercises, you may use the hint "Arithmetic" for relations involving addition, subtraction, and multiplication of elements of an integral domain.

15.50 Prove theorem (15.72), $abs.x = abs(-x)$.

15.51 Prove Triangle inequality (15.73), $abs(x + y) \leq abs.x + abs.y$.

15.52 Prove theorem (15.74), $abs(abs.x) = abs.x$.

15.53 Prove theorem (15.75), $abs(x \cdot y) = abs.x \cdot abs.y$.

15.54 Prove theorem (15.76), $-(abs.x + abs.y) \leq x + y \leq abs.x + abs.y$.

Exercises on operator Divides

In these exercises, you may use the hint "Arithmetic" for relations involving addition, subtraction, and multiplication of elements of an integral domain.

15.55 Prove theorem (15.78), $c \mid c$.

15.56 Prove theorem (15.79), $c \mid 0$.

15.57 Prove theorem (15.80), $1 \mid b$.

15.58 Prove theorem (15.81), $c \mid 1 \ \Rightarrow \ c = 1 \ \lor \ c = -1$.

15.59 Prove theorem (15.82), $d \mid c \land c \mid b \ \Rightarrow \ d \mid b$.

15.60 Prove theorem (15.83), $b \mid c \wedge c \mid b \equiv b = c \vee b = -c$.

15.61 Prove theorem (15.84), $b \mid c \Rightarrow b \mid c \cdot d$.

15.62 Prove theorem (15.85), $b \mid c \Rightarrow b \cdot d \mid c \cdot d$.

15.63 Prove theorem (15.86), $1 < b \wedge b \mid c \Rightarrow \neg(b \mid (c + 1))$.

15.64 Prove $b \mid c \equiv b \mid -c$.

15.65 Prove that if $d \mid b$ and $d \mid c$ then $d \mid (b + c)$.

15.66 Prove that if $b \neq 0$ and $d \mid b$, then $abs.d \leq abs.b$.

15.67 Complete the proof of theorem (15.87) by showing that the theorem holds for negative integers as well as positive integers. Do this by extending algorithm (15.88) to apply to negative as well as positive integers.

15.68 Prove Corollary (15.89).

Exercises on greatest common divisors

15.69 Prove Symmetry (15.96), b **gcd** $c = c$ **gcd** b.

15.70 Prove Associativity (15.97), $(b$ **gcd** $c)$ **gcd** $d = b$ **gcd** $(c$ **gcd** $d)$.

15.71 Prove (15.98), b **gcd** $b = abs.b$.

15.72 Prove Zero (15.99), 1 **gcd** $b = 1$.

15.73 Prove (15.100), 0 **gcd** $b = abs.b$.

15.74 Prove (15.101), b **gcd** $c = (abs.b)$ **gcd** $(abs.c)$.

15.75 Prove (15.102), b **gcd** $c = b$ **gcd** $(b + c) = b$ **gcd** $(b - c)$.

15.76 Prove (15.103), $b = a \cdot c + d \Rightarrow (b$ **gcd** $c = c$ **gcd** $d)$.

15.77 Prove Distributivity (15.104), $0 \leq d \Rightarrow d \cdot (b$ **gcd** $c) = (d \cdot b)$ **gcd** $(d \cdot c)$.

15.78 What is n **gcd** $(n + 1)$, for $n \geq 0$?

15.79 Extend Euclid's algorithm (page 318) to find the greatest common divisor of any two integers.

15.80 Suppose the conditional statement **if** $x > y \rightarrow x := x - y \ [\!] \ y > x \rightarrow y := y - x$ **fi** of Euclid's algorithm (page 318) is replaced by

$$x, y := x \uparrow y - x \downarrow y, x \downarrow y \quad.$$

Prove that the algorithm still truthifies R.

15.81 Here are two ways to compute the **gcd** of three integers. (i) Extend the iterative algorithm that follows inductive definition (15.107) for computing b **gcd** c to compute the **gcd** of three integers, all together. (ii) Use that iterative algorithm twice, using $gcd(b, c, d) = b$ **gcd** $(c$ **gcd** $d)$. Which do you prefer, and for what reasons?

15.82 Prove that if b is positive and composite, then it has a divisor d that satisfies $1 < d^2 \leq b$.

15.83 Prove that if $b \perp c$, then $(b - c) \, \mathbf{gcd} \, (b + c)$ equals 1 or 2.

Exercises on primes

15.84 Prove Theorem (15.113), $b \perp c \, \wedge \, c \,|\, (b \cdot d) \; \Rightarrow \; c \,|\, d$.

Exercises on congruences

In these exercises, you may use the hint "Arithmetic" for relations involving addition, subtraction, and multiplication of elements of an integral domain.

15.85 Prove that congruence relation $\overset{m}{=}$ is an equivalence relation.

15.86 Prove Addition (15.120), $b \overset{m}{=} c \; \equiv \; b + d \overset{m}{=} c + d$.

15.87 Prove Negation (15.121), $b \overset{m}{=} c \; \equiv \; -b \overset{m}{=} -c$.

15.88 Prove Multiplication (15.122), $b \overset{m}{=} c \; \Rightarrow \; b \cdot d \overset{m}{=} c \cdot d$.

15.89 Prove Powers (15.123), $b \overset{m}{=} c \; \Rightarrow \; b^n \overset{m}{=} c^n$ (for $n \geq 0$).

15.90 Prove Cancellation (15.124), $d \perp m \; \Rightarrow \; (b \cdot d \overset{m}{=} c \cdot d \; \equiv \; b \overset{m}{=} c)$.

15.91 Prove Cancellation (15.125), $b \cdot d \overset{m \cdot d}{=} c \cdot d \; \equiv \; b \overset{m}{=} c$ (for $d \geq 0$).

15.92 Prove theorem (15.126), $d \perp m \; \Rightarrow \; (\exists x \,|: d \cdot x \overset{m}{=} b)$.

15.93 Prove theorem (15.127), $d \perp m \; \Rightarrow \; (d \cdot x \overset{m}{=} b \, \wedge \, d \cdot y \overset{m}{=} b \; \Rightarrow \; x \overset{m}{=} y)$.

Chapter 16

Combinatorial Analysis

T his chapter concerns *combinatorial analysis*: the branch of mathematics that deals with *permutations* of a set or bag and *combinations* of a set. These ideas lead to *binomial coefficients* and the *Binomial theorem*. The first two sections of this chapter introduce the theory, with just enough examples to make clear the points being made. The third section illustrates the power of the theory through a variety of examples.

16.1   Rules of counting

RULES OF SUM AND PRODUCT

Three basic rules used in counting are the *rule of sum, rule of product,* and *rule of difference*. Stated in terms of sets and their cross products, these rules are straightforward.

(16.1)   **Rule of sum.** The size of the union of n (finite) pairwise disjoint sets is the sum of their sizes.

$(\forall i \mid 0 \leq i < j < n : S_i \cap S_j = \emptyset) \Rightarrow$
$\#(\cup i \mid 0 \leq i < n : S_i) = (\Sigma i \mid 0 \leq i < n : \#S_i)$

(16.2)   **Rule of product.** The size of the cross product of n sets is the product of their sizes.

$\#(S_0 \times \cdots \times S_{n-1}) = (\Pi i \mid 0 \leq i < n : \#S_i)$

(16.3)   **Rule of difference.** The size of a set with a subset of it removed is the size of the set minus the size of the subset.

$T \subseteq S \Rightarrow \#S - \#T = \#(S - T)$

Applying these rules in concrete situations requires identifying the sets involved. Here is an example. Suppose a child can draw 4 different faces (a set of size 4) and 2 different hats (a set of size 2). Then the rule of sum tells us the child can draw 6 different faces or hats, and the rule of product tells us that the child can draw $4 \cdot 2 = 8$ different combinations of faces with hats on them.

As another example, we calculate the number of different license plates if each license plate is to contain three letters followed by two digits. This number is the size of the cross product of three sets with 26 elements each and two sets with 10 elements each, or, according to the rule of product, $26 \cdot 26 \cdot 26 \cdot 10 \cdot 10$.

Permutations of a Set

A *permutation* of a set of elements (or of a sequence of elements) is a linear ordering of the elements. For example, two permutations of the set $\{5, 4, 1\}$ are $1, 4, 5$ and $1, 5, 4$. A permutation of a sequence of letters is called an *anagram*. An anagram of TUESDAY NOVEMBER THIRD, the day of the 1992 American presidential elections, is MANY VOTED BUSH RETIRED.

How many different permutations of a set of n elements are there? For the first element of the permutation, we choose from a set of n elements. For the second element of the permutation, we choose from the set of $n-1$ remaining elements. For the third, $n-2$, and so on. Thus, there are $n \cdot (n-1) \cdot (n-2) \cdots 1$, or $n!$, different permutations.

Sometimes, we want to construct a permutation of only r (say) of the n items. Such an r-*permutation* of a set of size n can be constructed as follows. For the first element, choose from n elements; for the second, from the $n-1$ remaining elements, ... , and for the last, from $n-r+1$ elements. Thus, there are

$$
\begin{aligned}
& n \cdot (n-1) \cdot \cdots \cdot (n-r+1) \\
= \quad & \langle \text{Multiply numerator and denominator by } (n-r)! \rangle \\
& \frac{n \cdot (n-1) \cdot \cdots \cdot (n-r+1) \cdot (n-r) \cdot (n-r-1) \cdot \cdots \cdot 1}{(n-r) \cdot (n-r-1) \cdot \cdots \cdot 1} \\
= \quad & \langle \text{Definition of } n! \text{ and } (n-r)! \rangle \\
& n!/(n-r)!
\end{aligned}
$$

different permutations. This number occurs frequently enough to give it a name.

(16.4) $P(n, r) = n!/(n-r)!$.

(16.5)   **Theorem.** The number of r-permutations of a set of size n equals $P(n, r)$.

We have: $P(n, 0) = 1$, $P(n, 1) = n$, $P(n, n-1) = n!$, and $P(n, n) = n!$. (Remember that $0! = 1$.)

For example, the number of 3-permutations of the 4-letter word BYTE is $P(4, 3)$, which is $4!/(4-3)! = 4! = 24$. The number of 2-permutations of BYTE is $P(4, 2) = 4!/(4-2)! = 4 \cdot 3 = 12$. These 2-permutations are: BY, BT, BE, YB, YT, YE, TB, TY, TE, EB, EY, and ET. There is one 0-permutation, the empty sequence; note that $P(n, 0) = 1$.

PERMUTATIONS WITH REPETITION OF A SET

Consider forming an r-permutation of a set but allowing each element to be used more than once. Such a permutation is called an *r-permutation with repetition*. For example, here are all the 2-permutations with repetition of the letters in SON: SS, SO, SN, OS, OO, ON, NS, NO, NN. Given a set of size n, in constructing an r-permutation with repetition, for each element we have n choices. The following theorem follows trivially from this observation and the rule of product.

(16.6)   **Theorem.** The number of r-permutations with repetition of a set of size n is n^r.

PERMUTATIONS OF A BAG

There will be fewer permutations of a bag than of a set of the same size, because the bag may have equal elements and because the transposition of equal elements in a permutation does not yield a different permutation. To illustrate the difference, we list below all the permutations of the set $\{S, O, N\}$ and the bag $\{\!| M, O, M |\!\}$. Although the set and the bag are the same size, the set has more permutations.

> SON, SNO, OSN, ONS, NSO, NOS
> MOM, MMO, OMM

The following theorem gives the number of permutations of a bag. Note that if the bag is really a set (i.e. each of the elements occurs once), the formula is equivalent to the number of permutations of a set.

(16.7)   **Theorem.** The number of permutations of a bag of size n with k distinct elements occurring $n_1, n_2, \ldots, n_k$ times is

$$\frac{n!}{n_1! \cdot n_2! \cdot \cdots \cdot n_k!}.$$

Proof. We prove the theorem by induction on k, the number of distinct elements in the bag.

Base case $k = 0$. The bag is empty, so there is 1 permutation: the empty sequence. The numerator of the fraction of the theorem is $0!$, which is 1. The denominator is a product of $k = 0$ values, which is also 1. Hence the expression reduces to $1/1 = 1$, and the theorem holds.

Inductive case. Assume the inductive hypothesis that the theorem holds for a bag B with n elements that consist of distinct elements e_i occurring n_i times, $1 \le i \le k$, and prove that the theorem holds for a bag constructed by adding n_{k+1} copies of a new value e_{k+1} to B. For the moment, assume that these copies of e_{k+1} are distinct and add them one at a time to B. The first copy can be inserted into each permutation in $n+1$ different places, the second in $n+2$ different places, and so on, giving

$$(n + n_{k+1}) \cdot \ \cdots \ \cdot (n+1) \cdot \frac{n!}{n_1! \cdot n_2! \cdot \ \cdots \ \cdot n_k!} \quad ,$$

that is,

(16.8) $\dfrac{(n + n_{k+1})!}{n_1! \cdot n_2! \cdot \ \cdots \ \cdot n_k!}$,

different permutations. (For example, to the bag $\{\!\{O\}\!\}$ with the single permutation O, adding two copies of M yields the permutations $M_1 M_2 O$, $M_2 M_1 O$, $M_1 O M_2$, $M_2 O M_1$, $O M_1 M_2$, $O M_2 M_1$.)

Since the n_{k+1} copies of e_{k+1} have been considered distinct, some permutations are counted more than once. Consider two permutations to be equivalent if removing the distinction between the copies of e_{k+1} makes the permutations the same, and partition the permutations into equivalence classes. (In the example given above, $M_1 M_2 O$, $M_2 M_1 O$ are in the same equivalence class and represent the permutation MMO.) Since there are $n_{k+1}!$ permutations of n_{k+1} distinct elements, each equivalence class contains $n_{k+1}!$ permutations. Therefore, to find the number of permutations, divide (16.8) by $n_{k+1}!$.   □

As an example, we compute the number of permutations of the letters in the word MISSISSIPPI. There are 11 letters. M occurs 1 time; I, 4 times; S, 4 times; and P, 2 times. Therefore, the number of permutations is

$$\frac{11!}{1! \cdot 4! \cdot 4! \cdot 2!} = 34650 \quad .$$

Combinations of a Set

An *r-combination* of a set is a subset of size r. A permutation is a sequence; a combination is a set.

For example, the 2-permutations of the set consisting of the letters in SOHN are

$$\text{SO, SH, SN, OH, ON, OS, HN, HS, HO, NS, NO, NH}$$

while the 2-combinations are

$$\{S, O\}, \{S, H\}, \{S, N\}, \{O, H\}, \{O, N\}, \{H, N\} \quad .$$

We now derive a formula for the number of r-combinations of a set of size n. For this purpose, the following notation will come in handy.

(16.9) **Definition.** The *binomial coefficient* $\binom{n}{r}$, which is read as " n choose r ", is defined by

$$\binom{n}{r} = \frac{n!}{r! \cdot (n-r)!} \qquad (\text{for } 0 \le r \le n) .$$

We have: $\binom{n}{0} = 1$, $\binom{n}{1} = n$, $\binom{n}{n-1} = n$, and $\binom{n}{n} = 1$. The reason for the term *binomial coefficient* will become clear later, on page 346.

The r-permutations of a set of size n can be generated by first generating the r-combinations and then generating the permutations of each r-combination —i.e. to construct an r-permutation, first choose the r elements to be used and then construct a permutation of them. Since each r-combination has $r!$ permutations, we have $P(n,r) = r! \cdot \binom{n}{r}$:

(16.10) **Theorem.** The number of r-combinations of n elements is $\binom{n}{r}$.

For example, suppose a student has to answer 6 of 9 questions on an exam. The number of ways in which this obligation can be discharged is

$$\binom{9}{6} = \frac{9!}{6! \cdot 3!} = \frac{9 \cdot 8 \cdot 7}{3 \cdot 2 \cdot 1} = 84 .$$

We can relate the number of r-combinations of a set of size n to the number of permutations of a certain bag. Consider Theorem (16.7) for the case of a bag with only two distinct elements. Thus, $n_2 = n - n_1$, and the formula of the theorem reduces to $\frac{n!}{n_1! \cdot (n-n_1)!} = \binom{n}{n_1}$. Comparing this case to Theorem (16.10) gives the following theorem.

(16.11) **Theorem.** The number $\binom{n}{r}$ of r-combinations of a set of size n equals the number of permutations of a bag that contains r copies of one object and $n - r$ copies of another.

Combinations with repetition of a set

An *r-combination with repetition* of a set S of size n is a bag of size r all of whose elements are in S. An r-combination of a set is a subset of that set; an r-combination with repetition of a set is a bag, since its elements need not be distinct. For example, the 2-combinations with repetition of the letters of SON are the bags

$$\{S,S\}, \ \{S,O\}, \ \{S,N\}, \ \{O,O\}, \ \{O,N\}, \ \{N,N\} .$$

On the other hand, the 2-permutations with repetition of SON are

$$\text{SS, SO, SN, OS, OO, ON, NS, NO, NN} .$$

We want to find a formula for the number of r-combinations with repetition of a set S of size n. To do so, we reduce the problem of finding this number to a problem whose solution is known. Let the elements of S be $e_1, \ldots, e_n$. Any r-combination with repetition can be represented by a permutation of its elements in which all the e_1's come first, then the e_2's, and so on:

$$e_1, e_1, \ldots, e_1, e_2, e_2, \ldots e_2, \ldots, e_n, e_n, \ldots e_n \quad .$$

Call this the *canonical* representation of the r-combination. In the canonical representation, if we distinguish the boundaries between distinct elements using a bar $|$, then we do not need to use the e_i at all; we can replace them, say, by the symbol x. For example, we show below a permutation with 3 e_1's, 1 e_2, no e_3's, and 4 e_4's; and below it, we show its representation using x for all the e_i and $|$ as a separator.

e_1	e_1	e_1		e_2			e_4	e_4	e_4	e_4
x	x	x	$|$	x	$|$	$|$	x	x	x	x

We have established a one-to-one correspondence between r-combinations with repetition of a set of size n and permutations of r x's and $n-1$ bars. Hence, we have the following.

$\qquad$ no. of r-combinations with repetition of a set of size n

$= \qquad$ ⟨The above one-to-one correspondence⟩

$\qquad$ no. of permutations of r x's and $n-1$ bars

$= \qquad$ ⟨Theorem (16.7), with $n, k, n_1, n_2 := r + n - 1, 2, r, n - 1$⟩

$\qquad \dfrac{(n+r-1)!}{r! \cdot (n-1)!}$

$= \qquad$ ⟨Definition (16.9)⟩

$\qquad \dbinom{n+r-1}{r}$

We have proved the following theorem.

(16.12) **Theorem.** The number of r-combinations with repetition of a set of size n is $\binom{n+r-1}{r}$.

Here is an application of Theorem (16.12). Suppose 7 businessmen stop at a fast-food restaurant, where each gets either a burger, a cheeseburger, or a fishwich. How many different orders are possible? The answer is the number of 7-combinations with repetition of a set of 3 objects (burger, cheeseburger, or fishwich). So n of Theorem (16.12) is 3 and r is 7. By the theorem, the number is $\binom{3+7-1}{7} = \frac{9!}{7! \cdot 2!} = 36$.

The equivalence of three different statements

In combinatorial analysis, the following three ways of expressing a certain number crop up often. It is useful to know that they are the same, because the set of techniques that can be used to solve any single problem is thus enlarged.

(16.13) **Theorem.** The following three numbers are equal.
  (a) The number of integer solutions of the equation
$$x_1 + x_2 + \cdots + x_n = r \,, \text{ where } 0 \le x_i \text{ for } 1 \le i \le n \,.$$
  (b) The number of r-combinations with repetition of a set of size n.
  (c) The number of ways r identical objects can be distributed among n different containers.

Proof. We show the equality of (a) and (c) by giving a one-to-one correspondence between the solutions of (a) and ways of (c). Given a solution of (a), let container X_i, $1 \le i \le n$, contain x_i objects. In total, the n containers contain r objects. Hence, a solution of (a) is mapped into a way in which r objects can be distributed among n containers. This mapping is one-to-one and onto, so (a) and (c) are equal.

We now show the equality of (b) and (c). Consider an r-combination with repetition of the set $\{x_1, \ldots, x_n\}$, where each x_i occurs n_i times (say) in the combination. We translate this combination into a distribution of r identical objects v (say) into n distinct containers X_i (say) as follows: place n_i copies of v into container X_i, for all i. To each such combination there exists such a distribution, and vice versa. This establishes a one-to-one correspondence between the r-combinations with repetition of n objects and the ways of distributing r identical objects among n containers, so (b) and (c) are equal.

Since (a) = (c) and (b) = (c), by transitivity (a) = (b). □

16.2 Properties of n choose r

Earlier, we defined the binomial coefficient $\binom{n}{r} = \frac{n!}{r! \cdot (n-r)!}$, for $0 \le r \le n$. We now discuss some properties of $\binom{n}{r}$.

Theorem (16.14) below follows trivially from $\binom{n}{r} = \frac{n!}{r! \cdot (n-r)!}$, since the RHS is symmetric in r and $n - r$ (i.e. replacing r by $n - r$ yields an equal expression). It is unfortunate that this symmetry is not apparent in the notation $\binom{n}{r}$. In the literature, this gives rise to the statement and proof of many theorems that would have been obvious had the symmetry been recognized and exploited. It would have been better to define, say,

$\widehat{C}(b,c)$ to be the number $\frac{(b+c)!}{b! \cdot c!}$ of combinations of $b + c$ objects taken b (or c) at a time and to note immediately that $\widehat{C}$ is symmetric in its arguments. However, the notation $\binom{n}{r}$ is too entrenched in mathematics to change.

Theorems for n choose r

(16.14) **Symmetry** : $\binom{n}{r} = \binom{n}{n-r}$

(16.15) **Absorption** : $\binom{n}{r} = \frac{n}{r} \cdot \binom{n-1}{r-1}$ (for $0 < r \le n$)

(16.16) **Absorption** : $r \cdot \binom{n}{r} = n \cdot \binom{n-1}{r-1}$ (for $0 < r \le n$)

(16.17) $(n - r) \cdot \binom{n}{r} = n \cdot \binom{n-1}{r}$ (for $0 \le r < n$)

(16.18) **Addition** : $\binom{n}{r} = \binom{n-1}{r} + \binom{n-1}{r-1}$ (for $0 < r < n$)

(16.19) $\binom{r+n+1}{n} = (\Sigma k \mid 0 \le k \le n : \binom{r+k}{k})$ (for $0 \le n,\ 0 \le r$)

(16.20) $2^n = (\Sigma r \mid 0 \le r \le n : \binom{n}{r})$ (for $0 \le n$)

(16.21) $\binom{n}{r} \cdot \binom{r}{k} = \binom{n}{k} \cdot \binom{n-k}{r-k}$ (for $0 \le k \le r \le n$)

Proofs of theorems (16.15)–(16.17) are left as exercises. Theorem (16.17) is unpleasant to prove by induction, but it can be proved very elegantly using Symmetry (16.14) and Absorption (16.16).

Theorem (16.18) can be proved in at least three ways. It can be proved by induction, but two better ways exist. It can be proved most easily by adding theorems (16.16) and (16.17) together (see Exercise 16.60). Finally, it can be proved using a *combinatorial argument*. A combinatorial argument relies on the interpretation of $\binom{n}{r}$ and $P(n, r)$ as the number of r-combinations and the number of permutations of a set of size n, instead of simply relying on the formulas $\binom{n}{r} = n!/((r! \cdot (n-r)!)$ and $P(n,r) = n!/(n-r)!$. Relying on such interpretations can sometimes result in shorter proofs. We now give a combinatorial proof of (16.18).

Proof. Choose some element e of a set S of n elements. Using predicate $C(s, r, S)$ to mean "s is an r-combination of S", we calculate the number of r-combinations of S as follows:

$$\binom{n}{r}$$

$$= \quad \langle \text{Definition of } C(s,r,S) \rangle$$
$$(\Sigma s \mid C(s,r,S) : 1)$$

$=$ $\langle\, true \equiv e \notin s \vee e \in s \,\rangle$

$(\Sigma s \mid (e \notin s \vee e \in s) \wedge C(s,r,S) : 1)$

$=$ $\langle$Range split$\rangle$

$(\Sigma s \mid e \notin s \wedge C(s,r,S) : 1) + (\Sigma s \mid e \in s \wedge C(s,r,S) : 1)$

$=$ $\langle e \notin s \Rightarrow C(s,r,S) = C(s,r,S-\{e\})$

$e \in s \Rightarrow C(s,r,S) = C(s-\{e\}, r-1, S-\{e\}) \,\rangle$

$(\Sigma s \mid e \notin s \wedge C(s,r,S-\{e\}) : 1) +$

$(\Sigma s \mid e \in s \wedge C(s-\{e\}, r-1, S-\{e\}) : 1)$

$=$ $\langle$Definition of $C(s,r,S)$, twice$\rangle$

$\binom{n-1}{r} + \binom{n-1}{r-1}$ $\square$

Theorem (16.21) can be proved quite simply:

$\binom{n}{r} \cdot \binom{r}{k}$

$=$ $\langle$Definition (16.9), twice$\rangle$

$\dfrac{n!}{r! \cdot (n-r)!} \cdot \dfrac{r!}{k! \cdot (r-k)!}$

$=$ $\langle$Cancel $r!$; Rearrange$\rangle$

$\dfrac{n!}{k! \cdot (r-k)! \cdot (n-r)!}$

$=$ $\langle$Multiply numerator and denominator by $(n-k)!\rangle$

$\dfrac{n!}{k! \cdot (n-k)!} \cdot \dfrac{(n-k)!}{(r-k)! \cdot (n-r)!}$

$=$ $\langle$Definition (16.9), twice — $n-r = (n-k) - (r-k)\rangle$

$\binom{n}{k} \cdot \binom{n-k}{r-k}$

Theorems (16.19)–(16.21) are three of many identities concerning the sum and product of binomial coefficients. This is only the beginning of a rich theory of binomial coefficients, which is not only elegant but useful in combinatorics and probability theory. One can even define $\binom{n}{r}$ for n a negative number or real number and r any natural number, which allows the expression of many more useful identities. See [16] for a full exploration.

FIGURE 16.1. PASCAL'S TRIANGLE

row 0					1					
row 1				1		1				
row 2				1	2	1				
row 3			1	3	3	1				
row 4		1	4	6	4	1				
row 5	1	5	10	10	5	1				
row 6	1	6	15	20	15	6	1			

$\cdots$ $\cdots$

Row n has $n+1$ values $\binom{n}{r}$ for $0 \leq r \leq n$.

PASCAL'S TRIANGLE

One interesting way of listing all binomial coefficients $\binom{n}{r}$ is *Pascal's triangle* of Fig. 16.1, named after Blaise Pascal, who wrote an influential treatise on the triangle (see Historical note 16.1). Row n of the triangle contains the $n+1$ values $\binom{n}{0}, \binom{n}{1}, \ldots, \binom{n}{n}$. This can be seen as follows. The two sides of the triangle are all 1's, since $\binom{n}{0} = \binom{n}{n} = 1$. Any other element of the triangle, $\binom{n}{r}$ for $0 < r < n$, is the sum of the two values just above it —since, by theorem (16.18), $\binom{n}{r} = \binom{n-1}{r} + \binom{n-1}{r-1}$.

Pascal's triangle has many surprising properties. For example, consider the hexagon of values $4, 5, 15, 20, 10, 6$ that surrounds the third element (10) in row 5:

$$
\begin{array}{ccc}
4 & 6 & \\
5 & 10 & 10 \\
15 & 20 &
\end{array}
$$

Both ways of multiplying alternate numbers of this hexagon give the same result: $4 \cdot 15 \cdot 10 = 5 \cdot 20 \cdot 6 = 600$. This property holds for any such hexagon of Pascal's triangle.

THE BINOMIAL THEOREM

Finally, we find out why the number $\binom{n}{r}$ is called a binomial coefficient. For natural number n and variables x and y, consider multiplying $x+y$ by itself n times:

$$
(x+y)^n = \underbrace{(x+y) \cdot (x+y) \cdot \cdots \cdot (x+y)}_{n \text{ factors}} .
$$

The expression $(x+y)^n$ is called a *binomial*, because it is a polynomial in two variables (x and y). This binomial can be expanded to

$$
(16.22) \quad (x+y)^n = c_0 \cdot x^0 \cdot y^n + c_1 \cdot x^1 \cdot y^{n-1} + \cdots + c_n \cdot x^n \cdot y^0
$$
$$
= (\Sigma k \mid 0 \le k \le n : c_k \cdot x^k \cdot y^{n-k})
$$

We want to determine the coefficients c_k. In order to understand the rule we use for calculating c_k, consider the case $n = 3$, so that the product is

$$
\begin{aligned}
& (x+y) \cdot (x+y) \cdot (x+y) \\
= \quad & \langle \text{Distributivity (15.5)} \rangle \\
& (x^2 + x \cdot y + y \cdot x + y^2) \cdot (x+y) \\
= \quad & \langle \text{Distributivity (15.5)} \rangle \\
& (x^3 + x \cdot y \cdot x + y \cdot x^2 + y^2 \cdot x) + (x^2 \cdot y + x \cdot y^2 + y \cdot x \cdot y + y^3)
\end{aligned}
$$

HISTORICAL NOTE 16.1. Blaise Pascal (1623–1662)

By the age of 12, and before reading Euclid's *Elements* (see Historical note 15.1), Pascal had already proved several of Euclid's propositions in geometry. At 16, Pascal wrote a significant work on conic sections, including what has been called one of the most beautiful theorems in geometry. Pascal invented and built the first calculating machine, before he was 21, to help his father in his calculations; along with Fermat, young Pascal created the theory of probability; and he is well known for his work on the cycloid. No wonder that Niklaus Wirth named his programming language *Pascal*.

Yet, Pascal is far better known for his religious and philosophical writings, and the general reader is more likely to have come across Pascal's *Pensées* and *Provincial Letters* than his mathematical works. Some think that Pascal wasted his mathematical genius on too many other things —E.T. Bell calls him "perhaps the greatest [mathematical] might-have-been in history".

Pascal was not physically well, and, from the age of 17 to the end of his life, he suffered from stomach trouble, insomnia and, later, incessant headaches. This pain and suffering, together with his family's deeply religious bent, was enough to turn his views inward. He spent the last 8 years of his life in a monastery, where much of his philosophical writings were done.

Pascal used probability theory (which he developed) in his own life. Probability theory deals in *expectations*. The expectation of a gamble is the value of the prize multiplied by the probability of winning the prize. In his *Pensées*, Pascal argued that the value of eternal happiness to be won by leading a religious life is infinite. Therefore, no matter how small the probability of winning eternal happiness, the expectation is infinite (infinity times a positive number, no matter how small, is infinity), so it pays to lead such a life. Convinced?

$$= \quad \langle \text{Collect terms} \rangle$$
$$x^3 + 3 \cdot x^2 \cdot y + 3 \cdot x \cdot y^2 + y^3$$

Look at the third formula in this calculation. There are three terms that equal $x^2 \cdot y$, so the final coefficient of $x^2 \cdot y$ is 3. Each of these terms comes from choosing an x from two of the original terms $(x + y)$ (which automatically chooses y from the other term). The coefficient of x^3 is calculated in the same way: it is the number of ways of choosing three x's from the three terms, or 1.

In the general case $(x + y)^n$, then, coefficient c_k is the number of ways in which k x's can be chosen from the n available factors (choosing k x's automatically chooses $n - k$ y's). The number of ways of choosing k elements from n elements is the number of combinations of n objects taken k at a time, or $\binom{n}{k}$. Thus, we have proved, in a combinatorial fashion, the Binomial theorem.

(16.23) **Binomial theorem.** [1] For $n \geq 0$,

$$(x + y)^n = (\Sigma k \mid 0 \leq k \leq n : \binom{n}{k} \cdot x^k \cdot y^{n-k}) .$$

Since $\binom{n}{k} = \binom{n}{n-k}$, the coefficients of $x^k \cdot y^{n-k}$ and $x^{n-k} \cdot y^k$ are equal. Thus, the coefficients of x^0 and x^n are equal, as are the coefficients of $x^{n-1} \cdot y$ and $x \cdot y^{n-1}$. Note that the binomial coefficients $c_0, \dots, c_n$ of $(x + y)^n$ are the numbers in row n of Pascal's triangle!

The Binomial theorem can also be proved by induction on n, as shown in Exercise 16.63. However, the combinatorial proof is shorter and simpler.

16.3   Examples of counting

We now give examples, drawn from a variety of contexts, to illustrate applications of the theory presented in the previous sections. We begin with examples concerning the rule of sum and the rule of product and advance from there.

RULES OF SUM AND PRODUCT

(16.24) **Example.** The twelfth-grade class has 55 boys and 56 girls. What is the total number of students in the class, and how many different possible boy-girl pairs are there?

There are two sets: the boys and the girls. The rule of sum tells us that there are 111 students. The rule of product tells us that the number of different pairs is the size of the cross product of the set of boys and the set of girls, which is $55 \cdot 56 = 3080$.   □

(16.25) **Example.** Suppose you can pass your language requirement in College by (i) gaining proficiency in French, German, or Japanese or (ii) gaining minimal qualification: (take two semesters of French, German, Japanese, or Italian) and (take two semesters of Korean or Hindi). (Above, we use parentheses to eliminate any possible ambiguity.) In how many different ways can the language requirement be satisfied?

---

[1] We tried to write a historical note on John Binomial but were unable to find sufficient material. Even Moriarity's long treatise on Binomial's theorem, which won Moriarity the Mathematical Chair at a small university (according to Sherlock Holmes's friend Watson [13]), provided little help in our researches.

The set P of ways in which proficiency can be gained has 3 elements. Consider the set S of ways in which minimal qualification can be satisfied. Each element of S is a pair whose first element is French, German, Japanese, or Italian and whose second element is Korean or Hindi. The rule of product asserts that $\#S = 4 \cdot 2 = 8$. The rule of sum then indicates that the number of ways the language requirement can be satisfied is $\#P + \#S = 11$.   □

(16.26) **Example.** One bag contains a red ball and a black ball; a second bag contains a red ball, a green ball, and a blue ball. A person first chooses a bag and then selects a ball, at random. In what fraction of the cases will a red ball be selected?

If the first bag is chosen, then there are 2 possible selections (a red ball is selected or a black ball is selected). If the second bag is chosen, then there are 3 possible selections. The rule of sum tells us that there are $2 + 3 = 5$ possible selections. In the same way, we see that there are 2 ways of selecting a red ball. Therefore, in 2/5 of the cases, a red ball will be chosen.

For the reader who knows something about probability, note that this problem has nothing to do with the probability that a red ball will be selected.   □

(16.27) **Example.** How many functions $f : S \to T$ from S to T are there for finite sets S and T?

Let $S = \{s_1, \ldots, s_{\#s}\}$. A function from S to T is constructed by giving for each s_i a value $f.s_i$ in T. For each s_i, there are $\#T$ different choices. By the rule of product, there are $\#T^{\#S}$ different functions from S to T. □

PERMUTATIONS

(16.28) **Example.** How many permutations of the letters are there in the word LIE? BRUIT? CALUMNY? FACETIOUSLY [2] ?

According to Theorem (16.5), the number of permutations of a set of size n is $n!$. Therefore, the number of permutations of the letters of LIE is $3! = 6$; of BRUIT, $5! = 120$; of CALUMNY, $7! = 5040$, of FACETIOUSLY, $11! = 39916800$.   □

(16.29) **Example.** How many one-to-one functions are there from a finite set S to a finite set T?

---

[2] *Facetiously* is unusual in that it has all the vowels a, e, i, o, u, and y in it, in order. If you don't consider y to be a vowel, then *facetious* will do. *Sequoia* has all the vowels except y. Doug McIlroy treats these problems abstemiously.

Let $S = \{s_1, \ldots, s_{\#S}\}$. Function f is one-to-one (see Def. (14.41)) if it satisfies $x \neq y \Rightarrow f.x \neq f.y$. Therefore, the sequence $f.s_1, f.s_2, \ldots, f.s_{\#S}$ has to be a $\#S$-permutation of T. The answer is thus the number of such permutations: $P(\#T, \#S)$. □

(16.30) **Example.** Besides taking CS courses, a computer science major has to take a course in each of the following: math, physics, English, history, art, and music. The order in which the student takes these requirements does not matter. How many different choices of order does the student have?

The number of permutations of the 6 topics is $6! = 720$. □

(16.31) **Example.** Three couples sit in one row at the movies. The women want to sit together and the men want to sit together. In how many ways may they be seated?

View the men as a single object and the women as a single object. There are $2! = 2$ different permutations of these two objects. For each of these two permutations, we count the number of ways the men can be seated and the number of ways the women can be seated; then, by the rule of product, multiply them. There are $3! = 6$ ways of seating the women and $3! = 6$ ways of seating the men. This gives $2 \cdot 6 \cdot 6 = 72$ different ways of seating the couples. □.

(16.32) **Example.** Suppose 6 people are to be seated at a round table. In how many ways may they be seated?

The number of permutations of the six people is $6! = 720$. However, since the table is round, one cannot distinguish which person is first, and we have made that distinction in counting permutations. For any ordering, there are 6 permutations of the people in that order, not counting who is first (e.g. ABCDEF, BCDEFA, CDEFAB, DEFABC, EFABCD, FABCDE, ABCDEF). Hence, we derive the number of ways to be seated by dividing the number of permutations by 6: $720/6 = 120$. □

(16.33) **Example.** How many permutations of the letters of ALGORITHM have the A and L together (in either order)? How many have the A and the L separated by at least one letter?

Think of AL as a single letter; then (AL)GORITHM has 8 letters, and the number of permutations is $8! = 40320$. Now, there are $2! = 2$ permutations of AL, so the total number of permutations of ALGORITHM with A and L together and in either order is $2 \cdot 40320 = 80640$.

The total number of permutations of ALGORITHM is $9! = 362880$, and 80640 of them have the A and the L together. Therefore, $9! - 80640 = 282240$ permutations of ALGORITHM have the A and the L separated.□

Permutations of a bag

(16.34) **Example.** A coin is tossed 5 times, landing H (heads) or T (tails) each time, to form an *outcome*. One possible outcome is HHTTT. How many possible outcomes are there? How many outcomes have 1 H? How many outcomes contain at most 1 H?

By the rule of product, there are $2^5 = 32$ possible outcomes. An outcome with 1 H is a permutation of a bag with 1 H and 4 T's. By Theorem (16.7), there are $\frac{5!}{1! \cdot 4!} = 5$ different outcomes with 1 H. Similarly, there is $\frac{5!}{0! \cdot 5!} = 1$ outcome with no H's. By the rule of sum, there are $5 + 1 = 6$ outcomes with at most 1 H.   □

(16.35) **Example.** How many paths are there in the plane from the point (0,0) to the point (5,4), where each step of the path consists of moving one unit to the right (R) or one unit up (U)? Two such paths are shown in Fig. 16.2.

Each such path consists of 5 steps R and 4 steps U, in some order. Hence, each such path is a permutation of the bag containing 5 R's and 4 U's. The number of such permutations is $\frac{9!}{4! \cdot 5!} = 126$.   □

(16.36) **Example.** This is an example of a combinatorial proof of a theorem from number theory. Let $n = 2 \cdot k$, for some k, $k \geq 0$. Prove that $n!/2^k$ is an integer.

Consider a bag with distinct elements x_i for $1 \leq i \leq k$, each of which occurs twice in the bag. By Theorem (16.7), the number of permutations of the bag (which is an integer) is $\frac{n!}{(2!)^k} = n!/2^k$.   □

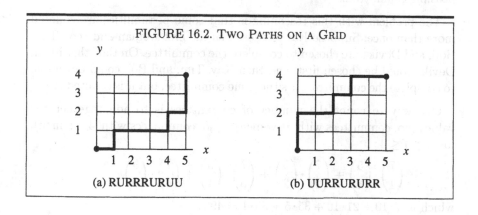

FIGURE 16.2. Two Paths on a Grid

(a) RURRRURUU   (b) UURRURURR

COMBINATIONS OF A SET

(16.37) **Example.** The chair has to select a committee of 5 from a faculty of 25. How many possibilities are there? How many possibilities are there if the chair should be on the committee?

The answer is the number of 5-combinations of a set of size 25; by Theorem (16.10), this is $\binom{25}{5} = \frac{25!}{5! \cdot 20!} = 53130$. If the chair has to be on the committee, then the other 4 members are chosen from a set of 24, so the answer is $\binom{24}{4} = \frac{24!}{4! \cdot 20!} = 10626$.   □

(16.38) **Example.** Suppose Gerry and John insist that they be on exactly the same committees in a department of 24. How many ways are there to choose a committee of 5?

Either Gerry and John are on a committee or they are not. We count the number of possible committees in each case. If Gerry and John are on a committee, then the other three members are chosen from a set of 22, so the number of committees is $\binom{22}{3} = 1540$. Committees of 5 that do not include Gerry and John are chosen from a set of 22; hence, there are $\binom{22}{5} = 26334$ different committees without Gerry and John. By the rule of sum, the answer is $1540 + 26334 = 27874$.   □

(16.39) **Example.** In a faculty of five men and seven woman, a committee of 4 with at least one woman is to be formed. How many possibilities are there?

Here is unsound reasoning. First choose the woman —there are $\binom{7}{1} = 7$ possibilities. Then, for the other 3 out of 11 people, there are $\binom{11}{3} = 165$ possibilities. Hence, by the product rule, there are $7 \cdot 165 = 1155$ different possible committees.

The problem with this reasoning is that some possibilities are counted more than once. Suppose Kay is chosen as the first woman and then Tim, Bob, and Devika are chosen to complete the committee. On the other hand, Devika could be chosen first, and then Kay, Tim, and Bob could be chosen to complete the committee. It's the same committee, but it is counted twice.

One way to count the number of committees is to add together the values (no. committees with i women) · (no. committees with $4 - i$ men), for $1 \le i \le 4$:

$$\binom{7}{1} \cdot \binom{5}{3} + \binom{7}{2} \cdot \binom{5}{2} + \binom{7}{3} \cdot \binom{5}{1} + \binom{7}{4} \cdot \binom{5}{0}$$

which is $7 \cdot 10 + 21 \cdot 10 + 35 \cdot 5 + 35 \cdot 1 = 490$.

There is an easier way to solve this problem. The total number of committees is $\binom{12}{4} = 495$. The number of committees of size 4 without a woman is $\binom{5}{4} = 5$. By the rule of difference, the number of committees with at least one woman is $495 - 5 - 490$. □

(16.40) **Example.** How many subsets does a set of size n have?

By Theorem (16.10), a set of size n has $\binom{n}{r}$ subsets of size r. The number of subsets is therefore $(\Sigma r \mid 0 \le r \le n : \binom{n}{r})$, and by theorem (16.20), this is 2^n.

Here is an alternative way of arriving at the solution. Each element is either in a subset or it is not (two possibilities). By the rule of product, there are 2^n possibilities. □

APPLYING COMBINATORIAL ANALYSIS TO POKER

A deck of playing cards has 52 cards. There are four suits: spades, hearts, diamonds, and clubs. In each suit, there are 13 cards of different value: 2, 3, 4, 5, 6, 7, 8, 9, 10, Jack (J), Queen (Q), King (K), and Ace (A). The basic idea of poker is for players to be given 5 cards from the deck; the players bet according to whether they expect to have the best hand, and the player with the best hand wins. "Best" is determined according to the list of hands given in Table 16.1, with the first hand listed being the best. Note that there are $\binom{52}{5} = 2,598,960$ different hands.

In betting, it helps to have a good idea of the chances of a hand being a winner, and this depends on the chances that someone has a better hand.

TABLE 16.1. POKER HANDS

Royal Flush: The cards 10, J, Q, K, A of one suit.

Straight flush: Five cards from the same suit, in sequence, with an Ace treated as coming before 2.

Four of a kind: (e.g. four 6's and one other card.)

Full house: Three cards of one value and two of another (e.g. three Jacks and two tens).

Flush: Five cards of the same suit (but not a straight or royal flush).

Straight: Five cards from at least two suits, in sequence, with the Ace coming either before the 2 or after the King.

Three of a kind: Three cards of one value and two other cards of different values.

Two pair : Two cards of one value, two cards of another value, and a fifth card of a third.

One pair: Two cards of one value and three other cards of differing value.

Combinatorial analysis can be used to determine such chances. The key to solving counting problems in poker is to separate the problem into pieces that can be solved using our counting techniques.

(16.41) **Example.** How many different royal flushes are there? Straight flushes?

In each suit, there is one royal flush: 10, J, Q, K, A. Since there are four suits, there are four royal flushes. A straight flush is determined by its highest card, which can have value 5, 6, ..., King. Thus, there are 9 straight flushes in each suit. Since there are 4 suits, by the rule of sum there are $9+9+9+9 = 36$ straight flushes. The chance of getting a straight flush or royal flush, then, is only 40 in 2,598,960, or 1 in 64974.   □

(16.42) **Example.** How many different three-of-a-kind hands are there?

We calculate the number of ways to choose the three cards with equal value:

$$\begin{aligned}
&\quad \text{(ways to choose the triple)}\\
&= \quad \langle \text{A card is made up of a value and a suit}\rangle\\
&\quad \text{(ways to choose the value)} \cdot \text{(ways to choose 3 suits)}\\
&= \quad \langle \text{There are 13 possible values; there are 4 suits}\rangle\\
&\quad 13 \cdot \binom{4}{3}\\
&= \quad \langle \text{Arithmetic}\rangle\\
&\quad 52
\end{aligned}$$

We now calculate the number of ways to choose the two other cards. Their values have to be different from the value of the other three.

$$\begin{aligned}
&\quad \text{(ways to choose the pair with different values)}\\
&= \quad \langle \text{A card is made up of a value and a suit}\rangle\\
&\quad \text{(ways to choose 2 values)} \cdot\\
&\quad \text{(ways to choose 2 suits with repetition)}\\
&= \quad \langle \text{There are 12 possible values; there are 4 suits}\rangle\\
&\quad \binom{12}{2} \cdot 4^2\\
&= \quad \langle \text{Arithmetic}\rangle\\
&\quad 1056
\end{aligned}$$

The number of three-of-a-kind hands is therefore $52 \cdot 1056 = 54912$.   □

16.4 The pigeonhole principle

The pigeonhole principle is usually stated as follows. [3]

(16.43) If more than n pigeons are placed in n holes, at least one hole
will contain more than one pigeon.

The pigeonhole principle is obvious, and one may wonder what it has to
do with computer science or mathematics. To find out, let us try to place
it in a more abstract setting.

The first point to note is that with more than n pigeons and with n
holes, the average number of pigeons per hole is greater than one. The
second point to note is that the statement "at least one hole contains more
than one pigeon" is equivalent to "the maximum number of pigeons in any
hole is greater than one".

Therefore, if we abstract away from pigeons and holes and just talk
about a bag S of real numbers (the number of pigeons in each hole), we
can restate the pigeonhole principle more mathematically. Let $av.S$ denote
the average of the elements of bag S and let $max.S$ denote the maximum.
Then the pigeonhole principle is:

(16.44) $av.S > 1 \;\Rightarrow\; max.S > 1$.

But this form of the principle can be generalized to the following. Provided
S is nonempty,

(16.45) **Pigeonhole principle.** $av.S \leq max.S$.

It is easy to prove that (16.45) implies (16.44) (see Exercise 16.67) but
the implication in the other direction does not hold. Hence, the generalized
pigeonhole principle is indeed more general. Second, we do not have to
accept the principle as intuitively true, for the proof of (16.45) is very
simple, given the definitions of average and maximum (see Exercise 16.68).

Frequently, the piegeonhole principle is applied to a bag of integers. In
this case, the maximum element in the bag is an integer, but the average
need not be. So, we can claim that the maximum is at least the smallest
integer that is not smaller than the average: [4]

(16.46) **Pigeonhole principle.** $\lceil av.S \rceil \leq max.S$.

The rest of this section illustrates applications of the pigeonhole principle.
In each example, the major task is to identify the bag S of numbers that

---

[3] The pigeonhole principle is also called the Dirichlet box principle, after Leje-
une Dirichlet, who first stated it, in the 1800s.

[4] See the footnote on page 320 for a definition of the ceiling function, $\lceil x \rceil$.

is used in the pigeonhole principle. Once S has been identified, the rest is easy.

(16.47) **Example.** Prove that in a room of eight people, at least two of them have birthdays on the same day of the week.

Let bag S contain, for each day of the week, the number of people in the room whose birthday is on that day. The number of people is 8 and the number of days is 7. Therefore,

$$
\begin{aligned}
&max.S \\
\geq\quad & \langle\text{Pigeonhole principle (16.46)} - S \text{ contains integers}\rangle \\
&\lceil av.S \rceil \\
=\quad & \langle\, S \text{ has 7 values that sum to } 8\,\rangle \\
&\lceil 8/7 \rceil \\
=\quad & \langle\text{Definition of ceiling}\rangle \\
& 2
\end{aligned}
$$

$\square$

(16.48) **Example.** A drawer contains ten black and ten white socks. How many socks must one take out (without looking at their color) to be sure that a matched pair has been selected?

In choosing two socks, two different-colored ones may be chosen, so at least three have to be taken out of the drawer. We use the pigeonhole principle to conclude that three is enough. Let b and c denote the number of black socks and white socks chosen, so that $S = \{b, c\}$. We have $b+c = \#S \geq 3$, and therefore

$$
\begin{aligned}
&max.S \\
\geq\quad & \langle\text{Pigeonhole principle (16.46)} - S \text{ contains integers}\rangle \\
&\lceil av.S \rceil \\
\geq\quad & \langle\, av.S \geq 3/2\,\rangle \\
&\lceil 1.5 \rceil \\
=\quad & \langle\text{Definition of } ceiling\,\rangle \\
& 2
\end{aligned}
$$

Thus, selecting 3 socks ensures that at least one of b and c is at least 2, and a matching pair is chosen.

(16.49) **Example.** Suppose Cornell has 51 computer science courses and that they are assigned numbers in the range 1..100. Prove that at least two courses have consecutive numbers.

Let the course numbers be c_i, for $1 \leq i \leq 51$. There are 51 distinct numbers c_i, so there are 51 distinct numbers $c_i + 1$. Each of the 102 numbers c_i or $c_i + 1$ is in the range 1..101. Define bag S by

$$
S = \{i \mid 1 \leq i \leq 101 : \text{number of } c_j \text{ and } c_k + 1 \text{ that equal } i\,\}.
$$

The average of the values in S is $102/101$, which is greater than 1. Hence, by the Pigeonhole principle (16.46), the maximum of the values is at least 2. Hence, for some i, j, and k we have $c_j = c_k + 1 = i$. Hence, c_k and $c_k + 1$ are consecutive course numbers.   □

(16.50) **Example.** Prove that in a group T of 85 people; at least four have the same initial letter of their last name.

Consider the bag S of 26 natural numbers defined by

$$\{c \mid c \in A..Z : (\Sigma p \mid p \in T \land p\text{'s last name begins with } c : 1)\} \quad .$$

The average of the numbers in S is $85/26$, which is greater than 3. By the pigeonhole principle the maximum of the numbers in S is greater than 3, so it is at least 4. Note that we really need the generalized pigeonhole principle to solve this problem.   □

(16.51) **Example.** Prove that if 101 integers are selected from the set $T = \{1, 2, \ldots, 200\}$, then there are two integers such that one divides the other.

Each selected integer x (say) may be written in the form $x = 2^k \cdot y$, where $0 \le k$ and y is odd, so that y is one of the 100 odd integers in $\{1, 2, \ldots, 200\}$. Since 101 integers are selected and there are only 100 different numbers y, by the pigeonhole principle, two of the selected numbers have the form $2^k \cdot y$ and $2^j \cdot y$ for some y, $0 \le k < j$. Then $2^k \cdot y$ divides $2^j \cdot y$.   □

Exercises for Chapter 16

16.1 Suppose the campus bookstore has 10 texts on FORTRAN and 25 on Pascal. How many different FORTRAN or Pascal texts can a student buy? How many choices are there to choose a pair of FORTRAN-Pascal books?

16.2 Suppose 15 bits are used to describe the address of a memory location in a computer, each bit being 0 or 1. How many different locations can there be?

16.3 Eight men are auditioning for the lead male role and six women for the lead female role in a play. How many different choices does the director have to fill the roles?

16.4 In early versions of the programming language BASIC, a variable name could be a single letter (A, B, ... , Z) or a single letter followed by a digit (0, 1, ... 9). Determine how many different variable names there are —identify the rule(s) you use to determine this number.

16.5 Eight Democrats and 7 Republicans are vying for their parties' nominations for president. How many different possibilities are there for President? How many different ways can a Democrat oppose a Republican in the final election?

16.6 Suppose Ford Taurus cars came in 3 models, 3 engine sizes, 2 transmission types, and 10 colors. How many distinct Tauruses can be manufactured?

16.7 The fast-food place serves hamburgers with or without mustard, ketchup, pickles, lettuce, and onions. How many different kinds of hamburgers are there?

16.8 There are five roads from Podunk to Kalamazoo and 3 from Kalamazoo to Central City. How many different ways are there to drive from Podunk to Central City?

16.9 One measure of security for a combination lock is the number of possible combinations. Suppose a combination lock requires selecting three numbers, each between 1 and 30 (inclusive). How many locks with different combinations can be made? If the numbers in a combination have to be different, how many different locks can be made?

16.10 How many nonnegative integers consisting of one to three digits are divisible by 5? Leading zeros are not allowed.

16.11 How many nonnegative integers consisting of one to three different digits are divisible by 5? Leading zeros are not allowed.

16.12 A red die and a black die are thrown. (A die is a cube with six sides, numbered $1, 2, 3, 4, 5, 6$.) How many different outcomes are there? How many outcomes sum to 2, 3, or 10?

16.13 One bag contains a red, a black, and a white ball; a second bag contains a red, a black, a green, and a white ball. Suppose a person chooses a bag and then selects two balls from it, at random. In what fraction of the cases are a red and a black ball selected, in that order?

16.14 How many different functions are there from the set { sun, no-sun } to the days of the week?

16.15 How many different functions are there from the days of the week to the set { sun, no-sun }.

16.16 How many different one-to-one functions are there from the set { sun, no-sun } to the days of the week?

16.17 How many different one-to-one functions are there from the days of the week to the set { sun, no-sun }?

16.18 A function $f : S \to T$ is *partial* if $f.s$ need not be defined for all s in S. If S and T are finite, how many partial functions exist from S to T?

Exercises on permutations of a set

16.19 How many permutations are there in each of the following words? LOT, LUCK, MAYBE, KISMET, DESTINY, RANDOMLY.

16.20 Six friends sit together in a row at the movies. One is a doctor and must sit on the aisle to allow for easy exit in case of an emergency. How many ways may the six people be seated?

16.21 How many permutations of ALGORITHM have the A, the L, and the G together (in any order)?

16.22 Ten people line up for a photograph. Juris, John, Gerry, and Bob want to stand together. How many different ways of lining the ten people up are there?

16.23 Write all the 3-permutations of a, b, c, d.

16.24 How many ways can 3 letters from the word ALGORITHM be chosen and written in a row? Five letters?

16.25 Suppose license plates are constructed using three letters from the word ITHACA followed by three digits. How many license plates can be constructed if (a) the letters have to be different, (b) the letters and digits need not be different?

16.26 A palindrome is a sequence of letters that reads the same backwards and forwards. Assuming there are 26 letters, determine how many palindromes there are of length 0, 1, 2, 3, and 4. Determine a general formula for the number of palindromes of length n.

16.27 In how many ways can a test with twenty true-false questions be answered (assuming all twenty questions are answered)?

16.28 In how many ways can a test with twenty true-false questions be answered if a student leaves some answers blank?

16.29 How many ways can 8 people be seated around a round table, if rotations are not considered different? If Mary and John, who are among the 8 people, do not want to sit together, how many ways are there?

16.30 Prove that $P(n, 2) + P(n, 1) = n^2$ for $n \geq 2$.

16.31 Prove that $P(n + 1, 3) = n^3 - n$ for $n \geq 2$.

16.32 Prove that $P(n + 1, 2) = P(n, 2) + 2 \cdot P(n, 1)$ for $n \geq 2$.

16.33 Prove that $P(n + 1, i) = P(n, i) + i \cdot P(n, i - 1)$ for $1 \leq i \leq n$. Mathematical induction is not needed; instead, look at the previous exercise.

16.34 Prove that $P(n, n) = P(n, n - 1)$ for $1 \leq n$.

Exercises on permutations of a bag

16.35 Determine the number of permutations of the bag consisting of the letters in the word EEE (a very big-shoe size). Use Theorem (16.7). Then write down all the permutations. Do the same for the words ERE and EAR.

16.36 For each of the words NOON and MOON, determine the number of permutations of the bag consisting of its letters. Use Theorem (16.7). Then write down all the permutations.

16.37 A coin is tossed 7 times, each time landing H (heads) or T (tails) to form a possible outcome. One possible outcome is HHTHHTH.

  (a) How many possible outcomes are there?

(b) How many outcomes have exactly 5 heads?

(c) How many outcomes have at least 6 heads?

(d) How many outcomes have at least 1 head?

16.38 How many different paths are there in the xy plane from $(1, 5)$ to $(7, 10)$ if a path consists of steps that go one unit to the right (an increase of 1 in the x-direction) or one unit up (an increase of one unit in the y-direction)?

16.39 How many different paths are there from $(-3, -2, -1)$ to $(3, 2, 1)$ if each step consists of a unit increase in one of the three dimensions?

16.40 A chessboard is an 8×8 grid of squares. A rook can move horizontally or vertically. In how many ways can a rook travel from the upper right corner to the lower left corner if all its steps are either to the left or down?

16.41 A byte consists of eight bits, each bit being a 0 or a 1.

(a) How many bytes are there?

(b) How many bytes contain exactly two 0's?

(c) Less than two 0's?

(d) At least two 0's?

16.42 How many two-byte (16-bit) strings of 0's and 1's contain

(a) seven 1's?

(b) At least fourteen 1's?

(c) At least two 1's?

(d) At most two 1's?

Exercises on combinations of a set

16.43 An urn contains red-colored numbers $1, 2, 3, 4, 5, 6$, blue-colored numbers $1, 2, 3, 4, 5$, green-colored numbers $1, 2, 3, 4$, and yellow-colored numbers $1, 2, 3$. How many different combinations of 4 red numbers, 3 blue numbers, 2 green numbers, and 1 yellow number can be selected from the urn?

16.44 An urn contains red-colored numbers $1, 2, 3, 4, 5, 6$, blue-colored numbers $1, 2, 3, 4, 5$, green-colored numbers $1, 2, 3, 4$, and yellow-colored numbers $1, 2, 3$. How many different combinations of 1 red number, 2 blue numbers, 3 green numbers, and 3 yellow numbers can be selected from the urn?

16.45 Sam and Tim, on a faculty of 10, refuse to be on the same committee. How many five-person committees can be formed?

16.46 A faculty of ten consists of 6 men and 4 women.

(a) How many committees of size 4 can be formed that have at least one man?

(b) How many committees of size 4 can be formed that have at least one woman?

(c) How many committees of size 4 can be formed that have at least one man and one woman?

(d) How many committees of two men and two women can be formed?

16.47 The student council consists of eight women and seven men.

(a) How many committees of 4 contain two women and two men?

(b) How many committees of size 4 have at least one woman?

(c) How many committees of size 4 can be formed that have at least one man and one woman?

General exercises on permutations and combinations

16.48 How many four-of-a-kind hands are there in poker?

16.49 How many poker hands are full houses?

16.50 How many poker hands are flushes?

16.51 How many poker hands are straights?

16.52 How many poker hands are two-pair hands?

16.53 How many poker hands are one-pair hands?

16.54 How many poker hands contain no pairs and are not flushes or straights of any kind? Having determined this, make a table of types of hands, the number of hands of that type, and the chances of being dealt that hand (to the nearest integer). For example, the chances of getting a straight flush are 1 in 64974.

16.55 While toasting, each of the n people at a party clink glasses once with all the others. How many "clinks" are there?

16.56 In how many combinations can the President of the U.S. invite 15 Senators from different States to the White House? (There are 50 States in the U.S. and two Senators per State.)

Exercises on $\binom{n}{r}$

16.57 Prove Absorption (16.15), $\binom{n}{r} = \frac{n}{r} \cdot \binom{n-1}{r-1}$ for $0 < r \le n$.

16.58 Prove Absorption (16.16), $r \cdot \binom{n}{r} = n \cdot \binom{n-1}{r-1}$ for $0 < r \le n$.

16.59 Prove theorem (16.17), $(n-r) \cdot \binom{n}{r} = n \cdot \binom{n-1}{r}$, for $0 \le r < n$. Hint: Apply Absorption (16.16) between two applications of Symmetry (16.14).

16.60 Prove Addition (16.18), $\binom{n}{r} = \binom{n-1}{r} + \binom{n-1}{r-1}$ for $0 < r < n$, by adding (16.16) and (16.17).

16.61 Prove theorem (16.19), $\binom{r+n+1}{n} = (\Sigma k \mid 0 \le k \le n : \binom{r+k}{k})$ for $0 \le n,\ 0 \le r$, by induction using Addition (16.18).

16.62 Prove theorem (16.20), $2^n = (\Sigma r \mid 0 \le r \le n : \binom{n}{r})$ for $0 \le n$, by induction using Addition (16.18).

16.63 Prove Binomial theorem (16.23) by induction on n.

Exercises on the principle of inclusion/exclusion

This set of exercises concerns the principle of inclusion/exclusion. This principle is slightly more advanced; hence, its relegation to exercises. Consider a set B of N objects, and let $p = p_0, \ldots, p_{r-1}$ be a list of properties that these objects might have. For example, a property could be having the color red or being an even integer. We want to develop a formula for the number of objects that have none of these properties.

Let q be a subsequence of sequence p. Let $N.q$ be the number of elements of B that have at least the properties in q. For example, $N(p_0 p_2)$ is the number of elements of B that have at least properties p_0 and p_2, while $N.p$ is the number of elements that have all the properties. Let $\overline{N}$ be the number of elements of B that do not have any of the properties of p. Then, we have the following theorem:

(16.52) $\overline{N} = (\Sigma i \mid 0 \leq i < r :$
$$(-1)^i \cdot (\Sigma q \mid q \text{ a subsequence of length } i \text{ of } p : N.q)) \quad .$$

16.64 Suppose twelve balls are painted as follows: Two are unpainted. Two are painted red, one is painted white, and one is painted blue. Two are painted red and white and one is painted white and blue. Three are painted red, white, and blue.

Let the properties be $p_0 = $ red, $p_1 = $ white, and $p_2 = $ blue. Write down $N.q$ for all subsequences q of p. Then verify that Theorem (16.52) holds in this case.

16.65 Use Theorem (16.52) to find the number of integers in the range $1..100$ that are not divisible by 3 or 5. Note that the number of integers in the range $1..n$ that are divisible by i is $\lfloor n/i \rfloor$. Check your answer by making a list of the 100 integers, crossing out those that are divisible by 3 or 5, and counting the rest.

16.66 Use Theorem (16.52) to find the number of integers in the range $1..200$ that are not divisible by any of the integers 2, 3, and 5.

Exercises on the pigeonhole principle

16.67 Prove that generalized Pigeonhole principle (16.45) implies (16.44).

16.68 Prove generalized Pigeonhole principle (16.45).

16.69 Suppose five distinct integers are selected from the set $\{1, 2, 3, 4, 5, 6, 7, 8\}$. Prove that at least one pair that has the sum 9.

Chapter 17

Recurrence Relations

In Sec. 12.2, we introduced inductive definitions, like the following definitions of exponentiation and the Fibonacci numbers.

$$b_0 = 1, \qquad b_n = b \cdot b^{n-1} \qquad \text{(for } n \geq 1\text{)}$$
$$F_0 = 0, \qquad F_1 = 1, \qquad F_n = F_{n-1} + F_{n-2} \qquad \text{(for } n \geq 2\text{)}$$

We found closed-form expressions for some of these inductive definitions, but not for others. For example, we found $b_n = (\Pi i \mid 1 \leq i \leq n : i)$, but we did not find a closed-form expression for F_n. In this chapter, we investigate techniques for finding closed-form expressions for inductive definitions. We restrict our attention to definitions that can be written as *linear recurrence relations*, or *linear difference equations*, as they are sometimes called. We give two "cookbook methods" for finding closed-form expressions of a large class of linear difference equations. The first method is based on *characteristic polynomials*; the second, on *generating functions*.

17.1 Homogeneous difference equations

For the moment, we deal only with the inductive part of an inductive definition; the constraints will be dealt with later. [1] Consider the recurrence relation

$$a_0 \cdot x_n + a_1 \cdot x_{n-1} + \cdots + a_k \cdot x_{n-k} = 0 \qquad \text{(for } n \geq k\text{)}$$

or

(17.1) $\quad (\Sigma i \mid 0 \leq i \leq k : a_i \cdot x_{n-i}) = 0 \quad ,$

for a function x, where the a_i are constants and $a_0 \neq 0$. (Throughout this chapter, we use x_i to denote the application of function x to argument i.) Dividing both sides of (17.1) by a_0, we arrive at a form in which $a_0 = 1$. All our examples have $a_0 = 1$.

Expression (17.1) is called an *order-k, homogeneous, linear difference equation* with constant coefficients (HDE, for short). It is called order-k because it can be viewed as defining x_n in terms of k values $x_{n-1}, \ldots, x_{n-k}$.

---

[1] We use the term *constraint* for a base case, or *boundary condition*.

It is called homogeneous because its RHS is 0. And, it is called linear because the exponents of all the x_i are 1.

As an example, the relation $x_n = 2 \cdot x_{n-1}$ (for $n \geq 1$) can be written as an order-1 HDE, with $a_0 = 1$ and $a_1 = -2$:

(17.2) $x_n - 2 \cdot x_{n-1} = 0$ (for $n \geq 1$) .

An HDE has many solutions. For example, $x_n = 0$ (for all n) and $x_n = 3 \cdot 2^n$ (for all n) are both solutions of (17.2). A key ingredient for finding all solutions of HDE (17.1) is its *characteristic polynomial*

$$a_0 \cdot \lambda^k + a_1 \cdot \lambda^{k-1} + \cdots + a_{k-1} \cdot \lambda + a_k \quad ,$$

or

(17.3) $(\Sigma i \mid 0 \leq i \leq k : a_i \cdot \lambda^{k-i})$.

Comparing (17.1) and (17.3), we see that the characteristic polynomial is constructed from the LHS of an HDE by replacing function application x_{n-i} by λ^{k-i}. For example, the characteristic polynomial for (17.2) is $\lambda - 2$.

Recall from high-school math that (17.3) with $a_0 \neq 0$ is called a *polynomial of degree* k in λ. The *roots* of this polynomial are values that, when substituted for λ in (17.3), result in an expression with value 0. A degree k polynomial (17.3) has k roots, call them $r_1, \ldots, r_k$. Thus, (17.3) can be written in the form $a_0 \cdot (\lambda - r_1) \cdot \ldots \cdot (\lambda - r_k)$. Further, if m of the roots are the same, that root is called a root of multiplicity m. For example, the polynomial $\lambda - 2$ has one root, $r_1 = 2$, of multiplicity 1. Also, the polynomial $\lambda^3 - 7 \cdot \lambda^2 + 15 \cdot \lambda - 9$ can be rewritten as $(\lambda - 1) \cdot (\lambda - 3) \cdot (\lambda - 3)$. Therefore, its roots are 1, 3, and 3; 1 is a root of multiplicity 1 and 3 is a root of multiplicity 2. Finally, the roots of a polynomial $a \cdot \lambda^2 + b \cdot \lambda + c$ are given by the *quadratic formula*

(17.4) $\dfrac{-b \pm \sqrt{b^2 - 4 \cdot a \cdot c}}{2 \cdot a}$.

The following theorem describes some solutions of HDE (17.1).

(17.5)   **Theorem.** Let r be any root of characteristic polynomial (17.3) of HDE (17.1). Then, $x_n = r^n$ is a solution of the HDE.

Proof. We substitute r^n for x_n (for all n) in the LHS of HDE (17.1) and calculate to show that the LHS equals the RHS (i.e. 0).

$$(\Sigma i \mid 0 \leq i \leq k : a_i \cdot r^{n-i})$$
$$= \quad \langle \text{Arithmetic} \rangle$$
$$(\Sigma i \mid 0 \leq i \leq k : a_i \cdot r^{n-k} \cdot r^{k-i})$$
$$= \quad \langle \text{Factor out } r^{n-k} \rangle$$

$$r^{n-k} \cdot (\Sigma i \mid 0 \leq i \leq k : a_i \cdot r^{k-i})$$
$$= \quad \langle \text{Textual substitution} \rangle$$
$$r^{n-k} \cdot (17.3)[\lambda := r]$$
$$= \quad \langle r \text{ is a root of } (17.3) \rangle$$
$$r^{n-k} \cdot 0$$
$$= \quad \langle \text{Arithmetic} \rangle$$
$$0 \qquad\qquad\qquad\qquad\qquad\qquad\qquad\qquad\qquad \square$$

As an example, order-1 HDE (17.2), $x_n - 2 \cdot x_{n-1} = 0$, has characteristic polynomial $\lambda - 2$, which has the one root $r_1 = 2$. Therefore, one solution of HDE (17.2) is $x_n = 2^n$ for all n.

If all k roots of the characteristic polynomial have multiplicity 1, then Theorem (17.5) yields k solutions to the HDE. However, if some root has multiplicity greater than 1, then the theorem yields fewer solutions. The following, more general, theorem yields k solutions no matter what the multiplicity of the roots. We will use this more general theorem. However, its proof is best done using calculus and is outside the scope of this text.

(17.6) **Theorem.** Let r be any root of multiplicity m of characteristic polynomial (17.3) for HDE (17.1). Then, for each j, $0 \leq j < m$,

$$x_n = n^j \cdot r^n \quad \text{(for all } n)$$

is a solution of the HDE.

Thus, the characteristic equation gives k solutions of the HDE. We now show that a linear combination of two solutions is also a solution.

(17.7) **Theorem.** Let $x = s1$ and $x = s2$ be two solutions of the HDE. (This means, for example, that $x_n = s1_n$ for $n \geq 0$.) Then the function f defined by $f_n = b_1 \cdot s1_n + b_2 \cdot s2_n$ (for $n \geq 0$) is also a solution.

Proof. We substitute f for x in the LHS of HDE (17.1) and calculate to show that the LHS is 0.

$$(\Sigma i \mid 0 \leq i \leq k : a_i \cdot x_{n-i})[x := f]$$
$$= \quad \langle \text{Textual substitution} \rangle$$
$$(\Sigma i \mid ...: a_i \cdot f_{n-i})$$
$$= \quad \langle \text{Definition of } f \rangle$$
$$(\Sigma i \mid ...: a_i \cdot (b_1 \cdot s1_{n-i} + b_2 \cdot s2_{n-i}))$$
$$= \quad \langle \text{Distributivity (15.51)} \rangle$$
$$(\Sigma i \mid ...: b_1 \cdot a_i \cdot s1_{n-i} + b_2 \cdot a_i \cdot s2_{n-i})$$
$$= \quad \langle \text{Distributivity (8.15)} \rangle$$
$$(\Sigma i \mid ...: b_1 \cdot a_i \cdot s1_{n-i}) + (\Sigma i \mid ...: b_2 \cdot a_i \cdot s2_{n-i})$$
$$= \quad \langle \text{Distributivity of } \cdot \text{ over } \Sigma \text{ (15.51)} \rangle$$
$$b_1 \cdot (\Sigma i \mid ...: a_i \cdot s1_{n-i}) + b_2 \cdot (\Sigma i \mid ...: a_i \cdot s2_{n-i})$$
$$= \quad \langle \text{Function } s1 \text{ and } s2 \text{ are solutions of HDE (17.1)} \rangle$$

$$
\begin{aligned}
&b_1 \cdot 0 + b_2 \cdot 0 \\
=\quad & \langle \text{Arithmetic} \rangle \\
&0
\end{aligned}
$$
                                                                                    □

In the same way, a linear combination of k solutions is also a solution. Further, we have the following theorem, which shows how to construct all solutions of HDE (17.1). (Its proof is beyond the scope of this text.)

(17.8) **Theorem.** Let $x = s1$, ... , $x = sk$ be the k solutions of HDE (17.1) given by Theorem (17.6). Then every solution of the HDE is a linear combination of the sj's, i.e. for arbitrary constants b_j, the function f that is defined as follows is a solution:

$$
f_n = (\Sigma j \mid 1 \le j \le k : b_j \cdot sj_n) \quad (\text{for } n \ge 0) \quad .
$$

Here is an example of the use of this theorem. We showed above that the single solution given by Theorem (17.8) of HDE $x_n - 2 \cdot x_{n-1} = 0$ is $x_n = 2^n$ (for $n \ge 0$). Therefore, by Theorem (17.8), all closed-form solutions of this HDE are given by $x_n = b_1 \cdot 2^n$ (for $n \ge 0$), where b_1 is an arbitrary constant.

CLOSED-FORM SOLUTIONS OF INDUCTIVE DEFINITIONS

We just showed how to find all solutions of an HDE. Many inductive definitions consist of such an HDE together with one or more constraints. These constraints help determine one particular solution. Such an inductive definitions can be solved as follows:

(17.9) **Method for solving an inductive definition based on an HDE.**
  (a) Rewrite the recurrence relation as an HDE.
  (b) Construct the characteristic polynomial of the HDE and find its roots.
  (c) Find the k solutions of the HDE according to Theorem (17.6).
  (d) Write down the general solution, using Theorem (17.8).
  (e) Use the base cases as constraints to determine the desired particular solution from the general solution.

(17.10) **Example.** Find a closed-form solution of inductive definition $x_0 = 2$, $x_n = 2 \cdot x_{n-1}$ (for $n \ge 1$).

The recurrence relation of this inductive definition equals the order-1 HDE $x_n - 2 \cdot x_{n-1} = 0$. Its characteristic polynomial is $\lambda - 2$, whose single root is 2. By Theorems (17.6) and (17.8), the general solution is $x_n = b_1 \cdot 2^n$ (for $n \ge 0$). Using the constraint $x_0 = 2$, we manipulate the general solution with $n := 0$ to calculate b_1:

$$x_0 = b_1 \cdot 2^0$$
$$= \quad \langle \text{Constraint } x_0 = 2 \rangle$$
$$2 = b_1 \cdot 2^0$$
$$= \quad \langle 2^0 = 1 \,;\, \text{Arithmetic} \rangle$$
$$2 = b_1$$

Therefore, the closed-form solution is $x_n = 2^{n+1}$ (for $n \geq 0$).   □

(17.11) **Example.** A bank pays an annual interest of 6 percent, which
is compounded monthly (that is, each month it pays .5 percent
interest). If Pat starts with an initial deposit of \$100, how much
will she have after one year?

Let p_n be the amount of money Pat has in the bank after n months, so
$p_0 = 100$. We have the recurrence relation $p_n = p_{n-1} + .005 \cdot p_{n-1}$ (for
$n > 0$), or $p_n = 1.005 \cdot p_{n-1}$. We can write this recurrence relation as the
HDE $p_n - 1.005 \cdot p_{n-1} = 0$. Its characteristic polynomial is $\lambda - 1.005$,
which has the single root 1.005, so the general solution of the HDE is
$p_n = b_1 \cdot 1.005^n$. Substituting the constraint $p_0 = 100$ into the general
solution yields $100 = b_1 \cdot 1.005^0$, so $b_1 = 100$. Thus, we have calculated
the solution $p_n = 100 \cdot 1.005^n$ (for $n \geq 0$). After 12 months, Pat has
$p_{12} = 100 \cdot 1.005^{12} = 106.17$ dollars.   □

(17.12) **Example.** Find a closed-form solution of the definition of Fibonacci numbers: $F_0 = 0$, $F_1 = 1$, and $F_n = F_{n-1} + F_{n-2}$
(for $n \geq 2$).

The inductive part of this definition can be written as the HDE $F_n -
F_{n-1} - F_{n-2} = 0$ (for $n \geq 2$). Its characteristic polynomial is $\lambda^2 - \lambda - 1$.
Using quadratic formula (17.4), we find the roots of this polynomial and
then write the polynomial as

$$(\lambda - \frac{1 + \sqrt{5}}{2}) \cdot (\lambda - \frac{1 - \sqrt{5}}{2}) \;,$$

or $(\lambda - \phi) \cdot (\lambda - \hat{\phi})$. (Recall from page 225 that $\phi = (1 + \sqrt{5})/2$ is the *golden
ratio* and $\hat{\phi}$ is its twin.) Therefore, the roots are $r_1 = \phi$ and $r_2 = \hat{\phi}$. By
Theorem (17.6), two solutions are $F_n = \phi^n$ and $F_n = \hat{\phi}^n$. By Theorem
(17.8), the general solution of the HDE is $F_n = b_1 \cdot \phi^n + b_2 \cdot \hat{\phi}^n$.

We use the base cases $F_0 = 0$ and $F_1 = 1$ to construct the required
particular solution. Substituting 0 for n in the general solution yields

$$F_0 = b_1 \cdot \phi^0 + b_2 \cdot \hat{\phi}^0$$
$$= \quad \langle F_0 = 0 \,;\, X^0 = 1 \text{ (for any } X \text{)}, \text{ twice} \rangle$$
$$0 = b_1 \cdot 1 + b_2 \cdot 1$$
$$= \quad \langle \text{Arithmetic} \rangle$$
$$b_1 = -b_2$$

Similarly, substituting 1 for n we obtain

$$1 = b_1 \cdot \phi + b_2 \cdot \hat{\phi} \quad .$$

Solving these two equations for b_1 and b_2 yields $b_1 = 1/\sqrt{5}$ and $b_2 = -1/\sqrt{5}$, so a closed-form expression is $F_n = (\phi^n - \hat{\phi}^n)/\sqrt{5}$.   □

(17.13) **Example.** Find a closed-form solution of the HDE $x_n - x_{n-2} = 0$ (for $n \geq 2$), subject to the constraints $x_1 = 2$ and $x_2 = 4$.

The HDE has the characteristic polynomial $\lambda^2 - 1$. This polynomial has roots 1 and -1. Hence, by Theorem (17.6), 1^n and $(-1)^n$ are solutions of the HDE. By Theorem (17.8), the general solution is $x_n = b_1 \cdot 1^n + b_2 \cdot (-1)^n$. Using constraint $x_1 = 2$, we derive the equation $2 = b_1 - b_2$. Using constraint $x_2 = 4$, we derive $4 = b_1 + b_2 \cdot (-1)^2 = b_1 + b_2$. Solving these two equations for b_1 and b_2, we get $b_1 = 3$ and $b_2 = 1$. Therefore, the closed-form solution is $x_n = 3 + (-1)^n$ for all $n \geq 0$.   □

(17.14) **Example.** Find a closed-form solution of the HDE $x_n - 2 \cdot x_{n-1} + x_{n-2}$ (for $n \geq 2$), subject to the constraints $x_1 = 1$ and $x_2 = 2$.

The HDE has the characteristic polynomial $\lambda^2 - 2 \cdot \lambda + 1$. This polynomial equals $(\lambda - 1) \cdot (\lambda - 1)$, so it has root 1 with multiplicity 2. According to Theorem (17.6), 1^n and $n \cdot 1^n$ are solutions of the HDE. By Theorem (17.8), the general solution is $x_n = b_1 \cdot 1^n + b_2 \cdot n \cdot 1^n$, i.e. $x_n = b_1 + b_2 \cdot n$. Constraint $x_1 = 1$ yields the equation $1 = b_1 + b_2$; Constraint $x_2 = 2$ yields $2 = b_1 + 2 \cdot b_2$. Solving these two equations for b_1 and b_2 yields $b_1 = 0$ and $b_2 = 1$, so the closed-form solution is $x_n = n$ for $n \geq 0$.   □

DEALING WITH COMPLEX ROOTS

A root of the characteristic polynomial of an HDE may be a *complex number* (defined below). We now show how to deal with such roots. We emphasize immediately that *the theory developed thus far holds for these cases* —Theorems (17.6) and (17.8) still provide the theory for solving HDEs. However, to apply the theory, we need to know how to manipulate complex numbers. Complex numbers are not central to this text, so our treatment is brief.

The roots of a polynomial may involve square roots of negative numbers, like $\sqrt{-1}$ and $\sqrt{-3}$. For example, quadratic formula (17.4) has a negative square root if $b^2 < 4 \cdot a \cdot c$. As with the square roots of positive numbers, this new kind of number satisfies $(\sqrt{x})^2 = x$, so

$$(\sqrt{-E})^2 = -E \quad .$$

Beyond that, all the usual rules of arithmetic hold. For example, $\sqrt{-E} = \sqrt{-1 \cdot E} = \sqrt{-1} \cdot \sqrt{E}$. The value $\sqrt{-1}$ has a special status in that it is given a name: $\mathbf{i}$. Thus, $\mathbf{i}^2 = -1$.

A complex number has the form $x + \mathbf{i} \cdot y$, where x and y are real numbers. Number x is called the *real part* of $x + \mathbf{i} \cdot y$ and y is called the *imaginary part*. If $y = 0$, then the number is real, so the real numbers are a subset of the complex numbers. The conventional rules of arithmetic apply. For example,

$$x + \mathbf{i} \cdot y + z + \mathbf{i} \cdot w = (x + z) + \mathbf{i} \cdot (y + w) \quad .$$

And, we multiply complex numbers as follows:

$$
\begin{aligned}
& (x + \mathbf{i} \cdot y) \cdot (z + \mathbf{i} \cdot w) \\
= \quad & \langle \text{Distributivity} \rangle \\
& x \cdot (z + \mathbf{i} \cdot w) + \mathbf{i} \cdot y \cdot (z + \mathbf{i} \cdot w) \\
= \quad & \langle \text{Distributivity} \rangle \\
& x \cdot z + \mathbf{i} \cdot x \cdot w + \mathbf{i} \cdot y \cdot z + \mathbf{i}^2 \cdot y \cdot w \\
= \quad & \langle \text{Symmetry, Associativity; } \mathbf{i}^2 = -1 \rangle \\
& (x \cdot z - y \cdot w) + \mathbf{i} \cdot (x \cdot w + y \cdot z)
\end{aligned}
$$

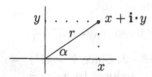 A complex number $x + \mathbf{i} \cdot y$ can be depicted as a point in a two-dimensional plane, as illustrated in this paragraph. Real part x measures a distance along the horizontal axis; imaginary part y, a distance along the vertical axis. As illustrated, the complex number can also be defined by its *magnitude* r and its *angle* α, which satisfy [2]

$$r = \sqrt{x^2 + y^2}, \qquad tan.\alpha = y/x \quad .$$

The pair $\langle r, \alpha \rangle$ is called the polar-coordinate representation of the complex number, and r and α are called polar coordinates. We also have $x = r \cdot cos.\alpha$ and $y = r \cdot sin.\alpha$. Thus,

$$x + \mathbf{i} \cdot y = r \cdot (cos.\alpha + \mathbf{i} \cdot sin.\alpha) \quad .$$

De Moivre's theorem comes in handy for manipulating HDEs whose characteristic polynomial has complex roots. We state De Moivre's theorem without proof. For $n \geq 0$,

(17.15) **De Moivre:** $(cos.\alpha + \mathbf{i} \cdot sin.\alpha)^n = cos(n \cdot \alpha) + \mathbf{i} \cdot sin(n \cdot \alpha)$

---

[2] We assume knowledge of trigonometric functions tan, sin, and cos.

We now illustrate how to deal with complex numbers that arise in solving HDEs. Most of the calculations are left as exercises.

(17.16) Example. Find a closed-form solution of the inductive definition

$$x_0 = 0, \quad x_1 = 1, \quad x_n = -x_{n-2} \quad (\text{for } n \geq 2) \ .$$

HDE: $x_n + x_{n-2} = 0$

Charact. polynomial: $\lambda^2 + 1$, or $(\lambda - \mathbf{i}) \cdot (\lambda + \mathbf{i})$

Roots r_1 and r_2: $\mathbf{i}$ and $-\mathbf{i}$

General solution: $x_n = b_1 \cdot \mathbf{i}^n + b_2 \cdot (-\mathbf{i})^n = (b_1 + b_2 \cdot (-1)^n) \cdot \mathbf{i}^n$

Constraint equation 1: $0 = b_1 + b_2$

Constraint equation 2: $1 = (b_1 - b_2) \cdot \mathbf{i}$

Constants b_1 and b_2: $-\mathbf{i}/2$ and $\mathbf{i}/2$

Closed-form solution: $x_n = (-1 + (-1)^n) \cdot \mathbf{i}^{n+1}/2$

The closed-form solution contains the complex number $\mathbf{i}$. One way to eliminate it is to use case analysis. Another way is to use polar coordinates and De Moivre's theorem. In this example, we use the first method; in the next example, we use the second method. We look at three cases: $n + 1$ odd, $n + 1$ divisible by 4, and $n + 1$ divisible by 2 but not by 4. These cases are chosen in such a way that, in each case, $\mathbf{i}$ cancels out of the closed-form solution.

Case $odd(n + 1)$**:** Here, $-1 + (-1)^n = -1 + 1 = 0$, so $x_n = 0$.

Case $4 \mid (n + 1)$**:** Here, n is odd, so $-1 + (-1)^n = -2$. Since $\mathbf{i}^4 = (\mathbf{i}^2)^2 = (-1)^2 = 1$, the closed-form expression for x_n reduces to $x_n = -1$.

Case $2 \mid (n + 1)$ **but** $4 \nmid (n + 1)$**:** As in the previous case, $-1 + (-1)^n = -2$. One can also show that $\mathbf{i}^{n+1} = -1$, so the closed-form expression for x_n reduces to $x_n = 1$.

Hence, the closed-form solution can be written as

$$x_n = \begin{cases} 0 & \text{if } odd(n+1) \\ -1 & \text{if } 4 \mid (n+1) \\ 1 & \text{if } 2 \mid (n+1) \wedge 4 \nmid (n+1) \ . \end{cases}$$

So the sequence $x_0, x_1, x_2, \ldots$ is $0, 1, 0, -1, 0, 1, 0, -1, \ldots$. It is a repeating sequence, with period 4. $\qquad\qquad\square$

(17.17) Example. Find a closed-form solution of

$$x_n = x_{n-1} - x_{n-2} \quad (\text{for } n \geq 2)$$

under the constraints $x_0 = 0$ and $x_1 = 1$.

HDE: $\qquad\qquad\qquad\qquad$ $x_n - x_{n-1} + x_{n-2} = 0$

Charact. polynomial: $\qquad$ $\lambda^2 - \lambda + 1$

Root r_1: $\qquad\qquad\qquad$ $(1 + \mathbf{i} \cdot \sqrt{3})/2 \quad = cos.\frac{\pi}{3} + \mathbf{i} \cdot sin.\frac{\pi}{3}$

Root r_2: $\qquad\qquad\qquad$ $(1 - \mathbf{i} \cdot \sqrt{3})/2 \quad = cos.\frac{\pi}{3} - \mathbf{i} \cdot sin.\frac{\pi}{3}$

General solution: $\qquad\quad$ $x_n = b_1 \cdot r_1^n + b_2 \cdot r_2^n$

Constraint equation 1: $\quad$ $0 = b_1 + b_2$

Constraint equation 2: $\quad$ $1 = b_1 \cdot \frac{1+\mathbf{i}\cdot\sqrt{3}}{2} + b_2 \cdot \frac{1-\mathbf{i}\cdot\sqrt{3}}{2}$

Constants b_1 and b_2 $\quad$ $1/\sqrt{-3}, \; -1/\sqrt{-3}$

The closed-form solution is then

$$x_n = b_1 \cdot r_1^n + b_2 \cdot r_2^n$$
$$= \quad \langle \text{Above definition of } b_1, \; b_2, \; r_1 \text{ and } r_2 \rangle$$
$$x_n = \tfrac{1}{\sqrt{-3}} \cdot (cos.\tfrac{\pi}{3} + \mathbf{i} \cdot sin.\tfrac{\pi}{3})^n - \tfrac{1}{\sqrt{-3}} \cdot (cos.\tfrac{\pi}{3} - \mathbf{i} \cdot sin.\tfrac{\pi}{3})^n$$
$$= \quad \langle \text{De Moivre's theorem, twice} \rangle$$
$$x_n = \tfrac{1}{\sqrt{-3}} \cdot (cos.\tfrac{n \cdot \pi}{3} + \mathbf{i} \cdot sin.\tfrac{n \cdot \pi}{3}) -$$
$$\tfrac{1}{\sqrt{-3}} \cdot (cos.\tfrac{n \cdot \pi}{3} - \mathbf{i} \cdot sin.\tfrac{n \cdot \pi}{3})$$
$$= \quad \langle \text{Arithmetic} \rangle$$
$$x_n = \tfrac{2}{\sqrt{-3}} \cdot \mathbf{i} \cdot sin.\tfrac{n \cdot \pi}{3}$$
$$= \quad \langle \mathbf{i}/\sqrt{-3} = \mathbf{i}/(\sqrt{3} \cdot \mathbf{i}) = 1/\sqrt{3} \rangle$$
$$x_n = \tfrac{2}{\sqrt{3}} \cdot sin.\tfrac{n \cdot \pi}{3}$$

The closed-form solution does not refer to $\mathbf{i}$, even though complex numbers were used in the manipulations that led to the solution. Substituting small values for n in the closed-form solution, we see that $x_0, x_1, x_2, \ldots$ is

$$0, 1, 1, 0, -1, -1, 0, 1, 1, 0, -1, -1, \ldots \quad .$$

This is a repeating sequence, with period 6. $\qquad\qquad\qquad\qquad$ $\square$

17.2 Nonhomogeneous difference equations

A *nonhomogeneous* difference equation (NDE, for short) has the form

$$a_0 \cdot x_n + a_1 \cdot x_{n-1} + \cdots + a_k \cdot x_{n-k} = f_n \quad (\text{for } n \geq k)$$

or

$$(17.18) \quad (\Sigma i \mid 0 \leq i \leq k : a_i \cdot x_{n-i}) = f_n \quad (\text{for } n \geq k) \quad ,$$

for some function f over the natural numbers. Comparing HDE (17.1) to NDE (17.18), we see that an HDE is an NDE for which $f_n = 0$ (for $n \geq 0$).

(17.19) **Example of an NDE.** $x_n - 2 \cdot x_{n-1} = 1$ (for $n \geq 1$). Here, $f_n = 1$ (for $n \geq 0$).   □

(17.20) **Example of an NDE.** $x_n - x_{n-1} = f_n$ (for $n \geq 1$), where f satisfies $f_n = f_{n-1}$ (for $n \geq 1$). Note that f is a constant function.   □

Finding a closed-form solution of an NDE with constraints involves first finding a general solution of the NDE. In turn, finding a general solution, will require finding some arbitrary solution p (say) of the NDE —it doesn't matter which one. We now show that, at least in some cases, finding an arbitrary solution can be done fairly easily. We give the method, without giving the theory behind it.

(17.21) **Method for finding an arbitrary solution of an NDE.** Suppose function f of the NDE has one of the forms given in the left column of Table 17.1. Then choose the corresponding trial solution p in the right column. The c_i of the trial solution are constant symbols. To determine their values, substitute the trial solution into the NDE and calculate.

If f is itself a solution of the HDE-form of the initial NDE and the calculation does not produce a solution, then, instead of p of Table 17.1, try the trial solutions given by $n \cdot p_n$, $n^2 \cdot p_n$, ...

(17.22) **Example.** Find some solution of

$$x_n - 2 \cdot x_{n-1} = 1 \quad (\text{for } n \geq 1) \ .$$

Here, $f_n = 1$ (for $n \geq 0$), so Table 17.1 suggests trying the solution $p_n = c_0$ for some constant c_0. We substitute p for x and calculate.

$$\begin{aligned}
&(x_n - 2 \cdot x_{n-1} = 1)[x := p] \\
= \quad &\langle \text{Textual substitution} \rangle \\
&p_n - 2 \cdot p_{n-1} = 1 \\
= \quad &\langle \text{Definition of } p\colon p_n = c_0 \rangle
\end{aligned}$$

TABLE 17.1. TRIAL SOLUTIONS FOR AN NDE

Function f	Trial solution p
$f_n = (\Sigma i \mid 0 \leq i \leq r : C_i \cdot n^i)$	$p_n = (\Sigma i \mid 0 \leq i \leq r : c_i \cdot n^i)$
$f_n = (\Sigma i \mid 0 \leq i \leq r : C_i \cdot n^i) \cdot d^n$	$p_n = (\Sigma i \mid 0 \leq i \leq r : c_i \cdot n^i) \cdot d^n$

The C_i are constants; the c_i are constant identifiers that denote values that are to be determined. For example, if $f_n = 5 + 6 \cdot n^2$, then $p_n = c_0 + c_1 \cdot n + c_2 \cdot n^2$, $C_0 = 5$, $C_1 = 0$, and $C_2 = 6$.

$$c_0 - 2 \cdot c_0 = 1$$
$$= \quad \langle \text{Rearrange} \rangle$$
$$c_0 = -1$$

Therefore, one solution is $x_n = p_n = -1$ (for all n).   □

(17.23) **Example.** Find some solution of

$$x_n - 2 \cdot x_{n-1} = 6 \cdot n, \quad \text{for } n \geq 1 \quad .$$

Here, $f_n = 6 \cdot n$ (for $n \geq 0$), so Table 17.1 suggests the trial solution p given by $p_n = c_0 + c_1 \cdot n$. We substitute p for x in the NDE and calculate.

$$(x_n - 2 \cdot x_{n-1} = 6 \cdot n)[x := p]$$
$$= \quad \langle \text{Textual substitution} \rangle$$
$$p_n - 2 \cdot p_{n-1} = 6 \cdot n$$
$$= \quad \langle \text{Definition of } p \colon p_n = c_0 + c_1 \cdot n \rangle$$
$$c_0 + c_1 \cdot n - 2 \cdot (c_0 + c_1 \cdot (n-1)) = 6 \cdot n$$
$$= \quad \langle \text{Rearrange} \rangle$$
$$2 \cdot c_1 - c_0 = (c_1 + 6) \cdot n$$

For the last equation to hold for all n, both sides must be 0, so $c_1 = -6$ and $c_0 = -12$. Hence, one solution of the NDE is $x_n = p_n = -12 - 6 \cdot n$ (for $n \geq 0$).   □

(17.24) **Example.** Find some solution of

$$x_n - 2 \cdot x_{n-1} = 2^n, \quad (\text{for } n \geq 1) \quad .$$

Here, $f_n = 2^n$, so Table 17.1 suggests the trial sequence p given by $p_n = c_0 \cdot 2^n$ (for $n \geq 0$). Substituting p for x in the NDE and manipulating yields

$$c_0 \cdot 2^n - c_0 \cdot 2^n = 2^n \quad ,$$

which has no solution. The problem is that $f_n = 2^n$ is itself a solution of the NDE. According to Method (17.21), we try the trial solution p given by $p_n = c_0 \cdot n \cdot 2^n$.

$$(x_n - 2 \cdot x_{n-1} = 2^n)[x := p]$$
$$= \quad \langle \text{Textual substitution} \rangle$$
$$p_n - 2 \cdot p_{n-1} = 2^n$$
$$= \quad \langle \text{Definition of } p \colon p_n = c_0 \cdot n \cdot 2^n \rangle$$
$$c_0 \cdot n \cdot 2^n - 2 \cdot c_0 \cdot (n-1) \cdot 2^{n-1} = 2^n$$
$$= \quad \langle \text{Divide both sides by } 2^n \rangle$$
$$c_0 \cdot n - c_0 \cdot (n-1) = 1$$

$$= \quad \langle \text{Arithmetic} \rangle$$
$$c_0 = 1$$

Hence, one solution of the NDE is $x_n = p_n = 2^n$ (for $n \geq 0$). $\square$

The following theorem tells us why an arbitrary solution of the NDE is important in finding a general solution of it.

(17.25) **Theorem.** Let p be an arbitrary solution of an NDE and let g be the general solution of the corresponding HDE. Then function pg defined by $pg_n = p_n + g_n$ (for $n \geq 0$) is the general solution of the NDE.

Proof. We show that pg is a solution by substituting it for x in the LHS of (17.18) and manipulating to show that it equals the RHS:

$$(\Sigma i \mid 0 \leq i \leq k : a_i \cdot x_{n-i})[x := pg]$$
$$= \quad \langle \text{Textual substitution} \rangle$$
$$(\Sigma i \mid 0 \leq i \leq k : a_i \cdot pg_{n-i})$$
$$= \quad \langle \text{Definition of } pg \rangle$$
$$(\Sigma i \mid 0 \leq i \leq k : a_i \cdot (p_{n-i} + g_{n-i}))$$
$$= \quad \langle \text{Distributivity} \rangle$$
$$(\Sigma i \mid 0 \leq i \leq k : a_i \cdot p_{n-i}) + (\Sigma i \mid 0 \leq i \leq k : a_i \cdot g_{n-i})$$
$$= \quad \langle p \text{ is one solution of the NDE;}$$
$$\quad\quad g \text{ is a solution of the corresponding HDE} \rangle$$
$$f_n + 0$$
$$= \quad \langle \text{Arithmetic} \rangle$$
$$f_n$$

Hence, pg is a solution. The proof that pg is the general solution is beyond the scope of this text. $\square$

(17.26) **Example.** Find a closed-form solution of the inductive definition

$$x_0 = 3, \quad x_n - 2 \cdot x_{n-1} = 1 \quad (\text{for } n \geq 1) \quad .$$

A general solution of the corresponding HDE was determined in Example (17.10) to be $g_n = b_1 \cdot 2^n$ (for $n \geq 0$). A solution of the NDE was determined in Example (17.22) to be $p_n = -1$. Therefore, a general solution of the NDE is

$$pg_n = g_n + p_n = b_1 \cdot 2^n - 1 \quad .$$

Using the constraint $x_0 = 3$, we get $b_1 = 4$, so the closed-form solution is $x_n = 4 \cdot 2^n - 1$. $\square$

17.3   Generating functions

We present a theory of *generating functions* for sequences of numbers. This little theory turns out to be extremely useful in finding closed-form solutions of inductive definitions. Often, the generating function for an inductive definition can be shown to be equal to the sum of a few generating functions of forms given in the theorems mentioned below, and theorem (17.34) then gives us the generating function for the inductive definition in a form that lets us read off a closed-form solution.

The generating function $G(z)$ for a finite sequence $x = x_0$, x_1, x_2, $\ldots$, $x_{\#x-1}$ of real numbers is the polynomial

(17.27) $G(z) = (\Sigma i \mid 0 \le i < \#x : x_i \cdot z^i)$.

For example, the generating function for the sequence $9, 2, 4$ is $9 + 2 \cdot z + 4 \cdot z^2$. The generating function is not used to evaluate the sum at a given point z . Think of it instead as a new kind of mathematical entity, whose main purpose is to give a different representation for a sequence (or a function over the natural numbers). Thus,

$$9, 2, 4 \quad \text{and} \quad 9 + 2 \cdot z + 4 \cdot z^2$$

are simply two different representations for the same sequence. Furthermore, the second representation affords techniques for analyzing and manipulating sequences, as we shall see.

Although generating functions are defined as sums, it turns out that they can have very different —and simple— forms. We see this in the following theorem.

(17.28)   **Theorem.** For $n \ge 0$, the generating function for the sequence of n binomial coefficients $\binom{n}{0}$, $\binom{n}{1}$, $\ldots$, $\binom{n}{n}$ is $(z+1)^n$.

Proof. Binomial theorem (16.23) (see page 348) is

$$(x+y)^n = (\Sigma k \mid 0 \le k \le n : \binom{n}{k} \cdot x^k \cdot y^{n-k})   .$$

We can use this theorem to compute the generating function by noticing that the substitution $x, y := z, 1$ in its RHS yields the desired generating function:

$$
\begin{aligned}
&((x+y)^n = (\Sigma k \mid 0 \le k \le n : \binom{n}{k} \cdot x^k \cdot y^{n-k}))[x, y := z, 1] \\
=\quad &\langle \text{Textual substitution; } 1^{n-k} = 1 \rangle \\
&(z+1)^n = (\Sigma k \mid 0 \le k \le n : \binom{n}{k} \cdot z^k) \\
=\quad &\langle \text{Definition of } G(z) \rangle \\
&(z+1)^n = G(z)
\end{aligned}
$$
□

We now turn to generating functions for infinite sequences. The generating function $G(z)$ for an infinite sequence x_0, x_1, x_2, ... (or function x whose domain is the natural numbers) is the infinite polynomial

(17.29) $G(z) = (\Sigma i \mid 0 \leq i : x_i \cdot z^i)$.

For example, the generating function for the sequence $2, 4, 6, 8, \ldots$ is

$$2 + 4 \cdot z + 6 \cdot z^2 + 8 \cdot z^3 + \ldots = (\Sigma i \mid 0 \leq i : 2 \cdot (i+1) \cdot z^i) \ .$$

Having introduced a new entity, the generating function, we analyze its properties. We start off by finding a simple form for a certain sequence.

(17.30) **Theorem.** For c a constant, $c \neq 0$, the generating function $G(z)$ for the sequence $c^0, c^1, c^2, c^3, \ldots$ is $1/(1 - c \cdot z)$.

Proof. We manipulate the definition of $G(z)$.

$$\begin{aligned}
& G(z) = (\Sigma i \mid 0 \leq i : c^i \cdot z^i) \\
= \quad & \langle \text{Subtract 1 from both sides —remember, } c^0 \cdot z^0 = 1 \rangle \\
& G(z) - 1 = (\Sigma i \mid 1 \leq i : c^i \cdot z^i) \\
= \quad & \langle \text{Change of dummy (8.22)} \rangle \\
& G(z) - 1 = (\Sigma i \mid 0 \leq i : c^{i+1} \cdot z^{i+1}) \\
= \quad & \langle \text{Factor out } c \cdot z \text{ from the sum in the RHS} \rangle \\
& G(z) - 1 = c \cdot z \cdot (\Sigma i \mid 0 \leq i : c^i \cdot z^i) \\
= \quad & \langle \text{Definition of } G(z) \rangle \\
& G(z) - 1 = c \cdot z \cdot G(z) \\
= \quad & \langle \text{Arithmetic} \rangle \\
& G(z) = 1/(1 - c \cdot z)
\end{aligned}$$
 $\square$

The calculation in this proof illustrates a generally useful technique for manipulating generating functions. Subtracting the first term of the generating function in the RHS in the first step allowed a change of dummy and a factoring step; together, these resulted in the generating function reappearing in the RHS, and this in turn allowed the summation to be eliminated completely.

Try this technique in proving the following theorems.

(17.31) **Theorem.** For c and d constants, the generating function $G(z)$ for the sequence $d \cdot c^0, d \cdot c^1, d \cdot c^2, d \cdot c^3, \ldots$ is $d/(1 - c \cdot z)$.

(17.32) **Theorem.** For d a constant, the generating function $G(z)$ for the sequence $0 \cdot d, 1 \cdot d, 2 \cdot d, 3 \cdot d, \ldots$ is $d \cdot z/(1 - z)^2$.

(17.33) **Theorem.** The generating function for the sequence x defined by $x_n = n \cdot (n - 1)$ is $2 \cdot z^2/(1 - z)^3$.

(17.34) **Theorem.** Let $G(z)$ be the generating function for the sequence $g_0, g_1, g_2, \ldots$. Let $H(z)$ be the generating function for the sequence $h_0, h_1, h_2, \ldots$. Then $GH(z) = G(z) + H(z)$ is the generating function for the sequence $g_0 + h_0, g_1 + h_1, g_2 + h_2, \ldots$.

We have one final theorem in our theory of generating functions. This theorem gives the generating function for a function that is defined only for integers that are at least k, where k is some natural number.

(17.35) **Theorem.** Let $y_0, y_1, y_2, \ldots$ be a sequence with generating function $G(z)$. Let k be a natural number, and let function f be defined by $f_i = y_{i-k}$ for $i \geq k$. Then the generating function for f is $z^k \cdot G(z)$.

Proof. The generating function for f is $(\Sigma n \mid k \leq n : f_n \cdot z^n)$. We manipulate this generating function.

$$
\begin{aligned}
& (\Sigma n \mid k \leq n : f_n \cdot z^n) \\
= \quad & \langle \text{Definition of } f \rangle \\
& (\Sigma n \mid k \leq n : y_{n-k} \cdot z^n) \\
= \quad & \langle \text{Change of dummy (8.22)} \rangle \\
& (\Sigma n \mid 0 \leq n : y_n \cdot z^{n+k}) \\
= \quad & \langle \text{Factor out } z^k \rangle \\
& z^k \cdot (\Sigma n \mid 0 \leq n : y_n \cdot z^n) \\
= \quad & \langle \, G(z) \text{ is the generating function for } y_0, y_1, \ldots \rangle \\
& z^k \cdot G(z) \qquad\qquad\qquad\qquad\qquad\qquad\qquad\qquad\qquad \square
\end{aligned}
$$

GENERATING FUNCTIONS FOR HDEs

An HDE of form (17.1) defines a sequence x_0, x_1, $\ldots$, so the general form of the generating function for an HDE is

$$
G(z) = x_0 + x_1 \cdot z^1 + x_2 \cdot z^2 + \ldots = (\Sigma n \mid 0 \leq n : x_n \cdot z^n) \quad .
$$

We now show that the generating function for an HDE is a fraction whose numerator and denominator are polynomials in z. Such a function is called a *rational function* of z. Further, we show how to construct this form of the generating function. We start with an example of the construction.

(17.36) **Example.** Construct the rational-function form of the generating function for HDE $x_n - x_{n-2} = 0$ (for $n \geq 2$).

$$
\begin{aligned}
& x_n - x_{n-2} = 0 \\
= \quad & \langle \text{Multiply both sides by } z^n \rangle \\
& x_n \cdot z^n - x_{n-2} \cdot z^n = 0 \\
= \quad & \langle \text{This holds for all } n \text{. Since this is an order-2 HDE, sum}
\end{aligned}
$$

both sides from $n = 2$ on.$\rangle$

$(\Sigma n \mid 2 \leq n : x_n \cdot z^n) - (\Sigma n \mid 2 \leq n : x_{n-2} \cdot z^n) = 0$

$=$ $\langle$First sum is $G(z) - x_0 - x_1 \cdot z$; Change of dummy (8.22)$\rangle$

$G(z) - x_0 - x_1 \cdot z - (\Sigma n \mid 0 \leq n : x_n \cdot z^{n+2}) = 0$

$=$ $\langle$Distributivity —to factor out z^2 $\rangle$

$G(z) - x_0 - x_1 \cdot z - z^2 \cdot (\Sigma n \mid 0 \leq n : x_n \cdot z^n) = 0$

$=$ $\langle$Definition of $G(z)$ $\rangle$

$G(z) - x_0 - x_1 \cdot z - z^2 \cdot G(z) = 0$

$=$ $\langle$Rearrange terms$\rangle$

$G(z) - z^2 \cdot G(z) = x_0 + x_1 \cdot z$

$=$ $\langle$Factor out $G(z)$; divide both sides by $1 - z^2$ $\rangle$

$G(z) = (x_0 + x_1 \cdot z) / (1 - z^2)$ $\qquad\qquad\qquad$ $\square$

The rational-function forms of the generating polynomials for HDEs of order 1, 2, and 3 are shown in Table 17.2. Thus, finding the rational-function form for the generating polynomial for a particular HDE of order 1, 2, or 3 is simply a matter of choosing the right formula from this table and sticking in appropriate values for the a_i and x_i.

From Table 17.2, you can probably guess what the rational-function form of the generating polynomial for an order-n HDE is. The proof-construction of this generating polynomial is not very difficult. The construction follows that of Example (17.36), and it relies heavily on the rules of manipulation of quantification that were introduced in Chap. 8, 9, and 15.

(17.37) **Theorem.** The generating function for HDE (17.1) is

$$G(z) = \frac{(\Sigma i \mid 0 \leq i < k : a_i \cdot z^i \cdot (\Sigma n \mid 0 \leq n < k - i : x_n \cdot z^n))}{(\Sigma i \mid 0 \leq i \leq k : a_i \cdot z^i)} .$$

Proof. We transform HDE (17.1).

$(\Sigma i \mid 0 \leq i \leq k : a_i \cdot x_{n-i}) = 0$

$=$ $\langle$Multiply both sides by z^n; Distributivity, to move z^n inside; sum over n, $k \leq n \rangle$

$(\Sigma n \mid k \leq n : (\Sigma i \mid .. : a_i \cdot x_{n-i} \cdot z^n)) = 0$

TABLE 17.2. GENERATING FUNCTIONS FOR HDEs OF ORDER 1, 2, AND 3

Order 1: $G(z) = \dfrac{a_0 \cdot x_0}{a_0 + a_1 \cdot z}$

Order 2: $G(z) = \dfrac{a_0 \cdot x_0 + (a_0 \cdot x_1 + a_1 \cdot x_0) \cdot z}{a_0 + a_1 \cdot z + a_2 \cdot z^2}$

Order 3: $G(z) = \dfrac{a_0 \cdot x_0 + (a_0 \cdot x_1 + a_1 \cdot x_0) \cdot z + (a_0 \cdot x_2 + a_1 \cdot x_1 + a_2 \cdot x_0) \cdot z^2}{a_0 + a_1 \cdot z + a_2 \cdot z^2 + a_3 \cdot z^3}$

$=$ 〈Interchange of dummies (8.19)〉
$$(\Sigma i \mid .. : (\Sigma n \mid k \le n : a_i \cdot x_{n-i} \cdot z^n)) = 0$$
$=$ 〈Change of dummy (8.22)〉
$$(\Sigma i \mid .. : (\Sigma n \mid k - i \le n : a_i \cdot x_n \cdot z^{n+i})) = 0$$
$=$ 〈Distributivity —to factor out $a_i \cdot z^i$ 〉
$$(\Sigma i \mid .. : a_i \cdot z^i \cdot (\Sigma n \mid k - i \le n : x_n \cdot z^n)) = 0$$
$=$ 〈Definition of $G(z)$ 〉
$$(\Sigma i \mid .. : a_i \cdot z^i \cdot (G(z) - (\Sigma n \mid 0 \le n < k - i : x_n \cdot z^n))) = 0$$
$=$ 〈Distributivity; move term to the RHS〉
$$(\Sigma i \mid .. : a_i \cdot z^i \cdot G(z)) = (\Sigma i \mid .. : a_i \cdot z^i \cdot (\Sigma n \mid .. : x_n \cdot z^n))$$
$=$ 〈Distributivity to factor out $G(z)$ 〉
$$G(z) \cdot (\Sigma i \mid .. : a_i \cdot z^i) = (\Sigma i \mid .. : a_i \cdot z^i \cdot (\Sigma n \mid .. : x_n \cdot z^n))$$
$=$ 〈Divide both sides by $(\Sigma i \mid .. : a_i \cdot z^i)$〉
$$G(z) = \frac{(\Sigma i \mid 0 \le i \le k : a_i \cdot z^i \cdot (\Sigma n \mid 0 \le n < k - i : x_n \cdot z^n))}{(\Sigma i \mid 0 \le i \le k : a_i \cdot z^i)}$$
$=$ 〈For $k = i$, the sum over n has an empty range〉
$$G(z) = \frac{(\Sigma i \mid 0 \le i < k : a_i \cdot z^i \cdot (\Sigma n \mid 0 \le n < k - i : x_n \cdot z^n))}{(\Sigma i \mid 0 \le i \le k : a_i \cdot z^i)}$$

Notice that the denominator is just the LHS of the HDE, with x_{n-i} replaced by z^i.   □

(17.38) **Example.** Construct the rational-function form of the generating
function for HDE $x_n - 2 \cdot x_{n-2} = 0$ (for $n \ge 2$).

This is an order-2 HDE, with $a_0 = 1$, $a_1 = 0$, and $a_2 = -2$. Using
Table 17.2, we find the generating function $(x_0 + x_1 \cdot z)/(1 - 2 \cdot z^2)$.   □

PARTIAL-FRACTION DECOMPOSITION

We now have the rational-function form of the generating function for an
order-k HDE. In this rational function, the numerator is of lower degree
than the denominator. Such a rational function has the following property.
Suppose the denominator can be put in the form

$$(1 - c_1 \cdot z) \cdot \ldots \cdot (1 - c_k \cdot z)$$

for distinct constants c_i. Then the rational function can be put in the form

$$\frac{d_1}{1 - c_1 \cdot z} + \frac{d_2}{1 - c_2 \cdot z} + \cdots + \frac{d_k}{1 - c_k \cdot z}$$

for constants d_i. This form is called the *partial-fraction decomposition* of
the rational function. A method for calculating the d_i is illustrated by an
example.

(17.39) **Example of a partial-fraction decomposition.** Given is

$$\frac{z}{(1-z)\cdot(1-2\cdot z)} = \frac{d_1}{1-z} + \frac{d_2}{1-2\cdot z} \quad.$$

To calculate d_1, first multiply both sides of the equation by $1-z$:

$$\frac{z}{(1-2\cdot z)} = d_1 + \frac{d_2\cdot(1-z)}{1-2\cdot z} \quad.$$

Then, in order to eliminate the term with d_2, set $z=1$ and simplify. This yields $\frac{1}{1-2\cdot 1} = d_1$, or $d_1 = -1$. Similarly, to calculate d_2, multiply both sides of the equation by $1-2\cdot z$, set z to $1/2$, and simplify; this yields $d_2 = 1$. □

The neat thing about the partial-fraction decomposition of the generating function of an inductive definition is that it is often a sum of generating functions of forms given in Theorems (17.30)–(17.33), so we get a closed-form expression for each term of the sequence. We now give an example of the calculation of a closed-form solution of an inductive definition.

(17.40) **Example.** Construct a closed form solution of

$$x_0 = 0, \quad x_1 = 2, \quad x_n = 2\cdot x_{n-2} \quad (\text{for } n \geq 2) \quad .$$

Step 0. The recurrence relation is equivalent to the HDE $x_n - 2\cdot x_{n-2} = 0$. The generating function for this HDE is $G(z) = \frac{x_0 + x_1\cdot z}{1 - 2\cdot z^2}$. Substituting the constraints $x_0 = 0$ and $x_1 = 2$ yields $(2\cdot z)/(1-2\cdot z^2)$.

Step 1. Since $1 - 2\cdot z^2 = (1+\sqrt{2}\cdot z)\cdot(1-\sqrt{2}\cdot z)$, the generating function has the partial-fraction decomposition

$$\frac{2\cdot z}{1 - 2\cdot z^2} = \frac{d_1}{1+\sqrt{2}\cdot z} + \frac{d_2}{1-\sqrt{2}\cdot z} \quad.$$

Solving for d_1 and d_2 yields $d_1 = -1/\sqrt{2}$ and $d_2 = 1/\sqrt{2}$.

Step 2. Using Theorem (17.31) twice, as well as Theorem (17.34), we see that

$$x_n = d_1\cdot(-\sqrt{2})^n + d_2\cdot(\sqrt{2})^n \quad (\text{for } n \geq 0).$$

Substituting for the d_i in this formula and simplifying, we get

$$x_n = \frac{(\sqrt{2})^n - (-\sqrt{2})^n}{\sqrt{2}} \quad (\text{for } n \geq 0).$$

For even n, this equation reduces to $x_n = 0$; for odd n, to $x_n = 2^{(n+1)/2}$. Thus, the sequence is $0, 2, 0, 4, 0, 8, \ldots$. □

The calculation performed in this example may seem like overkill, for a result that seems obvious, but it illustrates nicely all the steps used in calculating a closed-form solution. This method can be used as well when the closed-form solution is not so obvious, as we see later on.

GENERATING FUNCTIONS FOR NDEs

We discuss solving NDEs using generating functions. Consider the NDE

(17.41) $a_0 \cdot x_n + a_1 \cdot x_{n-1} + \cdots + a_k \cdot x_{n-k} = f_n$ (for $n \geq k$)

The proof of the following theorem is similar to the proof of Theorem (17.37) and is left to the reader.

(17.42) **Theorem.** Let $G(z)$ be the generating function for the homogeneous version of (17.41). Let $F(z) = (\Sigma n \mid k \leq n : f_n \cdot z^n)$ be the generating function for f. Then the generating function for NDE (17.41) is

$$G(z) + \frac{F(z)}{(\Sigma i \mid 0 \leq i \leq k : a_i \cdot z^i)} \quad .$$

This theorem can be used to solve some functions that are defined using an NDE, using techniques similar to those used for functions defined using HDEs. We give a simple example.

(17.43) **Example.** Find a closed-form solution of

$$x_0 = 3, \quad x_n - x_{n-1} = 2 \quad \text{(for } n \geq 1 \text{)}.$$

According to Table 17.2, the generating function of the homogeneous form of this inductive definition is $3/(1-z)$. Function f is $f_n = 2$ (for $n \geq 1$), so its generating function $F(z)$ is

$\quad (\Sigma z \mid 1 \leq n : 2 \cdot z^n)$
$=\quad \langle \text{Arithmetic —to prepare for use of Split off term} \rangle$
$\quad (\Sigma z \mid 1 \leq n : 2 \cdot z^n) + 2 - 2$
$=\quad \langle \text{Split off term (8.23) —backwards} \rangle$
$\quad (\Sigma z \mid 0 \leq n : 2 \cdot z^n) - 2$
$=\quad \langle (17.31) \rangle$
$\quad 2/(1-z) - 2$
$=\quad \langle \text{Arithmetic} \rangle$
$\quad 2 \cdot z/(1-z)$

Therefore, by Theorem (17.42), the generating function for the inductive definition is

$$\frac{3}{1-z} + \frac{2 \cdot z}{(1-z)^2}$$

$=$ ⟨Theorem (17.31); Theorem (17.32)⟩
   $(\Sigma n \mid 0 \leq n : 3 \cdot z^n) + (\Sigma n \mid 0 \leq n : 2 \cdot n \cdot z^n)$

$=$ ⟨Distributivity⟩
   $(\Sigma n \mid 0 \leq n : (3 + 2 \cdot n) \cdot z^n)$

Therefore, $x_n = 3 + 2 \cdot n$ (for $n \geq 0$).   □

(17.44) **Example.** Find a closed-form solution of

$$x_0 = 0, \ x_1 = 0, \ x_2 = 2,$$

$$x_n - 2 \cdot x_{n-1} - x_{n-2} + 2 \cdot x_{n-3} = \begin{cases} 1 & \text{if } n = 3 \\ 0 & \text{if } n > 3 \end{cases} .$$

The sequence $x_0, x_1, \ldots$ begins with $0, 0, 2, 5, 12, 25, 52$. It is difficult to guess a closed-form solution. But we can calculate one using our theory of generating functions. The recurrence relation is given by an NDE. According to Table 17.2, the generating function for the homogeneous form of the inductive definition is

$$\frac{2 \cdot z^2}{1 - 2 \cdot z - z^2 + 2 \cdot z^3} .$$

Function f is given by the sequence $0, 0, 0, 1, 0, 0, \ldots$, so its generating function $F(z)$ is z^3. Therefore, the generating function for the inductive definition is

$$\frac{2 \cdot z^2}{1 - 2 \cdot z - z^2 + 2 \cdot z^3} + \frac{z^3}{1 - 2 \cdot z - z^2 + 2 \cdot z^3}$$

$=$ ⟨The fractions have a common denominator, so combine;
   Factor z from numerator⟩

$$z \cdot \frac{2 \cdot z + z^2}{1 - 2 \cdot z - z^2 + 2 \cdot z^3}$$

$=$ ⟨Apply partial-fraction decomposition⟩

$$z \cdot \left(\frac{d_1}{1 - z} + \frac{d_2}{1 + z} + \frac{d_3}{1 - 2 \cdot z} \right)$$

Solving for the d_i yields $d_1 = -3/2$, $d_2 = -1/6$, and $d_3 = 5/3$. Therefore, the part within the parentheses is the generating function for

$$y_n = -\frac{3}{2} - \frac{(-1)^n}{6} + \frac{5 \cdot 2^n}{3} .$$

By Theorem (17.35), since this is multiplied by z^1, we have the generating function for x_n (for $x \geq 1$) as shown below, and we indicate also that $x_0 = 0$:

$$x_0 = 0, \quad x_n = -\frac{3}{2} - \frac{(-1)^{n-1}}{6} + \frac{5 \cdot 2^{n-1}}{3} \quad \text{(for } n > 0 \text{)} \ .$$

Verify that this defines a sequence that begins with $0, 0, 2, 5, 12, 25, 52$. $\square$

Exercises for Chapter 17

17.1 Prove that if r is a root of multiplicity $m > 1$ of characteristic polynomial (17.3) of HDE (17.1), then $x_n = n \cdot r^n$ is a solution of the HDE. Hint: If r is root of multiplicity greater than 1 of (17.3), then it is a root of the derivative of (17.3).

17.2 Find closed-form solutions of the following inductive definitions.

(a) $x_0 = 6$, $x_n = 2 \cdot x_{n-1}$ (for $n \geq 1$).
(b) $x_n = 2 \cdot x_{n-1}$ (for $n \geq 1$), with constraint $x_2 = 4$.
(c) $x_0 = 5$, $x_n = 5 \cdot x_{n-1}$ (for $n \geq 1$).
(d) $x_0 = 2$, $x_n = -5 \cdot x_{n-1}$ (for $n \geq 1$).
(e) $x_0 = 5$, $x_n = 3 \cdot x_{n-1}$ (for $n \geq 1$).
(f) $x_0 = 4$, $x_n = x_{n-1}$ (for $n \geq 1$).

17.3 For each of the following HDE's find a closed-form solution subject to the given constraints.

(a) $x_n - 3 \cdot x_{n-1} + 2 \cdot x_{n-2} = 0$, $\quad x_1 = 1$ and $x_3 = 1$
(b) $x_n - 3 \cdot x_{n-1} + 2 \cdot x_{n-2} = 0$, $\quad x_1 = 3$ and $x_3 = 9$.
(c) $x_n + 2 \cdot x_{n-1} + x_{n-2} = 0$, $\quad x_1 = 3$ and $x_2 = 0$.
(d) $x_n + 4 \cdot x_{n-1} + 4 \cdot x_{n-2} = 0$, $\quad x_0 = 0$ and $x_1 = 2$.
(e) $x_n - 3 \cdot x_{n-1} + 3 \cdot x_{n-2} - x_{n-3} = 0$, $\quad x_0 = 0$, $x_3 = 3$, and $x_5 = 10$.
(f) $x_n + 2 \cdot x_{n-1} - 15 \cdot x_{n-2} = 0$, $\quad x_0 = 0$ and $x_1 = 1$.
(g) $x_n - 8 \cdot x_{n-1} + 16 \cdot x_{n-2} = 0$, $\quad x_0 = 0$ and $x_1 = 8$.
(h) $x_n - 3 \cdot x_{n-1} + 3 \cdot x_{n-2} - x_{n-3} = 0$, $\quad x_0 = 0$, $x_1 = 0$, and $x_2 = 1$.

17.4 Let x_n be the number of subsets of the integers $0..(n-1)$ that do not contain consecutive integers (for $0 \leq n$). For example, $\{1, 4, 6\}$ is such a subset for $n = 7$, while $\{0, 1, 4, 6\}$ is not. Find an inductive definition for x_n and, from it, find a closed-form solution.

17.5 Consider the sequence $0, 1, 1/2, 3/4, 5/8, \ldots$ in which each value x_n (except the first two) is the average of the preceding two. Find a closed-form solution for x_n.

17.6 Suppose n parking spaces in a row are to be filled completely, with no empty places. A big car takes two spaces; a compact car takes one. Give an inductive definition of the number of ways in which the n spaces can be filled. Find a closed-form solution.

17.7 Find an inductive definition for the number of sequences of zeros, ones, and twos of length n that do not contain consecutive zeros. Then find a closed-form solution.

17.8 A particle starts at position 0 and moves in a particular direction. After one minute, it has moved 4 inches. Thereafter, the distance it travels during minute n, for $n > 1$, is twice the distance it traveled during minute $n - 1$. Define inductively the distance d_n the particle has traveled after n minutes. Then find a closed-form solution.

Exercise on dealing with complex roots

17.9 Show that i and $-i$ are the roots of $\lambda^2 + 1$.

17.10 Solve the pair of equations $0 = b_1 + b_2$ and $1 = b_1 \cdot i - b_2 \cdot i$ for b_1 and b_2.

17.11 Find closed-form solutions to the following HDE's.

(a) $x_0 = 0$, $x_1 = 1$, $x_n = -2 \cdot x_{n-2}$ for $n \geq 2$.
(b) $x_0 = 0$, $x_1 = 1$, $x_n = -x_{n-1} - x_{n-2}$ for $n \geq 2$.
(c) $x_0 = 0$, $x_1 = 1$, $x_n = 2 \cdot x_{n-1} - 2 \cdot x_{n-2}$ for $n \geq 2$.
(d) $x_0 = 0$, $x_1 = 1$, $x_n = -2 \cdot x_{n-1} - 2 \cdot x_{n-2}$ for $n \geq 2$.
(e) $x_0 = 0$, $x_1 = 1$, $x_n = 3 \cdot x_{n-1} - 3 \cdot x_{n-2}$ for $n \geq 2$.
(f) $x_0 = 0$, $x_1 = 1$, $x_n = -3 \cdot x_{n-1} - 3 \cdot x_{n-2}$ for $n \geq 2$.

Exercises on NDEs

17.12 Use Method (17.21) to find a particular solution of the following NDEs.

(a) $x_n - 3 \cdot x_{n-1} = 3 \cdot n + 2$.
(b) $x_n - 3 \cdot x_{n-1} = 2^n$.
(c) $x_n - 3 \cdot x_{n-1} = 4 \cdot 2^n + 3$.
(d) $x_n - x_{n-1} - 2 \cdot x_{n-2} = 1$.
(e) $x_n - x_{n-1} - 2 \cdot x_{n-2} = 1 + n$.
(f) $x_n + 2 \cdot x_{n-1} - 15 \cdot x_{n-2} = 10 + 6 \cdot n$.
(g) $x_n - 4 \cdot x_{n-1} - 4 \cdot x_{n-2} = 2^n$.
(h) $x_n - 5 \cdot x_{n-1} + 6 \cdot x_{n-2} = 2 \cdot n$.
(i) $x_n - 5 \cdot x_{n-1} + 6 \cdot x_{n-2} = 3^n$.
(j) $x_n - 2 \cdot x_{n-1} + x_{n-2} = 2^n$.

Exercises on generating functions

17.13 Prove Theorem (17.31).

17.14 Prove Theorem (17.32).

17.15 Prove Theorem (17.33).

17.16 Prove Theorem (17.34).

17.17 Give the generating functions for the following sequences.

(a) $1, 0, 0, 0, \ldots$.

(b) $0, 0, 2, 0, 0, 0, \ldots$.

(c) $0, -2, 0, 0, 0, 0, \ldots$.

17.18 Find a rational-function form of the generating function for the sequence $x_n = n^2$, for $n \geq 0$.

17.19 Use Table 17.2 to find the rational-function form of the generating function for the following HDEs.

(a) $x_n - 2 \cdot x_{n-1} = 0$.

(b) $x_n + x_{n-1} = 0$.

(c) $x_n - x_{n-1} = 0$.

(d) $x_n + 5 \cdot x_{n-1} = 0$.

(e) $x_n - 5 \cdot x_{n-1} = 0$.

(f) $F_n - F_{n-1} - F_{n-2} = 0$.

(g) $x_n - 2 \cdot x_{n-1} + x_{n-2} = 0$.

(h) $x_n - x_{n-1} + x_{n-2} = 0$.

(i) $x_n + 3 \cdot x_{n-1} + x_{n-2} = 0$.

(j) $x_n + 3 \cdot x_{n-1} - x_{n-2} = 0$.

(k) $x_n - x_{n-1} - x_{n-2} + x_{n-3} = 0$.

(l) $x_n - x_{n-3} = 0$.

17.20 Use the method given on page 380 to find a closed-form solution of the following inductive definitions.

(a) $x_0 = 2$, $x_n - 2 \cdot x_{n-1} = 0$ for $n \geq 1$.

(b) $x_0 = 4$, $x_n + x_{n-1} = 0$ for $n \geq 1$.

(c) $x_0 = 4$, $x_n - x_{n-1} = 0$ for $n \geq 1$.

(d) $x_0 = 2$, $x_n + 5 \cdot x_{n-1} = 0$ for $n \geq 1$.

(e) $x_0 = 2$, $x_n - 5 \cdot x_{n-1} = 0$ for $n \geq 1$.

(f) $F_0 = 0$, $F_1 = 1$, $F_n - F_{n-1} - F_{n-2} = 0$ for $n \geq 2$.

(g) $x_0 = 1$, $x_1 = 2$, $x_n + 3 \cdot x_{n-1} + x_{n-2} = 0$ for $n \geq 2$.

(h) $x_0 = 1$, $x_1 = 2$, $x_n + 3 \cdot x_{n-1} - x_{n-2} = 0$ for $n \geq 2$.

(i) $x_0 = 1$, $x_1 = 2$, $x_2 = 3$,
    $x_n - x_{n-1} - x_{n-2} + x_{n-3} = 0$ for $n \geq 3$.

(j) $x_0 = 1$, $x_1 = 2$, $x_2 = 3$, $x_n - x_{n-3} = 0$ for $n \geq 3$.

(k) $x_0 = 1$, $x_1 = 2$, $x_n - 2 \cdot x_{n-1} + x_{n-2} = 0$ for $n \geq 2$.

(l) $x_0 = 1$, $x_1 = 2$, $x_n - x_{n-1} + x_{n-2} = 0$ for $n \geq 2$.

(m) $x_0 = 0$, $x_1 = 1$, $x_n - 2 \cdot x_{n-1} - 15 \cdot x_{n-2} = 0$, for $n \geq 2$.

17.21 Consider the sequence $0, 1, 1/2, 3/4, 5/8, \ldots$ in which each value x_n (except the first two) is the average of the preceding two. Use generating functions to find a closed-form solution for x_n.

17.22 Prove Theorem (17.42).

Chapter 18

Modern Algebra

M *odern algebra* is the study of the structure of certain sets along with operations on them. An *algebra* is basically a model of a theory, as discussed near the beginning of Chap. 9. The algebras discussed here are *semigroups, monoids, groups,* and *boolean algebras.* They are useful throughout computer science and mathematics. For example, Chap. 8 was devoted to the study of quantification over an arbitrary abelian monoid. Semigroups and monoids find application in formal languages, automata theory, and coding theory. And, one boolean algebra is the standard model of the propositional calculus. Important in our study is not only the various algebras but their interrelationship. Thus, we study topics like *isomorphisms, homomorphisms,* and *automorphisms* of algebras. (Historical note 18.1 discusses the origin of these words.)

18.1   The structure of algebras

An algebra consists of two components:

- A set S (say) of elements, called the *carrier* of the algebra.

- Operators defined on the carrier.

Each operator is a total function of type $S^m \to S$ for some m, where m is called the *arity* of the operator. The algebra is *finite* if its carrier S is finite; otherwise, it is *infinite*.

Operators of arity 0, called *nullary* operators, are functions of no arguments. For ease of exposition, we view the nullary operators as *constants* in the carrier. For example, we consider 1 to be a function that takes no arguments and returns the value one. Operators of arity 1 are *unary* operators; of arity 2, *binary* operators; of arity 3, *ternary* operators. (The conditional expression **if** b **then** c **else** d is a ternary operation). Unary operators are written in prefix form; binary operators in infix form.

Examples of algebras

(a) The set of even integers and the operator $+$ form an algebra.

(b) The set of even numbers together with the operations multiplication and division is not an algebra, because division is not a total function

---

HISTORICAL NOTE 18.1. MORPHING AND OTHER WORDS

Some inkling of the meaning of words like isomorphism, homomorphism, and automorphism can be gained by looking at their Greek roots. In Greek, *isos* means *equal*. The prefix *iso* is used in many English words, such as *isosceles* (having equal legs or sides), *isonomic* (equal in law or privilege), and *isobar* (a line on a weather map connecting places with equal barometric pressure). Prefix *homo* comes from the Greek *homos*, meaning *same*. We see it used in *homogenized* and *homosexual*, for example. And, prefix *auto* comes from the Greek word meaning *self*, as in autohypnosis and automobile.

Putting these three prefixes together with *morphic*, which is a combining form, again from the Greek, meaning *having a form or shape*, gives *isomorphism*, *homomorphism*, and *automorphism*. The change in the shape of the U.S. car industry in the past fifteen years is *not* what we mean by an automorphism.

Lately, *morph* has been used in another context. Programs have been written that produce a *morph* of two images: an image that is a combination of the two images. On the back cover are five morphs of pictures of the two authors. The first picture is Gries; the second, 70% Gries and 30% Schneider; the third, 50% Gries; the fourth, 30% Gries; and the last, pure Schneider. These morphs were produced by the Macintosh program Morph.

---

on the even integers (division by 0 is not defined).

(c)   The set {*false*, *true*} and operators $\vee$, $\wedge$, and $\neg$ is an algebra. This is a *finite* algebra, because the set is finite.   □

We often want to discuss a class of algebras that have the same properties. To aid in this discussion, we present algebras in a standard form. For example, algebra (a) above is described by $\langle S, + \rangle$, where S is the set of even integers, and algebra (c) above is described by $\langle \mathbb{B}, \vee, \wedge, \neg \rangle$. We use $\langle S, \Phi \rangle$ to denote an algebra with carrier S and list of operators Φ.

The *signature* of an algebra consists of the name of its carrier and the list of types of its operators. For example, the algebra $\langle \mathbb{B}, \vee, \wedge, \neg \rangle$ has the signature

$$\langle \mathbb{B}, \mathbb{B} \times \mathbb{B} \to \mathbb{B}, \mathbb{B} \times \mathbb{B} \to \mathbb{B}, \mathbb{B} \to \mathbb{B} \rangle$$

Two algebras are said to have the same signature if (i) they have the same number of operators and (ii) corresponding operators have the same types (modulo the name of the carrier).

For example, algebras $\langle \mathbb{B}, \vee, \wedge, \neg \rangle$ and $\langle \mathcal{P}S, \cap, \cup, \sim \rangle$ for some set S have the same signature. Algebra $\langle \mathcal{P}S, \sim, \cup, \cap \rangle$ has a different signature, since $\sim$ is of arity 1 and $\vee$ and $\cap$ are of arity 2.

In a particular class of algebras, some constants may be distinguished because they satisfy certain properties. The properties that crop up most often are those of being an *identity*, a *zero*, and an *inverse*. We have seen these terms in earlier chapters, but for completeness we repeat the definitions here.

(18.1)   **Definition.** An element 1 in S is a *left identity* (or *unit*) of binary operator $\circ$ over S if $1 \circ b = b$ (for b in S);

1 is a *right identity* if $b \circ 1 = b$ (for b in S); and

1 is an *identity* if it is both a left and a right identity.

(18.2)   **Theorem.** If c is a left identity of $\circ$ and d is a right identity of $\circ$, then $c = d$. (Hence, in this case, all left and right identities are equal and $\circ$ has a unique identity.)

Proof. c
        $=$ $\langle d$ is a right identity$\rangle$
              $c \circ d$
        $=$ $\langle c$ is a left identity$\rangle$
              d ☐

(18.3)   **Definition.** An element 0 in S is a *left zero* of binary operator $\circ$ over S if $0 \circ b = 0$ (for b in S);

0 is a *right zero* if $b \circ 0 = 0$ (for b in S); and

0 is a *zero* if it is both a left and a right zero.

An algebra can have more than one left zero. For example, consider algebra $\langle \{b, c\}, \circ \rangle$ with operator $\circ$ defined below. Both b and c are left zeros —and both are right identities!

$$b \circ b = b \qquad c \circ b = c$$
$$b \circ c = b \qquad c \circ c = c$$

The proof of the following theorem is left to the reader.

(18.4)   **Theorem.** If c is a left zero of $\circ$ and d is a right zero of $\circ$, then $c = d$. (Hence, in this case, all left and right zeros are equal and $\circ$ has a unique zero.)

(18.5)   **Definition.** Let 1 be the identity of binary operator $\circ$ on S. Then b has a *right inverse* c with respect to $\circ$ and c has a *left inverse* b with respect to $\circ$ if $b \circ c = 1$. Elements b and c are called *inverses* of each other if $b \circ c = c \circ b = 1$.

Examples of inverses

(a) In algebra $\langle \mathbb{Z}, + \rangle$, 0 is an identity. As shown in Chap. 15, every element b in $\mathbb{Z}$ has an inverse, denoted by $-b$.

(b) In algebra $\langle \mathbb{R}, \cdot \rangle$, 1 is an identity. Every element of $\mathbb{R}$ except 0 has an inverse with respect to $\cdot$.

(c) Consider the set F of functions of arity 1 over a set S, and let $\bullet$ be function composition: $(f \bullet g).b = f(g.b)$. Then the function I given by $I.b = b$ (for b in F) is an identity. By Theorem (14.44), every onto function has a right inverse. By Theorem (14.45), every one-to-one function has a left inverse. And, by Theorem (14.46), every one-to-one and onto function has an inverse.   □

The proof of the following theorem is left to the reader.

(18.6)   **Theorem.** Let li be a left inverse and ri be a right inverse of element b with respect to an associative operator $\circ$. Then $li = ri$ and b has a unique inverse.

For a finite algebra, a binary operation can be given as a table, much like a truth table for boolean expressions, as shown in Table 18.1 —and much like the multiplication table used in elementary school.

Many properties of $\circ$ can be read directly from Table 18.1. For example, $\circ$ has a right identity if some column of entries is the same as the leftmost column; it has a left identity if some row of entries is the same as the top row. The operator has a right zero if all entries in a column equal the value that heads it. The operator is symmetric iff the table is symmetric about its principal diagonal.[1] And an equation $b \circ x = c$ has a solution for x iff the row for b contains an entry c.

---

[1] The principal diagonal of a square matrix $b[1..n, 1..n]$ is the list of elements $b[1,1], b[2,2], \ldots, b[n,n]$.

TABLE 18.1. TABLE FOR $\circ$ AND A MULTIPLICATION TABLE

$\circ$	b	c	d	e		$\cdot$	1	2	3	4
b	b	c	d	e		1	1	2	3	4
c	c	d	e	b		2	2	4	6	8
d	d	e	b	c		3	3	6	9	12
e	e	b	c	d		4	4	8	12	16

SUBALGEBRAS

Algebras are differentiated by their structure. Structure refers to properties like the existence of an identity, but it also refers to the kinds of *subalgebras* an algebra has. To define the term subalgebra, we first need to define the term *closed*.

(18.7)   **Definition.** A subset T of a set S is *closed* under an operator if applying the operator to elements of T always produces a result in T.

Example of closed operators

(a) The set of even integers is closed under $+$ because the sum of two even integers is even.

(b) Subset $\{0,1\}$ of the integers is not closed under $+$ because $1+1$ is not in this subset.

(c) Subset $\{0,1\}$ of the integers is closed under $\uparrow$ (maximum) because the maximum of any two of these integers is one of the integers.   □

(18.8)   **Definition.** $\langle T, \Phi \rangle$ is a subalgebra of $\langle S, \Phi \rangle$ if (i) $\emptyset \subset T \subseteq S$ and (ii) T is closed under every operator in Φ.

The term *subalgebra* is sensible, because a subalgebra satisfies all the properties of an algebra: T is nonempty and T is closed under the operators. Note that we use the same symbol f (say) for a function over S and the restriction of f to T. This overloading of function names simplifies notation and rarely leads to confusion.

Examples

(a) Algebra $\langle \mathbb{N}, + \rangle$ is a subalgebra of $\langle \mathbb{Z}, + \rangle$ because $\mathbb{N} \subseteq \mathbb{Z}$ and $\mathbb{N}$ is closed under $+$.

(b) $\langle \{0,1\}, + \rangle$ is not a subalgebra of $\langle \mathbb{Z}, + \rangle$ because $\{0,1\}$ is not closed under $+$.

(c) Algebra $\langle \{0,1\}, \cdot \rangle$ is a subalgebra of $\langle \mathbb{N}, \cdot \rangle$.

(d) Any algebra is a subalgebra of itself.   □

ISOMORPHISMS AND HOMOMORPHISMS

In what follows, for expository simplicity, we deal only with algebras that have nullary, unary, and binary operators. The extension to algebras with operators of higher arity is obvious. Well, that's what they say.

We discuss two ways to characterize structural similarities of two algebras. The first characterization is that they look the same: they have the same signature, their carriers are the same size, and their operators and constants have the same properties. They are essentially the same algebra, but for the names of operators and elements of the carrier. Two such algebras are called *isomorphic*. Isomorphism is formally defined as follows.

(18.9)   **Definition.** Let algebras $A = \langle S, \Phi \rangle$ and $\hat{A} = \langle \hat{S}, \hat{\Phi} \rangle$ have the same signature. A function $h : S \to \hat{S}$ is an *isomorphism* from A to $\hat{A}$ if:

(a)   Function h is one-to-one and onto.

(b)   For each pair of corresponding nullary operators (constants) c in Φ and $\hat{c}$ in $\hat{\Phi}$, $h.c = \hat{c}$.

(c)   For each pair of corresponding unary operators $\sim$ in Φ and $\hat{\sim}$ in $\hat{\Phi}$, $h(\sim b) = \hat{\sim} h.b$ (for b in S).

(d)   For each pair of corresponding binary operators $\circ$ in Φ and $\hat{\circ}$ in $\hat{\Phi}$, $h(b \circ c) = h.b \,\hat{\circ}\, h.c$.

A and $\hat{A}$ are *isomorphic*, and $\hat{A}$ is the *isomorphic image* of A under h.

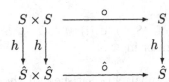

Property (18.9d) is sometimes depicted as the *commuting diagram* to the left. Each downward arrow represents application of h to a value of S; the upper horizontal arrow represents application of $\circ$ to a value of $S \times S$; and the lower one represents application of $\hat{\circ}$ to a value of $\hat{S} \times \hat{S}$. This diagram is said to *commute* if traveling right and then down yields the same value as traveling down and then right, i.e. if $h(b \circ c) = h.b \,\hat{\circ}\, h.c$. Such commuting diagrams are used frequently to illustrate isomorphisms and homomorphisms.

In the following example, algebra A is isomorphic to itself. An isomorphism from A to A is called an *automorphism*. Exercise 18.10 asks you to prove that h defined by $h.b = -b$ (for b in $\mathbb{Z}$) is an automorphism of $\langle \mathbb{Z}, + \rangle$. Thus, there are other automorphisms besides the identity function.

Example of an automorphism. Let $A = \hat{A} = \langle S, \Phi \rangle$. Let h be the identity function on S: $h.b = b$ for b in S. Obviously, A and $\hat{A}$ have the same signatures. We show that each of the points (a)–(d) of Definition (18.9) hold. Hence, $\hat{A}$ is the isomorphic image of A.

(a)   The identity function is one-to-one and onto.

(b)   For each nullary operator c in Φ, $h.c = c$, so h satisfies point (b).

(c)   For each unary operator $\sim$ in Φ, $h(\sim b) = \sim b = \sim (h.b)$. Hence, h satisfies point (c).

(d)   For each binary operator $\circ$ in Φ, $h(b \circ c) = b \circ c = h.b \circ h.c$. $\square$

Example of an isomorphism. Let $A = \langle \mathbb{B}, \vee \rangle$ and $\hat{A} = \langle \mathbb{B}, \wedge \rangle$. A and $\hat{A}$ have the same signature. Define $h : \mathbb{B} \rightarrow \mathbb{B}$ by $h.b = \neg b$. Function h is one-to-one and onto, so (a) of Definition (18.9) holds. Since A has no nullary and no unary operators, points (b) and (c) hold. Finally, we show that h satisfies (d). Thus, A is isomorphic to $\hat{A}$. For b and c in $\mathbb{B}$, we have:

$$
\begin{array}{ll}
& h(b \vee c) \\
= & \langle \text{Definition of } h \rangle \\
& \neg(b \vee c) \\
= & \langle \text{De Morgan (3.47b)} \rangle \\
& \neg b \wedge \neg c \\
= & \langle \text{Definition of } h, \text{ twice} \rangle \\
& h.b \wedge h.c \qquad\qquad\qquad\qquad \Box
\end{array}
$$

Example of an isomorphism. Let $A = \langle \mathbb{N}, + \rangle$ and $\hat{A} = \langle even, + \rangle$, where $even$ is the set of even natural numbers. A and $\hat{A}$ have the same signature. Define $h : \mathbb{N} \rightarrow even$ by $h.b = 2 \cdot b$ (for b in $\mathbb{N}$). Point (a) is satisfied, since h is one-to-one and onto. Points (b) and (c) are satisfied, since there are no nullary or unary operators. We prove below that point (d) holds. Thus, A is isomorphic to $\hat{A}$. For b and c in $\mathbb{N}$, we have,

$$
\begin{array}{ll}
& h(b + c) \\
= & \langle \text{Definition of } h \rangle \\
& 2 \cdot (b + c) \\
= & \langle \text{Distributivity} \rangle \\
& 2 \cdot b + 2 \cdot c \\
= & \langle \text{Definition of } h, \text{ twice} \rangle \\
& h.b + h.c \qquad\qquad\qquad\qquad \Box
\end{array}
$$

Example of an isomorphism —the slide rule. Let $A = \langle \mathbb{R}^+, \cdot \rangle$ and $\hat{A} = \langle \mathbb{R}, + \rangle$, where $\mathbb{R}^+$ is the set of positive real numbers. A and $\hat{A}$ have the same signature. Define $h : \mathbb{R}^+ \rightarrow \mathbb{R}$ by $h.r = log.r$ for $r > 0$, so that $h^{-1}.r = 2^r$. Function h is one-to-one and onto, so point (a) holds. There are no nullary and no unary operators, so (b) and (c) hold. We show that point (d) holds. For b and c in $\mathbb{R}^+$, we have,

$$
\begin{array}{ll}
& h(b \cdot c) \\
= & \langle \text{Definition of } h \rangle \\
& log(b \cdot c) \\
= & \langle \text{Property of logarithms} \rangle \\
& log.b + log.c \\
= & \langle \text{Definition of } h, \text{ twice} \rangle \\
& h.b + h.c
\end{array}
$$

So, A is isomorphic to $\hat{A}$. This isomorphism is the basis of the slide rule —which was obsoleted by the pocket calculator and computer. Using the slide rule, two numbers are multiplied by adding lengths that correspond to their logarithms.   □

We observed earlier that isomorphic algebras were the same, except for the renaming of elements and operators. How do we really know that the definition of isomorphism has this property? The following three theorems give some evidence. The proofs of part of the first theorem and of the second and third theorem are left to the reader.

(18.10) **Theorem.** An isomorphism maps identities to identities, zeros to zeros, and inverses to inverses.

Proof. Consider an isomorphism h from an algebra A to $\hat{A}$. Let 1 be a right identity of operator $\circ$ of A. For all $\hat{b}$ in the carrier of $\hat{A}$, we have

$$\hat{b}\,\hat{\circ}\,h.1$$
$$=\quad \langle\, h.b = \hat{b}, \text{ for some } b \text{ of } A - h \text{ is onto}\rangle$$
$$h.b\,\hat{\circ}\,h.1$$
$$=\quad \langle\, h \text{ commutes with } \circ, \text{ point (d) of Def. (18.9)}\rangle$$
$$h(b\circ 1)$$
$$=\quad \langle\, 1 \text{ is a right identity}\rangle$$
$$h.b$$
$$=\quad \langle\, h.b = \hat{b} \text{ —see above}\rangle$$
$$\hat{b}$$

Hence, a right identity 1 is mapped into a right identity $h.1$. The rest of the theorem is proved in a similar fashion and is left to the reader.   □

(18.11) **Theorem.** If $\hat{A}$ is an isomorphic image of A, then A is an isomorphic image of $\hat{A}$.

(18.12) **Theorem.** Let C be a set of algebras. The relation "A is isomorphic to $\hat{A}$" is an equivalence relation.

HOMOMORPHISMS

We now study a second relation between algebras, the *homomorphism*. This relation relaxes the requirement that function h of an isomorphism be one-to-one and onto.

(18.13) **Definition.** Let algebras $A = \langle S, \Phi \rangle$ and $\hat{A} = \langle \hat{S}, \hat{\Phi} \rangle$ have the
same signature. A function $h : S \rightarrow \hat{S}$ is a *homomorphism* from
A to $\hat{A}$ if it satisfies:

(a) For each pair of corresponding nullary operators c in Φ and
$\hat{c}$ in $\hat{\Phi}$, $h.c = \hat{c}$.

(b) For each pair of corresponding unary operators $\sim$ in Φ and
$\hat{\sim}$ in $\hat{\Phi}$, $h(\sim b) = \hat{\sim} h.b$ (for b in S).

(c) For each pair of corresponding binary operators $\circ$ in Φ and
$\hat{\circ}$ in $\hat{\Phi}$, $h(b \circ c) = h.b \hat{\circ} h.c$ (for b in S).

An isomorphism, then, is a homomorphism that is also one-to-one and onto.
Let $h.S$ denote the range of function h: $h.S = \{s \mid s \in S : h.s\}$ With an
isomorphism h, S and $h.S$ have the same size. With a homomorphism
h, $h.S$ may be smaller than S, but the structure of the sets is similar
with respect to the properties of the operators because of the commutative
nature of h, which is implied by (a)–(c) of Definition (18.13).

Examples of homomorphisms

(a) Function $h.b = 5 \cdot b$ is a homomorphism from algebra $\langle \mathbb{N}, + \rangle$ to itself.
There are no unary operators, and $h(b+c) = 5 \cdot (b+c) = 5 \cdot b + 5 \cdot c = h.b + h.c$ (for b and c in $\mathbb{N}$). Actually, for any integer k (including
0), $h.b = k \cdot b$ is a homomorphism from $\langle \mathbb{N}, + \rangle$ to itself.

(b) Let $\oplus$ be the function defined by $b \oplus c = (b + c) \bmod 5$. Then
$h.b = b \bmod 5$ is a homomorphism from $\langle \mathbb{N}, \oplus \rangle$ to $\langle 0..4, \oplus \rangle$.

(c) Function $h : seq(char) \rightarrow \mathbb{N}$ defined by $h.z = \#z$ is a homomorphism
from $\langle seq(char), \hat{\ }, \epsilon \rangle$ to $\langle \mathbb{N}, +, 0 \rangle$.   □

(18.14) **Theorem.** Let h be a homomorphism from $A = \langle S, \Phi \rangle$ to
$\hat{A} = \langle \hat{S}, \hat{\Phi} \rangle$. Then $\langle h.S, \hat{\Phi} \rangle$ is a subalgebra of $\hat{A}$, called the *homomorphic image of A under h*.

Proof. We show that $\langle h.\hat{S}, \hat{\Phi} \rangle$ satisfies Definition (18.8) of a subalgebra.

(a) Since $h : S \rightarrow \hat{S}$, $h.S \subseteq \hat{S}$.

(b) We show that $h.S$ is closed under each binary operator $\hat{\circ}$ in $\hat{\Phi}$. Let
$\hat{b}$ and $\hat{c}$ be in $h.S$. Then there exist values b, c in S that satisfy
$h.b = \hat{b}$ and $h.c = \hat{c}$. We have:

$$\hat{b} \hat{\circ} \hat{c}$$
$$= \quad \langle \text{Existence of } b \text{ and } c \rangle$$
$$h.b \hat{\circ} h.c$$
$$= \quad \langle h \text{ is a homomorphism (Point (c) of Def. (18.13))} \rangle$$
$$h(b \circ c)$$

Hence, $\hat{b} \hat{\circ} \hat{c}$ is in $h.S$ and $h.S$ is closed under $\hat{\circ}$. Similarly, $h.S$
is closed under all the nullary and unary operators of $\hat{\Phi}$.   □

18.2   Group theory

We explore three varieties of algebras, of increasingly complex structure: semigroups, monoids, and groups. Actually, we spend the most time on groups, since they are the most interesting and have so many applications.

(18.15)   **Definition.** A *semigroup* is an algebra $\langle S, \circ \rangle$ where $\circ$ is a binary associative operator.

Examples of semigroups

(a) $\langle S, \char94 \rangle$ is a semigroup, where S is the set of finite strings over a nonempty alphabet and $\char94$ is string catenation.

(b) $\langle S, \cdot \rangle$ is a semigroup, where $S = \{ r \in \mathbb{R} \mid 0 \leq r \leq 1 \}$ and $\cdot$ is multiplication.

(c) $\langle S, \uparrow \rangle$, where S is any nonempty subset of the real numbers and $b \uparrow c$ is the maximum of b and c.

(d) $\langle \{b, c\}, \circ \rangle$, where $\circ$ is defined by $b \circ b = c \circ b = b$ and $b \circ c = c \circ c = c$. This is a finite semigroup (since S is finite).

(e)   Let R be the set of binary relations over some set. Since relation product $\circ$ is associative, $\langle R, \circ \rangle$ is a semigroup.   □

Let T be a subset of carrier S of semigroup $\langle S, \circ \rangle$. Suppose T is closed under $\circ$. Then algebra $\langle T, \circ \rangle$ is called a *subsemigroup* of $\langle S, \circ \rangle$. Note that a subsemigroup is a semigroup — T is closed under $\circ$ and, since $\circ$ is associative, its restriction to T is associative.

(18.16)   **Definition.** A *monoid* $\langle S, \circ, 1 \rangle$ is a semigroup $\langle S, \circ \rangle$ with an identity 1. If $\circ$ is also symmetric, the monoid is called *abelian* (after Niels Abel; see Historical note 8.1 on page 144). A subalgebra of a monoid that contains the identity of the monoid is called a *submonoid*.

A submonoid $\langle T, \circ, 1 \rangle$ of monoid $\langle S, \circ, 1 \rangle$ is itself a monoid, for the following reasons. We know already that $\langle T, \circ \rangle$ is a semigroup. Also, the identity of $\circ$ is in T, so $\langle T, \circ, 1 \rangle$ satisfies the requirements of a monoid.

The abelian monoid was the basis for our study of quantification in Chap. 8. Here are some other examples of monoids.

Examples of monoids

(a) $\langle S, \hat{}, \epsilon \rangle$ is a monoid, where S is the set of finite strings over a nonempty alphabet, $\hat{}$ is string catenation, and the empty string ϵ is the identity. This monoid is not abelian.

(b) $\langle S, \cdot, 1 \rangle$ is a monoid, where $S = \{r \in \mathbb{R} \mid 0 \le r \le 1\}$ and $\cdot$ is multiplication.

(c) $\langle \mathbb{R}, \uparrow \rangle$, where $b \uparrow c$ is the maximum of b and c, is not a monoid, since $\uparrow$ has no identity in $\mathbb{R}$.

(d) $\langle \mathbb{N}, \uparrow, 0 \rangle$ is a monoid. Note that $0 \uparrow b = b \uparrow 0 = b$ for b in $\mathbb{N}$.

(e) Let R be the set of binary relations over some set S. Then $\langle R, \circ, \imath_S \rangle$ is a monoid (but not an abelian monoid), where $\circ$ is relation product. The identity is the identity relation $\imath_S$. □

Any semigroup $\langle S, \circ \rangle$ can be made into a monoid $\langle S \cup \{c\}, \circ, c \rangle$ for c a fresh element that is defined to satisfy $c \circ b = b \circ c = b$ for all elements of $S \cup \{c\}$. For example, operator $\uparrow$ can be extended to $\mathbb{R} \cup \{-\infty\}$ by $r \uparrow -\infty = -\infty \uparrow r = r$ for all elements of $\mathbb{R} \cup \{-\infty\}$, so that $\uparrow$ has an identity. One must be wary of this extension, however, because other properties of the reals $\mathbb{R}$ may not hold for $\mathbb{R} \cup \{-\infty\}$. For example, $1 + b > b$ does not hold for $b = -\infty$.

We come now to the most significant class of algebras of this section, the *group*. A group is monoid in which each element has an inverse. Thus, we have the following definition.

(18.17) **Definition.** A *group* is an algebra $\langle S, \circ, 1 \rangle$ in which

   (a) $\circ$ is a binary, associative operator,

   (b) $\circ$ has the identity 1 in S,

   (c) Every element b (say) of S has an inverse, which we write as b^{-1}.

A *symmetric, commutative,* or *abelian group* is an abelian monoid in which every element has an inverse.

Examples of groups

(a) The *additive group of integers* $\langle \mathbb{Z}, +, 0 \rangle$ is a group. The inverse b^{-1} of b is $-b$.

(b) Let K be the set of multiples of 5. Then $\langle K, +, 0 \rangle$ is a group. The inverse b^{-1} of b is the element $-b$.

(c) Let $n > 0$ be an integer. Define $\oplus$ for operands b and c in $0..(n-1)$ by $b \oplus c = (b+c) \bmod n$. Then $M_n = \langle 0..(n-1), \oplus, 0 \rangle$ is a group, called the *additive group of integers modulo* n.

(d) $\langle \mathbb{R}, \cdot, 1 \rangle$ has identity 1 but is not a group, because 0 has no inverse.

(e) $\langle \mathbb{R}^+, \cdot, 1 \rangle$ is a group. The inverse r^{-1} of r in $\mathbb{R}^+$ is $1/r$. □

By Theorem (18.2), the identity of a group is unique. By Theorem (18.6), the inverse of each element is unique. The additional property that each element have an inverse provides enough structure for us to prove a number of theorems about groups. Theorem (18.18) below establishes that the inverse of the inverse of an element is that element. Theorems (18.19) allow us to cancel, as we do with addition and multiplication (e.g. $b + d = c + d \equiv b = c$). Theorem (18.20) tells us the unique solution to $b \circ x = c$. Finally, Theorems (18.21) and (18.22) indicate that $\circ$ with one of its operands held fixed is one-to-one and onto. Note that all but one of the theorems fall into pairs that are symmetric with respect to operator $\circ$, even though symmetry need not be a property of $\circ$.

Theorems for groups

(18.18) $b = (b^{-1})^{-1}$

(18.19) **Cancellation:** $b \circ d = c \circ d \equiv b = c$

$\qquad\qquad\qquad\quad\; d \circ b = d \circ c \equiv b = c$

(18.20) **Unique solution:** $b \circ x = c \equiv x = b^{-1} \circ c$

$\qquad\qquad\qquad\qquad\quad\; x \circ b = c \equiv x = c \circ b^{-1}$

(18.21) **One-to-one:** $b \neq c \equiv d \circ b \neq d \circ c$

$\qquad\qquad\qquad\quad\; b \neq c \equiv b \circ d \neq c \circ d$

(18.22) **Onto:** $(\exists x \mid : b \circ x = c)$

$\qquad\qquad\quad\; (\exists x \mid : x \circ b = c)$

We give the proof of (18.18). In our proofs, as done earlier in the text, associativity is handled implicitly.

$$
\begin{aligned}
&(b^{-1})^{-1} \\
= \quad &\langle\, 1 \text{ is the identity of the group} \rangle \\
&1 \circ (b^{-1})^{-1} \\
= \quad &\langle \text{Definition of (right) inverse} \rangle \\
&b \circ b^{-1} \circ (b^{-1})^{-1} \\
= \quad &\langle \text{Definition of (right) inverse} \rangle \\
&b \circ 1 \\
= \quad &\langle\, 1 \text{ is the identity of the group} \rangle \\
&b
\end{aligned}
$$

There are different, but equivalent, definitions of groups. For example, Exercise 18.16 concerns showing that only a left identity and a left inverse for each element are required.

We can define integral powers b^n of an element b of a group $\langle S, \circ, 1 \rangle$:

(18.23) $b^0 = 1$,

$\qquad b^n = b^{n-1} \circ b \qquad$ (for $n > 0$)

$\qquad b^{-n} = (b^{-1})^n \qquad$ (for $n > 0$)

Thus, for $n \geq 0$, b^n is b composed with itself n times and b^{-n} is b^{-1} composed with itself n times. The proofs of the following theorems are easy and are left to the reader.

Properties of powers of group elements

(18.24) $b^m \circ b^n = b^{m+n}$ (for m and n integers)

(18.25) $(b^m)^n = b^{m \cdot n}$ (for m and n integers)

(18.26) $b^n = b^p \;\equiv\; b^{n-p} = 1$

(18.27) **Definition.** The *order* of an element b of a group with identity 1 (say), written *ord.b*, is the least positive integer m such that $b^m = 1$ (or ∞ if no such integer exists).

(18.28) **Theorem.** The order of each element of a finite group is finite.

Proof. Let n be the size of the group, and consider an arbitrary element b. By Pigeonhole principle (16.43), for the sequence $b^0, b^1, b^2, \ldots, b^n$ there is a least positive integer k, $k < n$, such that

$\qquad b^0, b^1, \ldots, b^{k-1}$ are all distinct and

$\qquad b^k$ equals one of those earlier values.

This is because the sequence has $n+1$ elements but the group has only n distinct values. Consider whether $b^k = b^i$, for $0 < i < k$. We have,

$\qquad b^k = b^i$

$= \qquad \langle$ Definition of b^k and b^i $\rangle$

$\qquad b^{k-1} \circ b = b^{i-1} \circ b$

$= \qquad \langle$ Cancellation (18.19)$\rangle$

$\qquad b^{k-1} = b^{i-1}$

$= \qquad \langle\, b^0, \ldots, b^{k-1}$ are all distinct$\rangle$

$\qquad$ *false*

Thus, $b^k \neq b^i$ for $0 < i < k$. Since b^k equals one of $b^0, \ldots, b^{k-1}$, we have $b^k = b^0 = 1$. $\qquad\qquad\qquad\qquad\qquad\qquad\qquad\qquad\qquad\qquad$ □

Example of orders of elements. Consider the additive group M_6 of integers modulo 6: for b and c in $0..5$, $b \oplus c = (b + c) \bmod 6$. We give the orders of each of the elements.

$ord.0 = 1$ (note that $0 \bmod 6 = 0$)
$ord.1 = 6$ (note that $(1 + 1 + 1 + 1 + 1 + 1) \bmod 6 = 0$)
$ord.2 = 3$ (note that $(2 + 2 + 2) \bmod 6 = 0$)
$ord.3 = 2$ (note that $(3 + 3) \bmod 6 = 0$)
$ord.4 = 3$ (note that $(4 + 4 + 4) \bmod 6 = 0$)
$ord.5 = 6$ (note that $(5 + 5 + 5 + 5 + 5 + 5) \bmod 6 = 0$)   □

Subgroups

Subalgebra $\hat{G} = \langle T, \circ, 1 \rangle$ of group $G = \langle S, \circ, 1 \rangle$ is called a *subgroup* of G if $\hat{G}$ is a group. We investigate the structure of groups and their subgroups.

Not every subalgebra of a group is a subgroup. For example, consider the additive group of integers $\langle \mathbb{Z}, +, 0 \rangle$. The inverse of any integer b is $-b$. Hence, the subalgebra $\langle \mathbb{N}, +, 0 \rangle$ is not a subgroup, since the inverse of 1 is not in $\mathbb{N}$.

On the other hand, consider $n > 0$ and the additive group M_n of integers modulo n,

$$M_n = \langle 0..n - 1, \oplus, 0 \rangle, \text{ where } b \oplus c = (b + c) \bmod n \quad,$$

Here, the inverse of an integer i, $0 \le i < n$, is $n - i$, since $(i + (n - i)) \bmod n = n \bmod n = 0$. Thus, the inverse of an element is in the group, which is enough to prove that a subalgebra of M_n is a subgroup. In fact, we will be able to prove that a subalgebra of any finite group is a group. Thus, we have the following two theorems. (The proof of the first theorem is left to the reader, since it is almost trivial.)

(18.29) **Theorem.** A subalgebra of a group is a group iff the inverse of every element of the subalgebra is in the subalgebra.

(18.30) **Theorem.** A subalgebra of a finite group is a group.

Proof. Let $\langle T, \circ, 1 \rangle$ be the subalgebra. Then T is closed under $\circ$ and, by definition, 1 is in T. We prove that the inverse of every element of T is in T. Since T is finite, the order m (say) of an element b is finite. Then $b^{m-1} \circ b = b^m = 1$, so the inverse of b is b^{m-1}. Element b^{m-1} is in T, since it is formed from the $(m - 1)$-fold composition of b and b is in T. □

The next three theorems give specific ways of constructing a subgroup. The first deals with an element of the group and its powers. The second requires a homomorphism of the group in order to construct the subgroup.

The third shows how to construct a subgroup from two subgroups. The proofs of the first two theorems are easy enough to be left as exercises.

(18.31) **Theorem.** Let b be an element of a group $\langle S, \circ, 1 \rangle$. Let set S_b consist of all powers of b (including negative powers). Then $\langle S_b, \circ, 1 \rangle$ is a subgroup of $\langle S, \circ, 1 \rangle$.

(18.32) **Theorem.** A homomorphic image of a group (monoid, semigroup) is a group (monoid, semigroup).

(18.33) **Theorem.** Let $G1 = \langle S1, \circ, 1 \rangle$ and $G2 = \langle S2, \circ, 1 \rangle$ be two subgroups of a group G. Then their intersection $\hat{G} = \langle S1 \cap S2, \circ, 1 \rangle$ is a subgroup G.

Proof. We first show that $S1 \cap S2$ is closed under $\circ$, so that $\hat{G}$ is an algebra. We have,

$$b \in S1 \cap S2 \land c \in S1 \cap S2$$
$$= \quad \langle \text{Definition of } \cap \ (11.21) \rangle$$
$$b \in S1 \land b \in S2 \land c \in S1 \land c \in S2$$
$$= \quad \langle S1 \text{ and } S2 \text{ are closed under } \circ \rangle$$
$$(b \circ c) \in S1 \land (b \circ c) \in S2$$
$$= \quad \langle \text{Definition of } \cap \ (11.21) \rangle$$
$$(b \circ c) \in S1 \cap S2$$

We now prove that $\hat{G}$ satisfies the three parts of Definition (18.17) of a group.

(a) $\circ$ restricted to $S1 \cap S2$ is a binary associative operator, since $S1 \cap S2$ is closed under $\circ$ and $\circ$ is associative on G.
(b) Since $G1$ and $G2$ are subgroups, they contain the identity of the group; therefore, so does $\hat{G}$.
(c) Let b be in $S1 \cap S2$. It is therefore in $S1$ and in $S2$, so its inverse is in $S1$ and in $S2$, so its inverse is in $S1 \cap S2$. □

We now show how a subgroup can be used to partition a group. This partition will tell us the relation between the sizes of groups and subgroups.

(18.34) **Definition.** Let $\langle \hat{S}, \circ, 1 \rangle$ be a subgroup of group $\langle S, \circ, 1 \rangle$, and let b be an element of group S. Then the set

$$\hat{S} \circ b = \{x \mid x \in \hat{S} : x \circ b\}$$

is called a *right coset* of $\hat{S}$. The number $index.\hat{S}$ of distinct right cosets $\hat{S} \circ b$ (for b in S) is called the *index* of $\hat{S}$.

(18.35) **Example.** We give below a table of subgroups and their cosets of the additive group M_6 of integers modulo 6. Remember here that operator $\circ$ is addition modulo 6.

Subgroup $\hat{S}$	Cosets of $\hat{S}$
0	$\hat{S} \circ 0 = \{0\}$, $\hat{S} \circ 1 = \{1\}$, $\hat{S} \circ 2 = \{2\}$, $\hat{S} \circ 3 = \{3\}$, $\hat{S} \circ 4 = \{4\}$, $\hat{S} \circ 5 = \{5\}$
0,3	$\hat{S} \circ 0 = \hat{S} \circ 3 = \{0,3\}$, $\hat{S} \circ 1 = \hat{S} \circ 4 = \{1,4\}$, $\hat{S} \circ 2 = \hat{S} \circ 5 = \{2,5\}$
0,2,4	$\hat{S} \circ 0 = \hat{S} \circ 2 = \hat{S} \circ 4 = \{0,2,4\}$, $\hat{S} \circ 1 = \hat{S} \circ 3 = \hat{S} \circ 5 = \{1,3,5\}$
0,1,2,3,4,5	$\hat{S} \circ i = \{0,1,2,3,4,5\}$ (for all i, $0 \le i < 6$) $\qquad \square$

Consider again a subgroup $\langle \hat{S}, \circ, 1 \rangle$ of $\langle S, \circ, 1 \rangle$. Since $\hat{S} \circ 1 = \hat{S}$, $\hat{S}$ is itself a right coset of $\hat{S}$. Moreover, Theorem Cancellation (18.19), $x \circ b = y \circ b \equiv x = y$, indicates that the elements $x \circ b$ of a coset (for all x in $\hat{S}$) are distinct. Therefore, we have

(18.36) **Theorem.** The size of a right coset $\hat{S} \circ b$ equals the size of $\hat{S}$.

Further, suppose $\hat{S} \circ b$ and $\hat{S} \circ c$ have an element d (say) in common, so that $x \circ b = y \circ c = d$, where x and y are in $\hat{S}$. Then, for any element $s \circ b$ in $\hat{S} \circ b$ we have

$$\begin{aligned} & s \circ b \\ = \quad & \langle \text{Identity; Inverse} \rangle \\ & s \circ x^{-1} \circ x \circ b \\ = \quad & \langle \text{Assumption } x \circ b = y \circ c \rangle \\ & s \circ x^{-1} \circ y \circ c \end{aligned}$$

Hence, any element of $\hat{S} \circ b$ is in $\hat{S} \circ c$. Similarly, any element of $\hat{S} \circ c$ is in $\hat{S} \circ b$. Hence, either $\hat{S} \circ b$ and $\hat{S} \circ c$ are disjoint or they are equal. We have proved that

(18.37) **Theorem.** The right cosets of subgroup $\langle \hat{S}, \circ, 1 \rangle$ of group $\langle S, \circ, 1 \rangle$ partition S. If the group is finite, then each coset has the same number $\#\hat{S}$ of elements, and $\#S = \#\hat{S} \cdot (index.\hat{S})$.

Since the right cosets of a subgroup partition S, they determine an equivalence relation. Two elements of S are equivalent under this relation iff they are in the same right coset.

(18.38) **Corollary.** The size of a finite group is a multiple of the size of each of its subgroups.

What can we extract from this theorem? Suppose the number of elements of a group is p, where p is a prime. Then the only positive divisors of p are 1 and p, so that the only two subgroups of the group are the group consisting of the identity of the group and the group itself. Also, if the size of the group is $p \cdot q$ for primes p and q, then the subgroups can have only

HISTORICAL NOTE 18.2. Pierre Fermat (1601-1665)

E.T. Bell calls Fermat the greatest mathematician of the seventeenth century. Fermat, along with Pascal (see Hist. note 16.1 on p. 347), created the theory of probability. He invented analytic geometry, independently of Descartes (see Hist. note 14.1 on p. 268), and was the first to apply it to three-dimensional space. He was one of the creators of calculus. But his greatest contributions were in number theory, where one analyzes the integers (for example) and derives theorems such as $b^p \stackrel{p}{=} b$ for p a prime. The numbers $2^{2^n} + 1$ for n an integer are called the Fermat numbers, after Fermat.

Amazingly, Fermat was a completely amateur mathematician. His higher education prepared him to be a magistrate, and he spent his working life in the service of the state —the last 17 as a King's councillor in Toulouse. For recreation, he pursued mathematics.

Fermat is the instigator of one of the great mysteries of mathematics. In reading a book on number theory, he wrote in the margin, "On the contrary, it is impossible to separate a cube into two cubes, a fourth power into two fourth powers, or, generally, any power above the second into two powers of the same degree; I have discovered a truly marvelous demonstration, which this margin is too narrow to contain" (quoted from E.T. Bell's *Men of Mathematics*). This theorem, called "Fermat's last theorem", is written as

For $n > 2$, no positive integers x, y, z exist such that $x^n + y^n = z^n$.

For 356 years, no one was able to prove Fermat's last theorem. Some found it too remote even to try. The great David Hilbert (see Hist. note 6.1 on p. 111), for example, said that it would take three years of intensive study before he could begin to prove the theorem, and he didn't have that much time to spend on a probable failure. Prizes were offered for a proof (or disproof), but no one came forward to claim them.

But lo and behold, in June 1993, Princeton Professor Andrew Wiles announced his proof during a lecture at Cambridge. The proof is long and difficult, and it has been estimated that only .1 % of working mathematicians could understand it. So the proof has to be read carefully by several mathematicians and refereed and finally published before its correctness can be assured. (See the introduction to Chap. 19 for an example of a published "proof" of a famous theorem whose incorrectness took ten years to discover.) Nevertheless, the celebrating began almost immediately in many places.

---

sizes 1, p, q, and $p \cdot q$. This theory is borne out in Example (18.35) of the additive integers modulo 6. This group is of size 6, and its subgroups are of sizes 1, 2, 3, and 6. Neat stuff!

Evidence of the applicability of group theory is given by proof of a theorem from number theory, due to Fermat (See Historical note 18.2).

(18.39) **Theorem.** If b is an integer and p a prime, then $b^p \overset{p}{=} b$.[2]

Proof. We look at the three cases $b = 0$, $1 \le b < p$, and $p \le b$.

Case $b = 0$. The theorem holds trivially in this case.

Case $1 \le b < p$. Consider the multiplicative group mod p: $\langle 1..p-1, \otimes, 1 \rangle$, where $b \otimes c = (b \cdot c) \bmod p$. The group has $p-1$ elements. For any element b, consider the subgroup consisting of powers of b. By Corollary (18.38), the size of this subgroup, which is the order of b, is a divisor of $p-1$. We have,

$$b^p \overset{p}{=} b$$
$$= \quad \langle \text{Definition (18.23)} \rangle$$
$$b^{p-1} \otimes b \overset{p}{=} b$$
$$= \quad \langle \text{The order of } b \text{ is a divisor of } p-1 \rangle$$
$$b^{ord.b \cdot k} \otimes b \overset{p}{=} b \qquad (\text{for some } k)$$
$$= \quad \langle (18.25) \rangle$$
$$(b^{ord.b})^k \otimes b \overset{p}{=} b \qquad (\text{for some } k)$$
$$= \quad \langle (18.27) \rangle$$
$$1^k \otimes b \overset{p}{=} b \qquad (\text{for some } k)$$
$$= \quad \langle 1 \text{ is the identity} \rangle$$
$$b \overset{p}{=} b \quad \text{—Reflexivity of } \overset{p}{=}$$

Case $p \le b$. Let c and r be the remainder when dividing b by p: $b = c \cdot p + r$ where $0 \le r < p$. Then

$$b^p \overset{p}{=} b$$
$$= \quad \langle \text{Definition of } X \overset{p}{=} Y \rangle$$
$$b^p \bmod p = b \bmod p$$
$$= \quad \langle b = c \cdot p + r \rangle$$
$$(c \cdot p + r)^p \bmod p = (c \cdot p + r) \bmod p$$
$$= \quad \langle \text{Property of } \mathbf{mod}, \text{twice} \rangle$$
$$r^p \bmod p = r \bmod p$$
$$= \quad \langle \text{Definition of } X \overset{p}{=} Y \rangle$$
$$r^p \overset{p}{=} r$$

Since $0 \le r < p$, from $r^p \overset{p}{=} r$ and the argument in the previous paragraph, we conclude $b^p \overset{p}{=} b$. $\square$

---

[2] On page 325, $b \overset{p}{=} c$ is defined as $p \mid (c - b)$, or p divides $c - b$. It is also equal to $b \bmod p = c \bmod p$.

CYCLIC GROUPS

(18.40) **Definition.** A group is *cyclic* if it contains some element whose
powers are the carrier of the group. An element whose powers form
the carrier is called a *generator* of the group and is said to *generate*
the group.

For example, in the additive group $\langle \mathbb{Z}, +, 0 \rangle$ of integers, 1 generates the
group, while in the additive group $M_6 = \langle 0..5, \oplus, 0 \rangle$ of integers modulo
6, not only 1 but also 5 generates the group. These two examples are
not chosen randomly; rather, they are fundamental in the study of cyclic
groups, as the following theorem shows.

(18.41) **Theorem.** An infinite cyclic group is isomorphic to the additive
group of integers. A finite cyclic group with n elements is isomorphic to the additive group of integers modulo n.

Proof. Let group $\langle S, \circ, b^0 \rangle$ be cyclic, where element b generates the group.
Therefore, every element of S can be written as b^m for some $m \geq 0$.

We handle the infinite case first. Let $\langle \mathbb{Z}, +, 0 \rangle$ be the additive group of
integers. Note that $\langle S, \circ, b^0 \rangle$ and $\langle \mathbb{Z}, +, 0 \rangle$ have the same signature. Define
function $h : S \to \mathbb{Z}$ by $h(b^m) = m$ (for m an integer). We prove that h
is an isomorphism from $\langle S, \circ, b^0 \rangle$ to $\langle \mathbb{Z}, +, 0 \rangle$ (see Definition (18.9)).

(a) Function h is one-to-one, because $h(b^m) = h(b^n) \equiv m = n$. It is
obviously onto.

(b) $h.(b^0) = 0$.

(c) The group has no unary operators, so point (c) of Definition (18.9) is
trivially satisfied.

(d) We show that h commutes with the binary operator of the group.

$$h(b^m \circ b^n)$$
$$= \quad \langle \text{Definition of } \circ \rangle$$
$$h(b^{m+n})$$
$$= \quad \langle \text{Definition of } h \rangle$$
$$m + n$$
$$= \quad \langle \text{Definition of } h, \text{ twice} \rangle$$
$$h(b^m) + h(b^n)$$

Now for the finite case. Let element b of finite group $\langle S, \circ, b^0 \rangle$ generate S,
and let n denote the size of S. Then the elements of S are $b^0, b^1, \ldots b^{n-1}$.
Let $M_n = \langle 0..(n-1), \oplus, 0 \rangle$ be the additive group of integers modulo n.
Thus, $m \oplus p = (m+p) \bmod n$. Define $hn : S \to 0..(n-1)$ by $hn(b^m) = m$.
We leave to the reader to show that hn is an isomorphism from $\langle S, \circ, b^0 \rangle$
to M_n.   □

A corollary of Theorem (18.41) is that two cyclic groups of the same order are isomorphic.

Not all groups are cyclic. But we do have the following theorem.

(18.42) Theorem. A subgroup of a cyclic group is cyclic.

Proof. By Theorem (18.41), the theorem is proved if we show that any subgroup of $\langle \mathbb{Z}, +, 0 \rangle$ is cyclic and that any subgroup of the additive group $M_n = \langle 0..n - 1, \oplus, 0 \rangle$ of integers modulo n is cyclic.

Case $\langle \mathbb{Z}, +, 0 \rangle$. Let S be a subgroup of this group. Let m be the least positive integer in S. Then, by the definition of a group, all multiples of m are in S. We show that every element of S is a multiple of m, so that m generates S and S is cyclic.

Let k be in S. Write k as $k = q \cdot m + r$ where $0 \leq r < m$. The following calculation shows that $r = 0$, so that k is a multiple of m.

$$
\begin{aligned}
& k = q \cdot m + r \\
= \quad & \langle \text{Arithmetic} \rangle \\
& k + (-q) \cdot m = r \\
\Rightarrow \quad & \langle k \in S \text{ and } (-q) \cdot m \in S, \text{ since it is a multiple of } m; \\
& \quad S \text{ is closed under addition} \rangle \\
& r \in S \\
= \quad & \langle 0 \leq r < m \text{ and } m \text{ is the smallest positive integer in } S \rangle \\
& r = 0
\end{aligned}
$$

Case M_n. Since this case is similar to the previous case, it is left to the reader. $\qquad\square$

As an example, consider again the additive group $M_6 = \langle 0..5, \oplus, 0 \rangle$ of integers modulo 6, as discussed in Example (18.35). According to our analysis, subgroups have sizes that divide 6, so the subgroups have sizes 1, 2, 3, and 6.

Groups of transformations

A *transformation* ϕ from a set S to a set T is simply a function $\phi : S \to T$. The term *transformation* comes from dealing with functions that transform, for example, the two-dimensional plane, in some fashion. For example, the transformation

$$halve(x, y) = (x/2, y/2)$$

halves the distance of every point (x, y) from the origin. Another transformation of the plane is a 90-degree clockwise rotation, which takes vertex b of Fig. 18.1 into vertex c, c into d, and so on. And the transformation

V defined by $V(x,y) = (-x,y)$ reflects points around the vertical axis $x = 0$. In this section, we are interested in one-to-one transformations of a set into itself, and we reserve the word *transformation* for them.

As we know from Chap. 14, the *composition* $\phi \bullet \tau$ of two transformations ϕ and τ is the transformation defined by

$$(\phi \bullet \tau).p = \phi(\tau.p) \quad (\text{for all } p).$$

Consider a nonempty set T of one-to-one onto transformations that is closed under $\bullet$. Let I be the identity transformation. Since each transformation ϕ in T is one-to-one and onto, it has an inverse ϕ^{-1}. Suppose the inverses are in T as well. Then $\langle T, \bullet, I \rangle$ is a group. First, $\bullet$ is a binary associative operator. The identity $I = \phi \bullet \phi^{-1}$ is in the group, since T is closed under $\bullet$. Hence, the properties of a group are satisfied.

As an example of an interesting group of transformations, consider the *symmetries of the square*, as depicted in Fig. 18.1. In general, a symmetry of a geometrical figure is a one-to-one transformation that takes the figure into itself and preserves distance. By *preserves distance*, we mean the following. For any points p and q of the figure and for ϕ a symmetry, the distance between p and q equals the distance between $\phi.p$ and $\phi.q$.

A symmetry of the square takes a vertex of the square into a vertex, since any other transformation would not preserve distance. Let us construct a symmetry ϕ, with $\phi.b = b$ where b is the upper right vertex of the square of Fig. 18.1. Since ϕ is a symmetry, ϕ does not take point d into either c or e, because then the distance between b and d would be greater than the distance between $\phi.b$ and $\phi.d$. Hence, $\phi.d = d$. Either of $\phi.c, \phi.e = c, e$ and $\phi.c, \phi.e = e, c$ makes ϕ a symmetry, so there are two symmetries of the square with $\phi.b = b$. In the same way, we can see that, for each vertex, there are two symmetries that take b into that vertex, so there are 8 symmetries of the square. We can describe them as follows.

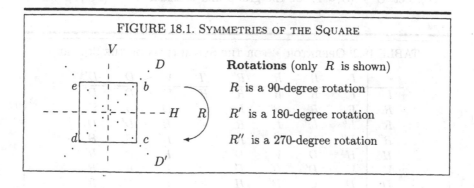

FIGURE 18.1. SYMMETRIES OF THE SQUARE

Rotations (only R is shown)

R is a 90-degree rotation

R' is a 180-degree rotation

R'' is a 270-degree rotation

I :   The identity transformation
R, R', R'' :   Rotations of 90, 180, and 270 degrees
H, V :   Reflections about the horizontal and vertical axes
D, D' :   Reflections about the two diagonals

Table 18.2 contains the composition table for these eight symmetries. The composition of two symmetries is itself a symmetry. For example, $R \cdot R' = R''$ —making a 180-degree rotation followed by a 90-degree rotation is the same as making a 270-degree rotation. Further, each symmetry has an inverse. Hence, these eight symmetries of a square form a group.

We give further examples of groups of transformations. In each case, the group operator is $\cdot$ and the identity is the identity transformation I.

Examples of groups of transformations

(a) The set of all one-to-one and onto transformations of the two-dimensional plane form a group.

(b) The set of rotations of the plane form a group. The identity transformation is a rotation of 0 degrees, and the product of two transformations of degrees p and q is a rotation of degree $p + q$.

(c) The set $\{I, H\}$ of symmetries of the square forms a group, since $H \cdot H = I$.

(d) The set of symmetries of the cube form a group.

(e) Let S be any set. Then the set G of one-to-one onto transformations of S is a group. $\square$

Theorem (18.43) indicates that a study of groups could be restricted to groups of transformations, since any group is essentially a group of transformations! This result is due to Arthur Cayley (see Historical note 18.3).

(18.43) **Theorem.** Any group is isomorphic with a group of transformations.

Proof. Let $G = \langle S, \circ, 1 \rangle$ be the group and consider $\hat{G} = \langle T, \cdot, I \rangle$ where

TABLE 18.2. OPERATOR $\cdot$ FOR THE SYMMETRIES OF THE SQUARE

$\cdot$	I	R	R'	R''	H	V	D	D'
I	I	R	R'	R''	H	V	D	D'
R	R	R'	R''	I	D'	D	H	V
R'	R'	R''	I	R	V	H	D'	D
R''	R''	I	R	R'	D	D'	V	H
H	H	D	V	D'	I	R'	R	R''
V	V	D'	H	D	R'	I	R''	R
D	D	V	D'	H	R''	R	I	R'
D'	D'	H	D	V	R	R''	R'	I

HISTORICAL NOTE 18.3. ARTHUR CAYLEY (1821–1895)

Cayley was one of the most productive mathematicians the world has seen. His *Collected Mathematical Papers* —13 volumes of about 600 pages each— contain 966 papers. His work has been of lasting value to mathematics and physics. The theory of invariants is due to Cayley. A topic of his creation that is more accessible to college students is the theory of matrices, including the way that we multiply matrices. Cayley is also responsible for the invention of the geometry of the space of n dimensions.

Cayley's mathematical genius showed itself early, and throughout his education he was far above the rest of the students in math. His undergraduate days were spent at Trinity College, Cambridge, and he was a Fellow at Trinity for three years thereafter —during which time he published 25 papers. To continue as a Fellow, he would have had to become a parson; instead, at the age of 25, he left Cambridge, studied law, and worked as a lawyer for 14 years. He did not give up mathematics, however, and his results were so good that, when he was 42, Cambridge offered him a professorship. He accepted, got married that same year, and spent much of the rest of his life pursuing math and university administration. (Largely through his efforts, women were at last admitted to Cambridge as students —in their own nunneries, of course.)

Cayley is admired as much for his character as for his mathematics. He was always strong, patient, steady, and unruffled. He never had much to say, but his opinions were usually accepted as final, for he reasoned things through thoroughly and was known for his impersonal judgement. Forsyth, his student and successor at Cambridge, closes his biography of Cayley with,

> But he was more than a mathematician. With a singleness of aim ... he persevered to the last in his nobly lived ideal. His life had a significant influence on those who knew him: they admired his character as much as they respected his genius: and they felt that, at his death, a great man had passed from the world.

T is a set of transformations defined as follows. For each element b of S, transformation $\phi_b : S \to S$ is given by $\phi_b.x = b \circ x$ (for x in S).

Note that G and $\hat{G}$ have the same signature. Define function $h : S \to T$ by $h.b = \phi_b$ for b in S. We leave to the reader the proofs of the following points, which establish that h is an isomorphism from G to $\hat{G}$.

(a) Each transformation is distinct: if $\phi_b = \phi_c$ then $b = c$. This proves that h is one-to-one. That h is onto is clear from its construction.

(b) Each transformation $\phi_b : S \to S$ is one-to-one and onto.

(c) $\phi_{b \circ c} = \phi_b \bullet \phi_c$.

(d) Let 1 be the identity of group G. Then ϕ_1 is the identity transformation, and for each b in S, the inverse of ϕ_b is $\phi_{b^{-1}}$. This shows that $\hat{G}$ is a group.

(e) h commutes with $\circ$: $h(b \circ c) = h.b \bullet h.c$.   □

GROUPS OF PERMUTATIONS

As our final topic under group theory, we investigate groups of *permutations*, where a permutation is a one-to-one transformation of a finite set onto itself. For example, two permutations ϕ and ϕ' of $\{0, 1, 2, 3\}$ are given by

$$\phi.0 = 1, \quad \phi.1 = 3, \quad \phi.2 = 2, \quad \phi.3 = 0 \quad \text{and}$$
$$\phi'.0 = 2, \quad \phi'.2 = 3, \quad \phi'.3 = 1, \quad \phi'.1 = 0 \ .$$

Try constructing the compositions of these two permutations: $\phi \cdot \phi'$ and $\phi' \cdot \phi$.

A permutation is simply a transformation, so a set of permutations forms a group under operator $\cdot$ if the set is closed under $\cdot$ and if the inverse of each permutation is in the set.

Permutation ϕ' above gives a circular arrangement of the elements. Such a permutation is *cyclic*. We could depict permutation ϕ' as

$$0 \rightarrow 2 \rightarrow 3 \rightarrow 1 \rightarrow 0 \ ,$$

where the first element 0 appears twice, at the beginning and end, and where $b \rightarrow c$ stands for $\phi'.b = c$. There is, however, a shorter representation for this permutation: (0231). The elements occur within parentheses, and each element is followed by its transform, except the last element in the list, whose transform is at the beginning of the list. Such a list could be called a *cycle*. In a cycle, it does not matter which element appears first, and the following four cyclic permutations are all equal: (0231), (1023), (3102), (2310) .

(18.44) **Theorem.** For a cyclic permutation $\phi = (b_0 b_1 \ldots b_{n-1})$ of n symbols, $\phi^k(b_i) = b_{k+i}$, (where all subscripts are modulo n).

Proof. We prove the theorem for $k \geq 0$ by induction on k .

Base case $k = 0$. We have $\phi^0(b_i) = b_i = b_{0+i}$, so the theorem holds.

Inductive case. Assume inductive hypothesis $\phi^k(b_i) = b_{k+i}$ and prove $\phi^{k+1}(b_i) = b_{k+1+i}$. We have,

$$\phi^{k+1}(b_i)$$
$$= \quad \langle \text{Definition of powers and } \cdot \rangle$$
$$\phi(\phi^k(b_i))$$
$$= \quad \langle \text{Inductive hypothesis} \rangle$$
$$\phi(b_{k+i})$$
$$= \quad \langle \text{Definition of } \phi \rangle$$
$$b_{k+i+1}$$

We leave the similar proof for negative k to the reader.   □

(18.45) **Corollary.** A cyclic permutation of n symbols has order n.

Proof. The corollary follows from $\phi^n(b_i) = b_{n+i} = b_i$, since all subscripts are taken modulo n, and since for $1 \le k < n$, $\phi^k(b_i) = b_{k+i} \ne b_{n+i}$. □

By definition, a cyclic permutation leaves unchanged elements that are not contained in it. For example, $(143).2 = 2$. Call two permutations *disjoint* if they have no elements in common. Then the composition of two disjoint cyclic permutations is symmetric. For example,

$$(04) \circ (132) = (132) \circ (04)$$
$$((04) \circ (132)).3 = (04).2 = 2$$
$$((132) \circ (04)).3 = (132).3 = 2 \ .$$

(18.46) **Theorem.** Any permutation can be written as a composition of disjoint cycles.

Proof. We present only an outline of the proof. Let ϕ be the permutation of a set S. For element b of S, $\phi_b = (b \ \phi.b \ \phi^2.b \ \ldots \ \phi^{ord.b-1}.b)$ is a cyclic permutation. Construct the set C of such cyclic permutations for all the elements of S:

$$C = \{b \mid b \in S : \phi_b\}$$

Each element of S appears in exactly one cyclic permutation of C, since cyclic permutations are considered equal up to rotation of the elements. (For example, $(123) = (231)$.) Thus, if $\phi^n.b = d$, then $\phi_b = \phi_d$. Construct the composition of the cyclic permutations in C —the order does not matter, since they are pairwise disjoint. This construction yields the desired result. □

For example, the identity permutation on the set $\{0, 1, 2, 3, 4\}$ is a product of five cyclic permutations of order 1: $(0) \bullet (1) \bullet (2) \bullet (3) \bullet (4) \bullet (5)$.

(18.47) **Theorem.** The order of any permutation is the least common multiple of the lengths of its disjoint cycles.

Proof. Let $\phi = \gamma_1 \bullet \cdots \bullet \gamma_k$ be a permutation, where the γ_i are disjoint cyclic permutations. Since the composition of two disjoint cyclic permutations is symmetric,

$$\phi^n = \gamma_1^n \bullet \cdots \bullet \gamma_k^n \ ,$$

for all integers n. Therefore $\phi^n = I$ (where I is the identity permutation) iff for every i, $1 \le i \le k$, γ_i^n is the identity transformation. By Corollary (18.45), $\phi^n = I$ iff n is a common multiple of the lengths of the γ_i. □

Let us use this theorem to study the symmetries of the square. These are depicted in Fig. 18.1 (page 407), and the composition table for this group is given in Table 18.2. It is a group of eight permutations. By Theorem (18.43) (any group is isomorphic to a group of transformations), we can represent this group as a group of permutations of the eight symbols that represent the elements of the group. For example, R corresponds to the permutation given by "multiplying" each element on the right by R. From Table 18.2, we see that this permutation is

$$(IRR'R'') \circ (HDVD') \quad .$$

Below, we give each of the eight permutations as a cyclic permutation; we omit the composition operator .

$$
\begin{array}{ll}
I: & (I)\,(R)\,(R')\,(R'')\,(H)\,(V)\,(D)\,(D') \\
R: & (IRR'R'')\,(HDVD') \\
R': & (IR')\,(RR'')\,(HV)\,(DD') \\
R'': & (IR''R'R)\,(HD'VD) \\
H: & (IH)\,(RD'R''D)\,(R'V) \\
V: & (IV)\,(RDR''D')\,(R'H) \\
D: & (ID)\,(RH)\,(R'D')\,(R''V) \\
D': & (ID')\,(RV)\,(R'D)\,(R''H)
\end{array}
$$

The symmetries of the cube form another interesting group. Exercise 18.54 asks you to show that the cube has 48 symmetries.

18.3 Boolean algebras

(18.48) **Definition.** A *boolean algebra* is an algebra $\langle S, \oplus, \otimes, \sim, 0, 1 \rangle$ in which

  (a) $\oplus$ and $\otimes$ are associative binary operators;

  (b) $\oplus$ and $\otimes$ are symmetric;

  (c) 0 and 1 are the identities of $\oplus$ and $\otimes$;

  (d) unary operator $\sim$ satisfies $b \oplus (\sim b) = 1$ and $b \otimes (\sim b) = 0$
        (for all b); $\sim b$ is called the *complement* of b;

  (e) $\otimes$ distributes over $\oplus$: $b \otimes (c \oplus d) = (b \otimes c) \oplus (b \otimes d)$;

  (f) $\oplus$ distributes over $\otimes$: $b \oplus (c \otimes d) = (b \oplus c) \otimes (b \oplus d)$.

It is tempting to try to see an abelian group embedded within a boolean algebra. After all, an abelian group has an associative, symmetric operator with an identity and an inverse for each element. Suppress this temptation, because it is does not work. In a group $\langle S, \oplus, 0 \rangle$, the inverse law says that $x \oplus b^{-1} = 0$ holds. Here, however, we have $b \oplus (\sim b) = 1$ instead of $b \oplus (\sim b) = 0$. Think of $\sim$ as the complement, and not the inverse, of b.

Examples of boolean algebras

  (a) $\langle \mathbb{B}, \vee, \wedge, \neg, false, true \rangle$ is a boolean algebra. It is our model for the propositional calculus. Thus, our discussion of boolean algebras may give us further information concerning our model of the propositional calculus.

(b) $\langle PS, \cup, \cap, \sim, \emptyset, S \rangle$ is a boolean algebra, where S is any nonempty set. We call this a *power-set algebra*.

(c) $\langle \{1, 2, 3, 6\}, \mathbf{lcm}, \mathbf{gcd}, \sim, 1, 6 \rangle$ is a boolean algebra. Here, **lcm** is the least common multiple, **gcd** is the greatest common divisor, and $\sim$ $x = 6/x$. We leave it to Exercise 18.59 to show that this is indeed a boolean algebra. Exercise 18.60 extends this as follows. Let π be a product of distinct primes. Then $\langle S, \mathbf{lcm}, \mathbf{gcd}, \sim, 1, \pi \rangle$ is a boolean algebra, where S is the set of divisors of π and $\sim x = \pi/x$.

(d) For n in $\mathbb{Z}^+$, let F_n be the set of functions of type $\mathbb{B}^n \to \mathbb{B}$, i.e. the set of boolean functions of n boolean arguments. Let s denote a sequence of n boolean values. Define $\oplus$, $\otimes$, and $\sim$ by $(f1 \oplus f2).s = f1.s \vee f2.s$, $(f1 \oplus f2).s = f1.s \wedge f2.s$, $(\sim f).s = \neg(f.s)$. Then $\langle F_n, \oplus, \otimes, \sim, f, t \rangle$ is a boolean algebra. The identity of $\oplus$ is the function f that always yields *false*, and the identity t of $\otimes$ always yields *true*. □

We now look at some basic theorems that can be proved for boolean algebras. Example (a) above tells us that a model of the propositional calculus restricted to $\wedge$, $\vee$, and $\neg$ is a boolean algebra, so many of the laws of propositional calculus will be theorems.

Theorems for boolean algebras

(18.49) **Idempotency:** $b \oplus b = b$, $b \otimes b = b$

(18.50) **Zero:** $b \oplus 1 = 1$, $b \otimes 0 = 0$

(18.51) **Absorption:** $b \oplus (b \otimes c) = b$, $b \otimes (b \oplus c) = b$

(18.52) **Cancellation:** $(b \oplus c = b \oplus d) \wedge (\sim b \oplus c = \sim b \oplus d) \equiv c = d$
$(b \otimes c = b \otimes d) \wedge (\sim b \otimes c = \sim b \otimes d) \equiv c = d$

(18.53) **Unique complement:** $b \oplus c = 1 \wedge b \otimes c = 0 \equiv c = \sim b$

(18.54) **Double complement:** $\sim(\sim b) = b$

(18.55) **Constant complement:** $\sim 0 = 1$, $\sim 1 = 0$

(18.56) **De Morgan:** $\sim(b \oplus c) = (\sim b) \otimes (\sim c)$
$\sim(b \otimes c) = (\sim b) \oplus (\sim c)$

Theorems for boolean algebras (continued)

(18.57) $b \oplus (\sim c) = 1 \equiv b \oplus c = b, \quad b \otimes (\sim c) = 0 \equiv b \otimes c = b$

The definition of a boolean algebra is symmetric in the pairs $(\oplus, 0)$ and $(\otimes, 1)$. By symmetric, we mean that exchanging $\oplus$ and $\otimes$ and also 0 and 1 gives exactly the same definition. For this reason, two boolean-algebra expressions are called *duals* of each other if each can be constructed from the other by this exchange. This notion is similar to the notion of a dual discussed in Sec. 2.3. Because of the symmetry, an expression (e.g. $x \otimes (\sim x) = 0$) is a theorem iff its dual (e.g. $x \oplus (\sim x) = 1$) is a theorem. This is why some of the theorems above are listed in pairs. We need to prove only one of each pair.

The proofs of the theorems are similar to the proofs of Chap. 3 and are therefore left to the reader. Also left to the reader is the proof of the following theorem.

(18.58) **Theorem.** A homomorphic image of a boolean algebra is a boolean algebra.

The rest of this section investigates the structure of boolean algebras. Recall from Example (b) on page 413 that $\langle \mathcal{P}S, \cup, \cap, \sim, \emptyset, S \rangle$ is a boolean algebra, called a power-set algebra. We will prove the startling fact that any finite boolean algebra is isomorphic to a power-set algebra. Thus, the only finite boolean algebras are the power-set algebras. Throughout the rest of this section, 0 denotes the identity of $\oplus$ and 1 the identity of $\otimes$.

We begin our investigation by noticing that the subsets of a set are partially ordered by the subset relation, and this subset relation can be defined in terms of intersection: $S1 \subseteq S2 \equiv S1 \cap S2 = S1$. Consider an arbitrary boolean algebra $\langle S, \oplus, \otimes, \sim, 0, 1 \rangle$. The properties of $\otimes$ are similar to the properties of $\cap$ in the power-set algebra. So let us define a similar relation for an arbitrary boolean algebra. .

(18.59) **Axiom:** $b \leq c \equiv b \otimes c = b$

(18.60) **Axiom:** $b < c \equiv b \leq c \wedge b \neq c$

We can immediately prove the following theorem.

(18.61) **Theorem.** Relation $\leq$ is a partial order.

Proof. A partial order is a reflexive, antisymmetric, and transitive relation. We have $b \leq b \equiv b \otimes b = b$, which holds by Idempotency (18.49), so $\leq$ is reflexive. We prove antisymmetry.

$$b \leq c \wedge c \leq b$$
$$= \quad \langle \text{Definition of } \leq, \text{ twice} \rangle$$

$$b \otimes c = b \ \wedge \ b \otimes c = c$$
$$= \quad \langle \text{Replacement (3.51)} \rangle$$
$$b \otimes c = b \ \wedge \ b = c$$
$$\Rightarrow \quad \langle \text{Weakening (3.76b)} \rangle$$
$$b = c$$

We leave the proof of transitivity to the reader.   □

We can also provide an alternative characterization of $\leq$, whose proof is left to Exercise 18.72.

(18.62) $\quad b \leq c \ \equiv \ b \oplus c = c$

Now, a power set of a set of size n contains the empty set and n singleton sets. If an arbitrary boolean algebra $\langle S, \oplus, \otimes, \sim, 0, 1 \rangle$ is to be isomorphic to a power-set algebra, it should have the equivalent of the empty set and the singleton sets. The equivalent of the empty set is the constant 0. We call the equivalent of the singleton sets *atoms*. The atoms are the elements that satisfy the following predicate *atom.a* .

(18.63) $\quad atom.a \ \equiv \ a \neq 0 \wedge (\forall b \colon S \mid 0 \leq b \leq a : 0 = b \vee b = a)$

The properties we expect of atoms are given in the following theorems. To get some understanding of them, think of $\otimes$ as intersection. The proofs of these theorems are straightforward and are left to the reader.

Properties of atoms of a boolean algebra

(18.64) $\quad atom.a \ \Rightarrow \ a \otimes b = 0 \vee a \otimes b = a$

(18.65) $\quad atom.a \wedge atom.b \wedge a \neq b \ \Rightarrow \ a \otimes b = 0$

(18.66) $\quad (\forall a \mid atom.a : a \otimes b = 0) \ \Rightarrow \ b = 0$

We can now prove the following theorem concerning the representation of elements of a finite boolean algebra.

(18.67) **Theorem.** Any element of a finite boolean algebra can be written uniquely: $b = y$ where y is a "sum" of atoms:

$$y = (\oplus a \mid atom.a \wedge a \otimes b \neq 0 : a) \quad .$$

Proof. Exercise 18.76 asks you to prove that $b \otimes y = y$. Exercise 18.77 asks you to prove that $b \otimes (\sim y) = 0$. Using these, we prove the desired result:

$$b$$
$$= \quad \langle \text{Identity of } \otimes \ (18.48c) \rangle$$
$$b \otimes 1$$
$$= \quad \langle (18.48d) \rangle$$

$$b \otimes (y \oplus (\sim y))$$
$$= \quad \langle \text{Distributivity } (18.48e) \rangle$$
$$(b \otimes y) \oplus (b \otimes (\sim y))$$
$$= \quad \langle \text{Exercise } 18.76; \text{ Exercise } 18.77 \rangle$$
$$y \oplus 0$$
$$= \quad \langle \text{Identity of } \oplus \ (18.48c) \rangle$$
$$y$$

The uniqueness of this representation is proved in Exercise 18.78.   □

Theorem (18.67) tells us that any member of a finite boolean algebra with n atoms can be written uniquely as a finite sum of the atoms. Also, any such sum is a member of the algebra. Hence the size of the algebra is the number of such sums. Each sum either contains atom a_i or it does not, for each i, $1 \leq i \leq n$. Hence, there are 2^n possible distinct sums, and we have the following theorem.

(18.68) **Theorem.** A boolean algebra with n atoms has 2^n elements.

Finally, we can prove that every finite boolean algebra is isomorphic to a power-set algebra.

(18.69) **Theorem.** A finite boolean algebra $A = \langle S, \oplus, \otimes, \sim, 0, 1 \rangle$ with n atoms is isomorphic to algebra $\hat{A} = \langle \mathcal{P}\hat{S}, \cup, \cap, \sim, \emptyset, S \rangle$, where $\hat{S} = 1..n$.

Proof. Let $a_1, \ldots, a_n$ be the atoms of A. By Theorem (18.68), A has 2^n elements. Note that A and $\hat{A}$ have the same signature. Define function $h : S \to \mathcal{P}\hat{S}$ by

(18.70) $h.b = \{i |: i \in S1\}$ if $b = (\oplus i \mid i \in S1 : a_i)$ $(S1 \subseteq 1..n)$.

Thus, the zero of A is mapped to the empty set and each other element $(\oplus i \mid i \in S1 : a_i)$ for some nonempty set $S1$ is mapped into the set containing the indices of the atoms. Since the representation of each element of A as a sum is unique, this mapping is well defined, one-to-one, and onto.

It remains to prove that h satisfies properties (18.9a) and (18.9d). We prove only (18.9d) for $\oplus$ and leave the rest to the reader, since they are similar. Suppose $b = (\oplus i \mid i \in S1 : a_i)$ and $c = (\oplus i \mid i \in S2 : a_i)$. Then

$$h(b \oplus c)$$
$$= \quad \langle \text{Definition } b \text{ and } c \rangle$$
$$h((\oplus i \mid i \in S1 : a_i) \oplus (\oplus i \mid i \in S2 : a_i))$$
$$= \quad \langle \text{Range split for idempotent } \star \ (8.18) \rangle$$
$$h(\oplus i \mid i \in S1 \vee i \in S2 : a_i)$$
$$= \quad \langle \text{Definition of } h \rangle$$
$$\{i \mid i \in S1 \vee i \in S2\}$$
$$= \quad \langle \text{Union } (11.20) \rangle$$

$$\{i \mid i \in S1\} \cup \{i \mid i \in S2\}$$
$$= \quad \langle \text{Definition of } h \,\rangle$$
$$h(\oplus i \mid i \in S1 : a_i) \cup h(\oplus i \mid i \in S2 : a_i)$$
$$= \quad \langle \text{Definition of } b \text{ and } c \,\rangle$$
$$h.b \cup h.c$$
□

Exercises for Chapter 18

18.1 Prove Theorem (18.4).

18.2 Prove Theorem (18.6).

18.3 Determine the properties of ∘ given by Table 18.1.

18.4 Give examples of algebras with carriers as small as possible and with one binary operator that have the following properties.

(a) There is an identity element.
(b) There is a zero element.
(c) There is an identity and a zero.
(d) There is an identity but no zero.
(e) There is a zero but no identity.
(f) The operator is not associative.
(g) The operator is not symmetric.
(h) There is a left identity, but no right identity.
(i) There is a left zero, but no right zero.
(j) Every element has an inverse.
(k) The carrier has at least two elements, every element has a left inverse, and only the identity has a right inverse.

18.5 Prove that if an operator has both a left identity and a right zero, they are the same element. (Hence, if the operator has an identity and a zero, they are the same.)

18.6 Prove the parts of Theorem (18.10) that were not proven in the text: An isomorphism maps zeros to zeros and inverses to inverses.

18.7 Prove that if $\hat{A}$ is isomorphic to A then A is isomorphic to $\hat{A}$.

18.8 Prove Theorem (18.12), that isomorphism is an equivalence relation.

18.9 Prove that function h of Example (b) on page 395 is a homomorphism.

18.10 Prove that h defined by $h.b = -b$ (for b in $\mathbb{N}$) is an automorphism of $\langle \mathbb{Z}, + \rangle$. That is, show that h is an isomorphism from $A = \langle \mathbb{Z}, + \rangle$ to $\hat{A} = \langle \mathbb{Z}, + \rangle$.

Exercises on semigroups, monoids, and groups

18.11 Prove that a subalgebra of a monoid is a monoid.

18.12 Prove Cancellation (18.19).

18.13 Prove Unique solution (18.20), $b \circ x = c \;\equiv\; x = b^{-1} \circ c$ and $x \circ b = c \;\equiv\; x = c \circ b^{-1}$

18.14 Prove One-to-one (18.21).

18.15 Prove Onto (18.22).

18.16 Show that $\langle S, \circ, 1 \rangle$ is a group if $\circ$ is a binary associative operator with a left identity 1 and every element has a left inverse. Hint: It will be advantageous to begin by proving a left-cancellation theorem, $d \circ b = d \circ c \;\equiv\; b = c$, by Mutual implication (3.80).

18.17 Consider algebra $A = \langle S, \circ, 1 \rangle$, with $\circ$ a binary associative operator. Assume that all equations $x \circ b = c$ have the same solution x and all equations $b \circ y = c$ have the same solution y. Prove that A is a group.

18.18 Let nonempty finite set S be closed under $\circ$. Suppose (i) $\circ$ is symmetric; (ii) $\circ$ is associative; and (iii) left cancellation holds: $d \circ b = d \circ c \;\equiv\; b = c$. Prove that $\circ$ has an identity 1 (say) in S, so that $\langle S, \circ, 1 \rangle$ is an (abelian) group.

18.19 Prove that, in a group, the equation $x \circ b \circ x \circ c \circ b = x \circ c \circ d$ has one and only one solution x.

18.20 Prove that in the operation table of a group, no two elements in a row are the same and no two elements in a column are the same. (This holds for an infinite as well as a finite group.)

18.21 Prove that in a group with an even number of elements, an element besides the identity is its own inverse.

18.22 Write down all groups of size 1, 2, 3, and 4 and show, by exhaustion, that they are all abelian.

18.23 Prove theorem (18.24), $b^m \circ b^n = b^{m+n}$. The proof requires induction.

18.24 Prove theorem (18.25), $(b^m)^n = b^{m \cdot n}$. The proof requires induction.

18.25 Prove theorem (18.26), $b^n = b^p \;\equiv\; b^{n-p} = 1$.

18.26 Give the orders of each of the elements of the following groups

  (a) The group of integers modulo 3.
  (b) The group of integers modulo 4.
  (c) The group of integers modulo 9.
  (d) The group of integers modulo 10.
  (e) The multiplicative group $\langle \mathbb{N}, \cdot, 1 \rangle$ of positive integers.
  (f) The additive group of multiples of 5.

18.27 Prove Theorem (18.29): A subalgebra of a group is a group iff the inverse of every element of the subalgebra is in the subalgebra.

18.28 Prove Theorem (18.31).

18.29 Prove that a group is symmetric iff the following condition holds: $(b \circ c)^n = b^n \circ c^n$ (for all elements b and c of the group and all natural numbers n).

18.30 Prove that in a group: $(b \circ y)^{-1} = b^{-1} \circ y^{-1}$.

18.31 Prove that in a group $\langle S, \circ, 1 \rangle$: $b \circ b = b \equiv b = 1$.

18.32 Prove that if $b \circ b = 1$ for all elements b of a group, then the group is abelian.

18.33 Prove Theorem (18.31), that the set of powers of an elements of a group form a subgroup.

18.34 Prove Theorem (18.32): the homomorphic image of a group (monoid, semigroup) is a group (monoid, semigroup).

18.35 We know that the additive group of integers is a group. Prove that the additive integers modulo n ($n > 0$) is a group without proving directly that it satisfies the group properties. Hint: Use Theorem (18.32).

18.36 Let element b of order m generate a finite cyclic group with identity 1. Prove that element c of order n generates the group iff $m \gcd n = 1$.

18.37 Prove that function hn of Theorem (18.41), defined in the finite case of the proof, is an isomorphism.

18.38 Indicate which of the following groups are cyclic:

  (a) $\langle 1..6, \otimes \rangle$, where $b \otimes c = (b \cdot c) \bmod 7$.
  (b) $\langle \{1, 3, 5, 7\}, \otimes \rangle$, where $b \otimes c = (b \cdot c) \bmod 8$.
  (c) $\langle \{1, 2, 4, 5, 7, 8\}, \otimes \rangle$, where $b \otimes c = (b \cdot c) \bmod 9$.

18.39 Consider the three symmetries of the square V, D, and R''. Compute the transformations $V \bullet D$, $(V \bullet D) \bullet R''$, $D \bullet R''$, and $V \bullet (D \bullet R'')$,

18.40 Compute the transformations $H \bullet R$, $(H \bullet R) \bullet D$, $D \bullet R$, $R \bullet D$, and $H \bullet (R \bullet D)$, where H, D, and R are three of the symmetries of the square.

18.41 Consider the set $\mathbb{R}$ of real numbers. Consider the set of transformations of the form $\phi.x = b \cdot x + c$ (for b and c real numbers). For example, $\phi.x = 3 \cdot x + 1/2$ is one such transformation. Each of the cases below puts restrictions on b and c, and thus on the set of transformations considered. Tell which sets of transformations are groups, and why.

  (a) b and c rational numbers.
  (b) b and c integers.
  (c) $b = 1$ and c an even integer.
  (d) $b = 1$ and c an odd integer.
  (e) b a rational number and $c = 0$.
  (f) $b \neq 0$ and c any real number.
  (g) $b = 0$ and c any number.

18.42 Suppose a set S of transformations is closed under composition. Show that if transformations ϕ_1 and ϕ_2 have inverses, then so does $\phi_1 \bullet \phi_2$.

18.43 Solve the equation $R \bullet X \bullet D = H$ for X, in the group of symmetries of the square.

18.44 Solve the equation $R \bullet X \bullet H = V$ for X, in the group of symmetries of the square.

18.45 Prove that if $\phi_1, \ldots, \phi_n$ are one-to-one, then so is $\phi_1 \bullet \phi_2 \bullet \ldots \bullet \phi_n$. Tell what its inverse is.

18.46 Do the set and operators given in Fig. 18.1 form a group? What is the identity of the group? Give the set of subgroups of this group. Give the generators of the group.

18.47 Find the order of each element of the group $\langle \{I, R, R', R''\}, \bullet \rangle$ of rotations of the square. Which elements generate the group?

18.48 Derive the group of symmetries of the isosceles triangle and compute its composition table. Identify its subgroups and give its generators.

18.49 Derive the group of symmetries of the equilateral triangle and compute its composition table. Identify its subgroups and give its generators.

18.50 Derive the group of rotations of a regular hexagon and give its generators.

18.51 Derive the group of symmetries of a rhombus (parallelogram with four equal sides).

18.52 Determine all the isomorphisms between the additive group M_4 of integers mod 4 and the group of rotations of the square.

18.53 Are any of the following groups isomorphic?

(a) The group of symmetries of an equilateral triangle.

(b) The group of symmetries of a square.

(c) The group of rotations of a regular hexagon.

(d) The additive group of integers mod 6.

18.54 Prove that there are 48 symmetries of the cube. List all the symmetries of the cube that take one of its vertices b (say) into itself.

18.55 Describes all symmetries of a wheel with k equally spaced spokes.

18.56 Prove that h of the proof of Theorem (18.41) commutes with $\circ$, i.e. prove $h(b^m \circ b^p) = h(b^m) + h(b^p)$.

18.57 Prove that hn of the proof of Theorem (18.41) commutes with $\circ$, i.e. prove $hn(b^m \circ b^p) = hn(b^m) \oplus hn(b^p)$ (for $0 \le m < n$ and $0 \le m < p$).

18.58 Prove the individual items (a)–(e) of Theorem (18.43).

Exercises on boolean algebras

18.59 Prove that $\langle \mathcal{P}\{1, 2, 3, 6\}, \text{lcm}, \text{gcd}, \sim, 1, 6 \rangle$ is a boolean algebra, where $\sim x = 6/x$.

18.60 Let π be a product of (at least one) distinct primes. Let S be the set of divisors of π and define $\sim b = \pi/b$ for b in S. Prove that $\langle S, \text{lcm}, \text{gcd}, \sim, 1, \pi \rangle$ is a boolean algebra.

18.61 Prove Idempotency (18.49), $b \oplus b = b$.

18.62 Prove Zero (18.50), $b \oplus 1 = 1$.

18.63 Prove Absorption (18.51), $b \oplus (b \otimes c) = b$.

18.64 Prove Cancellation (18.52), $(b \oplus c = b \oplus d) \wedge (\sim b \oplus c = \sim b \oplus d) \equiv c = d$.

18.65 Prove Unique complement (18.53), $b \oplus c = 1 \wedge c \otimes b = 0 \equiv c = \sim b$

18.66 Prove Double complement (18.54), $\sim(\sim b) = b$. Hint: Use Unique complement (18.53).

18.67 Prove Constant complement (18.55), $\sim 0 = 1$. Hint: Use Unique complement (18.53).

18.68 Prove De Morgan (18.56), $\sim(b \oplus c) = \sim b \otimes \sim c$. Hint: Use Unique complement (18.53).

18.69 Prove (18.57), $b \oplus (\sim c) = 1 \equiv b \oplus c = b$.

18.70 Prove theorem (18.58).

18.71 Prove the third part of Theorem (18.61), that $\leq$ is transitive.

18.72 Prove theorem (18.62), $b \leq c \equiv b \oplus c = c$.

18.73 Prove theorem (18.64), $atom.a \Rightarrow a \otimes b = 0 \vee a \otimes b = a$.

18.74 Prove theorem (18.65), $atom.a \wedge atom.b \wedge a \neq b \Rightarrow a \otimes b = 0$.

18.75 Prove theorem (18.66), $(\forall a \mid atom.a : a \otimes b = 0) \Rightarrow b = 0$.

18.76 Define y as in Theorem (18.67). Prove that $b \otimes y = y$.

18.77 Define y as in Theorem (18.67). Prove that $b \otimes \sim y = 0$.

18.78 Prove that the representation y of b given in Theorem (18.67) is unique.

Chapter 19

A Theory of Graphs

A graph is just a bunch of points with lines between some of them, like a map of cities linked by roads. A rather simple notion. Nevertheless, the theory of graphs has broad and important applications, because so many things can be modeled by graphs. For example, planar graphs — graphs in which none of the lines cross— are important in designing computer chips and other electronic circuits. Also, various puzzles and games are solved easily if a little graph theory is applied.

Graph theory is also a source of intriguing questions that are simple to state but hard to answer. For example, how many colors are needed to color countries on a map so that adjacent countries have different colors? A graph-theoretic proof that only four colors are needed was published in 1879 but was found to be wrong ten years later. It took almost 100 more years of work to finally decide that, indeed, only, four colors are required.

The proofs of some results in graph theory involve an algorithm to construct some object, like a path from one vertex to another. Therefore, this chapter contains a number of algorithms.

19.1 Graphs and multigraphs

We begin by looking at three kinds of graphs: *directed graphs* or *digraphs*, *undirected graphs*, and *multigraphs*.

(19.1) **Definition.** Let V be a finite, nonempty set and E a binary relation on V. Then $G = \langle V, E \rangle$ is called a *directed graph*, or *digraph*. An element of V is called a *vertex*; an element of E, an *edge*.

Digraphs are usually depicted using a diagram like the one in Fig. 19.1. In this digraph, E is the binary relation

(19.2) $\{\langle b,b \rangle, \langle b,c \rangle, \langle b,d \rangle, \langle c,e \rangle, \langle e,c \rangle, \langle e,d \rangle\}$

over the set $\{a,b,c,d,e\}$. In such a diagram, each vertex is shown as a small circle containing a name. Sometimes, a vertex is shown as a large dot with its element or label next to it, and sometimes the name of the vertex is omitted. An edge $\langle b,c \rangle$ is shown as an arrow from b to c. An edge

$\langle b, b \rangle$ is called a *self-loop*. A digraph with no self-loops is called *loop-free*. Thus, a digraph $\langle V, E \rangle$ is loop-free iff E is an irreflexive relation.

Vertex b is the *start vertex* and c the *end vertex* of edge $\langle b, c \rangle$. The edge is *incident on* b and c, and b and c are *adjacent* (to each other).

UNDIRECTED GRAPHS

In an undirected graph $\langle V, E \rangle$, E is a set of *unordered* pairs. For example, Fig. 19.2a shows an undirected graph whose edges are $\{a, b\}$, $\{a, c\}$, $\{b, c\}$, and $\{b, d\}$. The edges are given as sets, since no ordering is implied.

When E of a digraph is a symmetric relation, for each arrow from a vertex b to a vertex c there is also an arrow from c to b, since $\langle b, c \rangle$ is in the relation iff $\langle c, b \rangle$ is. If we replace each pair of arrows by a single, undirected, line between the two vertices, we get an undirected graph. Fig. 19.2b is a digraph; Fig. 19.2a is the corresponding undirected graph. This construction illustrates a one-to-one correspondence between undirected graphs and digraphs with symmetric relations.

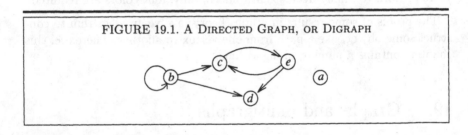

FIGURE 19.1. A DIRECTED GRAPH, OR DIGRAPH

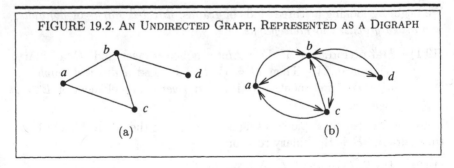

FIGURE 19.2. AN UNDIRECTED GRAPH, REPRESENTED AS A DIGRAPH

(a)   (b)

MULTIGRAPHS

A *multigraph* is a pair $\langle V, E \rangle$, where V is a set of vertices and E is a *bag* of undirected edges. [1] Thus, in a multigraph, there may be many edges between two given vertices. Every undirected graph is a multigraph, but not vice versa. A multigraph is shown in Fig. 19.3b. Multigraphs do not depict relations. Sometimes, in order to talk about edges and to represent them in the computer, we name the edges —by using just $\langle b, c \rangle$ for an edge between b and c, we cannot tell which edge is meant. That is why the edges in Fig. 19.3b are labeled.

THE DEGREE OF A VERTEX AND GRAPH

The *indegree* of a vertex of a digraph is the number of edges for which it is an end vertex; the *outdegree* is the number of edges for which it is a start vertex. The degree $deg.v$ of a vertex v is the sum of its indegree and outdegree. For example, in Fig. 19.1, vertex b has indegree 1, outdegree 3, and degree 4, while vertex d has indegree 2, outdegree 0, and degree 2. A vertex whose degree is 0 is called *isolated*. In Fig. 19.1, vertex a is isolated.

The degree $deg.v$ of a vertex of an undirected graph or multigraph is the number of edge ends attached to it. For example, vertex d in the undirected graph of Fig. 19.2a has degree 1, vertex b has degree 3, and each of the other two has degree 2.

From the definition of degree, we see that each edge contributes 1 to the degree of two vertices (which may be the same): 1 for each end of the edge. The following theorem and corollary follow readily from this observation.

(19.3)   **Theorem.** The sum of the degrees of the vertices of a digraph or multigraph equals $2 \cdot \#E$.

---

[1] One could also define multigraphs with directed edges, but we have no need to do so in this chapter.

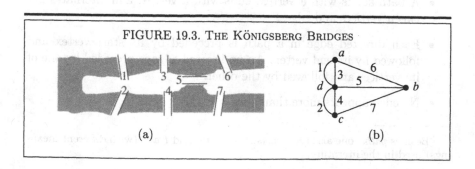

FIGURE 19.3. THE KÖNIGSBERG BRIDGES

(a)   (b)

(19.4)   **Corollary.** In a digraph or multigraph, the number of vertices of odd degree is even.

Finally, the degree of a graph is the maximum degree of its vertices.

PATHS

Graph theory had its beginnings with a puzzle that Leonhard Euler (see Historical note 19.1) saw in the city of Königsberg, Prussia, in 1736. Königsberg has a river flowing through it, which branches and joins to form islands. Seven bridges had been built over the river, as shown in Fig. 19.3a. Euler wondered whether it was possible to tour the city in a way that would traverse each bridge exactly once. We represent this situation as the multigraph in Fig. 19.3b. The vertices a, b, c, d are the four land masses: the two outside banks of the river and the two islands. The edges are the bridges that link the land masses. From the Königsberg bridges, Euler abstracted the following problem: which multigraphs have a "path" that contains each edge exactly once? We answer this question in Sec. 19.2, but first we must introduce some more concepts and terminology.

A *path* [2] of a multigraph or a digraph is the sequence of vertices and edges that would be traversed when "walking" the graph from one vertex to another, following the edges, but with *no edge traversed more than once.* (A directed edge can be traversed only in the forward direction.) A path that starts at vertex b and ends at vertex c is called a *b-c* path. Below are two paths of the graph of Fig. 19.1 and one non-path (because an edge is included twice). The second path "visits" vertex c twice:

(19.5)   a path : $\langle c, \langle c, e \rangle, e, \langle e, d \rangle, d \rangle$

a path : $\langle b, \langle b, c \rangle, c, \langle c, e \rangle, e, \langle e, c \rangle, c \rangle$

not a path : $\langle b, \langle b, c \rangle, c, \langle c, e \rangle, e, \langle e, c \rangle, c, \langle c, e \rangle, e \rangle$

One path of Fig. 19.3b is $\langle a, 6, b, 5, d, 1, a \rangle$.

In summary, a path has the following properties:

- A path starts with a vertex, ends with a vertex, and alternates between vertices and edges.

- Each directed edge in a path is preceded by its start vertex and followed by its end vertex. An undirected edge is preceded by one of its vertices and followed by the other.

- No edge appears more than once.

---

[2] Besides *paths*, one also sees *walks*, *traversals*, and *trails* (with different meanings) used in the literature.

The *length* of a path is the number of edges in it. Thus, path $\langle b, \langle b, c \rangle, c \rangle$ has length 1, while path $\langle b \rangle$ has length 0.

The notation used in (19.5) to describe paths in a digraph or graph is redundant. In a digraph, for any two vertices b and c, there is at most one edge from b to c. Similarly, in an undirected graph, there is at most one edge between two given vertices. Hence, there is no need to show the edges. For example, the first two paths in (19.5) could be presented simply as $\langle c, e, d \rangle$ and $\langle b, c, e, c \rangle$. Henceforth, we use this shorter representation of paths in digraphs and graphs (but not multigraphs). However, remember that a path contains the edges, too, and that the length of the path is the number of edges. For example, the length of path $\langle b \rangle$ is 0 and the length of path $\langle b, c, e \rangle$ is 2.

It is also possible to represent a nonempty path by the sequence of edges in the path. We call this the *edge-path* representation.

A *simple path* is a path in which no vertex appears more than once, except that the first and last vertices may be the same. For example, $\langle b \rangle$, $\langle b, b \rangle$, $\langle b, c, e, d \rangle$ and $\langle c, e, c \rangle$ are simple paths of Fig. 19.1, but $\langle b, c, e, c \rangle$ is not.

(19.6) **Theorem.** If a graph has a b-c path, then it has a simple b-c path.

Proof. Suppose there is a b-c path; call it p, and consider the following algorithm.

> **do** p not simple $\rightarrow$
>   let $p = x \,\hat{}\, \langle d \rangle \,\hat{}\, y \,\hat{}\, \langle d \rangle \,\hat{}\, z$;   (where $x \,\hat{}\, z \neq \langle \rangle$)
> $p := x \,\hat{}\, \langle d \rangle \,\hat{}\, z$
> **od**

We prove that execution of the loop changes p into a simple path using the loop invariant

$$P: \quad p \text{ is a } b\text{-}c \text{ path (expressed as the list of vertices)}$$

By assumption, P is initially *true*. Second, if p is not a simple path, then p has the form shown in the let statement, and execution of the body reduces the length of p while leaving it a b-c path. Therefore, the loop terminates. Finally, P together with the falsity of the loop condition gives the desired result. $\qquad\Box$

A path with at least one edge and with the first and last vertices the same is called a *cycle*.

(19.7) **Theorem.** Suppose all vertices of a loop-free multigraph have even degree. Suppose $deg.b > 0$ for some vertex b. Then some cycle contains b.

Proof. Let $\langle V, E \rangle$ be the graph. We prove the theorem by giving an algorithm that constructs a cycle containing b. The algorithm uses a sequence variable s that contains the part of the cycle constructed thus far, as an edge-path and a variable v that contains the last vertex of s.

solid lines are the path

The essential idea of the algorithm is this. Suppose a b-v path has been constructed, where $b \neq v$. Because the degree of v is even, at least one edge incident on v does not yet appear on the path. Adding it to the path increases the length of the path. Because all path are finite, this process cannot continue forever, and finally $v = b$.

Invariant P below captures the properties of s and v just before each iteration of the loop of the algorithm. In the invariant and algorithm, G' denotes graph G with the *edges* of path s removed.

P : (s is a b-v edge-path) $\land$
(the length of path s is not 0) $\land$
($\forall u{:}V \mid u \neq b \land u \neq v :$ degree of u in G' is even) $\land$
($b \neq v \Rightarrow$ the degrees of b and v in G' are odd)

The algorithm below begins by choosing some edge incident on b and placing it in path s. It is easy to see that the four conjuncts of P are truthified by the initialization.

$\{deg.b > 0\}$
Let e be an edge $\{b, c\}$ for some c;
$v, s := c, \langle e \rangle$;
do $b \neq v \rightarrow$ Let e be an edge $\{v, c\}$ in G' (for some c);
$\qquad\qquad v, s := c, s \mathbin{\hat{}} \langle e \rangle$
od
$\{s$ is a cycle$\}$

A key point in showing that each iteration maintains invariant P is that, as illustrated in Fig. 19.4(a), at the beginning of each iteration, there is an edge incident on v that is not in s. This is because the degree of v in graph G' is odd. Each iteration adds such an edge to s, as shown in Fig. 19.4(b), thereby removing it from graph G'. Therefore, at some point, the algorithm terminates. And, upon termination, $b = v$ so that path s is actually a cycle.   □

An undirected multigraph is *connected* if there is a path between any two vertices. A digraph is *connected* if making its edges undirected results in a connected multigraph. The graph of Fig. 19.1 on page 424 is unconnected; the graphs of Fig. 19.2 and Fig. 19.3b are connected.

REPRESENTATION OF DIGRAPHS

In order to process digraphs using a computer, we need some data structure to represent them. There are two standard ways of representing a digraph: the *adjacency-matrix* and the *adjacency-list* representations.

Let the digraph G be $\langle V, E \rangle$, where $V = 0..n-1$. That is, the n vertices are given by the natural numbers $0, 1, \ldots, n-1$. The adjacency matrix for G is a boolean matrix $b[0..n-1, 0..n-1]$, where

$$b[i,j] \equiv \langle i,j \rangle \in E \quad \text{(for } 0 \le i < n, 0 \le j < n)$$

Equivalently, we can represent the graph by a two-dimensional 0-1 array c, where

$$c[i,j] = \mathbf{if} \ \langle i,j \rangle \in E \ \mathbf{then} \ 1 \ \mathbf{else} \ 0$$

For example, if we represent the vertices a, b, c, d, e by $0, 1, 2, 3, 4$, the two possible adjacency matrices for the graph of Fig. 19.1 are

$$b = \begin{pmatrix} f & f & f & f & f \\ f & t & t & t & f \\ f & f & f & f & t \\ f & f & f & f & f \\ f & f & t & t & f \end{pmatrix} \qquad c = \begin{pmatrix} 0 & 0 & 0 & 0 & 0 \\ 0 & 1 & 1 & 1 & 0 \\ 0 & 0 & 0 & 0 & 1 \\ 0 & 0 & 0 & 0 & 0 \\ 0 & 0 & 1 & 1 & 0 \end{pmatrix}$$

With the adjacency-matrix representation, some operations are quite efficient. For example, the existence of a particular edge $\langle i,j \rangle$ can be tested in constant time using $b[i,j]$ (or $c[i,j] = 1$), and an edge can be added or deleted in constant time. However, some operations are more expensive than we would like. For example, determining the outdegree of vertex v requires time proportional to $\#V = n$, since every element of row v of the matrix has to be tested, even if the outdegree is 0. Further, an adjacency matrix requires n^2 bits, even for a graph with no edges.

The adjacency-list representation of graph $G = \langle 0..n-1, E \rangle$ consists of a one-dimensional array $d[0..n-1]$, where each element $d[i]$ is the se-

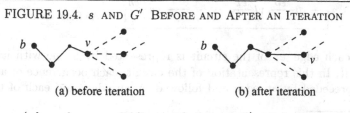

FIGURE 19.4. s AND G' BEFORE AND AFTER AN ITERATION

(a) before iteration   (b) after iteration

(edges of s are solid lines and edges of G' are dashed.)

quence of vertices adjacent to it. For example, representing again the vertices a, b, c, d, e by $0, 1, 2, 3, 4$, the adjacency list for the graph of Fig. 19.1 on page 424 is [3]

$$d = \begin{pmatrix} \langle \rangle \\ \langle 1, 2, 3 \rangle \\ \langle 4 \rangle \\ \langle \rangle \\ \langle 2, 3 \rangle \end{pmatrix}$$

In this representation, the space required is proportional to the number of edges plus the number of vertices. Also, counting the outdegree of a vertex takes time proportional to that outdegree, but counting the indegree of a vertex requires time proportional to the total number $\#E$ of edges (why?). So, an adjacency list is better for some operations than an adjacency matrix, and vice versa.

19.2   Three applications of graph theory

THE KÖNIGSBERG BRIDGES PROBLEM

We are now ready to tackle Euler's problem of the Königsberg bridges. By an *Euler path* of a multigraph, we mean a path that contains each edge of the graph (exactly once). If the first and last vertices are the same, the Euler path is called an *Euler circuit*.

We can characterize multigraphs with Euler circuits as follows.

(19.8)   **Theorem.** An undirected connected multigraph has an Euler circuit iff every vertex has even degree.

Proof. The proof is by mutual implication.

LHS $\Rightarrow$ **RHS**. Assume that the multigraph has an Euler circuit. We can view the Euler circuit as a ring, by identifying the first and last vertices of the circuit. We cannot show this in a linear fashion, but we can draw it:

An Euler circuit

Here, each edge e_i of the circuit is represented by a line with its name above it. In this representation of the circuit, each occurrence of a vertex v_i is preceded by an edge and followed by an edge, and each of the two

---

[3] Each array element, which is a sequence, can be implemented by a linked list or doubly-linked list. Such data structures are outside the scope of this text.

HISTORICAL NOTE 19.1. LEONHARD EULER (1707–1783)

Euler, a Swiss, is the most prolific mathematician of all time. It has been said that "Euler calculated without apparent effort, as men breathe or eagles sustain themselves in the wind." Euler had an intense mathematical curiosity —reading about an anchor being let out would set him to investigate the ship's motion in such circumstances, the Königsberg bridges led to the first insights into graph theory, and so on. He also was able to work anywhere, even with a child on his lap. And, he had a phenomenal memory and a great ability to calculate mentally. This was fortunate, because for the last seventeen years of his life he was blind, which did not slow down his mathematics. While blind, for example, he worked out in his head the basics for a calculable solution to the orbit of the moon, a problem which Newton had struggled with.

One reason for Euler's productivity was that he was *paid* to do science. He spent the periods 1727–1740 and 1766–1783 at the St. Petersburgh Academy (Russia), under the sponsorship of Catherine I and Catherine the Great, where, at 26, he was the leading mathematician. As such, Euler spent time at Court, but his main job was to do whatever science he wished. Euler had a similar position at the Berlin Academy under Frederick the Great from 1740 to 1766.

E.T. Bell [3] relates an amusing story concerning Euler and the French philosopher Denis Diderot's visit to Catherine the Great's Court. Diderot had been making a pest of himself, trying to convert courtiers to atheism. Catherine asked Euler how Diderot could be muzzled. Euler, who was deeply religious, had someone tell Diderot that a learned mathematician would give the Court an algebraic proof of the existence of God. With all watching, Euler advanced toward Diderot and pronounced solemnly and with deep conviction,

Sir, $\frac{a+b^n}{n} = x$, hence God exists; reply!

Diderot, who knew no mathematics, did not know how to respond, and his embarrassed and confused silence was met with laughter from all. Humiliated, Diderot asked Catherine's permission to return to France and left.

---

edges contributes 1 toward the degree of v_i. Since all edges appear in the circuit, the degree of each v_i is even.

RHS $\Rightarrow$ LHS. Assume that each vertex has even degree. Choose any vertex b of the multigraph. We prove that the graph has an Euler circuit by cases: either $deg.b = 0$ or $deg.b > 0$.

Case $deg.b = 0$. Since the multigraph is connected, it consists of a single vertex and no edges, so $\langle b \rangle$ is the Euler circuit.

Case $deg.b > 0$. By Theorem (19.7) on page 427, some cycle s (say) contains b, and the proof of that theorem shows how to construct s. We write an algorithm that changes cycle s into an Euler circuit by repeatedly adding edges to s, always leaving it a cycle, until s contains all edges. The invariant P of the loop of the algorithm is:

$P:$ s is a cycle of G; s is maintained as an edge-path.

In the following algorithm, let G' stand for graph G but with the edges of s removed.

Choose some cycle s;
do s does not contain all edges of G →
   Let c be a vertex in s whose degree in G'
      is greater than 0;
   Choose some cycle z of G' that begins with c
      (represent z as an edge-path);
   Write s as $x \hat{\ } y$ for some sequences x and y, where
      the last edge of x ends in vertex c;
   $s := x \hat{\ } z \hat{\ } y$
od

It is obvious that the initialization of the loop truthifies P. Second, from the falsity of the loop condition together with P, we see that upon termination s contains all the edges, so s is an Euler circuit.

We investigate the invariance of P under execution of the repetend when s does not contain all the edges —execution is illustrated in Fig. 19.5. We discuss each statement of the repetend, in turn.

- Since G is connected and s does not contain all the edges, some vertex c (say) of s has an edge e of G' incident on it. Therefore, a vertex c as indicated in the first statement exists.

- Consider the connected component of G' that contains such an edge e. Each vertex of this connected component has even degree, so by Theorem (19.7) on page 427, the connected component has a cycle with first vertex c. Construct such a cycle z (say).

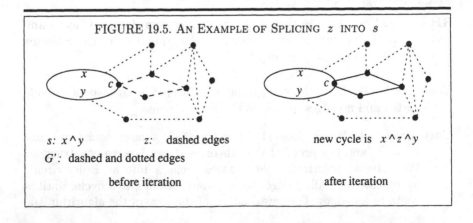

FIGURE 19.5. AN EXAMPLE OF SPLICING z INTO s

$s:$ $x \hat{\ } y$ $z:$ dashed edges   new cycle is $x \hat{\ } z \hat{\ } y$
$G':$ dashed and dotted edges

before iteration   after iteration

- The final two statements splice cycle z into cycle s by replacing c by z.

Finally, each iteration makes progress towards termination by adding at least one edge to s. Hence, the algorithm terminates.   □

The proof of the following corollary is left to the reader.

(19.9)   **Corollary.** An undirected connected multigraph with at least one edge has an Euler path (but no Euler circuit) iff exactly two vertices have odd degree.

We can apply the theorem and its corollary to Euler's problem of the Königsberg bridges. In Fig. 19.3b, all four vertices have odd degree. Hence, the graph has no Euler path or circuit.

 The graph to the left also does not have an Euler path or Euler circuit, because it has four vertices of odd degree. This graph is sometimes presented as a common misleading puzzle: you are asked to draw the edges of the graph, going through each edge once, without lifting your pencil from the paper. It cannot be done.

INSTANT INSANITY

Suppose we have four cubes. Each face of a cube is painted red (R), white (W), blue (B), or yellow (Y) as shown below. (To the left, we show the first cube and the colors on its six faces. To its right, we show the four cubes cut and flattened in the plane, in order to make their colors easy to see.)

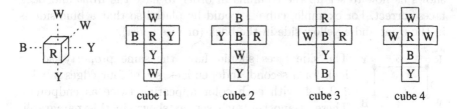

cube 1   cube 2   cube 3   cube 4

 The object of the game of instant insanity is to place the four cubes in a column so that the four colors appear on each of the four sides of the column. An exercise asks you to show that there are $41,472$ possibilities for arranging these cubes in a column, so simply trying all possibilities is an insane way to proceed. Let's see how graph theory helps us solve the puzzle without going insane.

Suppose the four cubes are stacked in front of us in a manner that solves the puzzle. Note that choosing the front face of a cube has also determined the back face. Similarly, choosing one side face has determined the opposite side face as well. A representation of the cube that highlights the fact that choosing a face automatically chooses the opposite face may help eliminate from consideration some of the useless configurations.

Consider the representation shown below. It is a multigraph with four colors R, W, B, and Y as vertices. For each pair of colors on opposite faces of cube i, there is an edge labeled i that joins those two colors (for $1 \le i \le 4$). For example, cube 1 contributes the edges $\langle W, Y \rangle$, $\langle Y, B \rangle$, and $\langle R, W \rangle$, all labeled 1.

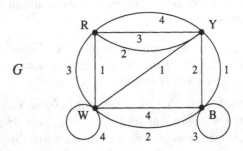

 Again, suppose the four cubes are stacked in front of us in a manner that solves the puzzle. There are four different colors facing us and four different colors on the opposite side, so each color appears twice on the eight front-and-back faces. Multigraph G above has four edges corresponding to the four pairs of front-and-back faces (one pair for each cube); they are labeled 1, 2, 3, and 4, and each color appears twice as an end point of these four edges. So we look in the multigraph for four such edges, and see the four edges as shown in this paragraph. Thus, this part of the multigraph shows us how to set up the columns in order to have the front and back faces correct. For example, cube 1 could be placed so that a blue side is in the front and a yellow side in the back (or vice versa).

The side faces should have the same property, so we look for a second, independent, set of four edges labeled $1, 2, 3, 4$ with each color appearing twice as endpoints. There is another such set, as shown in this paragraph. To solve the problem, then, first set up the cubes according to the diagram in the previous paragraph; then rearrange the cubes, taking care not to mess up their front and back faces, so that the sides are according to the diagram in this paragraph. We then have the stacked cubes as follows:

THE CELEBRITY PROBLEM

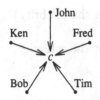

At a party of n people, there is a celebrity c: a person whom everyone knows and who knows no one. Let the people at the party be numbered $0, 1, \ldots, n - 1$. Consider the digraph with vertices $0..n - 1$ whose edges describe who knows whom: There is a directed edge from i to j iff i knows j. (We do not care whether someone knows themself, so self-loops may or may not be present.) The graph of this paragraph shows celebrity c; missing are edges between the other people. Let this graph be given by its adjacency matrix $b[0..n-1, 0..n-1]$: $b[i, j] \equiv$ "i knows j".

We want an efficient algorithm for finding the celebrity. It might be expected that the algorithm must look at (almost) all the elements of b and thus will take time proportional to n^2. However, the graph has the following nice property. Look at $b[i, j]$ for $i \neq j$. If $b[i, j]$ is *true*, then i is not the celebrity, since i knows someone. If $b[i, j]$ is *false*, then j is not the celebrity, since someone does not know j. Therefore, one test can eliminate one person from possible celebrity status. Perhaps we can write an algorithm that requires only $n - 1$ tests.

We now specify the algorithm more formally. The precondition Q states that the celebrity exists (all dummies range over $0..n - 1$):

$$Q : (\exists c \mid : c \text{ is the celebrity}) \quad ,$$

where

$$c \text{ is the celebrity} \equiv (\forall i \mid c \neq i : b[i, c] \land \neg b[c, i]) \quad .$$

The algorithm should store the celebrity in integer variable c, thus truthifying the postcondition that c is the celebrity. However, since the algorithm will presumably operate by iteratively determining who is *not* the celebrity, it may be better to phrase postcondition R as

$$R : 0 \leq c < n \land$$
$$(\forall i \mid 0 \leq i < n \land i \neq c : i \text{ is not the celebrity})$$

With this postcondition, we can replace n by a fresh variable j and get the loop invariant

$$P : 0 \leq j \leq n \land 0 \leq c < j \land$$
$$(\forall i \mid 0 \leq i < j \land i \neq c : i \text{ is not the celebrity})$$

The algorithm is then easily written as shown below. Within the repetend, if $b[j, c]$ holds, then j is not the celebrity, so increasing j maintains invariant P. On the other hand, if $b[j, c]$ is *false*, then c is not the celebrity, and the assignment $c, j := j, j + 1$ maintains P.

$c, j := 0, 1;$
do $j \neq n \rightarrow$
    if $\neg b[j, c] \rightarrow \{(\forall i \mid 0 \leq i < j + 1 \land i \neq c : i \text{ not the celebrity}\}$
           $j := j + 1$
    $[\!]\ \neg b[j, c] \rightarrow \{(\forall i \mid 0 \leq i < j + 1 \land i \neq j : i \text{ not the celebrity}\}$
           $c, j := j, j + 1$
  fi
od

This development illustrates that studying the properties of the objects to be manipulated by an algorithm may lead to an efficient algorithm. The more properties one knows, the better chance of developing a good algorithm. In this case, the property we used was that an element $b[i, j]$ determines that one person (either i or j) is not the celebrity.

19.3 Classes of graphs

We now focus our attention on graphs without self-loops. We look at various kinds of such graphs.

COMPLETE GRAPHS

The *complete graph* of n vertices, denoted by K_n, is an undirected, loop-free graph in which there is an edge between every pair of distinct vertices. (The use of K for these graphs comes from the German word for complete, *Komplett*.) Here are the graphs K_1, K_2, K_3, K_4, and K_5:

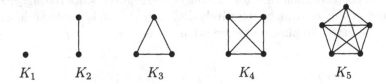

    K_1 K_2 K_3 K_4 K_5

Complete graph K_n has $n \cdot (n - 1)/2$ edges (see Exercise 19.7).

A *bipartite graph* is an undirected graph in which the set of vertices are partitioned into two sets X and Y such that each edge is incident on one vertex in X and one vertex in Y. (The prefix *bi* means *two* and the stem *partite* means *divided into parts or partition elements*.) The graph in this paragraph is bipartite, with $X = \{a, b, c\}$ and $Y = \{d, e, f\}$. Vertex f could be placed in either X or Y, since it is isolated.

(19.10) **Theorem.** A path of a bipartite graph is of even length iff its ends are in the same partition element.

Proof. The details of a proof by induction are left to the reader. The proof rests on the fact that every edge $\{b, c\}$ has its endpoints in different partition elements, so every path $\langle b, c, d \rangle$ has its endpoints in the same partition element and its middle vertex in the other partition element.   □

(19.11) **Corollary.** A connected graph is bipartite iff every cycle has even length.

A *complete bipartite graph* $K_{m,n}$ is a bipartite graph in which one partition element X has m vertices, the other partition element Y has n vertices, and there is an edge between each vertex of X and each vertex of Y. Below are complete bipartite graphs $K_{3,2}$ and $K_{3,3}$. $K_{3,3}$ is known as the *utility graph*. Think of the vertices $\{a, b, c\}$ as houses and $\{d, e, f\}$ as the source of three utilities: gas, water, and telephone. Is there a way to connect the three utilities to the three houses such that no utility line crosses another utility line? We answer this question in Sec. 19.6.

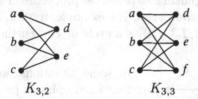

19.4   Subgraphs and morphisms

A multigraph or digraph $\langle V', E' \rangle$ is called a *subgraph* of $\langle V, E \rangle$ if $V' \subseteq V$, $E' \subseteq E$, and the endpoints of the edges of E' are in V'. Thus, to form a subgraph of a graph, simply choose some of the vertices of the graph and some of the edges incident on them. Note that any graph is a subgraph of itself. Below is a graph and, to its right, one of its subgraphs.

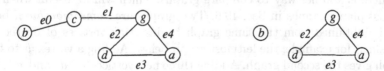

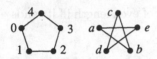

The two graphs in this paragraph look quite different. But actually, they are structurally identical. If we rename the vertices a, b, c, d, e of the second graph $0, 1, 2, 3, 4$, we can see that these two graphs have the same vertices and the same edges. Hence, they are the same graph, up to renaming of vertices. We call two such graphs *isomorphic* (See Historical note 18.1 on page 388 for a dissection of the word *isomorphism*.)

Here is a formal definition of graph isomorphism. Let two graphs G and G' have sets of vertices V and V', respectively. The graphs are isomorphic if there exists a one-to-one and onto function $f: V \to V'$ such that

$$(\forall v, w \mid: \langle v, w \rangle \text{ is an edge of } G \text{ iff } \langle f.v, f.w \rangle \text{ is an edge of } G')$$

Function f is called an *isomorphism* from G to G'.

The isomorphism that shows that the two graphs in the above paragraph are isomorphic is given by $f.0 = a$, $f.1 = b$, $f.2 = c$, $f.3 = d$, $f.4 = e$.

An isomorphism f "preserves edges": f takes an edge into an edge. Therefore, an isomorphism also preserves properties of a graph that depend on edges, like paths and cycles. For example, if $\langle 0, 1, 2, 3, 4 \rangle$ is a cycle of G, then $\langle f.0, f.1, f.2, f.3, f.4 \rangle$ is a cycle of G'. Further, $deg.b = deg(f.b)$ for all vertices b.

In some situations, we can use the fact that isomorphisms preserve various structural properties of graphs to tell quickly that two graphs are *not* isomorphic. For example, the two graphs in this paragraph are not isomorphic because the leftmost graph has two vertices of degree 3 while the rightmost has none.

It is much harder to prove that two graphs are isomorphic. In fact, all the known general graph algorithms for testing isomorphism take time that is exponential in the number of vertices of the graphs, which makes the test infeasible for large graphs. For restricted classes of graphs, faster algorithms exist.

There is another way to compare graphs, which will be useful when we discuss planar graphs in Sec. 19.6. Two graphs are *homeomorphic* if both can be obtained from the same graph by inserting vertices of degree 2. Consider, for example, the leftmost graph below. Adding a vertex b to this graph gives the second graph. Adding three new vertices c, d, and e gives the third graph. Hence, the second and third graphs are homeomorphic. In fact, all three graphs are homeomorphic to each other.

19.5 Hamilton circuits

A *Hamilton path* of a graph or digraph is a path that contains each vertex exactly once, except that the end vertices of the path may be the same. A *Hamilton circuit* is a Hamilton path that is a cycle. (In contrast, recall that an Euler circuit contains each *edge* exactly once.)

The name honors Sir William Hamilton (see Historical note 19.2), who developed a puzzle that he sold to a Dublin toy manufacturer. The puzzle consisted of a wooden dodecahedron, a solid figure with twelve identical pentagonal faces. [4] The twenty corner points were labeled with the names of cities, and edges modeled roads. The object of the puzzle was to figure out how to tour the cities, visiting each city once and returning to the starting city.

One way to solve the puzzle is to draw the 20 vertices of the dodecahedron and its edges as a graph, as shown to the left below. The game is then solved if a Hamilton path can be found in this graph. One solution is shown in the graph to the right below, where the edges that are not in the Hamilton circuit are dotted.

Hamilton paths or circuits appear in many applications. For example, suppose a salesman wants to find a least expensive (or shortest) route that visits a number of cities and ends back at their home city. Such a route is a Hamilton circuit. If a number is placed on each edge between cities to indicate the cost of taking that road or the length of that road, then summing the edges of a Hamilton path gives the cost of taking that path. The traveling-salesman problem, then, is to find a minimum-cost Hamilton circuit.

Unfortunately, there is no neat characterization of graphs that have Hamilton paths or circuits, as there is for graphs with Euler paths or circuits (see Theorem (19.8)). Also, there is no known efficient algorithm for finding a Hamilton path or circuit. The fastest known algorithms require time exponential in n for a graph with n vertices, and this quickly becomes

---

[4] *Dodeca* comes from the Greek word δωδεκα, which means twelve, and *hedron* from the Greek word εδρα, which means *face*.

HISTORICAL NOTE 19.2. Sir William Rowan Hamilton (1805–1865)

Hamilton is considered by far the greatest Irish scientist. His *Theory of Systems of Rays*, completed when he was 23, provides a scientific basis for optics that is still in use today. But Hamilton was prouder of his *quaternions*, a linear algebra of four-dimensional vectors, which he developed in order to have an algebra of rotations in three dimensions. The great step forward in this work was the realization (for the first time) that multiplication of vectors did not have to be symmetric (or commutative). Hamilton felt that this discovery was "as important for the middle of the nineteenth century as the discovery of [the calculus] was for the close of the seventeenth." In hindsight, this claim is rather overstated.

Hamilton received instruction from an uncle and did not attend school before he was eighteen. He loved languages, and, by thirteen, he had mastered one language for each year of his life, including Latin, Greek, Hebrew, Sanskrit, Hindustani, and Bengali. His interest in math started when he was twelve. By seventeen, he had mastered integral calculus and developed the germs of his great work on optics. In 1823, he entered Trinity College, Dublin. There, he won the highest honors and awards in classics and mathematics —and also completed the first draft of his *Theory of Systems of Rays*. In 1827, a Professor of Astronomy resigned to become Bishop of Cloyne. The committee to replace the professor passed over all applicants and offered the professorship to Hamilton, who had not even applied. Thus, at 22, Hamilton went from undergrad to professor.

Hamilton spent the last twenty years of his life almost exclusively on elaborating quaternions and their applications, believing that this work would stand as one of the greatest mathematical achievements of all time. This period of his life was beset by domestic difficulties and alcohol. He and his wife, who was a semi-invalid, were unable to control their servants, and their home degenerated into a pigsty. When he died, enough dirty dinner plates to supply a large household were found among the mess in the house, as well as some sixty huge manuscript books filled with mathematics. As Hamilton lay on his deathbed, the National Academy of Sciences of the U.S. (founded during the Civil War) elected him as the first foreign member.

infeasible as n grows large. Typical algorithms in use for the traveling-salesman problem find near-optimal, but not optimal, solutions.

We can, however, give a few fairly obvious hints on finding a Hamilton circuit in a loop-free graph $G = \langle V, E \rangle$:

- If G has a Hamilton circuit, then $deg.v \geq 2$ for all vertices v.

- If $deg.v = 2$, then both edges incident on v are in the circuit.

- If $deg.v > 2$ and two of the edges incident on v are in a Hamilton circuit, then the other edges incident on v are not in that Hamilton circuit.

THE KNIGHT'S TOUR

Chess is played on an 8×8 board of squares. A knight, one of the playing pieces, makes a move by moving two squares horizontally or vertically and then one square in a perpendicular direction. This is illustrated in the diagram to the left, with k marking an initial position of the knight and x marking its possible moves. A knight's tour on an $n \times n$ board begins at some square and visits each square of the board, returning to the beginning square —making only legal moves, of course. The problem is to determine values of n for which a knight's tour exists.

We model this problem using a graph. Let the vertices be the board squares. Draw an undirected edge between two vertices iff a knight on the square that one vertex represents can move (in one move) to the square that the other vertex represents. In the diagram to the left below, we show a 4×4 board; the graph in the middle is its representation. Because of the way a graph is constructed, a knight's tour exists on an $n \times n$ board iff its graph representation has a Hamilton circuit.

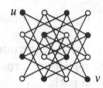

We have partitioned the board into white and black squares, so the graph has white and black vertices. Note that every edge joins a white vertex with a black vertex. Thus, the graph is bipartite and, by Theorem (19.11), every cycle has even length. Since a Hamilton circuit visits each vertex once and there are n^2 vertices, n^2 must be even for a knight's tour to exist. By theorem (4.8) on page 76, n is even as well. Hence, a knight's tour does not exist on an $n \times n$ board with n odd.

A 2×2 board has no knight's tour because the board is too small for the knight to make a move. So let us consider the 4×4 board shown above. The corresponding graph looks messy, but we can use the hints listed above to eliminate some of the edges from consideration for appearing in a Hamilton circuit. Since the upper left and lower right vertices u and v are of degree 2, the edges incident on them appear in any Hamilton circuit. Let the two vertices that are adjacent to u and v be named b and c. The edges $\langle v, b \rangle$, $\langle u, b \rangle$, $\langle v, c \rangle$ and $\langle u, c \rangle$ appear in any Hamilton path, so that any other edges incident on b and c do not and can therefore be eliminated. Eliminating these edges, and also edges that are determined from the upper right and lower left vertices in a similar manner, yields the graph to the right above. But this graph is not connected, so it cannot have a Hamilton circuit. Hence, the 4×4 board has no Hamilton circuit.

Exercise 19.13 asks you to construct a knight's tour for a 6×6 board. One can then show that for all even integers n, $6 \leq n$, a knight's tour exists on an $n \times n$ board. The inductive proof of this fact basically constructs knight's tours for small boards and pastes them together to construct knight's tours for larger boards.

Two classes of graphs with Hamilton circuits

We now give a class of directed graphs and a class of undirected graphs that contain Hamilton paths. For each, we give an algorithm that constructs a Hamilton path.

(19.12) **Theorem.** A complete digraph (a digraph with exactly one edge between each pair of vertices) has a Hamilton path.

Proof. Let the vertices of the digraph be $0..n-1$. The algorithm below constructs a Hamilton path, storing it in sequence variable s as the sequence of vertices in the path. The invariant of the algorithm is

$$P: \; 0 \leq i \leq n \; \wedge$$
$$\text{path } s \text{ has length } i-1 \text{ and contains vertices } 0..i-1.$$

The initialization truthifies P. The loop terminates since i is increased by each iteration and $0 \leq i \leq n$. Upon termination of the loop, when $i = n \wedge P$ holds, s is a Hamilton path.

```
i, s := 1, ⟨0⟩;
do i ≠ n →
    if ⟨s(i − 1), i⟩ is an edge          → s := s ⌢ ⟨i⟩
    ▯ ⟨i, s.0⟩ is an edge                → s := ⟨i⟩ ⌢ s
    ▯ ⟨s.0, i⟩ and ⟨i, s(i − 1)⟩ are edges  →
        Find j, 0 ≤ j < i, such that ⟨s.j, i⟩ and
            ⟨i, s(j + 1)⟩ are edges;
        s := s[0..j] ⌢ ⟨i⟩ ⌢ s[j + 1..]
    fi;
    i := i + 1
od
{ P is a Hamilton path }
```

To maintain invariant P, the repetend must add vertex i to path s. The diagram below illustrates the three ways of doing this. The first two guarded commands of the conditional statement within the loop append or prepend i to s. If neither is possible, then the third guard is true, since the digraph is complete (for each pair of vertices b, c (say), one of the edges $\langle b, c \rangle$ and $\langle c, b \rangle$ exists). In this case, a position j after which i should be inserted is found and i is inserted there. Exercise 19.14 asks you to write a binary-search algorithm to find j; the existence of this algorithm proves that j exists.   □

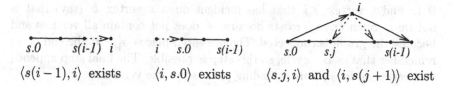

$\langle s(i-1), i \rangle$ exists $\langle i, s.0 \rangle$ exists $\langle s.j, i \rangle$ and $\langle i, s(j+1) \rangle$ exist

(19.13) Theorem. Let G be an undirected graph of n nodes. If $deg.b + deg.c \geq n-1$ for each pair b, c of vertices, then G has a Hamilton path.

Proof. We leave it to the reader to show that the graph is connected (Exercise 19.16). The algorithm below stores a Hamilton path in sequence variable s as the sequence of vertices in the path. The invariant of the loop of the algorithm is

$$P: 0 \leq i \leq n \wedge$$
$$\text{path } s \text{ has length } i - 1 \text{ and contains } i \text{ distinct vertices.}$$

The algorithm begins by storing an arbitrary vertex in s. Then, each iteration of the loop adds one vertex to s, until s has length $n - 1$ and, by the invariant, it is a Hamilton path.

$i, s := 1, \langle b \rangle;$ (for an arbitrary vertex b)
do $i \neq n \rightarrow$
    if $s(i-1)$ is adjacent to a vertex b (say) not in s
        $\rightarrow s := s \,\hat{}\, \langle b \rangle$
    [] $s.0$ is adjacent to a vertex b (say) not in s
        $\rightarrow s := \langle b \rangle \,\hat{}\, s$
    [] all vertices adjacent to $s.0$ and $s(i-1)$ are in s
        $\rightarrow$ Change s into a cycle on its vertices;
            Choose j such that $s.j$ has a vertex b (say)
            incident on it that is not in s;
            $s := s[j + 1..i - 1] \,\hat{}\, s[0..j];$
            $s := s \,\hat{}\, \langle b \rangle$
    fi;
    $i := i + 1$
od
$\{ P \text{ is a Hamilton path} \}$

We investigate the loop repetend in order to see how it maintains invariant P. It must add one vertex (and thus one edge) to path s. The first two guarded commands append and prepend a vertex to s. If the first two guards are *false*, then the guard of the third guarded command is true. The first step in this case is to change s into a cycle on its vertices. Lemma 19.14, given below, indicates that this can be done. The next step is to find a vertex $s.j$ that has incident on it a vertex b (say) that is not in s. Such an $s.j$ exists because s does not contain all vertices and because the graph is connected. The next step places $s[j]$ at the end of s; remember that s is a cycle, so this step is possible. The final step appends this vertex b to s, thus extending path s by one vertex. □

(19.14) **Lemma.** Consider an undirected graph that satisfies $deg.b + deg.c \geq n - 1$ for each pair of vertices b, c. Let s be a path of length $i - 1$ such that, for all vertices b that are not in s, neither $\langle s.0, b \rangle$ nor $\langle s(i-1), b \rangle$ are edges. Then s can be changed into a cycle.

Proof. An algorithm to change s into a cycle is given below. If $\langle s.0, s(i-1) \rangle$ is an edge, then s is already a cycle. Otherwise, some processing must be done. So assume $\langle s.0, s(i-1) \rangle$ is not an edge. That a j exists as required in this algorithm is proved as follows. Consider the following sets $T1$ and $T2$ of edges:

$$T1 = \{j \mid 2 \leq j < i - 1 \land \langle j - 1, s(i-1) \rangle \text{ is an edge}\}$$
$$T2 = \{j \mid 2 \leq j < i - 1 \land \langle s.0, j \rangle \text{ is an edge}\}$$

By the hypothesis of the theorem and the assumption that $\langle s.0, s(i-1) \rangle$ is not an edge, $1 + \#T1$ is the number of vertices adjacent to $s(i-1)$ and $1 + \#T2$ is the number of vertices adjacent to $s.0$. Thus, by the stated property of the graph

$$\#T1 + \#T2 \geq n - 1 - 2 = n - 3 \quad .$$

However, $T1 \cup T2 \subseteq 2..i - 2$, so $\#(T1 \cup T2) \leq i - 3 < n - 3$. Hence, $T1$ and $T2$ have an element in common.

```
{ Change s into a cycle on its vertices }
  if ⟨s.0, s(i − 1)⟩ is not an edge then
    begin  Find an integer j, 0 ≤ j < i, such that
             ⟨s.j, s(i − 1)⟩ and ⟨s.0, s(j + 1)⟩ are edges;
           s := s[0..j] ^ reverse(s[j + 1..i − 1])
    end
```

Finally, the assignment $s := s[0..j] \; \hat{} \; reverse(s[j + 1, i - 1])$ changes s into a cycle. We see this in the following diagram, which shows the path before and after this assignment. By changing s as indicated, the edge $\langle s.j, s(j+1) \rangle$ is deleted and the two edges $\langle s.j, s(i-1) \rangle$ and $\langle s.j, s.0 \rangle$ are inserted. □

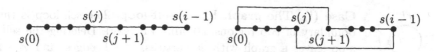

The proof of the following corollary is left to the reader.

(19.15) **Corollary.** Let G be an undirected loop-free graph of n nodes. If the degree of each vertex is at least $(n-1)/2$, then G has a Hamilton path.

19.6   Planar graphs

A graph is *planar* if it can be drawn in the plane without any edges crossing. Below, we show several ways of drawing the complete graph K_4. In the leftmost one, two edges cross, but in the other three, no edges cross. Hence, K_4 is a planar graph.

 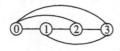

 A planar graph separates the points in the plane into *regions*. Consider two points that are not points of the edges of the graph. These two points are in the same region iff it is possible to draw a (curved) line segment that joins them and does not touch an edge. In the graph in this paragraph, the regions are labeled B, C, and D. Each *interior* region (e.g. B and C) is characterized by the cycle that forms its boundary. In addition, there is an *exterior* region (D), which contains all the points that are not bounded by some cycle.

Euler proved the following about planar graphs.

(19.16) **Theorem.** For a connected planar graph with v vertices, e edges, and r regions, $r = e - v + 2$.

Proof. The proof is by induction on the number of edges of the graph.

Base case $e = 0$. Since the graph is connected, there is one vertex and one (exterior) region. Substituting into the formula $r = e - v + 2$, we get $1 = 0 - 1 + 2$, which is valid.

Inductive case $e > 0$. We assume the inductive hypothesis that the theorem holds for an arbitrary graph of $e-1$ edges and prove that it holds for a graph of e edges (and r regions and v vertices). We distinguish three cases: (i) there is a self-loop, (ii) there is a vertex of degree 1, and (iii) there is no self-loop and all vertices have degree at least 2.

Case (i). The graph has a self-loop. The self-loop is the boundary of two regions $r0$ and $r1$ (say). Deleting the self-loop leaves a graph with v vertices, $e - 1$ edges, and $r - 1$ regions, since $r0$ and $r1$ are merged into a single region. By the inductive hypothesis, the first formula in the manipulation below is *true* . Hence, so is the second formula, and the theorem holds in this case.

$$r - 1 = (e - 1) - v + 2$$
$$= \quad \langle\text{Arithmetic}\rangle$$
$$r = e - v + 2$$

Case (ii). There is a vertex c (say) of degree 1. The points of the plane on either side of the edge belong to the same region $r1$ (say). Deleting the edge and also vertex c leaves a connected graph with $v - 1$ vertices, $e - 1$ edges, and r regions. By the inductive hypothesis, the first formula in the manipulation below is *true* . Hence, so is the second formula, and the theorem holds in this case.

$$r = (e - 1) - (v - 1) + 2$$
$$= \quad \langle\text{Arithmetic}\rangle$$
$$r = e - v + 2$$

Case (iii). Each vertex has degree at least 2 **and there are no self-loops.** Hence, there is an edge $\langle b, c \rangle$ on a boundary, as shown in this paragraph. This edge borders on two regions. Deleting the edge results in a graph of $r - 1$ regions, v vertices, and $e - 1$ edges. The manipulation in case (i) shows that the theorem holds in this case as well.   □

(19.17) **Theorem.** In a loop-free, connected, planar graph with $v > 2$ vertices, e edges, and r regions, $e \leq 3 \cdot v - 6$. In addition, if the graph is bipartite, then $e \leq 2 \cdot v - 4$.

Proof. Each boundary contains at least three edges, for the following reason. First, the graph is loop-free, so there are no boundaries consisting of 1 edge. Second, the graph is not a multigraph, so there are no boundaries consisting of 2 edges.

Since each edge is on the boundary of at most 2 regions, there are at least 3/2 edges per region, i.e. $3 \cdot r / 2 \leq e$. Hence, $3 \cdot r \leq 2 \cdot e$ holds. Using Euler's Theorem (19.16), we derive the first formula. First, we have

$$3 \cdot r \leq 2 \cdot e$$
$$= \quad \langle\text{Theorem (19.16)}, \ r = e - v + 2 \rangle$$
$$3 \cdot (e - v + 2) \leq 2 \cdot e$$
$$= \quad \langle\text{Distributivity of } \cdot \text{ over } + \text{ and } - \rangle$$

$$3 \cdot e - 3 \cdot v + 3 \cdot 2 \leq 2 \cdot e$$
$$= \quad \langle \text{Predicate calculus} \rangle$$
$$e \leq 3 \cdot v - 6$$

Now suppose that the graph is bipartite. Then each cycle is even, so the number of edges that bound each region is at least 4. Since each edge is on the boundary of at most two regions, we have $4 \cdot r \leq 2 \cdot e$. The theorem now follows by a manipulation similar to the one in the previous case. $\square$

This theorem gives us a simple way to determine that some graphs are nonplanar, for its contrapositives are:

$$e > 3 \cdot v - 6 \ \Rightarrow \ \text{the loop-free connected graph is nonplanar}$$

$$e > 2 \cdot v - 4 \ \Rightarrow \ \text{the loop-free bipartite connected graph}$$
$$\text{is nonplanar.}$$

For example, we can prove

(19.18) **Theorem.** Complete graph K_5 and complete bipartite graph $K_{3,3}$ are nonplanar.

Proof. K_5 has 10 edges and 5 vertices. Since $10 > 3 \cdot 5 - 6 = 9$, K_5 is nonplanar. $K_{3,3}$ has 9 edges and 6 vertices. Since $9 > 2 \cdot 6 - 4 = 8$, $K_{3,3}$ is nonplanar. $\square$

Obviously, if a graph contains K_5 or $K_{3,3}$ as a subgraph, then the graph is not planar. What about the converse of this statement: if a graph is nonplanar, does it contain K_5 or $K_{3,3}$? Let's look at an example.

Consider graph G on the left below. This graph is not planar, but does it contain K_5 or $K_{3,3}$? We can almost see $K_{3,3}$ in G, but somehow vertices g and h are in the way. Deleting vertex h and the edges incident on it yields the subgraph in the middle below. However, we cannot delete vertex g and the edges incident on it, because that would result in a graph that is a proper subgraph of $K_{3,3}$. However, if we remove g from G but splice together the two edges incident on it, we arrive at the graph to the right, which is $K_{3,3}$. Thus, $K_{3,3}$ is homeomorphic (see page 438) to a subgraph of G.

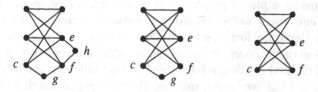

In 1930, K. Kuratowski proved the following characterization of the class of planar graphs. It is not at all obvious that one should even consider characterizing the planar graphs so simply, so this theorem is quite amazing. The proof of this theorem is beyond the scope of this text.

(19.19) **Theorem.** A graph is planar iff it does not contain a subgraph that is homeomorphic to K_5 or $K_{3,3}$.

Kuratowski's theorem provides a nice characterization of planar graphs, but not an efficient algorithm for testing planarity of graphs. How quickly might we expect to be able to test a given graph for planarity? Theorem (19.17), $e \leq 3 \cdot v - 6$, gives us hope that we might be able to do it in time proportional to the number v of vertices, even though the graph might have $n \cdot (n-1)/2$ edges. Why? If we start counting edges and get past $3 \cdot v - 6$, then we know that the graph is not planar and do not have to process it any further. On the other hand, if $e \leq 3 \cdot v - 6$, then the number of edges is linear in v, so the amount of data to process is linear in v.

A linear-time algorithm to test for planarity and construct a planar representation if it exists was first developed by Robert Tarjan and John E. Hopcroft in 1971. This and related work on algorithms and computational complexity formed the basis for their receiving the ACM Turing Award in 1986. Their planarity algorithm is quite complex and is outside the scope of this text.

DUAL GRAPHS AND MAP COLORING

Given a planar graph G, we construct its *dual* graph G_D as follows. Let the r regions of G be $R_1, \ldots, R_r$. Then the dual G_D of G has the r vertices $R_1, \ldots, R_r$. Graph G_D has an edge $\langle R_i, R_j \rangle$ iff regions R_i and R_j are adjacent, i.e. if some edge of G is on the boundary of both. We show below a graph and its dual.

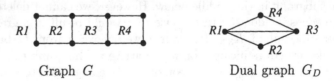

Graph G Dual graph G_D

Now consider a planar graph G to be a map, where regions represent countries and edges define boundaries between countries. In the above graph G, there are four countries. One can see that the countries can be colored so that adjacent countries have different colors exactly when the vertices of its dual graph can be colored so that adjacent vertices have different colors. Thus, we have translated the problem of coloring a map into the problem of coloring the vertices of a graph. The above graph requires only three colors, but the graph below require four colors.

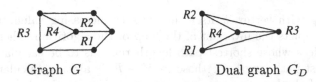

Graph G Dual graph G_D

In the introduction to this chapter, we mentioned the famous four-color problem: can the countries of a map be colored using only four colors so that adjacent countries have different colors? The relation between planar graphs and their duals shows how this problem becomes one of coloring the vertices of a planar graph with four colors so that adjacent vertices have different colors. This is an example of a problem that is easy to state, has an interesting application, but is hard to prove. In fact, for 100 years, many mathematicians worked on the 4-color problem, before K. Appel and W. Haken solved it in 1977, and they needed computers to solve it.

19.7   Shortest paths and spanning trees

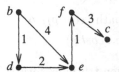

In some applications, nonnegative *weights* are associated with edges of a digraph or graph to represent the cost of traversing or using that edge. To the left is a weighted digraph —the numbers on the edges are not edge names but edge weights. For example, the cost of traversing or using edge $\langle b, e \rangle$ is 4. We now investigate a few basic algorithms that process graphs with edge weights: shortest-path algorithms and spanning-tree algorithms.

DIJKSTRA'S SHORTEST-PATH ALGORITHM

Given is a weighted, connected digraph $G = \langle V, E \rangle$ with vertices $V = 0..n-1$. The weights of the edges are given by a function w: for each edge $\langle b, c \rangle$, $w(b, c) > 0$ is its weight, and $w(b, c) = 0$ if there is no edge $\langle b, c \rangle$. The cost $cost.p$ of a path p is the sum of the weights of its edges. In the graph in the introductory paragraph of this section, path $\langle b, d, e, f \rangle$ has cost 4 and path $\langle b, e, f \rangle$ has cost 5.

We denote by $min(b, c)$ the minimum cost over all b-c paths:

$$min(b, c) = (\downarrow p \mid p \text{ is a } b\text{-}c \text{ path} : cost.p)$$

Note that if all the edge weights are 1, a minimum-cost path is also a shortest path —i.e. one with the fewest number of edges.

We now present an algorithm that, given two vertices b and c, computes $min(b, c)$. The algorithm iteratively computes $min(b, v)$ for vertices

v, starting with vertex b itself, in order of increasing value $min(b, v)$. Fig. 19.6 depicts the invariant of the loop of the algorithm. There is a set F of vertices whose shortest-path length remains to be calculated, while the shortest-path length for those in $V - F$ is already calculated. Each iteration of the loop moves one vertex across the frontier, by removing it from F, thus making progress towards termination.

We now give precise details about the invariant of this algorithm. First, PF explains the relation between vertices b and F.

$$PF : \quad V - F \neq \emptyset \Rightarrow b \in V - F$$

Next, an array $L[0..n-1]$ of natural numbers is used to maintain information about lengths of shortest paths. For elements v of $V - F$, $L[v]$ is the length of the shortest b-v path.

Consider v in F. A b-v path that has only one vertex in F (vertex v) is called a b-v *xpath* (for want of a better name). Thus, all the vertices of an xpath, except the last, are in $V - F$. For elements v of F, $L[v]$ is the length of the shortest b-v xpath (∞ if no xpath to v exists).[5] So, the invariant PL that describes array L is as follows.

$$PL : \quad (\forall v \mid v \in V - F : L[v] = (\downarrow p \mid p \ a \ b\text{-}v \ \text{path} : cost.p)) \ \wedge$$
$$(\forall v \mid v \in F : L[v] = (\downarrow p \mid p \ a \ b\text{-}v \ \text{xpath} : cost.p))$$

The algorithm can now be presented. The initialization truthifies invariants PF and PL. Each iteration removes some vertex u from F and then updates L in order to reestablish the invariants. Since each iteration removes one element from F, at some point $c \notin F$ and the loop terminates.

---

[5] For ∞, we can use the number $W \cdot n$, where W is the largest edge weight, since $W \cdot n$ is greater than any path cost.

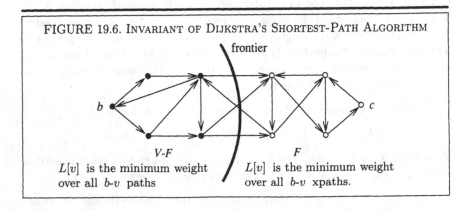

FIGURE 19.6. Invariant of Dijkstra's Shortest-Path Algorithm

frontier

b c

V-F

$L[v]$ is the minimum weight over all b-v paths

F

$L[v]$ is the minimum weight over all b-v xpaths.

for all vertices v **do** $L[v] := \infty$;
$L[b], F := 0, V$;
do $c \in F \rightarrow$
    Let u be the vertex of F with minimum L value;
    $F := F - \{u\}$;
    for each v adjacent to u **do** $L[v] := L[v] \downarrow (L[u] + w(u, v))$;
od

We now prove that at each iteration $L[u]$ is the shortest path length from b to u, so that invariant PL is maintained when u is removed from F. We show that the definition of $L[v]$ for v in F implies the desired result.

$$(\forall v \mid v \in F : L[v] = (\downarrow p \mid p \text{ a } b\text{-}v \text{ xpath} : cost.p))$$
$$= \quad \langle L[u] \text{ is the minimum of the } L[v] \text{ in question} \rangle$$
$$(\forall v \mid v \in F : L[u] \leq (\downarrow p \mid p \text{ a } b\text{-}v \text{ xpath} : cost.p))$$
$$= \quad \langle \text{Interchange of quantification} \rangle$$
$$L[u] \leq (\downarrow p, v \mid v \in F \wedge p \text{ a } b\text{-}v \text{ xpath} : cost.p)$$
$$= \quad \langle \text{Any } b\text{-}v \text{ path has a } b\text{-}v \text{ xpath as a prefix} \rangle$$
$$L[u] \leq (\downarrow p, v \mid v \in F \wedge p \text{ a } b\text{-}v \text{ path} : cost.p)$$
$$= \quad \langle \text{Instantiation, with } v := u \rangle$$
$$L[u] \leq (\downarrow p \mid \; : p \text{ a } b\text{-}u \text{ path} : cost.p)$$

Since $L[u]$ is actually the cost of some b-u path, the result follows.

Exercise 19.27 concerns keeping track of the vertices in F whose L value is not ∞, in order to reduce the time needed to find the minimum L value.

This algorithm can be extended to store in $L[v]$ the shortest path length for *all* vertices v simply by changing the loop test to $F \neq \emptyset$. Secondly, it can be easily modified to compute least-cost paths themselves, and not just the least costs (see Exercise 19.28).

The algorithm can also be used to compute shortest paths in an undirected graph; just represent each undirected edge with cost k by two directed edges, each with cost k.

Spanning trees

Trees can be defined in various ways. For example, on page 233 we defined binary trees inductively. Here, we give a definition of a tree that is helpful in graph theory. A *tree* is a loop-free connected graph that contains no cycles. The leftmost graph in this paragraph is a tree; the rightmost one is not a tree, because it contains a cycle. We now explore some fairly obvious properties of trees.

(19.20) **Theorem.** Each pair of vertices of a tree is connected by a unique simple path.

Proof. Since a tree is a connected graph, there exists at least one path between any two vertices. We outline the proof by contradiction that there cannot be more than one path between two vertices. Suppose two paths exist between two vertices. Then some of the edges of the two paths can be used to construct a cycle, contradicting the fact that a tree has no cycle.□

(19.21) **Theorem.** A tree with at least two vertices has at least two vertices of degree 1.

Proof. Since the tree has no cycle, there is at least one vertex c (say) of degree 1. Since a tree is connected, there is at least one edge incident on c. Construct the longest path possible with c as the start vertex. A longest path exists because a tree has no cycles. The last vertex of this longest path has degree 1 and is different from c. Hence, at least two vertices have degree 1.   □

(19.22) **Theorem.** For a tree $\langle V, E \rangle$, $\#V = 1 + \#E$.

Proof. The proof is by induction on the number of edges in the tree.

Base case $\#E = 0$. Since the graph is connected and there are no edges, there is only one (isolated) vertex, so $1 = \#V = 1 + 0 = 1 + \#E$.

Inductive case $\#E > 0$. We assume the inductive hypothesis that the theorem holds for graphs of $n-1$ edges, $0 < n$, and prove that it holds for a graph of n edges. Consider a tree with n edges and v (say) vertices. By Theorem (19.21), the tree has a vertex of degree 1. Deleting that vertex and the single edge incident on it yields a tree with $n-1$ edges and $v-1$ vertices. Applying the inductive hypothesis and calculating yields the result.   □

The following theorem provides several characterizations of trees. The proof is left to the reader.

(19.23) **Theorem.** The following statements are equivalent (where $G = \langle V, E \rangle$ is a loop-free graph).
   (a) G is a tree.
   (b) G is connected and the removal of any edge yields two trees.
   (c) G contains no cycles and $\#V = 1 + \#E$.
   (d) G is connected and $\#V = 1 + \#E$.
   (e) G has no cycles, and adding any edge introduces exactly one cycle.

A *spanning tree* of a connected graph is a tree that is a subgraph and that contains all the vertices of the graph. In the middle below is a graph; on either side of it is a spanning tree of the graph.

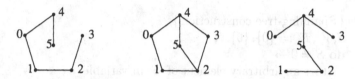

A spanning tree of a graph contains the minimum number of edges to keep the graph connected.

A GENERAL SPANNING-TREE ALGORITHM

We present an algorithm for constructing a spanning tree of a connected graph. Let $G = \langle V, E \rangle$ be the given connected graph; the spanning tree will be $\langle V, E' \rangle$; i.e. the edges of the spanning tree will be stored in a set variable E'. The algorithm is *nondeterministic*, in that during execution an arbitrary choice of an edge to place in E' is made at each iteration of the loop, and a different choice can lead to a different spanning tree. Later, we show how this choice can be restricted in three different ways.

Variable V' contains the vertices of the spanning tree as it is being constructed, so that at all times $\langle V', E' \rangle$ is a tree. (Alternatively, one could have a bit attached to each vertex of V to indicate whether or not it is in V'.)

$$P1: \quad V' \subseteq V \ \wedge \ E' \subseteq E \ \wedge \ \langle V', E' \rangle \text{ is a tree}$$

Variable S contains a set of vertices, to keep track of which parts of the graph still need to be explored in building the spanning tree. The following predicate *reachable.v* is used to explain how set S is used.

$$reachable.v \ \equiv \ \text{there is a path } \langle u \rangle \ \hat{} \ x \ \hat{} \ \langle v \rangle \text{ where } u \in S$$
$$\text{and vertices of } x \ \hat{} \ \langle v \rangle \text{ are in } V - V'.$$

We now give the second invariant $P2$, which is illustrated in Fig. 19.7.

$$P2: \ S \subseteq V' \ \wedge \ (\forall v \mid v \in V - V' : reachable.v)$$

Invariant $P2$ is used in proving that upon termination, all the vertices of the graph are in the spanning tree $\langle V', E' \rangle$. Suppose S is empty (which is *true* upon termination of the algorithm). Then no vertex v satisfies *reachable.v*. $P2$ then implies that no vertex is in $V - V'$, so all vertices are in V'.

Now consider the algorithm below. The first assignment truthifies $P1$ and $P2$ —arbitrarily, vertex 0 is placed in V, and, because the graph is connected, each other vertex is reachable along some path that includes 0 as its first vertex.

(19.24)   { Spanning-tree construction }
$\quad$ $S, V', E' := \{0\}, \{0\}, \emptyset;$
$\quad$ **do** $S \neq \emptyset \rightarrow$
$\qquad$ Store an arbitrary element of S in variable u;
$\qquad$ **if** $(\exists v \mid \langle u, v \rangle$ is an edge $: v \notin V') \rightarrow$
$\qquad\qquad$ $S, V', E' := S \cup \{v\}, V' \cup \{v\}, E' \cup \{\langle u, v \rangle\}$
$\qquad$ [] $(\forall v \mid \langle u, v \rangle$ is an edge $: v \in V') \rightarrow$ $\ S := S - \langle u \rangle$
$\qquad$ **fi**
$\quad$ **od**

Next, we show that the loop terminates. Consider pair $\langle \#V - \#V', \#S \rangle$. Each iteration either decreases $\#V - \#V'$ or leaves $\#V - \#V'$ unchanged and decreases $\#S$. Hence, $\langle \#V - \#V', \#S \rangle$ is decreased, lexicographically speaking. Since the pair is bounded below by $\langle 0, 0 \rangle$, the loop terminates.

FIGURE 19.7. One Iteration of Spanning-Tree Construction

The graph to the left illustrates the invariant. Solid edges are in E' (in the spanning tree), and the rest of the edges of the graph are drawn with dashed lines. The vertices in the dotted circle (all of which are drawn with unfilled circles) are in S. Every vertex v that is not in the spanning tree is reachable from a vertex in S along a path whose vertices (except the first) are not in the spanning tree.

Suppose the node labeled u is chosen during an iteration. The graph to the right illustrates the state after the iteration. An edge leaving u has been placed in the spanning tree, and the vertex at the other end of the edge has been placed in S.

We show that each iteration maintains invariant $P1$. That $V' \subseteq V$ and $E' \subseteq E$ are maintained is obvious. The conjunct "$\langle V', E' \rangle$ is a tree" is maintained because an edge $\langle u, v \rangle$ is placed in E' only if v is not yet in the tree. Also, each element of S remains a member of V' because the statement that adds an element to S adds it also to V'.

It is easily seen that the first guarded command maintains invariant $P2$ (see Fig. 19.7 for an illustration). And the second guarded command does also.

BREADTH-FIRST AND DEPTH-FIRST SPANNING TREE ALGORITHMS

Spanning-tree algorithm (19.24) is nondeterministic in two ways. First, the choice of vertex u is arbitrary; second, the choice of edge incident on u to add to the spanning tree is arbitrary. We limit the choice of vertex u in two different ways to arrive at two different spanning-tree algorithms.

The breadth-first algorithm. Replace variable S of algorithm (19.24) by a sequence variable s, where the two are related by $v \in S \equiv v \in s$. For u, always choose the first element of s, and always *append* vertices to s. Thus, the algorithm is written as follows.

```
{ Breadth-first spanning-tree construction }
s, V', E' := ⟨0⟩, {0}, ∅;
do s ≠ ε →
    u := head.s;
    if (∃v | ⟨u, v⟩ is an edge : v ∉ V') →
            s, V', E' := s ^ ⟨v⟩, V' ∪ {v}, E' ∪ {⟨u, v⟩}
    ▯ (∀v | ⟨u, v⟩ is an edge : v ∈ V') →  s := tail.s
    fi
od
```

This algorithm first adds to the spanning tree all possible edges that are incident on the first vertex added to the tree, then all possible edges that are incident on the second edge added, and so on. As an example, consider the graph in the middle below, and suppose its uppermost vertex is the first vertex placed in the spanning tree. To the left, we show the spanning tree constructed by the breadth-first construction, with the edges marked in the order in which they are added to the tree.

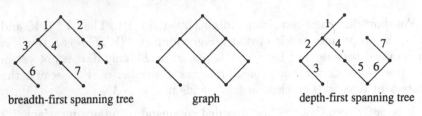

breadth-first spanning tree   graph   depth-first spanning tree

The depth-first algorithm. Replace set S of algorithm (19.24) by sequence variable s, where the two are related by $v \in S \equiv v \in s$. For u, always choose the first element of s, and always *prepend* vertices to s. The diagram to the right above shows the spanning tree constructed by the depth-first algorithm applied to the graph in the middle, with the edges labeled in the order of insertion into the tree.

```
{ Depth-first spanning-tree construction }
s, V', E' := ⟨0⟩, {0}, ∅;
do s ≠ ε →
  u := head.s;
  if (∃v | ⟨u, v⟩ is an edge : v ∉ V') →
       s, V', E' := ⟨v⟩ ^ s, V' ∪ {v}, E' ∪ {⟨u, v⟩}
  ⟦ (∀v | ⟨u, v⟩ is an edge : v ∈ V') →   s := tail.s
  fi
od
```

PRIM'S MINIMAL SPANNING TREE ALGORITHM

Consider a complex of computers in a business, which are to be linked together so that the computers can communicate with each other. The cost of installing a transmission line between two computers can depend on various factors, such as the distance between them and whether they are in the same room or building. Suppose we construct a weighted graph to model this complex of computers. The vertices represent the computers, the edges represent the transmission lines, and the weights on the edges are the costs of building those transmission lines. To keep costs to a minimum, we think of constructing the communication lines modeled by a *minimum (weight) spanning tree*: a tree whose sum of the edge weights is a minimum.

As another example, consider the problem of linking cities with pipelines for oil or gas. Here, it is best to keep costs down by building pipelines according to a minimum spanning tree.

Robert Clay Prim's algorithm for computing a minimum spanning tree is a version of our original nondeterministic algorithm (19.24) for computing a spanning tree. In our version of Prim's algorithm, we still use set variable S, but the choice of edge $\langle u, v \rangle$ to add to the tree is restricted.

In algorithm (19.24), consider all edges $\langle u', v' \rangle$ where u' is in S and v' is in $V - V'$. Among all such edges, choose an edge $\langle u, v \rangle$ with minimum weight. This choice determines the u to use in the statement "Store an arbitrary element of S in variable u" of algorithm (19.24). This choice also determines the vertex v to add to E' within the repetend of the loop of (19.24). Since the only change in the algorithm is to restrict nondeterminism, the modified algorithm still produces a spanning tree.

We have to show that Prim's algorithm constructs a *minimum* spanning tree. To that end, we introduce an additional loop invariant P'.

$\qquad P'$: Some minimum spanning tree of $\langle V, E \rangle$ contains $\langle V', E' \rangle$

P' is true initially, since the graph has a minimum spanning tree and $E' = \emptyset$. And if P' is true upon termination, then $\langle V, E' \rangle$ *is* a minimum spanning tree since it is a tree and has $\#V - 1$ edges. So it remains only to show that each iteration maintains P', thus showing that P' is *true* after the iteration.

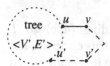 At the beginning of the iteration, as per P', some minimum spanning tree T (say) has $\langle V', E' \rangle$ as a subgraph. Suppose T does not contain the edge $\langle u, v \rangle$ that is added by the iteration. We show how to change T into a minimum spanning tree that does contain $\langle u, v \rangle$. Since T is connected, there is a path in T that connects u to v, where the first edge out of v' is $\langle u', v' \rangle$ (say) and $v' \notin V'$. Because of the way u is chosen in the iteration, we have $w.\langle u, v \rangle \leq w.\langle u', v' \rangle$. Replacing edge $\langle u', v' \rangle$ in T by $\langle u, v \rangle$ therefore leaves T a spanning tree but does not increase its cost, so this replacement constructs the desired minimum tree containing $\langle u, v \rangle$.

KRUSKAL'S MINIMUM SPANNING TREE ALGORITHM

Prim's algorithm always maintains a tree, and at each iteration it adds one edge to the tree. Kruskal's algorithm, on the other hand, starts with the graph $\langle V, \emptyset \rangle$ and seems to add edges to it in a rather indiscriminate fashion.

```
{ Kruskal's Algorithm }
E' := ∅;
do #E' ≠ #V − 1 →
   Let ⟨u, v⟩ be an edge of minimum weight whose
      addition to E' does not create a cycle in graph ⟨V, E'⟩ ;
   E' := E' ∪ {⟨u, v⟩}
od
```

We take as loop invariant simply

P : Graph $\langle V, E' \rangle$ has no cycle

Since E' ends up with $\#V - 1$ vertices and no cycles, by Theorem (19.23), $\langle V, E' \rangle$ is a spanning tree. Our only task, then, is to show that upon termination $\langle V, E' \rangle$ is a *minimum* spanning tree. To this end, we consider another invariant $P1$.

$P1$: Some minimum spanning tree of $\langle V, E \rangle$ contains $\langle V, E' \rangle$

$P1$ is true initially, since the graph has a minimum spanning tree and $E' = \emptyset$. And if $P1$ is true upon termination, then $\langle V, E' \rangle$ *is* a minimum spanning tree since it is a tree and has $\#V - 1$ edges. So it remains only to show that each iteration maintains $P1$.

At the beginning of the iteration, as per $P1$, some minimum spanning tree T (say) has $\langle V, E' \rangle$ as a subgraph. Suppose T does not contain the edge $\langle u, v \rangle$ that is added by the iteration. We show how to change T into a minimum spanning tree that does contain $\langle u, v \rangle$.

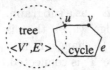 Since T does not contain $\langle u, v \rangle$, adding $\langle u, v \rangle$ to T creates a cycle. At least two edges of this cycle are not in $\langle V, E' \rangle$. Edge $\langle u, v \rangle$ is not, because it is being added during the iteration. And another edge of the cycle e (say) is not in T because adding $\langle u, v \rangle$ to $\langle V, E' \rangle$ does not create a cycle. By the way in which $\langle u, v \rangle$ is chosen, $w.\langle u, v \rangle \le w.e$. Therefore, replacing e in T by $\langle u, v \rangle$ does not increase the cost T and leaves it a spanning tree. Hence this replacement leaves T a minimum spanning tree.

Exercises for Chapter 19

19.1 Prove Corollary (19.4), the number of vertices of odd degree of a digraph or graph is even.

19.2 Prove that, in any group of two or more people, there are always two people with exactly the same number of friends in the group. Hint: We assume that if X is a friend of Y, then Y is a friend of X. Use an undirected graph to represent the friendship relation.

19.3 Prove Corollary (19.9).

19.4 Does there exist an undirected connected multigraph with an even number of vertices and an odd number of edges that has an Euler circuit? If yes, show one; if no, explain why.

19.5 (a) Under what conditions does the complete graph K_n have an Euler circuit?

(b) Under what conditions does the complete bipartite graph $K_{m,n}$ have an Euler circuit?

19.6 Show that there are $41,472$ ways of arranging 4 cubes in a column.

19.7 Prove by induction that complete graph K_n has $n \cdot (n-1)/2$ edges.

19.8 Prove Theorem (19.10).

19.9 Prove Corollary (19.11).

19.10 The k-cube is the graph whose vertices are labeled by k-bit numbers; two vertices are joined by an edge iff their labels differ in exactly one bit. For example, the 1-cube, 2-cube, and 3-cube are shown below. Show that the k-cube has 2^k vertices, has $k \cdot 2^{k-1}$ edges, and is bipartite.

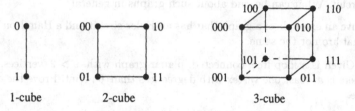

19.11 Show that the following two graphs are isomorphic.

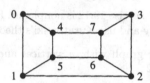

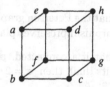

19.12 Show that the following three graphs are isomorphic. The leftmost graph is known as the *Petersen Graph*.

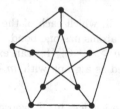

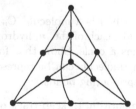

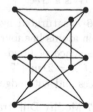

19.13 Construct a knight's tour for a 6×6 board.

19.14 Write a binary search algorithm for the following problem. Given is a complete digraph whose vertex set is $0..n-1$ (for some $n > 0$). Path s contains the vertices $0..i-1$ for some $i < n$. It is given that $\langle s.0, i \rangle$ and $\langle i, s(\#s-1) \rangle$ are edges. Write a binary search algorithm that truthifies

$$0 \le j < i \wedge \langle s.j, i \rangle \text{ and } \langle i, s(j+1) \rangle \text{ are edges} \quad .$$

The binary search should be a single loop, and its loop invariant should be

$$P1: \quad 0 \le j < k < i \wedge \langle s.j, i \rangle \text{ and } \langle i, s.k \rangle \text{ are edges} .$$

19.15 Prove Corollary (19.15).

19.16 Prove (by contradiction) that if a loop-free graph of n vertices satisfies $(\forall b, c \mid b \neq c : deg.b + deg.c \geq n - 1)$, then the graph is connected.

19.17 Under what conditions does the complete bipartite graph $K_{m,n}$ have a Hamilton circuit?

19.18 Find a Hamilton path in the Petersen graph of Exercise 19.12.

19.19 (a) Give an example of a graph that has an Euler circuit but no Hamilton circuit.

  (b) Give an example of a graph that has a Hamilton circuit but no Euler circuit.

  (c) Give an example of a graph in which an Euler circuit is also a Hamilton circuit. What can be said about such graphs in general?

  (d) Give an example of a graph that has an Euler circuit and a Hamilton circuit that are not the same.

19.20 Given is a loop-free, connected, planar graph with $v > 2$ vertices. Prove that there is at least one vertex with degree less than 6. Hint: Prove the contrapositive.

19.21 Give an example of a loop-free, connected, non-planar graph in which $e \leq 3 \cdot v - 6$.

19.22 Prove that the Petersen graph of Exercise 19.12 is nonplanar. Hint: Use the fact that isomorphism preserves planarity and use Kuratowski's theorem (19.19).

19.23 Show that if a connected, planar graph with v vertices and e edges is isomorphic to its dual, then $e = 2 \cdot v - 2$.

19.24 Prove Theorem (19.23).

19.25 Prove of disprove: If a graph $G = \langle V, E \rangle$ is loop-free and $\#V = 1 + \#E$, then G is a tree.

19.26 A saturated hydrocarbon is a molecule $C_m H_n$ in which each of the m carbon atoms has four bonds, each of the n hydrogen atoms has one bond, and no sequence of bonds forms a cycle. Show that for $m \geq 1$, $C_m H_n$ can exist only if $n = 2 \cdot m + 2$. (The molecule can be represented as a graph with $m + n$ vertices and with edges representing bonds.)

19.27 Modify Dijkstra's algorithm to keep track of the vertices in F whose L value is not ∞.

19.28 Modify Dijkstra's shortest-path algorithm to compute the shortest path and not just its cost.

Chapter 20

Infinite Sets

I n this chapter, we investigate infinite sets. We learn that "infinity" comes in different sizes. For example, there are more real numbers than integers, even though both sets are infinite . We also learn that the smallest infinity is the size of the set of natural numbers and that there are just as many even nonnegative numbers as there are natural numbers.

In our study, we come across some techniques that are useful in theoretical computer science, like Cantor diagonalization. We also discover that there are more possible tasks to perform than programs, so that many tasks simply cannot be performed on a computer.

20.1   Finite versus infinite sets

We begin our investigation of infinite sets by defining the terms finite and infinite.

(20.1)   **Definition.** Set S is *finite,* with cardinality or size n for some natural number n, if there exists a one-to-one, onto function $f : (0..n-1) \rightarrow S$; otherwise, S is *infinite.*

For example, to show that the set $\{0, 4, 2\}$ is finite, we exhibit the function f with $f.0 = 0$, $f.1 = 4$, and $f.2 = 2$ and conclude by definition (20.1) that the cardinality of this set is 3.

The proofs of the following four theorems are left to the reader.

(20.2)   **Theorem.** Every subset of a finite set is finite.

(20.3)   **Theorem.** The union of two finite sets is finite.

(20.4)   **Theorem.** The cross product of two finite sets is finite.

(20.5)   **Theorem.** The power set of a finite set is finite.

Perhaps more interesting is to prove that

(20.6)   **Theorem.** The set $\mathbb{N}$ of natural numbers is infinite.

Proof. According to Definition (20.1), we have to show that there is no one-to-one, onto function $f : (0..n-1) \rightarrow \mathbb{N}$, for any natural number n. Any one-to-one function $f : (0..n-1) \rightarrow \mathbb{N}$ has a range of n values. Therefore, its range cannot contain all the $n+1$ values in $0..n$, which is a subset of $\mathbb{N}$. Therefore f cannot be onto $\mathbb{N}$.   □

Showing the absence of an object, as done in the previous proof, is sometimes difficult, so we seek a different characterization of the infinite sets. To this end, we prove (see Exercise 20.5) the following theorem.

(20.7)   **Theorem.** If S is infinite, then there exists a one-to-one function $f : \mathbb{N} \rightarrow S$.

With the help of this theorem, we give a neat characterization of the infinite sets (see Exercise 20.6).

(20.8)   **Theorem.** S is infinite iff there is a one-to-one function that maps S into a proper subset of itself.

As an example, the set $\mathbb{N}$ of natural numbers is infinite, since the function $f.i = 2 \cdot i$ maps $\mathbb{N}$ onto the even natural numbers and the even naturals are a proper subset of $\mathbb{N}$.

This characterization of infinite sets may at first be disconcerting. If five men and five women are together in a room and the women leave the room, half as many people remain in the room. But if all the naturals are in a box and the odd naturals fall out through a hole, then the number of naturals in the box does not change!

The following theorems are readily proved using the characterization of infinite sets given in Theorem (20.8).

(20.9)   **Theorem.** If a subset of a set is infinite, then so is the set.

(20.10)   **Theorem.** Let $f : S \rightarrow T$ be one-to-one. If S is infinite, then so is T.

(20.11)   **Theorem.** If S is infinite, then so is its power set $\mathcal{P}S$.

(20.12)   **Theorem.** If S is infinite, then so is $S \cup T$.

(20.13)   **Theorem.** If S is infinite and $T \neq \emptyset$, then $S \times T$ is infinite.

20.2   The cardinality of an infinite set

Is infinity simply infinity, or are there different kinds of infinities? We answer the latter question affirmatively. In fact, we show that there is an unending sequence of increasingly bigger infinite sets, so there are an infinite number of infinities!

Consider the set C of real numbers [1] in the range $0..1$. That is, $C = \{r : \mathbb{R} \mid 0 \le r \le 1\}$. In decimal notation, a real number in C has the form

$$.d_0 d_1 d_2 d_3 \cdots .$$

where each d_i is a digit between 0 and 9. We now prove that C is bigger than $\mathbb{N}$.

(20.14) **Theorem.** $C = \{r : \mathbb{R} \mid 0 \le r \le 1\}$ is bigger than $\mathbb{N}$.

Proof. There are at least as many elements in C as there are natural numbers. To see this, consider the one-to-one function that maps each natural $i_0 i_1 \cdots i_n$ (in decimal notation) into the real number

$$.i_n \cdots i_1 i_0 0\, 0\, 0 \cdots .$$

We show that C is bigger than $\mathbb{N}$ by proving that no one-to-one function $f : \mathbb{N} \to C$ is onto. Thus, no matter how we map $\mathbb{N}$ to C, some number in C is left over. Consider any such one-to-one function f and write down the correspondence as shown below. We write $(f.j)_k$ for digit k of the decimal representation of $f.j$. Each line j shows $f.j$ in its decimal representation.

$$
\begin{aligned}
f.0 &= .(f.0)_0\ (f.0)_1\ (f.0)_2\ \cdots \\
f.1 &= .(f.1)_0\ (f.1)_1\ (f.1)_2\ \cdots \\
f.2 &= .(f.2)_0\ (f.2)_1\ (f.2)_2\ \cdots \\
f.3 &= .(f.3)_0\ (f.3)_1\ (f.3)_2\ \cdots \\
&\cdots
\end{aligned}
$$

Construct a real number $e = .e_0 e_1 e_2 \cdots$ as follows. Make e different from $f.0$ by defining $e_0 = 2$ if $(f.0)_0 = 1$ and $e_0 = 1$ otherwise. In the same way, for each natural i, make e different from $f.i$ by defining $e_i = 2$ if $(f.i)_i = 1$ and $e_i = 1$ otherwise. Since e is different from every $f.i$, f is not onto. □

This result was first proved by Georg Cantor, in a series of groundbreaking papers on the theory of sets beginning in 1874. The technique used to construct e different from all the $f.i$ is known as the *Cantor diagonalization technique*. The real numbers $f.i$ in the diagram above can be viewed as a matrix with an infinite number of rows and an infinite number of

---

[1] C is sometimes called a *continuum*. Some of the numbers in C are *rational*, for example, $.3333\cdots = 1/3$ and $.5000\cdots = 1/2$. Others are *irrational*, in that the infinite sequence of d_i's does not have a repeating part. In representing real numbers in C, care must be taken to use only unique representations. Exercise 20.16 asks you to prove that $.3000\cdots$ and $.2999\cdots$ are equal, so only one of these two representations should be chosen.

HISTORICAL NOTE 20.1. Georg F.L.P. Cantor (1845–1918)

As a boy of 15, Cantor was determined to become a mathematician, but his father forced him to study engineering, a more promising career. The sensitive Georg tried very hard to follow his father's wishes, but in the end, his father gave in, and Cantor was allowed to study math. In 1867, at the age of 22, Cantor completed his Ph.D. in Berlin under Kummer, Weierstrass and Kronecker.

Cantor's most brilliant work was done between the ages of 29 and 39. This work included groundbreaking but controversial work on set theory, in particular, the theory of infinite sets, which we just touch on in this chapter. Part of the controversy surrounding his work dealt with the fact that his work was non-constructive. His old teacher Kronecker believed that one should use only constructive reasoning —a thing existed only if one had an algorithm for constructing it in a finite number of steps. Cantor's work did not have this constructive flavor, and Kronecker, considering the work illegitimate and dangerous, repeatedly attacked the work and Cantor. Further, Kronecker blocked Cantor from professional positions that a mathematician of Cantor's stature should rightly have had, and Cantor spent most of his career at the then third-rate University of Halle.

The sensitive Cantor suffered very much under Kronecker's attacks and was often depressed. In 1884, a complete breakdown forced him into a mental clinic, the first of several times this was to happen. Cantor died in a mental hospital at the age of 73.

---

columns. The diagonalization technique consists of constructing e so that each digit e_i differs from the diagonal element $(f.i)_i$ of the matrix. The diagonalization technique is used extensively in the theory of computability.

We want to be able to compare the sizes of infinite sets. In keeping with tradition, we call such a size the *cardinality* of the set, and the size is called a *cardinal number*. In Chap. 11, we defined the notation $\#S$ to be the cardinality of a finite set S. We now use $\#S$ for the cardinality of S even when S is infinite. For finite S, $\#S$ is a natural number; but for infinite S, $\#S$ is a different kind of object, something entirely new. We will not say what this object is; we simply name it. Later, we investigate some properties of $\#S$ for infinite S.

The cardinality of $\mathbb{N}$ is denoted by $\aleph_0$, i.e. $\#\mathbb{N} = \aleph_0$.[2]

Without actually defining $\#S$ for infinite sets S, we define the comparison operators $\#S < \#T$, $\#S \leq \#T$, etc. for cardinalities $\#S$, $\#T$.

---

[2] $\aleph$, read "aleph", is the first letter of the Hebrew alphabet.

(20.15) **Definition.** $\#S \leq \#T$ (and also $\#T \geq \#S$) if there exists a one-to-one function $f : S \to T$.

$\#S = \#T$ if there exists a one-to-one, onto function $f : S \to T$.

$\#S < \#T$ (and also $\#T > \#S$) if $\#S \leq \#T$ but $\#S \neq \#T$.

The next two theorems seem rather obvious. However, their proofs are too complex for this text and are not included.

(20.16) **Theorem. (Schröder-Bernstein).** For sets S and T,
$$\#S = \#T \;\equiv\; \#S \leq \#T \land \#T \leq \#S.$$

(20.17) **Theorem. (Zermelo).** For sets S and T,
$$\#S \leq \#T \;\lor\; \#T \leq \#S.$$

The following theorems are left as exercises.

(20.18) **Theorem.** The relation $\#S = \#T$ is an equivalence relation.

(20.19) **Theorem.** For any two sets S and T, exactly one of $\#S < \#T$, $\#S = \#T$, and $\#S > \#T$ holds.

(20.20) **Theorem.** The relation $\#S \leq \#T$ over cardinal numbers $\#S$ and $\#T$ is a linear order.

We now prove an important result that relates the cardinalities of a set S and its power set $\mathcal{P}S$ —i.e. the set of all subsets of S.

(20.21) **Theorem.** For all sets S, $\#S < \#(\mathcal{P}S)$.

Proof. The function $f : S \to \mathcal{P}S$ given by $f.x = \{x\}$ is one-to-one. Therefore, $\#S \leq \#(\mathcal{P}S)$. We prove $\#S < \#(\mathcal{P}S)$ by proving that no one-to-one function $g : S \to \mathcal{P}S$ is onto.

Let $g : S \to \mathcal{P}S$ be one-to-one. Consider the subset u of S defined by

$$x \in u \;\equiv\; x \notin g.x \quad \text{(for all x in S).}$$

We show that for all y in S, $u \neq g.y$, which means that g is not onto. For arbitrary y in S, we have

$$\begin{aligned}
&u = g.y \\
=\quad & \langle \text{Definition of } u \text{ and Extensionality (11.4)} \rangle \\
& (\forall x \mid x \in S : x \notin g.x \equiv x \in g.y) \\
\Rightarrow\quad & \langle \text{Instantiation (9.13) —note that } y \in S \rangle \\
& y \notin g.y \equiv y \in g.y \\
=\quad & \langle (3.15), \neg P \equiv P \equiv false \rangle \\
& false
\end{aligned}$$

Since $u = g.y \Rightarrow false$ equivales $u \neq g.y$, we have our result.   □

Using this theorem, we see that $\#\mathbb{N} < \#(\mathcal{P}\mathbb{N}) < \#(\mathcal{P}(\mathcal{P}\mathbb{N})) < \cdots$. Thus, there are an infinite number of infinite sets with increasing cardinality, so there are an infinite number of infinities.

Addition of cardinal numbers has not been defined. It can be defined as follows. Let s and t be two cardinal numbers, and let S and T be two disjoint sets satisfying $s = \#S$ and $t = \#T$. Define $s + t = \#(S \cup T)$. Addition of cardinal numbers is commutative and associative, and ordering $\leq$ of cardinal numbers is preserved by addition. Addition of cardinal numbers does have some strange properties. For example, if T is infinite and $\#S \leq \#T$, then $\#S + \#T = \#T$. Further investigation of arithmetic of cardinal numbers is beyond the scope of this text.

20.3   Countable and uncountable sets

Theorem (20.7) says that for any infinite set S, there is a one-to-one function $f : \mathbb{N} \to S$. Therefore, by the definition of $\leq$ for cardinal numbers, no infinite set is smaller than $\mathbb{N}$, and we have:

(20.22) **Theorem.** If set S is infinite, then $\aleph_0 \leq \#S$.

It is also easy to see that

(20.23) **Theorem.** If S is finite, then $\#S < \aleph_0$.

Proof. If S is finite, there is a one-to-one, onto function $f : 0..(\#S - 1) \to S$, and its inverse f^{-1} is a one-to-one function $f^{-1} : S \to \mathbb{N}$ that is not onto. Hence, $\#S < \aleph_0$.   □

As a smallest infinite set, $\mathbb{N}$ is of special interest, which has led to the introduction of special terminology for it.

(20.24) **Definition.** Set S is *countable*, or *denumerable*, if $\#S \leq \aleph_0$ —i.e. if S is finite or equal in cardinality to $\mathbb{N}$. If S is not countable, it is *uncountable*.

(20.25) **Definition.** An *enumeration* of a set S is a one-to-one, onto function $f : (0..n-1) \to S$ (for some natural n) or $f : \mathbb{N} \to S$. If f is onto but not one-to-one, it is called an *enumeration with repetition of S*.

We sometimes use sequence notation to indicate an enumeration of a set, writing it in the form $\langle f.0, f.1, f.2, \cdots \rangle$. This form makes clear that an enumeration provides an ordering $\prec$ of the elements of S. For any two different elements x and y of S, there exist unique integers i and j that satisfy $f.i = x$ and $f.j = y$, and we define $x \prec y \equiv i < j$.

It is fairly easy to prove (see Exercise 20.20) that

(20.26) **Theorem.** A set is countable iff there exists an enumeration of it.

Examples of enumerations

(a)   The empty set has one enumeration, the empty function.

(b)   The set $\{b, c, d\}$ has six enumerations, which we describe as sequences: $\langle b, c, d \rangle$, $\langle b, d, c \rangle$, $\langle c, b, d \rangle$, $\langle c, d, b \rangle$, $\langle d, b, c \rangle$, and $\langle d, c, b \rangle$.

(c)   The set of squares of natural numbers has the enumeration $f.i = i^2$, which can be viewed as the sequence $\langle 0, 1, 4, 9, 16, \cdots \rangle$. □

The following theorem will be used to show that a number of different sets are countable.

(20.27) **Theorem.** Suppose the elements of a set can be placed in an infinite matrix m (say) with a countable number of rows and a countable number of columns:

$$
\begin{matrix}
m_{00} & m_{01} & m_{02} & m_{03} & m_{04} & \cdots \\
m_{10} & m_{11} & m_{12} & m_{13} & m_{14} & \cdots \\
m_{20} & m_{21} & m_{22} & m_{23} & m_{24} & \cdots \\
\cdots
\end{matrix}
$$

Then the set is countable.

Proof. We exhibit an enumeration function —a one-to-one function D whose domain is the natural numbers and whose range are the pairs $\langle i, j \rangle$ for i, j natural numbers.

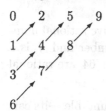

Enumeration function D is exhibited to the left in this paragraph. This diagram indicates that $D_0 = \langle 0, 0 \rangle$, $D_1 = \langle 1, 0 \rangle$, $D_2 = \langle 1, 0 \rangle$, $D_3 = \langle 2, 0 \rangle$, and so on. It is readily seen that each element of the matrix is in the range of D, so that D is onto. It is also clear that D is one-to-one, since each natural number i is assigned a different matrix element $D.i$. The rest of this proof defines D more formally.

Let D_k be the diagonal beginning at $\langle k, 0 \rangle$ in the diagram of the previous paragraph.

$$\langle \langle k, 0 \rangle, \langle k - 1, 1 \rangle, \cdots, \langle 1, k - 1 \rangle, \langle 0, k \rangle \rangle \quad .$$

For example,

$$D_0 = \langle \langle 0,0 \rangle \rangle \quad,$$
$$D_1 = \langle \langle 1,0 \rangle, \langle 0,1 \rangle \rangle \quad,$$
$$D_2 = \langle \langle 2,0 \rangle, \langle 1,1 \rangle, \langle 0,2 \rangle \rangle \quad.$$

Construct the infinite sequence of pairs by catenating the elements of the sequences $D_0, D_1, D_2, \ldots$, in that order. This infinite sequence is function D described in the diagram above.

We now define the one-to-one function $g : \mathbb{N} \times \mathbb{N} \to \mathbb{N}$ that gives the ordering of pairs according to this infinite sequence. Each set D_k contains $k + 1$ elements, and the number of elements in $D_0, D_1, \ldots, D_{i+j-1}$ is therefore $1 + 2 + \cdots + i + j = (i + j)(i + j + 1)/2$. Since the points on D_k are in the order $\langle k, 0 \rangle, \langle k - 1, 1 \rangle, \langle k - 2, 2 \rangle, \ldots, \langle 0, k \rangle$, the position in the ordering of $\langle i, j \rangle$ is $g(i,j)$. Thus, the one-to-one function is

$$g(i,j) = j + \frac{(i+j)(i+j+1)}{2}$$

and its inverse is the enumeration.   □

We can now easily prove that

(20.28) Theorem. The set $\mathbb{N} \times \mathbb{N}$ is countable.

Proof. Construct a matrix m with an infinite number of rows and an infinite number of columns, where element m_{ij} is the pair $\langle i, j \rangle$. Matrix m contains the set $\mathbb{N} \times \mathbb{N}$. By Theorem (20.27), $\mathbb{N} \times \mathbb{N}$ is countable.   □

In some situations, it is easy to construct an enumeration with repetition for a set, while an enumeration (without repetition) is more difficult to construct. However, we can prove that

(20.29) Theorem. A set with an enumeration with repetition also has an enumeration.

Let us investigate the cardinality of the set of nonnegative rational numbers, i.e. the set of values i/j where i is a natural number and j is a positive natural number. For example, $5/3$, $1/2$, and $64/64$ are rational numbers.

(20.30) Theorem. The set of nonnegative rationals is countable —its cardinality is the same as that of $\mathbb{N}$.

Proof. Consider the infinite matrix $r[0.., 1..]$ shown below, which contains all pairs $\langle i, j \rangle$ where i is a natural and j a positive natural number.

$$\langle 0,1 \rangle \quad \langle 0,2 \rangle \quad \langle 0,3 \rangle \quad \langle 0,4 \rangle \quad \cdots$$
$$\langle 1,1 \rangle \quad \langle 1,2 \rangle \quad \langle 1,3 \rangle \quad \langle 1,4 \rangle \quad \cdots$$
$$\langle 2,1 \rangle \quad \langle 2,2 \rangle \quad \langle 2,3 \rangle \quad \langle 2,4 \rangle \quad \cdots$$
$$\cdots$$

This set of pairs has an enumeration f (say), since it is a matrix. Each element $r_{ij} = \langle i, j \rangle$ can be mapped into the rational number i/j, say $h(i,j) = i/j$. Therefore, the function $h \bullet f$ is an enumeration with repetition of the non-negative rational numbers, which means that the set of non-negative rational numbers is countably infinite. (Function $h \bullet f$ is an enumeration with repetition because matrix r contains an infinite number of representations for many of the rationals. For example, the first row contains an infinite number of representations of 0, and the entries $\langle 1, 1 \rangle, \langle 2, 2 \rangle, \langle 3, 3 \rangle, \cdots$ all represent the number 1.)   □

Next, consider the set $seq(\mathbb{N})$ of all finite sequences of natural numbers. For example, $\langle 3, 2, 69, 1000 \rangle$ and $\langle 0, 0, 0, 0 \rangle$ are members of $seq(\mathbb{N})$.

(20.31) **Theorem.** The set $seq(\mathbb{N})$ is countable.

Proof. The idea is to consider each sequence as the sequence of exponents of the prime factorization of a natural number. We exhibit a one-to-one function $f : seq(\mathbb{N}) \rightarrow \mathbb{N}$. Let $p = p.0, p.1, \ldots$ be the prime numbers, listed in increasing order. Define function f by

$$f(s) = (\Pi i \mid 0 \leq i < \#s : p.i^{s.i+1})$$

By the Fundamental theorem of arithmetic, (15.114), if sequences s and t are different, then $f.s \neq f.t$. Hence, f is one-to-one, as required.   □

We leave to the reader the proofs of the following theorems.

(20.32) **Theorem.** The union of a countable set of countable sets is countable.

(20.33) **Theorem.** If S and T are countable, then so is $S \times T$.

(20.34) **Theorem.** If sets S_i are countable for $0 \leq i \leq n$, then so is $S_0 \times S_1 \times \cdots \times S_n$.

We have seen several countable sets. There are, however, sets that are not countable. For example, on page 463, we showed that the continuum $C = \{r : \mathbb{R} \mid 0 \leq r \leq 1\}$ is not countable. We also know from theorem (20.21) that $\#\mathbb{N} < \#\mathcal{P}\mathbb{N}$. How are sets C and $\mathcal{P}\mathbb{N}$ related? We have the following theorem, whose proof is left to the exercises.

(20.35) **Theorem.** $\#C = \#(\mathcal{P}\mathbb{N})$.

The reader may wonder whether any infinite set S satisfies $\aleph_0 < S < \#C$. The existence or non-existence of such set S has yet to be proved, in spite of efforts by the best minds in mathematics. The statement that there is no such set S is known as the *continuum hypothesis*. It was first conjectured by Cantor, and it was included as the first of Hilbert's unsolved problems (see Historical note 6.1 on page 111). In 1938, Kurt Gödel (see Historical note 7.1 on page 129) showed that adding the continuum hypothesis to set theory as an axiom does not introduce an inconsistency into set theory. But then in 1963, Paul Cohen showed that postulating the non-existence of such a set does not introduce an inconsistency either! Thus, we have a deep, unresolved, question on our hands.

AN APPLICATION TO COMPUTING

One often wonders what is computable (can be computed by some program) and what can not. Some understanding of this question can be seen in the following.

Let us calculate how many programs can be written in any given language L (say). First, the set of characters used to write a program is finite. Second, any program of L is a finite sequence of such characters. By Exercise 20.32, the set of such finite sequences of characters drawn from a finite set is countable, and since L is a subset of all such sequences, L is countable.

Now consider all programs of L whose purpose is to print a real number. Well, in general we can't really print a real number because we can't print an infinite number of digits. So let us think of writing a program P_r for a real number $r = .r_0 r_1 r_2 \cdots$ that, given $i \geq 0$, prints digit r_i of r. Thus, by executing P_r enough times, we can print as long an approximation to r as we wish.

There are an uncountable number of such real numbers r, but there are only a countable number of programs in L. Therefore, for only a countable number of the real numbers r does there exist a program P_r. We conclude that there are more real numbers r than there are programs P_r to print them. We cannot perform all tasks on a computer.

Exercises for Chapter 20

20.1 Prove theorem (20.2): Every subset of a finite set S is finite. Hint: View the existing one-to-one function $f : (0..\#S - 1) \to S$ as a sequence and determine what has to be done to sequence f to arrive at a one-to-one function for a given subset of S.

20.2 Prove theorem (20.3): The union of two finite sets is finite.

20.3 Prove theorem (20.4): The cross product of two finite sets is finite.

20.4 Prove theorem (20.5): The power set of a finite set is finite.

20.5 Prove theorem (20.7): If S is infinite, then there exists a one-to-one function $f : \mathbb{N} \to S$. Hint: By Axiom of Choice (11.77), an element e_0 (say) can be chosen from S; let $f.0 = e_0$, delete e_0 from S, and repeat the process.

20.6 Prove theorem (20.8): S is infinite iff there is a one-to-one function f that maps S into a proper subset of itself. Hint: Use function f to construct a one-to-one function $g : S \rightarrow S - \{f.0\}$.

20.7 Prove that the set $seq(\{`b`, `c`\})$ of finite sequences of characters 'b' and 'c' is infinite, by exhibiting a one-to-one function of the set to a proper subset.

20.8 Prove that the set of Pascal programs is infinite, by exhibiting a one-to-one function of the set to a proper subset.

20.9 Prove that the set of real numbers is infinite, by exhibiting a one-to-one function from the set to a proper subset.

20.10 Prove theorem (20.9): If a subset of a set is infinite, then so is the set.

20.11 Prove theorem (20.10): Given $f : S \rightarrow T$ be one-to-one, if set S is infinite then so is set T.

20.12 Prove theorem (20.11): If S is infinite, then so is its power set $\mathcal{P}S$.

20.13 Prove theorem (20.12): If S is infinite, then so is $S \cup T$.

20.14 Prove theorem (20.13): If S is infinite and $T \neq \emptyset$, then $S \times T$ is infinite.

20.15 Prove that the intersection of two infinite sets is not necessarily infinite.

20.16 Prove that $.3000\cdots$ and $.2999\cdots$ are equal. Hint: Investigate the expression $10 \cdot .2999\cdots - .2999\cdots$.

20.17 Prove theorem (20.18): The relation $\#S = \#T$ is an equivalence relation.

20.18 Prove theorem (20.19): For any two sets S and T, exactly one of $\#S < \#T$, $\#S = \#T$, and $\#S > \#T$ holds.

20.19 Prove theorem (20.20): The relation $\#S \leq \#T$ over cardinal numbers $\#S$ and $\#T$ is a linear order.

Exercises on countability

20.20 Prove theorem (20.26): A set is countable iff there exists an enumeration of it.

20.21 Prove that $\mathbb{N} \times \mathbb{N}$ is countable by proving that the function $g : \mathbb{N} \times \mathbb{N} \rightarrow \mathbb{N}$ defined by $g(m, n) = 2^m \cdot (2 \cdot n + 1) - 1$ is one-to-one and onto.

20.22 Prove theorem (20.29): A set with an enumeration with repetition also has an enumeration. Hint: Consider the enumeration with repetition f to be an array (which is infinite if S is infinite and finite otherwise). Construct an enumeration by throwing out duplicates from f.

20.23 Prove theorem (20.32): The union of a countable set of countable sets is countable. Hint: Use a technique similar to that used in the proof of theorem (20.30) that the set of nonnegative rationals is countable.

20.24 Prove theorem (20.33): If S and T are countable, then so is $S \times T$. Hint: Use a technique similar to that used in the proof of theorem (20.30) that the set of nonnegative rationals is countable.

20.25 Prove theorem (20.34): If sets S_i are countable for $0 \leq i \leq n$, then so is $S_0 \times S_1 \times \cdots \times S_n$. Hint: Use mathematical induction.

20.26 Prove $\#C \leq \#(\mathcal{P}\mathbb{N})$. Hint: Represent each number r (say) in C as a binary fraction, $r = .r_0 r_1 r_2 \cdots$ where each r_i satisfies $0 \leq r_i < 2$.

20.27 Prove $\#C \geq \#(\mathcal{P}\mathbb{N})$. Hint: Don't map each subset into a binary fraction, as a kind of inverse of the mapping given in the previous exercise, because that will not yield a one-to-one function. Instead, map each subset of $\mathbb{N}$ into a *decimal* fraction.

20.28 Prove theorem (20.35), $\#C = \#(\mathcal{P}\mathbb{N})$, using the results of the previous two exercises.

20.29 Prove that $\mathbb{Z}$ is countable.

20.30 We have shown that the set of real numbers in the range 0..1 is uncountable. Prove that the set of real numbers in any range $a..b$, $a < b$, has the same cardinality as the set of real numbers in 0..1.

20.31 Prove that the sets $\{r : \mathbb{R} \mid 0 \leq r \leq 1\}$ and $\{r : \mathbb{R} \mid 0 < r < 1\}$ have the same cardinality.

20.32 Let S be a finite set. Show that the set $seq(S)$ is countable by giving an enumeration for it. Set $seq(S)$ is the set of all sequences of finite length, where the elements of the sequence are members of S.

20.33 Use theorem (20.32) to show that the following sets are countable.

(a) The set of all polynomials of degree n for a given $n > 0$ and with rational coefficients.
(b) The set of all polynomials with rational coefficients.
(c) The set of all $n \times m$ matrices with rational coefficients, for fixed m, n at least 0.
(d) The set of all infinite matrices with rational coefficients.

References

[1] Allen, L.E *WFF'N PROOF: The Game of Modern Logic.* Autotelic Instructional Material Publishers, New Haven, 1972.

[2] Roland C. Backhouse. *Program Construction and Verification.* Prentice-Hall International, Englewood Cliffs, N.J., 1986.

[3] Eric T. Bell. *Men of Mathematics.* Simon and Schuster, New York, 1937.

[4] Errett Bishop. *Foundations of Constructive Mathematics.* McGraw-Hill, New York, 1967.

[5] Errett Bishop and Douglas Bridges. *Constructive Analysis.* Springer-Verlag, Berlin, 1985. (A thoroughly rewritten edition of [4]).

[6] George Boole. *An Investigation of the Laws of Thought, on which are founded the Mathematical Theories of Logic and Probabilities.* 1854.

[7] Florian Cajori. *A History of Mathematical Notations*, Vols. I and II. The Open Court Publishing Company, La Salle, Illinois, 1928 and 1929.

[8] Robert L. Constable et al. *Implementing Mathematics with the Nuprl Proof Development System.* Prentice-Hall, Englewood Cliffs, 1986.

[9] Edsger W. Dijkstra. *A Discipline of Programming.* Prentice-Hall, Englewood Cliffs, 1976.

[10] Edsger W. Dijkstra and Carel S. Scholten. *Predicate Calculus and Program Semantics.* Springer-Verlag, New York, 1990.

[11] René Descartes. *Geometrie.* Leyden, 1637.

[12] *Encyclopædia Britannica.* Encyclopædia Britannica, Inc., Chicago, 1967.

[13] A. Conan Doyle. The sign of the four; or, the problem of the sholtos. *Lippincott's Monthly Magazine* 45 (1890), 147–223.

[14] Wim H.J. Feijen. Exercises in formula manipulation. In E.W. Dijkstra (ed.), *Formal Development of Programs*, Addison-Wesley, Menlo Park, 1990, pp. 139–158.

[15] Solomon Feferman et al (eds.). *Collected Works of Gödel, Vol I (1929–1936)*. 1986.

[16] Ronald L. Graham, Donald E. Knuth, and Oren Patashnik. *Concrete Mathematics: A Foundation for Computer Science*. Addison-Wesley, Menlo Park, 1989.

[17] Alan Foster and Graham Shute. *Propositional Logic: A Student Introduction*. Aston Educational Enquiry Monograph 4, University of Aston in Birmingham, 1976.

[18] Gerhard Gentzen. Untersuchungen über das logische Schliessen. *Mathematische Zeitschrift 39* (1935), 176–210, 405–431. ([42] contains an English translation.)

[19] Karl Imanuel Gerhardt (editor). *Die Philosophischen Schriften von G. W. Leibniz, Band VII, "Scientific Generalis. Characteristica"*, XIX and XX. Weidmannsche Buchh., Berlin, 1887.

[20] Kurt Gödel. On formally undecideable propositions of *Principia Mathematica* and related systems I (originally in German). *Monatshefte für Mathematik und Physik 38* (1931), 173–198.

[21] Eric C.R. Hehner. *The Logic of Programming*. Prentice-Hall International, London, 1984.

[22] C.A.R. (Tony) Hoare and Cliff B. Jones. *Essays in Computing Science*. Prentice-Hall International, London, 1989.

[23] Andrew Hodges. *Alan Turing: The Enigma*. Simon & Schuster, New York, 1983.

[24] Douglas R. Hofstadter. *Gödel, Escher, Bach: An Eternal Golden Braid*. Basic Books, New York, 1979.

[25] John Horgan. Claude E. Shannon. *IEEE Spectrum* (April 1992), 72–75.

[26] Donald E. Knuth. *The Art of Computer Programming, Vols. 1, 2, 3*. Addison-Wesley Publishing Co., Menlo Park, 1969.

[27] Donald E. Knuth. *The TEXbook*. Addison-Wesley Publishing Co., Menlo Park, 1984.

[28] Leslie Lamport. *LATEX: A Document Preparation System*. Addison-Wesley Publishing Co., Menlo Park, 1986.

[29] Clarence I. Lewis. *A Survey of Symbolic Logic*. Dover Publications, New York, 1960.

[30] Gottfried Wilhelm Leibniz. *Scientific Generalis. Characteristica*

[31] Robert Recorde. *The Whetstone of Witte*. London, 1557.

[32] Constance Reid. *Hilbert*. Springer-Verlag, Heidelberg, 1970.

[33] Bertrand A.W. Russell. Mathematical logic as based on the theory of types. *American J. of Mathematics 30* (1908), 222–262.

[34] Pamela Samuelson. Should program algorithms be patented? *Comm. of the ACM 33* 8 (August 1990), 23–27.

[35] Claude E. Shannon. A Symbolic analysis of relay and switching circuits. *AIEE Transactions 57* (1938), 713–723 (abstract of Shannon's Masters thesis.)

[36] Claude E. Shannon. *The Mathematical Theory of Communications*. University of Illinois Press, Urbana, 1949.

[37] Raymond M. Smullyan. *What Is the Name of This Book?* Prentice-Hall, Englewood Cliffs, 1978.

[38] Raymond M. Smullyan. *To Mock a Mockingbird*. Alfred A. Knopf, New York, 1985.

[39] Raymond M. Smullyan. *Forever Undecided*. Alfred A. Knopf, New York, 1987.

[40] Richard M. Stallman. *GNU Emacs Manual, Fifth Edition, Version 18*. Free Software Foundation, Cambridge, Mass., 1986.

[41] Richard M. Stallman and Simson Garfinkle. Against software patents. *Comm. of the ACM 35* 1 (January 1992), 17–22, 121.

[42] M.E. Szabo (ed.). *The Collected Papers of Gerhard Gentzen*. North-Holland, Amsterdam, 1969.

[43] Anne S. Troelstra and Dirk van Dalen. *Constructivism in Mathematics, an Introduction*. North-Holland, Amsterdam, 1988.

[44] Hao Wang. *Reflections on Kurt Gödel*. The MIT Press, Cambridge, Mass., 1987.

[45] Wayne A. Wickelgren. *How to Solve Problems*. W.H. Freeman and Company, San Francisco, 1974.

Index

Texts and Monographs in Computer Science

(continued from page ii)

Texts and Monographs in Computer Science

THEOREMS OF THE PROPOSITIONAL CALCULUS

EQUIVALENCE AND TRUE

(3.1) **Axiom, Associativity of** $\equiv$: $((p \equiv q) \equiv r) \equiv (p \equiv (q \equiv r))$

(3.2) **Axiom, Symmetry of** $\equiv$: $p \equiv q \equiv q \equiv p$

(3.3) **Axiom, Identity of** $\equiv$: $true \equiv q \equiv q$

(3.4) *true*

(3.5) **Reflexivity of** $\equiv$: $p \equiv p$

NEGATION, INEQUIVALENCE, AND FALSE

(3.8)   **Axiom, Definition of** *false*: $false \equiv \neg true$

(3.9)   **Axiom, Distributivity of** $\neg$ **over** $\equiv$: $\neg(p \equiv q) \equiv \neg p \equiv q$

(3.10) **Axiom, Definition of** $\not\equiv$: $(p \not\equiv q) \equiv \neg(p \equiv q)$

(3.11) $\neg p \equiv q \equiv p \equiv \neg q$

(3.12) **Double negation:** $\neg\neg p \equiv p$

(3.13) **Negation of** *false*: $\neg false \equiv true$

(3.14) $(p \not\equiv q) \equiv \neg p \equiv q$

(3.15) $\neg p \equiv p \equiv false$

(3.16) **Symmetry of** $\not\equiv$: $(p \not\equiv q) \equiv (q \not\equiv p)$

(3.17) **Associativity of** $\not\equiv$: $((p \not\equiv q) \not\equiv r) \equiv (p \not\equiv (q \not\equiv r))$

(3.18) **Mutual associativity:** $((p \not\equiv q) \equiv r) \equiv (p \not\equiv (q \equiv r))$

(3.19) **Mutual interchangeability:** $p \not\equiv q \equiv r \equiv p \equiv q \not\equiv r$

DISJUNCTION

(3.24) **Axiom, Symmetry of** $\vee$: $p \vee q \equiv q \vee p$

(3.25) **Axiom, Associativity of** $\vee$: $(p \vee q) \vee r \equiv p \vee (q \vee r)$

(3.26) **Axiom, Idempotency of** $\vee$: $p \vee p \equiv p$

(3.27) **Axiom, Distributivity of** $\vee$ **over** $\equiv$: $p \vee (q \equiv r) \equiv p \vee q \equiv p \vee r$

(3.28) **Axiom, Excluded Middle:** $p \vee \neg p$

(3.29) **Zero of** $\vee$: $p \vee true \equiv true$

(3.30) **Identity of** $\vee$: $p \vee false \equiv p$

(3.31) **Distributivity of** $\vee$ **over** $\vee$: $p \vee (q \vee r) \equiv (p \vee q) \vee (p \vee r)$

(3.32) $p \vee q \equiv p \vee \neg q \equiv p$

CONJUNCTION

(3.35) **Axiom, Golden rule**: $p \wedge q \equiv p \equiv q \equiv p \vee q$

(3.36) **Symmetry of** $\wedge$: $p \wedge q \equiv q \wedge p$

(3.37) **Associativity of** $\wedge$: $(p \wedge q) \wedge r \equiv p \wedge (q \wedge r)$

(3.38) **Idempotency of** $\wedge$: $p \wedge p \equiv p$

(3.39) **Identity of** $\wedge$: $p \wedge true \equiv p$

(3.40) **Zero of** $\wedge$: $p \wedge false \equiv false$

(3.41) **Distributivity of** $\wedge$ **over** $\wedge$: $p \wedge (q \wedge r) \equiv (p \wedge q) \wedge (p \wedge r)$

(3.42) **Contradiction:** $p \wedge \neg p \equiv false$

(3.43) **Absorption:** (a) $p \wedge (p \vee q) \equiv p$

   (b) $p \vee (p \wedge q) \equiv p$

(3.44) **Absorption:** (a) $p \wedge (\neg p \vee q) \equiv p \wedge q$

   (b) $p \vee (\neg p \wedge q) \equiv p \vee q$

(3.45) **Distributivity of** $\vee$ **over** $\wedge$: $p \vee (q \wedge r) \equiv (p \vee q) \wedge (p \vee r)$

(3.46) **Distributivity of** $\wedge$ **over** $\vee$: $p \wedge (q \vee r) \equiv (p \wedge q) \vee (p \wedge r)$

(3.47) **De Morgan:** (a) $\neg(p \wedge q) \equiv \neg p \vee \neg q$

   (b) $\neg(p \vee q) \equiv \neg p \wedge \neg q$

(3.48) $p \wedge q \equiv p \wedge \neg q \equiv \neg p$

(3.49) $p \wedge (q \equiv r) \equiv p \wedge q \equiv p \wedge r \equiv p$

(3.50) $p \wedge (q \equiv p) \equiv p \wedge q$

(3.51) **Replacement:** $(p \equiv q) \wedge (r \equiv p) \equiv (p \equiv q) \wedge (r \equiv q)$

(3.52) **Definition of** $\equiv$: $p \equiv q \equiv (p \wedge q) \vee (\neg p \wedge \neg q)$

(3.53) **Exclusive or:** $p \not\equiv q \equiv (\neg p \wedge q) \vee (p \wedge \neg q)$

(3.55) $(p \wedge q) \wedge r \equiv p \equiv q \equiv r \equiv p \vee q \equiv q \vee r \equiv r \vee p \equiv p \vee q \vee r$

IMPLICATION

(3.57) **Axiom, Definition of Implication:** $p \Rightarrow q \equiv p \vee q \equiv q$

(3.58) **Axiom, Consequence:** $p \Leftarrow q \equiv q \Rightarrow p$

(3.59) **Definition of implication:** $p \Rightarrow q \equiv \neg p \vee q$

(3.60) **Definition of implication:** $p \Rightarrow q \equiv p \wedge q \equiv p$

(3.61) **Contrapositive:** $p \Rightarrow q \equiv \neg q \Rightarrow \neg p$

(3.62) $p \Rightarrow (q \equiv r) \equiv p \wedge q \equiv p \wedge r$

(3.63) **Distributivity of** $\Rightarrow$ **over** $\equiv$: $p \Rightarrow (q \equiv r) \equiv p \Rightarrow q \equiv p \Rightarrow r$

(3.64) $p \Rightarrow (q \Rightarrow r) \equiv (p \Rightarrow q) \Rightarrow (p \Rightarrow r)$

(3.65) **Shunting:** $p \wedge q \Rightarrow r \equiv p \Rightarrow (q \Rightarrow r)$

(3.66) $p \wedge (p \Rightarrow q) \equiv p \wedge q$

(3.67) $p \wedge (q \Rightarrow p) \equiv p$

(3.68) $p \vee (p \Rightarrow q) \equiv true$

(3.69) $p \vee (q \Rightarrow p) \equiv q \Rightarrow p$

(3.70) $p \lor q \Rightarrow p \land q \equiv p \equiv q$

(3.71) **Reflexivity of** $\Rightarrow$: $p \Rightarrow p \equiv true$

(3.72) **Right zero of** $\Rightarrow$: $p \Rightarrow true \equiv true$

(3.73) **Left identity of** $\Rightarrow$: $true \Rightarrow p \equiv p$

(3.74) $p \Rightarrow false \equiv \neg p$

(3.75) $false \Rightarrow p \equiv true$

(3.76) **Weakening/strengthening:** (a) $p \Rightarrow p \lor q$

　　　　　　　　　　　　　　　　 (b) $p \land q \Rightarrow p$

　　　　　　　　　　　　　　　　 (c) $p \land q \Rightarrow p \lor q$

　　　　　　　　　　　　　　　　 (d) $p \lor (q \land r) \Rightarrow p \lor q$

　　　　　　　　　　　　　　　　 (e) $p \land q \Rightarrow p \land (q \lor r)$

(3.77) **Modus ponens:** $p \land (p \Rightarrow q) \Rightarrow q$

(3.78) $(p \Rightarrow r) \land (q \Rightarrow r) \equiv (p \lor q \Rightarrow r)$

(3.79) $(p \Rightarrow r) \land (\neg p \Rightarrow r) \equiv r$

(3.80) **Mutual implication:** $(p \Rightarrow q) \land (q \Rightarrow p) \equiv (p \equiv q)$

(3.81) **Antisymmetry:** $(p \Rightarrow q) \land (q \Rightarrow p) \Rightarrow (p \equiv q)$

(3.82) **Transitivity:** (a) $(p \Rightarrow q) \land (q \Rightarrow r) \Rightarrow (p \Rightarrow r)$

　　　　　　　　　　　 (b) $(p \equiv q) \land (q \Rightarrow r) \Rightarrow (p \Rightarrow r)$

　　　　　　　　　　　 (c) $(p \Rightarrow q) \land (q \equiv r) \Rightarrow (p \Rightarrow r)$

LEIBNIZ AS AN AXIOM

(3.83) **Axiom, Leibniz:** $e = f \Rightarrow E_e^z = E_f^z$

(3.84) **Substitution:** (a) $(e = f) \land E_e^z \equiv (e = f) \land E_f^z$

　　　　　　　　　　 (b) $(e = f) \Rightarrow E_e^z \equiv (e = f) \Rightarrow E_f^z$

　　　　　　　　　　 (c) $q \land (e = f) \Rightarrow E_e^z \equiv q \land (e = f) \Rightarrow E_f^z$

(3.85) **Replace by** $true$: (a) $p \Rightarrow E_p^z \equiv p \Rightarrow E_{true}^z$

　　　　　　　　　　　　 (b) $q \land p \Rightarrow E_p^z \equiv q \land p \Rightarrow E_{true}^z$

(3.86) **Replace by** $false$: (a) $E_p^z \Rightarrow p \equiv E_{false}^z \Rightarrow p$

　　　　　　　　　　　　 (b) $E_p^z \Rightarrow p \lor q \equiv E_{false}^z \Rightarrow p \lor q$

(3.87) **Replace by** $true$: $p \land E_p^z \equiv p \land E_{true}^z$

(3.88) **Replace by** $false$: $p \lor E_p^z \equiv p \lor E_{false}^z$

(3.89) **Shannon:** $E_p^z \equiv (p \land E_{true}^z) \lor (\neg p \land E_{false}^z)$

(4.1) $p \Rightarrow (q \Rightarrow p)$

(4.2) **Monotonicity of** $\lor$: $(p \Rightarrow q) \Rightarrow (p \lor r \Rightarrow q \lor r)$

(4.3) **Monotonicity of** $\land$: $(p \Rightarrow q) \Rightarrow (p \land r \Rightarrow q \land r)$

(4.4) **Deduction:** To prove $P \Rightarrow Q$, assume P and prove Q.

(4.5) **Case analysis:** If E^z_{true}, E^z_{false} are theorems, then so is E^z_P.

(4.6) **Case analysis:** $(p \vee q \vee r) \wedge (p \Rightarrow s) \wedge (q \Rightarrow s) \wedge (r \Rightarrow s) \Rightarrow s$

(4.7) **Mutual implication:** To prove $P \equiv Q$, prove $P \Rightarrow Q$ and $Q \Rightarrow P$.

(4.9) **Proof by contradiction:** To prove P, prove $\neg P \Rightarrow false$.

(4.12) **Proof by contrapositive:** To prove $P \Rightarrow Q$, prove $\neg Q \Rightarrow \neg P$

General Laws of Quantification

For symmetric and associative binary operator $\star$ with identity u.

(8.13) **Axiom, Empty range:** $(\star x \mid false : P) = u$

(8.14) **Axiom, One-point rule:** Provided $\neg occurs(\text{'}x\text{'}, \text{'}E\text{'})$,
$(\star x \mid x = E : P) = P[x := E]$

(8.15) **Axiom, Distributivity:** Provided each quantification is defined,
$(\star x \mid R : P) \star (\star x \mid R : Q) = (\star x \mid R : P \star Q)$

(8.16) **Axiom, Range split:** Provided $R \wedge S \equiv false$ and each
quantification is defined,
$(\star x \mid R \vee S : P) = (\star x \mid R : P) \star (\star x \mid S : P)$

(8.17) **Axiom, Range split:** Provided each quantification is defined,
$(\star x \mid R \vee S : P) \star (\star x \mid R \wedge S : P) = (\star x \mid R : P) \star (\star x \mid S : P)$

(8.18) **Axiom, Range split for idempotent** $\star$**:** Prov. each quant. is defined,
$(\star x \mid R \vee S : P) = (\star x \mid R : P) \star (\star x \mid S : P)$

(8.19) **Axiom, Interchange of dummies:** Provided each quantification
is defined, $\neg occurs(\text{'}y\text{'}, \text{'}R\text{'})$, and $\neg occurs(\text{'}x\text{'}, \text{'}Q\text{'})$,
$(\star x \mid R : (\star y \mid Q : P)) = (\star y \mid Q : (\star x \mid R : P))$

(8.20) **Axiom, Nesting:** Provided $\neg occurs(\text{'}y\text{'}, \text{'}R\text{'})$,
$(\star x, y \mid R \wedge Q : P) = (\star x \mid R : (\star y \mid Q : P))$

(8.21) **Axiom, Dummy renaming:** Provided $\neg occurs(\text{'}y\text{'}, \text{'}R, P\text{'})$,
$(\star x \mid R : P) = (\star y \mid R[x := y] : P[x := y])$

(8.22) **Change of dummy:** Provided $\neg occurs(\text{'}y\text{'}, \text{'}R, P\text{'})$, and f
has an inverse, $(\star x \mid R : P) = (\star y \mid R[x := f.y] : P[x := f.y])$

(8.23) **Split off term:** $(\star i \mid 0 \leq i < n+1 : P) = (\star i \mid 0 \leq i < n : P) \star P^i_n$

Theorems of the Predicate Calculus

Universal quantification

(9.2) **Axiom, Trading:** $(\forall x \mid R : P) \equiv (\forall x \mid: R \Rightarrow P)$

(9.3) **Trading:** (a) $(\forall x \mid R : P) \equiv (\forall x \mid: \neg R \vee P)$
(b) $(\forall x \mid R : P) \equiv (\forall x \mid: R \wedge P \equiv R)$
(c) $(\forall x \mid R : P) \equiv (\forall x \mid: R \vee P \equiv P)$

(9.4) **Trading:** (a) $(\forall x \mid Q \wedge R : P) \equiv (\forall x \mid Q : R \Rightarrow P)$
(b) $(\forall x \mid Q \wedge R : P) \equiv (\forall x \mid Q : \neg R \vee P)$
(c) $(\forall x \mid Q \wedge R : P) \equiv (\forall x \mid Q : R \wedge P \equiv R)$
(d) $(\forall x \mid Q \wedge R : P) \equiv (\forall x \mid Q : R \vee P \equiv P)$

(9.5) **Axiom, Distributivity of $\vee$ over $\forall$:** Prov. $\neg occurs(\text{'}x\text{'}, \text{'}P\text{'})$,
$P \vee (\forall x \mid R : Q) \equiv (\forall x \mid R : P \vee Q)$

(9.6) **Provided** $\neg occurs(\text{'}x\text{'}, \text{'}P\text{'})$, $(\forall x \mid R : P) \equiv P \vee (\forall x \mid : \neg R)$

(9.7) **Distributivity of $\wedge$ over $\forall$:** Provided $\neg occurs(\text{'}x\text{'}, \text{'}P\text{'})$,
$\neg(\forall x \mid : \neg R) \Rightarrow ((\forall x \mid R : P \wedge Q) \equiv P \wedge (\forall x \mid R : Q))$

(9.8) $(\forall x \mid R : true) \equiv true$

(9.9) $(\forall x \mid R : P \equiv Q) \Rightarrow ((\forall x \mid R : P) \equiv (\forall x \mid R : Q))$

(9.10) **Range weakening/strengthening:** $(\forall x \mid Q \vee R : P) \Rightarrow (\forall x \mid Q : P)$

(9.11) **Body weakening/strengthening:** $(\forall x \mid R : P \wedge Q) \Rightarrow (\forall x \mid R : P)$

(9.12) **Monotonicity of $\forall$:**
$(\forall x \mid R : Q \Rightarrow P) \Rightarrow ((\forall x \mid R : Q) \Rightarrow (\forall x \mid R : P))$

(9.13) **Instantiation:** $(\forall x \mid : P) \Rightarrow P[x := e]$

(9.16) P is a theorem iff $(\forall x \mid : P)$ is a theorem.

Existential Quantification

(9.17) **Axiom, Generalized De Morgan:**
$(\exists x \mid R : P) \equiv \neg(\forall x \mid R : \neg P)$

(9.18) **Generalized De Morgan:** (a) $\neg(\exists x \mid R : \neg P) \equiv (\forall x \mid R : P)$
(b) $\neg(\exists x \mid R : P) \equiv (\forall x \mid R : \neg P)$
(c) $(\exists x \mid R : \neg P) \equiv \neg(\forall x \mid R : P)$

(9.19) **Trading:** $(\exists x \mid R : P) \equiv (\exists x \mid : R \wedge P)$

(9.20) **Trading:** $(\exists x \mid Q \wedge R : P) \equiv (\exists x \mid Q : R \wedge P)$

(9.21) **Distributivity of $\wedge$ over $\exists$:** Provided $\neg occurs(\text{'}x\text{'}, \text{'}P\text{'})$,
$P \wedge (\exists x \mid R : Q) \equiv (\exists x \mid R : P \wedge Q)$

(9.22) **Provided** $\neg occurs(\text{'}x\text{'}, \text{'}P\text{'})$, $(\exists x \mid R : P) \equiv P \wedge (\exists x \mid : R)$

(9.23) **Distributivity of $\vee$ over $\exists$:** Provided $\neg occurs(\text{'}x\text{'}, \text{'}P\text{'})$,
$(\exists x \mid : R) \Rightarrow ((\exists x \mid R : P \vee Q) \equiv P \vee (\exists x \mid R : Q))$

(9.24) $(\exists x \mid R : false) \equiv false$

(9.25) **Range weakening/strengthening:** $(\exists x \mid R : P) \Rightarrow (\exists x \mid Q \vee R : P)$

(9.26) **Body weakening/strengthening:** $(\exists x \mid R : P) \Rightarrow (\exists x \mid R : P \vee Q)$

(9.27) **Monotonicity of $\exists$:**
$(\forall x \mid R : Q \Rightarrow P) \Rightarrow ((\exists x \mid R : Q) \Rightarrow (\exists x \mid R : P))$

(9.28) **$\exists$-Introduction:** $P[x := E] \Rightarrow (\exists x \mid : P)$

(9.29) **Interchange of quantifications:**
Provided $\neg occurs(\text{'}y\text{'}, \text{'}R\text{'})$ and $\neg occurs(\text{'}x\text{'}, \text{'}Q\text{'})$,
$(\exists x \mid R : (\forall y \mid Q : P)) \Rightarrow (\forall y \mid Q : (\exists x \mid R : P))$

(9.30) **Provided** $\neg occurs(\text{'}\hat{x}\text{'}, \text{'}Q\text{'})$,
$(\exists x \mid R : P) \Rightarrow Q$ is a theorem iff $(R \wedge P)[x := \hat{x}] \Rightarrow Q$ is a theorem

Table of Precedences

(a) $[x := e]$ (textual substitution)   (highest precedence)
(b) . (function application)
(c) unary prefix operators: $+$ $-$ $\neg$ $\#$ $\sim$ $\mathcal{P}$
(d) $**$
(e) $\cdot$ $/$ $\div$ **mod** **gcd**
(f) $+$ $-$ $\cup$ $\cap$ $\times$ $\circ$ $\bullet$
(g) $\downarrow$ $\uparrow$
(h) $\#$
(i) $\triangleleft$ $\triangleright$ $\char94$
(j) $=$ $<$ $>$ $\in$ $\subset$ $\subseteq$ $\supset$ $\supseteq$ $|$ (conjunctional, see page 29)
(k) $\vee$ $\wedge$
(l) $\Rightarrow$ $\Leftarrow$
(m) $\equiv$ (lowest precedence)

All nonassociative binary infix operators associate to the left, except $**$, $\triangleleft$, and $\Rightarrow$, which associate to the right.

The operators on lines (j), (l), and (m) may have a slash $/$ through them to denote negation —e.g. $b \not\equiv c$ is an abbreviation for $\neg(b \equiv c)$.

Greek letters and their Transliterations

Name	Sign	Tr.	Name	Sign	Tr.	Name	Sign	Tr.
Alpha	α	a	Iota	ι	i	Rho	ρ	r
Beta	β	b	Kappa	κ	k	Sigma	σ	s
Gamma	γ	g	Lambda	λ	l	Tau	τ	t
Delta	δ	d	Mu	μ	m	Upsilon	υ	y, u
Epsilon	ϵ	e	Nu	ν	n	Phi	ϕ	ph
Zeta	ζ	z	Xi	ξ	x	Chi	χ	ch
Eta	η	e	Omicron	o	o	Psi	ψ	ps
Theta	θ	th	Pi	π	p	Omega	ω	o

Types Used in this Text

integer	$\mathbb{Z}$	integers: $\ldots, -3, -2, -1, 0, 1, 2, 3, \ldots$
nat	$\mathbb{N}$	natural numbers: $0, 1, 2, \ldots$
positive	$\mathbb{Z}^+$	positive integers: $1, 2, 3, \ldots$
negative	$\mathbb{Z}^-$	negative integers: $-1, -2, -3, \ldots$
rational	$\mathbb{Q}$	rationals i/j for i, j integers, $j \neq 0$
reals	$\mathbb{R}$	real numbers
positive reals	$\mathbb{R}^+$	positive real numbers
bool	$\mathbb{B}$	booleans: *true*, *false*
sets	$set(t)$	set of elements of type t
bags	$bag(t)$	bag of elements of type t
sequences	$seq(t)$	sequence of elements of type t